GRAPES AND WINES - ADVANCES IN PRODUCTION, PROCESSING, ANALYSIS AND VALORIZATION

Edited by **António Manuel Jordão**
and **Fernanda Cosme**

Grapes and Wines - Advances in Production, Processing, Analysis and Valorization
http://dx.doi.org/10.5772/68059
Edited by António Manuel Jordão and Fernanda Cosme

Contributors

Violeta Carolina Niculescu, Nadia Paun, Roxana Ionete, Renato Vasconcelos Botelho, Gabriela Datsch Bennemann, Yohandra Reyes Torres, Alessandro Jefferson Sato, Anca Sipos, Manuel Oliveira, Jesus Heras-Roger, Carlos Díaz-Romero, Jacinto Darias-Martín, Domingo Ríos-Mesa, Anibal Jose-Coutinho, Patricia Avila, Jorge Ricardo-Da-Silva, Ana Alexandra Marta-Costa, João Rebelo, Lina Lourenço-Gomes, Sofia Helena Gouveia, Raul Ferrer-Gallego, Miquel Puxeu, Laura Martín, Enric Nart, Claudio Hidalgo, Imma Andorrà, Daniela Fracassetti, Ileana Vigentini, Ximena Besoain, Carmina Gisbert, Rosa María Peiró, Tania San Pedro Galán, Antonio Olmos, Rosa Adela Arroyo-Garcia, Eugenio Revilla, Alberto Bellido, Davis Carrasco, Anna Puig, Leonor Ruiz, Zenaida Guadalupe, Belén Ayestarán, Leticia Martínez-Lapuente, Juan Carlos De Haro Sanchez, Juan F. Rodriguez, Manuel Carmona Franco, Ángel Pérez Martínez, Francisco Espinosa, Esperanza Valdés, David Uriarte, Inmaculada Garrido, Esther Gamero, Maria-Henar Prieto, David Moreno, Antônio Teixeira, Janice Leivas, Jorge Tonietto

Published by InTech
Janeza Trdine 9, 51000 Rijeka, Croatia

Publishing Process Manager Martina Usljebrka
Technical Editor SPi Global
Cover InTech Design team

First published February, 2018
Printed in Croatia
Legal deposit, Croatia: National and University Library in Zagreb

Additional hard copies can be obtained from orders@intechopen.com

Grapes and Wines - Advances in Production, Processing, Analysis and Valorization, Edited by António Manuel Jordão and Fernanda Cosme
p. cm.
Print ISBN 978-953-51-3833-4
Online ISBN 978-953-51-3834-1

For EU product safety concerns:
IN TECH d.o.o., Prolaz Marije Krucifikse Kozulić 3, 51000 Rijeka, Croatia,
info@intechopen.com or visit our website at intechopen.com.

Meet the editors

António Manuel Jordão is currently an assistant professor of Oenology at Polytechnic Institute of Viseu, Agrarian Higher School and member of the Chemistry Research Centre, Vila Real. He is a graduate of Agro-Industrial Engineering and has a Master's degree in Food Science and Technology from Technical University of Lisbon. In 2006, he obtained a PhD degree in Agro-Industrial Engineering at Agronomy Higher Institute of Technical University of Lisbon. His research interests are phenolic compounds from grapes and wines, wine technology, and, in particular, the topic of wine aging process. He is a scientific editor of two books of enology. He is the author/coauthor of 40 articles in peer-reviewed scientific journals, 10 book chapters, and 65 communications in scientific congress. He is also a member of the editorial board of scientific journals, a supervisor of several Master's theses in enology and has developed consultancy services for wine companies and evaluation of research projects.

Maria Fernanda Gil Cosme Martins is a graduate of Agro-Industrial Engineering from Technical University of Lisbon and a PhD degree holder in Food Science at University of Trás-os-Montes and Alto Douro (UTAD). Currently, she is an assistant professor of Food Science/Oenology at UTAD and a member of the Chemistry Research Centre, Vila Real. Her main research lines include grape composition, wine technology, and wine stabilization and aging, namely fining, phenolic compounds, polysaccharides, and proteins. She is a scientific editor of 1 book of Oenology and author/coauthors of 22 articles in ISI journals, 4 articles in open access peer-reviewed international scientific journals, 10 book chapters, and 79 papers in conference proceedings. She has participated in 10 funded research projects and supervised PhD and MSc student thesis in the fields of Food Science and Oenology.

Contents

Preface

Grape and wine sector has probably inspired more research and publications than any other agricultural and food area. In fact, through their passion for vine, wine, and derived products, a great number of scientists not only have contributed to the development of viticultural practices but also have made advances in the winemaking technological process, in the wine aging conditions, in the valorization of wine products and their by-products, in the grape and wine characterization, and finally in the economic characterization of the wine sector. In addition, in the last years, the impact of climatic changes on viticulture practices has also been an important research line in order to analyze how the vine adapts and resists to new environmental conditions that affect several wine regions. Each applied development in grapes and wines has led to a better control of the physicochemical and sensory quality of the different wine products.

The main goal of vine and wine production is to ensure a high production level combined with a high quality standard of grapes and wine products. During the whole life of the vine cycle, as well as, during all stages of grape development, different factors affect the grapevine and the development and composition of the grapes, namely, climate, soil, wine-growing practices, and the genetic potential that each grape variety presents. In addition, wine production and preservation also involve numerous factors of microbiological, chemical, and technological innovation, which will contribute decisively to the wine quality and derived products. In addition, the grape and wine production is a very important economic activity and, as such, economic and social dimension is still a central theme for the sustainability of the sector. Thus, the determined evolution of the grape and wine industry demands persistent advancements in all parts of the production chain.

The goal of this book is to summarize in a concise manner the accumulated information about the most recent developments in grape and wine production, since the different origins and location of grape production, environmental factors that determine their production, the vine sanitation problems and the diverse cultural practices, the production, characterization and conservation of wines, the valorization of wine by-products, and finally the economic analysis of the wine sector.

This book is composed of three different general sections: (1) Viticulture and Environmental Conditions, (2) Wine Production and Characterization, and (3) Economic Analysis and Valorization of Wine Products. Inside these 3 general sections, 16 different chapters provide current research on different topics of recent advances on production, processing, analysis, and valorization of grapes and wines.

Thus, inside the first section (Viticulture and Environmental Conditions), Chapter 1 focuses on an example of the recovery of two ancient Spanish grapevine varieties taking into consideration several topics from genetic variability to *In vitro* conservation. Chapter 2 evaluates the anthocyanin and flavonol composition of wild grapes and determines whether some wild genotypes present some genetic characters of interest related to anthocyanin and flavonol accumulation during grape maturation. Chapter 3 discusses the impact of the main grapevine trunk diseases that are affecting *Vitis vinifera* in Chile, while Chapter 4 discusses the viticulture in warmer climates and the use of several mitigation techniques (irrigation, canopy shading, water nebulization, and kaolin coating) to counteract the worse effects of adverse weather conditions associated to the climate changes in a specific Portuguese wine-demarcated region. Other researchers discuss in Chapter 5 the elaboration and application of water balance indices and improving them by using a geographic information system in the wine grape–growing regions located in semiarid regions of Brazil. Finally, Chapter 6, analyzes the effects of the combination of water deficit and thinning on the constitution and composition of the *Tempranillo* grape variety, namely on the lipid peroxidation, oxidant and antioxidant enzyme activities and their interactions on phenolic profile, and chromatic characteristic of the wines.

In the second section (Wine Production and Characterization), several perspectives of wine production and characterization are exposed. Thus, Chapter 7 focuses on the chemical structures of the main polyphenols in grape and wines, followed by their identification and quantification, while in Chapter 8, the thematic of the application of mathematical models applicable to enology, in particular, the nonlinear models for white wine alcoholic fermentation process, is approached. Other researchers discuss in Chapter 9 an important subject in modern oenology, which is the elaboration of SO_2-free wines and the microbiological, physical, and chemical factors that influence it. In Chapter 10, particular attention is given to sparkling wines. The authors of this chapter analyze the influence of wine chemical compounds on the foaming properties of sparkling wines. Chapter 11 discusses the aromatic profile of wines, in particular the occurrence and analysis of sulfur compounds in wines, while in Chapter 12 the thematic of the chemical characterization of red wines produced in eight Atlantic islands from Canary and Cape Verde archipelago and also a comparative analysis with the red wines produced in other Atlantic islands (Azores and Madeira) are discussed. Finally, in Chapter 13, the authors approach the thematic of wine sensory analyses by the use of a new novel memory-based sensory approach to asses a large-scale typicality of mainland Portuguese red wines.

In the last section (Economic Analysis and Valorization of Wine Products), several standpoints of wine economy and the potential valorization of wine products are exposed. Thus, in Chapter 14, the authors analyze the economic performance of wineries (from two Portuguese wine regions), using indicators widely employed in economic and business literature, and assess the potential influence of size on a firm´s performance. The authors in Chapter 15 focus on the potential use of winery residues as nutrition for humans and as agricultural input, and finally other researchers discuss in Chapter 16 the potential utilization of grape seed oil for producing bio-based materials through environmentally friendly processes that could substitute petroleum-derived products.

All of these chapters are written by a group of international researchers, in order to provide up-to-date reviews, overviews, and summaries of current research on the different dimensions of grape and wine production. This book is not only for technicians actively engaged in the field but also for students attending technical schools and/or universities and other professionals that might be interested in reading and learning about some fascinating areas of grape and wine research and discovering the most recent tendencies of production, processing, analysis, and valorization of grape and wine products.

Finally, it was with great pleasure that we accepted the opportunity offered by InTechOpen Editors to assemble and edit this book. We are greatly indebted to the authors that generously shared their scientific knowledge and experience with others through their contribution to this book.

Prof. António Manuel Jordão (M.Sc., PhD)
Assistant Professor of Oenology
Polytechnic Institute of Viseu
Agrarian Higher School
Department of Food Industries, Portugal

Prof. Fernanda Cosme (PhD)
Assistant Professor of Food Science and Oenology
Chemical Research Centre of Vila Real (CQ-VR)
University of Trás-os-Montes and Alto Douro
School of Life Science and Environment, Portugal

Section 1

Viticulture and Environmental Conditions

Chapter 1

Recovering Ancient Grapevine Varieties: From Genetic Variability to In Vitro Conservation, A Case Study

Carmina Gisbert, Rosa Peiró, Tania San Pedro, Antonio Olmos, Carles Jiménez and Julio García

Additional information is available at the end of the chapter

http://dx.doi.org/10.5772/intechopen.71133

Abstract

A great number of varieties have been described in grapevine; however, few of them are currently in use. The increasing concern on varietal diversity loss has encouraged actions for recovering and preserving grapevine germplasm, which represents valuable resources for breeding as well as for diversification in grapevine-derived products. On the other hand, it is expected that this important crop, which is distributed in warm areas worldwide, will suffer the climate changes. Therefore, it is also convenient the identification of intravarietal variability and the recovery of accessions well adapted to particular environments. In this chapter, we will contribute to highlight the importance of recovering ancient materials, the usefulness of SSR markers to determine their molecular profile, the importance to analyze their virus status, and the possibilities that offer biotechnological tools for virus sanitation and in vitro storage as a complement of field preservation. In this context, we have evaluated different grapevine accessions and developed in vitro culture protocols for micropropagation, sanitation, and storage grapevine cultivars. In this work, we report the results obtained for the historic variety "Valencí Blanc" (or "Beba") and the historic and endangered variety "Esclafagerres" ("Esclafacherres" or "Esclafacherris").

Keywords: Valencí Blanc, Beba, Esclafagerres, Esclafacherres, virus, sanitation, varietal identification, in vitro culture

1. Introduction

Grapevine (*Vitis vinifera* L.) is a crop of major economic importance distributed in warm areas worldwide [1] with a wine production of 2910 million hectoliters in 2014 and 75,866 square kilometers dedicated to grapevine culture [2]. The majority of the world's wine-producing regions are found between the temperate latitudes of 30 and 50° in both hemispheres [3]. Grapes are mainly used for making wine but also can be eaten fresh as table grapes or used for making jam, juice, jelly, grape seed extract, raisins, vinegar, and grape seed oil. Approximately 71% of world grape production is used for wine, 27% as fresh fruit, and 2% as dried fruit. In the *Vitis* International Variety Catalog (VIVC; http://www.vivc.de/), supported by Biodiversity and the International Organization of Vine and Wine (OIV), there are around 24,500 accessions which include cultivars, breeding lines, and different *Vitis* species. Around 50% (12,679) of the varieties correspond to *Vitis vinifera* Linné Subsp. *vinifera* (or *sativa*), and 30% (7714) correspond to *Vitis* interspecific crossing. Around 25% of the cultivars were registered in France (5602), followed by the United States (2401) and Italy (2348) with approximately 10% each one. Spain has registered a total of 734 varieties, being most of them (631) *V. vinifera*. According to Lacombe [4], a total of 1902 grape varieties (both scions and rootstocks) are officially authorized for cultivation in at least one country of the European Union. Around 65% of these grape varieties are registered only in one country, meaning the responsibility to preserve these varieties is too focused. On the other hand, four varieties ("Cabernet Sauvignon," "Merlot," "Chardonnay Blanc," and "Sauvignon Blanc") were maintained in at least 60 different institutions. Nowadays, not only these cultivars but also "Syrah" (or "Shiraz") dominates vineyards worldwide [5]. Considering that most major wine-producing regions could become by 2050 unsuitable for currently grown cultivars [6, 7], the preservation of genetic variability and the selection and/or development of cultivars well adapted to upcoming climate changes are important. The long juvenile period of grapevine makes breeding a slow process; therefore, the knowledge of the raw material and their availability is very important to speed up breeding programs.

1.1. Grapevine: gain and loss of diversity

Vitis vinifera subsp. vinifera was domesticated in the Neolithic period (*ca* 8500–4000 BC) [8] from wild grapevines (*Vitis vinifera* L. subsp. *sylvestris* (Gmelin) Hegi) [9–12]. Grapevine domestication appears to have occurred between the seventh and the fourth millennia BC, in a geographical area between the Black Sea and Iran [8, 13–16], and the earliest evidence for large-scale winemaking was found in the North of Zagros Mountains and in the Caucasian region around 6000–5000 BC [17]. Cuttings of cultivated grapevines would have been spread by humans in the Near East, Middle East, and Central Europe. As a result, these areas may have constituted secondary domestication centers [18, 19] where spontaneous hybridizations among cultivars or local wild plants generated the pattern of admixture that is observed in current cultivars [19–23]. In consequence, genetic variability of grapevine has increased due to the contribution of different genetic pools in the process of grapevine spreading. The appearance of spontaneous mutations [24] and the selective pressure by humans which depended on the different uses of grapevine (fresh consumption, raisin, or wine production) [25] were also contributed to increase the genetic variability of this crop.

Along the years, genetic erosion has occurred in both cultivated and wild grapevines. Anthropogenic pressure on the wild natural habitats greatly decreased the wild grapevine populations that were also affected by the phylloxera aphid (*Daktulosphaira vitifoliae* Fitch) introduced from North America during the second part of the nineteenth century [26]. The phylloxera pest devastated the vineyards in all Europe. Since that time, grapevines need to be grafted onto phylloxera-resistant rootstocks. This fact reduced the number of grapevine cultivars used as scions which provoke genetic erosion [12]. On the other hand, the creation of denominations of origin (DO), each one including a reduced number of authorized varieties, has also contributed to reduce the varieties cultured in a specific area. Therefore, the preservation of grapevine minor cultivar and that on risk of disappearance together with *Vitis vinifera subsp. sylvestris* is a major stake in grapevine preservation.

1.2. Grapevine preservation

The importance of germplasm preservation focused on their putative use, in the present or in the future. It is the source of genes to face new pathogens or climate constraints. Genetic diversity of grapevine is maintained normally as living plants in the field [27, 28]. Several important ex situ grapevine collections exist like "The Domaine de Vassal" in Montpellier (France), the "Julius Kühn Institute" in Siebeldingen (Germany), and "La colección de vides de El Encín" in the IMIDRA (Instituto Madrileño de Investigación y Desarrollo Rural, Agrario y Alimentario) center of Alcalá de Henares (Spain). The French collection houses 7800 accessions *ca* 50 countries, representing 2300 different grape varieties, including wild species, rootstocks, hybrids, and mutants. Its transfer to the INRA Pech Rouge Experimental Unit (Gruissan, Aude) is under progress (https://www6.montpellier.inra.fr/vassal_eng/). The German collection of the Institute for Grapevine Breeding Geilweilerhof holds more than 3000 accessions of cultivars and wild species as well as important breeding lines. Beyond many others old and neglected cultivars from Germany, Switzerland, and Austria and rare germplasm from Eastern Europe can be found (https://www.julius-kuehn.de/en/grapevine-breeding/fields-of-activity/genetic-resources-and-information-centre-vine-and-wine/). The Spanish collection consists of 3532 accessions that are grouped into 852 rootstocks; 69 interspecific hybrids; 111 *Vitis* spp.; 1852 *V. vinifera* varieties, of which 1178 are for wine use and 674 for table use; and 648 *V. vinifera sylvestris* (http://www.madrid.org/coleccionvidencin/index.php?option=com_content&view=article&id=9&Itemid=2). In order to avoid the loss of the stored materials which are exposed to environmental disasters and pest attacks, the duplication of accessions for storage in different collections is a common strategy, although limited by budget constraints.

Another complementary strategy very useful in vegetative propagated plants is the in vitro preservation that offers the possibility to maintain plants under controlled and slow growing conditions and their micropropagation and transference to the field when need it. Although the first attempts to store grapevine under in vitro culture conditions were reported in the 1980s [29–31], this strategy is not usual in grapevine, although it is commonly used in other vegetative propagated crops like banana (preserved both through standard in vitro conditions and cryopreservation in the International Network for the Improvement of Banana and Plantain germplasm bank, in Leuven, Belgium). Cryopreservation is the storage of viable tissues, generally meristems or embryos, at ultralow temperature [32]. The success of in vitro conservation is tightly related to the choice of an adequate conservation method with the development of the corresponding methodology [33].

In grapevine, it is convenient to check by molecular markers that the variety to be stored really corresponds to it because homonymies (similar name for different cultivars) and/or synonymies (different names for a same cultivar) are commonly found. The identification of homonymies is important to avoid the loss of variability (loss of genotypes). On the contrary, the detection of synonymies avoids the maintenance of duplicated materials that do not contribute to increase variability but increase the cost. It is also very important to check the sanitary status of the plants, sanitize them if necessary, and provide suboptimal culture conditions that limit and slow down plant development, without causing physiological damage to the plant material. Grapevine can be infected by numerous viruses [34], and a high incidence of virus infection is commonly found in autochthonous cultivars [35, 36]. The EU Directive 2002/11/EC rules require that the initial plant material for vegetative propagation is free of *Grapevine fanleaf virus* (GFLV), *Arabis mosaic virus* (ArMV), *Grapevine fleck virus* (GFkV), *Grapevine leafroll–associated virus-1* (GLRaV-1), and *Grapevine leafroll–associated virus-3* (GLRaV-3). When samples to be storage are virus infected, different approaches to regenerated virus-free plants can be used. Since the middle late of the twentieth century, meristem culture and thermotherapy were applied with this aim for grapevine sanitation [37, 38]. Other techniques such as chemotherapy, electrotherapy, cryotherapy, and somatic embryogenesis were also reported [39–41].

For in vitro preservation under standard or limiting conditions, the development of protocols adjusted to the variety to be preserved is necessary. For this kind of storage, it is important to choose the adequate culture medium and environmental conditions in order to reduce the number of subcultures and hence minimize the cost and the putative errors that could arise in each subculture. Protocols for storage grapevine under in vitro culture have been reported by several authors [42, 43]. Recently, we reported the effectiveness of the MW medium to store a broad spectrum of grapevine cultivars, including endangered varieties, as well as the modifications of this medium (reduction of sucrose or elimination of indole-3-butyric acid (IBA) in the medium) to reduce growth in the faster growing varieties [44]. Cryopreservation protocols have also been developed for grapevine [45–47]. When developing cryopreservation protocols, the evaluation of the cryopreservation solution toxicity in the varieties to be preserved is required. Pre-culture in culture media that facilitate the dehydration of tissues (i.e., in medium with high content of sugars) is also a common step.

Finally, studies of genetic variability among varieties under conservation are important because they allow the detection of mutations and the study of relationships among them. The determination of molecular profiles is also of great importance in the development of core collections which represent the higher variability present in a whole collection in a reduced number of accessions. The core collections are very useful for breeders because studies on the core collection provide an overall view of the properties to be found in the whole collection [48]. In the following schema, the different steps to be carried out for germplasm storage are shown (**Figure 1**).

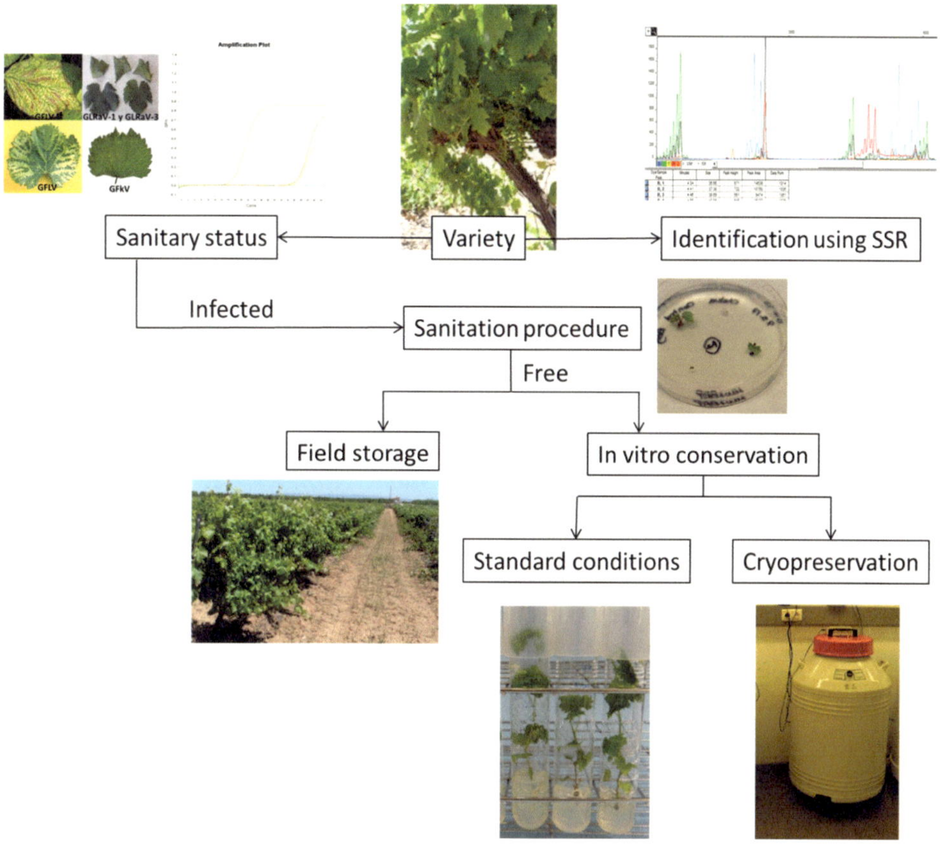

Figure 1. Scheme including the main steps required for germplasm storage.

2. Recovering ancient germplasm at the Comunitat Valenciana: the historic varieties "Valencí Blanc" and "Esclafagerres"

2.1. Richness of grapevine germplasm in the Comunitat Valenciana

The Comunitat Valenciana which includes the provinces of Alicante, Valencia, and Catellón is located in the Mediterranean coast of Spain and has been an important viticulture area since historic times to nowadays. For instance, in the archeological place of "L'Alt de Benimaquia" (Denia, Alicante), dated back to the end of the seventh century BC, significant quantities of vinification residues (tartaric acid and seeds) were found [49]. Important vineyards must be also located in the Requena-Utiel plateau (Valencia) as evidenced by the big presses dated in the fifth century BC found at Las Pilillas site [50]. Nowadays there exist three DO for wine production (DO Alicante (http://www.vinosalicantedop.org/), DO Valencia (http://www.dovalencia.info/), and DO Utiel-Requena (http://utielrequena.org/)), one DO for table grape (DO Uva de mesa del Vinalopó (http://uva-vinalopo.org/wp/)) and one protected geographical indication in Castellón (http://www.igpcastello.com/).

The richness in grapevine cultivars before the arrival to the phylloxera pest in the provinces of Alicante and mainly, in that of Valencia, is well documented [51, 52]. In a report about grapevine varieties cultured in Spain in 1889, it is pointed that more than 150 varieties were cultured in different locations of the Valencia province. These varieties included varieties with berries of black, white, and red color. In comparison with other provinces that also appeared in this report, Valencia was one of the richest [52]. The phylloxera aphids that devastated European vineyards invaded Spain in 1878 from three areas (Girona, Málaga, and Portuguese border). Its arrival to the Comunitat Valenciana, with the consequent loss of grapevine variability, occurred in 1912 when the aphids spread to Sagunto, Líria, and Requena [53].

In the context of the research project CGL2015–70843-R, we initiated different approaches in order to contribute in the recovering of ancient varieties in the risk of disappearance from Alicante and Valencia provinces. The objectives of this project include the analysis of grapevine germplasm diversity and the development of protocols for virus sanitation and in vitro conservation. Different prospections have been performed in order to rescue ancient varieties. The determination of SSR profiles is being useful to confirm the identity and to detect synonymies and homonymies that are very common in grapevine. The analysis of the genetic variability will identify accessions which may carry useful mutations for adaptation to specific environments. As occurred in other areas, grapevine cultivars were found commonly infected by the viruses GFLV, GLRaV, and GFkV. Sanitation of cultivars to be preserved in vitro is being carried out through meristem culture, although other alternative sanitation procedures are being developed [42, 54]. In this work, we report the SSR profiles of different accessions of the historic variety "Valencí Blanc" or "Beba" and the historic and endangered variety "Esclafagerres" (or "Esclafacherres/is"). The methodologies used for their sanitation and in vitro conservation are also summarized.

2.2. The historic varieties "Valencí Blanc" and "Esclafagerres"

The "Valencí Blanc" variety also known as "Beba" is a minor cultivar usually used as white table grape (**Figure 2A**). Despite in the past it was used for wine and raisin production [52, 55], today it is cultured as table grape for minor consume, and it is authorized for wine production in DO Ribera del Guadiana (Spain), where it is also named as "Eva" (http://riberadelguadiana.eu/esp/). The origin of this variety is unknown although it is proposed an oriental or North African origin [56]. The name "Valencí" ("Valensi" and "Balansí" in older reports) remembers to the name of Valencia City [57]. Oliver-Fuster (1980) cited by [58] proposed that this variety maybe was introduced by Balearic people who emigrate to Argelia. The most antique synonymy assigned for this variety is "Calop" [59] although other synonymies like "Ain el Kelp," "Tebourbi," "Panse the Provence" and "Grumer" are reported [56, 58–60]. In a report about grapevines cultivated in 1889 [52], the culture of "Grumer" is noted, among other 18 grapevine varieties in the Alicante province and in Albaida and Onteniente (locations nearest to the Alicante province). In the same report, the culture of "Valensí" in Alberique and Enguera (nearest to Valencia City) appeared. Recently, with the name of "Grumer," we have identified some accessions from the Alicante province that do not correspond to "Valencí" but grouped with "Muscat Istambul" [61]. Lacombe et al. [62] proposed as the origin of "Muscat Istambul" the cross of "Muscat of Alexandria" × "Valencí Blanc" which was confirmed by Mena et al. [63]. Therefore, it could be

Figure 2. Grapes of "Valencí Blanc" (A) and "Esclafagerres" (B) varieties.

easy to found this homonym. In the VIVC database, 71 synonymies for "Beba" appeared. However, some of these ("Chelva," "Hebén," "Mantúo," "Teta de Vaca," and "Uva de Planta") were rejected as they did not share the same SSR profile [64]. Probably other of those proposed are also false synonymies. On the other hand, in the report about grapevine varieties cultured in Spain in 1889 [52], the variety "Valencí" was included in the groups of cultivars with black and white berries. The comparison among the SSR profiles of some accessions of "Valencí" with white grapes ("Valencí Blanc") and with black grapes ("Valencí Tinto" or "Valencí Negre" in the Comunitat Valenciana) indicates that they are not mutant for berry color but resulting from different crosses. Comparing the SSR profile of the variety "Heben" (or "Gibi"), proposed as parent of "Valencí Blanc," with the SSR profile of the accession of "Valencí Tinto" hold in the VIVC database, no relationship between them was observed. However, one or two alleles were shared between "Valencí Blanc" and "Valencí Tinto." Therefore, the unknown parent from "Valencí Blanc" could be the parent of "Valencí Tinto." The name of "Valensi Chaselas" also appeared in the report of grapevines cultured in 1889, concretely as cultured in the Valencian location of Gandía.

"Esclafagerres" variety (**Figure 2B**), which name means that which bursts the jars, is also an ancient variety with white berries commonly grown on the Alicante and Valencia provinces. Some old references that mentioned the culture of this variety in the Alicante province are reported by several authors [65–67]. In DGAIC [52], the culture of "Esclafagerres" appeared in the Alicante and in the Valencian locations of Albaida, Onteniente, and Sagunto, where it was included among the varieties with white and also with black berries. This variety was usually mixed with other grapevine varieties like "Merseguera" for wine production. The "Esclafagerres" variety gives high yields (probably the meaning of the name is related to this) and has grapes with low sugar content despite it was commonly cultured under dry land.

In this work, we report the assays performed with both varieties in order to determine their genetic profiles and resume the strategies performed for virus sanitation and in vitro conservation.

2.3. Determining the SSR profiles

A total of 14 samples of "Valencí Blanc" from different locations (**Table 1**) and two samples of "Esclafagerres" from La Mata and Monforte del Cid were used for DNA extraction and SSR

Code	Origin	VVMD27		VVMD5		VVS2		VrZAG83		VrZAG79		VrZAG62		VrZAG64		VVMD7		VVMD24		VVMD32		VVMD25		VMC1b11		VVMD28		VVMD6		VVMD21	
		A1	A2	A1	A2	A1	A2	A1	A2	A1	A2	A1	A2	A1	A2	A1	A2	A1	A2	A1	A2	A1	A2	A1	A2	A1	A2	A1	A2	A1	A2
Vb-Pe0	Penàguila (Alicante)	179	187	233	237	136	144	197	197	242	246	189	205	134	158	241	247	206	208	254	270	255	255	185	189	243	257	209	211	248	254
Vb-Pe1	Penàguila (Alicante)	179	187	233	237	136	144	197	197	242	246	189	205	134	158	241	247	206	208	254	270	255	255	185	189	243	257	209	211	248	254
Vb-Pe2	Penàguila (Alicante)	179	187	233	237	136	144	197	197	242	246	189	205	134	158	241	247	206	208	254	270	255	255	185	189	243	257	209	211	248	254
Vb-Pe3	Penàguila (Alicante)	179	187	233	237	136	144	197	197	242	246	189	205	134	158	241	247	206	208	254	270	255	255	185	189	243	257	209	211	248	254
Vb-Pe4	Penàguila (Alicante)	179	187	233	237	136	144	197	197	242	246	189	205	134	158	241	247	206	208	254	270	255	255	185	189	243	257	209	211	248	254
Vb-Pe5	Penàguila (Alicante)	179	187	233	237	136	144	197	197	242	246	189	205	134	158	241	247	206	208	254	270	255	255	185	189	243	257	209	211	248	254
Vb-Pe10	Penàguila (Alicante)	179	187	233	237	136	144	197	197	242	246	189	205	134	158	241	247	206	208	254	270	255	255	185	189	243	257	209	211	248	254
Vb-Be1	Benirarres (Alicante)	179	187	233	237	136	144	197	197	242	246	189	205	134	158	241	247	206	208	254	270	255	255	185	189	243	257	209	211	248	254
Vb-Al1	Alicante	179	187	233	237	136	144	197	197	242	246	189	205	134	158	241	247	206	208	254	270	255	255	185	189	243	257	209	211	248	254
Vb-FF1	La Font de la Figuera (Valencia)	179	187	233	237	136	144	197	197	242	246	189	205	134	158	241	247	206	208	254	*254*	255	255	185	189	243	257	209	211	248	254
Vb-FA1	Fontanars dels Alforins (Valencia)	179	187	233	237	136	144	197	197	242	246	189	205	134	158	241	247	206	208	254	*254*	255	255	185	189	243	257	209	211	248	254
Vb-On1	Ontinyent (Valencia)	179	187	233	237	136	144	197	197	242	246	189	205	134	158	241	247	206	208	254	*254*	255	255	185	189	243	257	209	211	248	254
Vb-Ba1	Fuente del Maestre (Badajoz)	179	187	233	237	136	144	197	197	242	246	189	205	134	158	241	247	206	208	254	*254*	255	255	185	189	243	257	209	211	248	254
Vb-Cu1	Campillo de Altobuey (Cuenca)	179	187	233	237	136	144	197	197	242	246	189	205	134	158	241	247	206	208	254	*254*	255	255	185	189	243	257	209	211	248	254
Es-Ma1	La Mata (Alicante)	187	191	233	237	146	152	193	197	242	246	189	197	134	136	237	247	206	208	254	270	241	255	171	189	233	247	209	211	248	248
Es-Mo1	Monforte del Cid (Alicante)	187	191	233	237	146	152	193	197	242	246	189	197	134	136	237	247	206	208	254	270	241	255	171	189	233	247	209	211	248	248

Allele sizes are expressed as base pairs.

Table 1. SSR profiles for 14 accessions of "Valencí Blanc" (Vb-Pe0, Vb-Pe1, Vb-Pe2, Vb-Pe3, Vb-Pe4, Vb-Pe5, Vb-Pe10, Vb-Be1, Vb-Al1, Vb-FF1, Vb-FA1, Vb-On1, Vb-Ba1, Vb-Cu1) and two of "Esclafagerres" (Es-Ma1, Es-Mo1) varieties.

analysis. DNA was extracted from fully expanded leaves using the commercial DNeasy Plant Mini Kit (Qiagen) according to the manufacturers' instructions. DNA quality and quantity were assessed using gel electrophoresis and spectrophotometry. Fifteen SSR markers (VVS2, VVMD5, VVMD6, VVMD7, VVMD24, VVMD25, VVMD27, VVMD28, VVMD32, VrZAG62, VRzAG64, VrZAG79, VrZAG83, and VMC1b11) were analyzed using two sets of multiplex PCR reactions. Each multiplex was carefully assembled according to the compatibility of the SSRs during PCR and the molecular size of their amplicons. The forward primer of the SSR markers was labeled with one of the four fluorescent dyes: carboxyfluorescein (FAM), carboxytetramethylrhodamine (TAMRA), hexachloro-6-carboxyfluorescein (HEX), or 6-carboxytetramethyl rhodamine (ROX) [61]. Multiplex PCR was carried out in a total volume of 11.00 μL, using 1.25 μL of commercial Master Mix PCR Multiplex (Takara Multiplex Hot Short PCR, Takara), 20–40 ng of genomic DNA, 0.1 μL of Takara Taq Hot Start, and labeled multiplexed SSR primers (from 5.5 to 35.0 μmol). The amplification was performed in an ABI 9700 thermocycler, and the amplification conditions were 95°C for 14 min followed by 30 cycles of 95°C for 30 s, 55°C for 90 s, and 72°C for 60 s and a final extension of 72°C for 30 min. Previous to PCR fragment size determination, the multiplex PCR product was previsualized using gel electrophoresis. The electrophoresis was carried out on an ABI 3100 platform (Applied Biosystems, Foster City, CA, USA). For PCR fragment size determinations, 0.13 μL of an internal size standard (GeneScan™ 500 LIZ, Applied Biosystems) was mixed with 1.00 μL of PCR product and 10.87 μL of formamide. The mixture was heated at 94°C for 3 min and then cooled in icy water. The size of the SSR fragments was determined with the software package GeneScan 3.7 (Applied Biosystems).

The SSR profiles of the analyzed accessions are shown in **Table 1**. Whereas similar SSR profiles were found for both accessions of "Esclafacherre," some variability is found among "Valencí Blanc" accessions. Among the 15 SSRs analyzed, differences in the SSR VVMD32 were found: the accessions collected in the province of Alicante have two alleles in this loci VVMD32 (254; 270), whereas the rest of accessions from the provinces Badajoz, Cuenca, and Valencia have the allele 254 in homozygosity. We can consider that among the analyzed germplasm, there are two variants of "Valencí Blanc" as in other cases in which two plants showed identical SSR profiles for all the SSR markers studied except for one or two alleles. This could be attributable to slight clonal polymorphism [68]. They may have originated in a similar place and then spread to different areas. The accession of "Valencí Blanc" in the VIVC database (accession number 22710) and that reported by Lacombe et al. [62] had the same profile for the comparable SSRs (including the VVMD32) to the accessions from Alicante. Similarly, accessions from Alicante showed also identical SSR profile to two accessions of "Beba" from El Encín grapevine collection analyzed by Mena [64].

The comparison of the SSR profile of the "Esclafagerres" accessions to SSR profiles in the VIVC database (including 3265 accessions), those in the International Vitis database (including 3430 accessions), as well as those reported in several publications [62, 63, 69] did not match with any of the included varieties. No matches were found when the SSR profile of "Esclafagerres" was blasted to the Italian Vitis database. Therefore, this profile should correspond with that of the "Esclafagerres" variety which has not been reported before.

2.4. Virus analysis and sanitation

To analyze the putative virus infection in the original samples, the methodology described by López-Fabuel et al. [70] was used. Briefly, extracts were prepared from leaves 1/20 in phosphate buffered saline (PBS) buffer, pH 7.2, supplemented with 0.2% diethyldithiocarbamic acid (DIECA) and 2% polyvinylpyrrolidone-10 (PVP-10) in individual plastic bags with a heavy net (Plant Print Diagnostics). Total RNA was extracted from 200 μL of crude extract using an Ultraclean Plant RNA isolation kit (Mobio) following the manufacturer's instructions. The real-time multiplex reverse transcription polymerase chain reaction (RT-PCR) was performed for the simultaneous detection of ArMV, GFLV, GFkV, GLRaV-1, and GLRaV-3 using a StepOnePlus thermocycler (Applied Biosystems) and a reaction mixture containing 1× AgPath-ID One-step RT-PCR buffer (Ambion) and 1.5× AgPath-ID One-step RT-PCR enzyme mix (Ambion); 5 μL of sample; 400 nM of GFLV, ArMV, GFkV, and GLRaV-1 primers; 800 nM of GLRaV-3 primers; and 200 nM of each probe. The amplification protocol consisted of an RT step at 45°C for 10 min and a denaturation step at 95°C for 10 min, followed by 45 cycles of amplification (95°C, 15 s; 50°C, 15 s; and 60°C, 60 s). As positive controls viral isolates maintained at the Instituto Valenciano de Investigaciones Agrarias (IVIA) were included. When amplification was observed for a specific virus, it was confirmed by real-time uniplex RT-PCR using the corresponding primers.

Meristem culture was used in a previous project (RTA2011–00067-C04) to obtain virus-free plants of "Esclafagerres." In the context of the project CGL2015–70843-R, "Valencí Blanc" (sample Vb-Pe0), which resulted in infection with GFkV and GLRaV-3, was sanitized through both meristem culture and somatic embryogenesis. Meristems (n = 35) from plants of "Valencí Blanc" were extracted using a binocular lent and cultured in vitro on plates (90 × 15 mm) containing the medium MW, selected for "Monastrell" micropropagation [71] supplemented with 6-benzylaminopurine (BAP) at 1.8 μM. Low light conditions were used for the two first weeks of culture. Only 54.3% of meristems grew after 20 days of culture (**Figure 3A**), and two develop into plants after transferring to tubes with MW, 70 days after (**Figure 3B**). Damage of meristem during extraction and/or the composition of the culture medium that could need to be enriched with other nutrients is putatively the cause of the low and slow regeneration of meristems. The analysis for virus presence of these two plants was carried out as described before, and one of them (50%) resulted free for both viruses. Therefore, from this plant, clones were obtained for in vitro conservation. Despite the fact that meristem culture is an efficient technology for virus sanitation, it is needed to obtain an adequate size of the meristem in order to avoid virus transmission allowing meristem development.

The other methodology used for virus sanitation was the induction of somatic embryos as reported in Peiró et al. [41]. Briefly, seeds of "Valencí Blanc" were extracted from grapes and disinfected and cut previously to be cultured on the embryogenesis induction medium (EIM2) which contained TDZ (thidiazuron) at 0.9 μM. Thirteen percent of explants responded after 60 days of culture on this medium (**Figure 3C–D**). Despite in grapevine a high percentage of somatic embryos are not able to develop into normal plants [72–74], in our work, germination of somatic embryos occurred directly in the induction medium and grew correctly (**Figure 3E**).

Figure 3. Strategies for virus sanitation of "Valencí Blanc" (accession Vb-Pe0). (A) Meristems cultured on MW supplemented with 6-benzylaminopurine (BAP) at 1.8 μM and without indole butyric acid (IBA) after 20 days of culture. (B) Plants from meristems cultured in MW medium after 90 days of culture. (C) Cut seeds cultured in EIM2 medium at day 0. (D) Seeds with some somatic embryos after 45 days of culture. (E) Plants regenerated from somatic embryos.

Ten developed plants were analyzed for evaluating the success of virus cleaning. All plants regenerated from somatic embryos resulted virus-free for both GFkV and GLRaV-3 viruses. Therefore, 100% of sanitation was obtained. The result obtained for GFkV was expected because this virus is not seed transmitted [75]. With respect to GLRaV-3, it is not clear if it is present in seeds [34, 75], but we have found this virus in some regenerated plants resulting from somatic embryos induced from seeds of other infected grapevines, which would indicate its presence in the seeds [54]. Induction of somatic embryos from stamens or pistils was also reported in grapevine to cure plants of GLRaV [76, 77]. We also analyzed the SSR profile of regenerated virus-free plants in order to select those regenerated from mother tissues of the seeds, which will show the mother genotype. The 15 SSRs used for determining the SSR profile of "Valencí Blanc" accessions were used. One of ten analyzed plants showed the same SSR profile as the mother plant, that is, a 10% of regenerated plants were obtained from mother tissue and not from the embryo.

2.5. In vitro storage

Sanitized plants of "Valencí Blanc" and "Esclafagerres" are maintained in tubes with MW medium in an in vitro culture growth chamber under standard conditions (25± 2°C; 16 h light). The MW medium is adequate to storage a broad spectrum of grapevine cultivars including "Valencí Blanc" and "Esclafagerres" [44]. Both cultivars grew less than 4 cm after 40 days of culture in this medium. We consider that this speed of growth is acceptable to maintain these cultivars under standard conditions with small number of subcultures. A reduction of sugar or the elimination of IBA in the culture medium is used for maintaining cultivars that grew faster [44]. The higher the number of subcultures, the higher the cost and the higher the possibility to make nomenclature errors [78].

Another strategy to germplasm ex situ storage using in vitro culture is cryopreservation. With this methodology, the metabolism is greatly reduced, and few requirements are needed for the maintenance of tissue samples. Meristems of "Esclafagerres" from micropropagated virus-free plants maintained in vitro were used to initiate the cryopreservation assays using the methodology described in Gisbert et al. [79]. The first results indicated that 50 min of incubation in the plant vitrification solution 2 (PVS2) is adequate for recuperating cryopreserved meristems of "Esclafagerres" (**Figure 2**). Recently, Pathirana et al. [47] have reported a positive effect on grapevine regeneration when a pretreatment with salicylic acid was performed prior to cryopreservation. In both works [47, 79], the droplet vitrification protocol was used.

3. Conclusion

As a result of different actions performed in the context of the projects CGL2015-70843-R and RTA2011-00067-C04, a broad spectrum of grapevine varieties are being evaluated in order to determine the varietal identification and their variability and also their capacity for in vitro culture, plant regeneration, and germplasm storage. Different strategies for virus cleaning have been developed and applied to rescue virus-free plants. Among the analyzed materials, the historic varieties "Valencí Blanc" and "Esclafagerres" were sanitized and currently are maintained under in vitro culture conditions. Differences for the microsatellite VVMD32 were found among "Valencí Blanc" accessions, clustering the accessions from Alicante and that of other origins. The SSR profile for the variety "Esclafagerres" was firstly reported in the present work.

Acknowledgements

The authors thank the support of the projects CGL2015-70843-R and RTA2011-0020067-C04 both co-funded by FEDER Funds.

Author details

Carmina Gisbert[1]*, Rosa Peiró[1], Tania San Pedro[2], Antonio Olmos[2], Carles Jiménez[3] and Julio García[4]

*Address all correspondence to: cgisbert@btc.upv.es

1 Polytechnic University of Valencia (Universitat Politècnica de València), Valencia, Spain

2 Valencian Institute of Agrarian Investigations (IVIA), Valencia, Spain

3 Plant Health Service of the Generalitat Valenciana, Valencia, Spain

4 Conselleria of Agriculture, Environment, Climate Change and Rural Development, Alicante, Spain

References

[1] Dallakyan MV, Yesoyan SS, Harutyunyan FA, Melyan GH, Yesayan AH, Hovhannisyan NA. Application of microsatellite markers for varietal identification grape rootstocks: A preliminary analysis. Biological Journal of Armenia. 2013;**3**:136-138

[2] FAOSTAT [Internet]. 2017. Available from: http://www.fao.org/faostat/en/#data/QC [Accessed: 2017–06-15]

[3] Fraga H, Garcia de CAI, Malheiro AC, Santos JA. Modelling climate change impacts on viticultural yield, phenology and stress conditions in Europe. Global Change Biology. 2016;**22**:3774-3788. DOI: 10.1111/gcb.13382

[4] Lacombe T, Audeguin I, Boselli M, Bucchetti B, Cabello F, Chatelet P, Crespan M, d'Onofrio C, Eiras-Dias J, Ercisli S, Gardiman M, Grando MS, Imazio S, Jandurova O, Jung A, Kiss E, Kozma P, Maul E, Maghradze D, Martinez MC, Muñoz G, Pátková JK, Pejic I, Peterlunger E, Pitsoli D, Preiner D, Raimondi S, Regner F, Savin G, Savvides S, Schneider A, Spring JL, Szoke A, Veres A, Boursiquot JM, Bacilieri R, This P. Grapevine European catalogue: Towards a comprehensive list. Vitis. 2011;**50**:65-68

[5] Anderson K, Nelgen S. Global Wine Markets, 1961 to 2009: A Statistical Compendium. Adelaide: University of Adelaide Press; 2011. p. 468. DOI: 10.1017/UPO9780987073013

[6] Hannah L, Roehrdanz PR, Ikegami M, Shepard AV, Shaw MR, Tabor G, Zhi L, Marquet PA, Hijmans RJ. Climate change, wine, and conservation. Proceedings of the Natural Academy of Sciences in the United States of America. 2013;**110**:6907-6912. DOI: 10.1073/pnas.1210 127110

[7] van Leeuwen C, Schultz HR, de Cortazar-Atauri IG, Duchene E, Ollat N, Pieri P, Bois B, Goutouly JP, Quénol H, Touzard JM, Malheiro AC, Bavaresco L, Delrot S. Why climate change will not dramatically decrease viticultural suitability in main wine-producing areas by 2050. Proceedings of the Natural Academy of Sciences of the United States of America. 2013;**110**:E3051-E3052. DOI: 10.1073/pnas.1307927110

[8] McGovern PE, Glusker DL, Exner LJ, Voigt MM. Neolithic resinated wine. Nature. 1996;**381**:480-481. DOI: 10.1038/381480a0

[9] Rossetto M, McNally J, Henry RJ. Evaluating the potential of SSR flanking regions for examining taxonomic relationship in the *Vitaceae*. Theoretical and Applied Genetics. 2002;**104**:61-66. DOI: 10.1007/s001220200007

[10] Sefc KM, Steinkellner H, Lefort F, Botta R, da Câmara-Machado A, Borrego J, Maletic E, Glössl J. Evaluation of the genetic contribution of local wild vines to European grapevine cultivars. American Journal of Enology and Viticulture. 2003;**54**:15-21

[11] Crespan M. Evidence on the evolution of polymorphism of microsatellite markers in varieties of *Vitis vinifera* L. Theoretical and Applied Genetics. 2004;**108**:231-237. DOI: 10.1007/s00122-003-1419-5

[12] This P, Jung A, Boccacci P, Borrego J, Costantini L, Crespan M, Eisenheld C, Grando MS, Lacombe T, Lacou V, Meredith CP, Milani N, Ortiz J, Peterlunger E, Regner Z, Zulini L, Dettweiler E. Development of a common set of standard varieties and standardized method of scoring microsatellites markers for the analysis of grapevine genetic resources. Theoretical and Applied Genetics. 2004;**109**:1448-1458. DOI: 10.1007/s00122-004-1760-3

[13] La CC. Transcaucasie au Néolithique et au Chalcolithique. Oxford: British Archaeological Series; 1995. p. 240

[14] McGovern PE, Rudolph HM. The analytical and archaeological challenge of detecting ancient wine: Two case studies from the ancient Near East. In: McGovern PE, Fleming SJ, Katz SH, editors. The Origins and Ancient History of Wine. New York: Gordon and Breach; 1996. p. 57-67

[15] Zohary D. The mode of domestication of the founder crops of the southwest Asian agriculture. In: Harris DR, editor. The Origin and Spread of Agriculture and Pastoralism in Eurasia. London: University College London Press; 1996. p. 142-158. DOI: 10.1017/S0016672300034558

[16] Zohary D, Hopf M. Domestication of Plants in the Old World: The Origin and Spread of Cultivated Plants in West Asia, Europe, and the Nile Valley. 3rd ed. New York: Oxford University Press; 2000. p. 151-159

[17] McGovern PE. Ancient Wine: The Search of the Origin of the Viniculture. Princeton University Press: Princeton; 2003. p. 392

[18] Grassi F, Labra M, Imazio S, Spada A, Sgorbati S, Scienza A, Sala F. Evidence of a secondary grapevine domestication centre detected by SSR analysis. Theoretical and Applied Genetics. 2003;**107**:1315-1320. DOI: 10.1007/s00122-003-1321-1

[19] Arroyo-García R, Ruíz-García L, Bolling L, Ocete R, López MA, Arnold C, Ergul A, Söylemezoğlu G, Uzun HI, Cabello F, Ibáñez J, Aradhya MK, Atanassov A, Atanassov I, Balint S, Cenis JL, Costantini L, Goris-Lavets S, Grando MS, Klein BY, McGovern PE, Merdinoglu D, Pejic I, Pelsy F, Primikirios N, Risovannaya V, Roubelakis-Angelakis KA, Snoussi H, Sotiri P, Tamhankar S, This P, Troshin L, Malpica JM, Lefort F, Martinez-Zapater JM. Multiple origins of cultivated grapevine (*Vitis vinifera* L. Ssp. *sativa*) based on chloroplast DNA polymorphisms. Molecular Ecology. 2006;**15**:3707-3714. DOI: 10.1111/j.1365-294X.2006.03049.x

[20] Aradhya MK, Dangl GS, Prins BH, Boursiquot JM, Walker MA, Meredith CP, Simon CJ. Genetic structure and differentiation in cultivated grape, *Vitis vinifera* L. Genetics Research. 2003;**81**:179-192. DOI: 10.1017/S0016672303006177

[21] Bacillieri R, Lacombe T, Le Cunff L, Di Vecchi-Staraz M, Lacou V, Genna B, Péros JP, This P, Boursiquot JM. Genetic structure in cultivated grapevines is linked to geography and human selection. BMC Plant Biology. 2013;**13**:25. DOI: 10.1186/1471-2229-13-25

[22] Fernández L, Le Cunff L, Tello J, Lacombe T, Boursiquot JM, Fournier-Level A, Bravo G, Lalet S, Torregrosa L, This P, Martínez-Zapater JM. Haplotype diversity of *VvTFL1A* gene

and association with cluster traits ingrapevine (V. *vinifera*). BMC Plant Biology. 2014;**14**:209. DOI: 10.1186/s12870-014-0209-3

[23] Myles S, Chia JM, Hurwitz B, Simon C, Zhong GY, Buckler E, Ware D. Rapid genomic characterization of the genus *Vitis*. PLoS One. 2010;**5**:e8219. DOI: 10.1371/journal.pone.0008219

[24] Bowers JE, Boursiquot JM, This P, Chu K, Johansson H, Meredith C. Historical genetics: The parentage of chardonnay, Gamay, and other wine grapes of north eastern France. Science. 1999;**289**:1562-1565

[25] Branas J. Viticulture Générale. Paul Dehan: Montpellier; 1974. p. 990

[26] Arnold C, Gillet F, Gobat M. Situation de la vigne sauvage *Vitis vinifera* ssp. *silvestris* en Europe. Vitis. 1998;**37**:159-170

[27] Santana JC, Hidalgo E, De Lucas AI, Recio P, Ortiz JM, Martín JP, Yuste J, Arranz C, Rubio JA. Identification and relationships of accessions grown in the grapevine (*Vitis vinifera* L.) Germplasm Bank of Castilla y Léon (Spain) and the varieties authorized in the VQPRD areas of the region by SSR-marker analysis. Genetic Resources and Crop Evolution. 2008;**55**:573-583. DOI: 10.1007/s10722-007-9261-2

[28] Leão PCS, Motoike SY. Genetic diversity in table grapes based on RAPD and microsatellite markers. Pesquisa Agropecuária Brasileira. 2011;**46**:1035-1044. DOI: 10.1590/S0100-204X2011000900010

[29] Barlass M, Skene KGM. Long term storage of grape *in vitro*. Plant Genetics Research Newsletter. 1983;**53**:19-21

[30] Alleweldt G. Investigations on the cold storage of in vitro cultures grapevines. In: Proceedings Moethennesy Coll Améliorament de la Vigne et Culture *in vitro*; 23–24 April 1985; Paris. p. 31-38

[31] Galzy R. Les possibilite de conservation in vitro d'une collection de clones de vignes. Bulletin Organisation Internationale de la Vigne et du Vin. 1985;**650**:378-390

[32] Benson EE. Cryopreservation of phytodiversity: A critical appraisal of theory and practice. Critical Reviews in Plant Sciences. 2008;**27**:141-219. DOI: 10.1080/07352680802202034

[33] García-Águila L, de Feria M, Acosta K. Aspectos básicos de la conservación *in vitro* de germoplasma vegetal. Biotecnología Vegetal. 2007;**7**:67-79

[34] Gasparro M, Caputo AR, Forleo LR, Perniola R, Alba V, Milella RA, Antonacci D. Study of main grapevine viruses transmission in breeding programs. BIO Web of Conferences. 2016;**7**:01039. DOI: 10.1051/bioconf/20160701039

[35] Laimer M, Lemaire O, Herrbach E, Goldschmidt V, Minafra A. Resistance to viruses, phytoplasmas and their vectors in the grapevine in Europe: A review. Journal of Plant Patholoigy. 2009;**91**:7-23. DOI: 10.4454/jpp.v91i1.620

[36] Bertolini E, García J, Yuste A, Olmos A. High prevalence of viruses in table grape from Spain detected by real-time RT-PCR. European Journal of Plant Pathology. 2010;**128**:283-287. DOI: 10.1007/s10658-010-9663-4

[37] Doré C, Varoquaux F. Histoire et Amélioration de Cinquante Plantes Cultivées. Paris: Editions Quae; 2006. p. 812

[38] Engelbrecht DJ, Schwerdtfeger U. *In vitro* grafting of grapevine shoot apices as an aid to the recovery of virus-free clones. Phytophylactica. 1979;**11**:183-185

[39] Panattoni A, Luvisi A, Triolo E. Review. Elimination of viruses in plants: Twenty years of progress. Spanish Journal of Agricultural Research. 2013;**11**:173-188. DOI: 10.5424/sjar/2013111-3201

[40] Bayati SH, Shams-Bakhsh M, Moieni A. Elimination of grapevine virus A (GVA) by cryotherapy and electrotherapy. Journal of Agricultural Science and Technology. 2011;**13**: 443-450

[41] Peiró R, Gammoudi N, Yuste A, Olmos A, Gisbert C. Mature seeds for *in vitro* sanitation of the grapevine leafroll associated virus (GLRaV-1 and GLRaV-3) from grape (*Vitis vinifera* L.). Spanish Journal of Agricultural Research. 2015;**13**:e1005. DOI: 10.5424/sjar/2015132-7094

[42] De Carvalho-Silva RC, Gomes-Luis Z, Scherwinski-Pereira JE. Short-term storage in vitro and large-scale propagation of grapevine genotypes. Pesquisa Agropecuaria Brasileira. 2012;**46**:344-350. DOI: 10.1590/S0100-204X2012000300005

[43] Hassanen SA, Abido AIA, Aly MAM, Rayan GA. *In vitro* preservation of grapevine (*Vitis vinifera* L.) 'Muscat of Alexandria' and 'black Monukka' cultivars as genetic resource. African Journal of Basic and Applied Sciences. 2013;**5**:55-63. DOI: 10.5829/idosi.ajbas.2013.5.2.23313

[44] San Pedro T, Peiró R, Jimenez C, Olmos A, Gisbert C. Evaluation of conditions for *in vitro* storage of commercial and minor grapevine (*Vitis vinifera* L.) cultivars. The Journal of Horticultural Science & Biotechnology. 2017. DOI: 10.1080/14620316.2017.1352462

[45] Bettoni JC, Dalla-Costa M, Pereira-Gardin JP, Kretzschmar AA, Pathirana R. Cryotherapy: A new technique to obtain grapevine plants free of viruses. Revista Brasileira de Fruticultura. 2016;**38**:e-883. DOI: 10.1590/0100-29452016833

[46] Markovic Z, Chatelet P, Preiner D, Sylvestre I, Kontic JK, Engelmann F. Effect of shooting medium and source of material on grapevine (*Vitis vinifera* L.) shoot tip recovery after cryopreservation. Cryo Letters. 2014;**35**:40-47

[47] Pathirana R, McLachlan A, Hedderley D, Panis B, Carimi F. Pre-treatment with salicylic acid improves plant regeneration after cryopreservation of grapevine (Vitis spp.) by droplet vitrification. Acta Physiologiae Plantarum. 2016;**38**:12. DOI: 10.1007/s11738-015-2026-1

[48] van Hintum TJL, Brown AHD, Spillane C, Hodgkin T. Core collections of plant genetic resources [Internet]. Available from: http://www.bioversityinternational.org/fileadmin/_migrated/uploads/tx_news/Core_collections_of_plant_genetic_resources_614.pdf [Accessed: 2017-06-19]

[49] Díes-Cusí E, Gómez-Bellard C, Guérin P. El vino en los inicios de la cultura ibérica: nuevas excavaciones en L'Alt de Benimaquia Denia. Revista de Arqueología. 1993;**142**:16-27

[50] Martínez A. Testimonios de viticultura durante la antigüedad en la meseta de Requena-Utiel (Valencia). In: Pérez SC, Blázquez-Pérez J, editors. Vine and Wine Cultural Heritage. Madrid: Universidad Autónoma de Madrid; 2013. p. 177-188

[51] Sanz de Bremón E. Denia. 1881-1980. Valencia: JM Borja, I Mora; 1982. p. 235

[52] DGAIC [Dirección General de Agricultura, Industria Y Comercio]. Avance estadístico Sobre Cultivo Y producción de la Vid en España Formado Por la Junta Consultiva Agronómica. Madrid: Madrid Tipografía de L. Péant e Hijos; 1891. p. 195

[53] Piqueras J. La filoxera en España y su difusión espacial: 1878-1926. Cuadernos de Geografía. 2005;**77**:101-136

[54] San Pedro T, Gammoudi N, Peiró R, Olmos A, Gisbert C. Somatic embryogenesis from seeds in a broad sense of *Vitis vinifera* L. varieties: Rescue of true to type virus free plants. In: BMC Plant Biology. 2017 (Submitted)

[55] Valcárcel JA. Agricultura General. Tomo VIII. Valencia: Ediciones Josep Esteban y Cervera; 1971. p. 290-295

[56] Laiadi Z, Bentchikou MM, Bravo G, Cabello F, Martínez-Zapater JM. Molecular identification and genetic relationships of Algerian grapevine cultivars maintained at the germplasm collection of Skidda (Algeria). Vitis. 2009;**48**:25-32

[57] Favà-Agud X. Diccionari dels Noms de Ceps i Raïms: L'ampelonímia Catalana. Institut d'Estudis Catalans: Barcelona; 2001. p. 456

[58] García-Muñoz S, Lacombe T, De Andrés MT, Gaforio L, Muñoz-Organero G, Lacou V, This P, Cabello F. Grape varieties (*Vitis vinifera* L.) from the Balearic Islands: Genetic characterization and relationship with Iberian peninsula and Mediterranean Basin. Genetic Resources of Crop Evolution. 2012;**59**:589-605. DOI: 10.1007/s10722-011-9706-5

[59] Martí G. Binisalem: su Historia y su Entorno. Cort: Palma de Mallorca; 1978. p. 238

[60] Truel P. Catalogue des Variétés en Collection à la Station de Recherches Viticoles de Montpellier. Domaine de Vassal INRA: Marseillan; 1985. p. 129

[61] Peiró R, Crespo A, Soler J, Jiménez C, Cabello F, Gisbert C. Analysis of genetic variability in Moscatel germplasm. Spanish Journal of Agricultural Research. 2017 (submitted)

[62] Lacombe T, Boursiquot JM, Lacou V, Di Vecchi-Staraz M, Peros JM, This P. Large-scale parentage analysis in an extended set of grapevine cultivars (*Vitis vinifera* L.). Theoretical and Applied Genetics. 2013;**126**:401-414. DOI: 10.1007/s00122-012-1988-2

[63] Mena A, Martínez J, Fernández-González M. Recovery, identification and relationships by microsatellite analysis of ancient grapevine cultivars from Castilla-La Mancha: The largest wine growing region in the world. Genetic Resources and Crop Evolution. 2014;**61**:625-637. DOI: 10.1007/s10722-013-0064-3

[64] Mena A. Recuperación, caracterización y conservación de variedades de vid (*Vitis vinifera* L.) minoritarias de Castilla-La Mancha [thesis]. Toledo: Universidad de Castilla la Mancha; 2013

[65] Lagasca M. El Herbario del Real Colegio Alfonso XII de San Lorenzo de El Escorial. Madrid: Anales del Jardín Botánico de Madrid; 1811. p. 397

[66] López-Camuñas. Diamante del Vinicultor, Cosechero de Vino, Fabricante de Aguardientes, Vinagres, Licores, etc. Nueva Obra de Consulta Necesaria. Madrid: Imprenta de Moreno y Rojas; 1880. p. 552

[67] Abela E, Sainz de Andino EJ. El Libro del Viticultor. Tipografía de Manuel G. Hernández. Maxtor Librería: Valladolid; 1885. p. 326

[68] Lacou V, Lacombe T, Dechesne F, Siret R, Bruno JP, Dessup M, Dessup T, Ortigosa P, Parra P, Roux C, Santoni S, Varès D, Péros JP, Boursiquot JM, This P. High throughput analysis of grape genetic diversity as a tool for germplasm collection management. Theoretical and Applied Genetics. 2011;**121**:1233-1245. DOI: 10.1007/s00122-010-1527-y

[69] Jiménez-Cantizano A. Caracterización molecular del Banco de Germoplasma de vid del Rancho de la Merced [thesis]. Cádiz: Universidad de Cádiz; 2014

[70] López-Fabuel I, Wetzel T, Bertolini E, Bassler A, Vidal E, Torres LB, Yuste A, Olmos A. Real-time multiplex RT-PCR for the simultaneous detection of the five main grapevine viruses. Journal of Virological Methods. 2013;**188**:21-24. DOI: 10.1016/j.jviromet.2012.11.034

[71] San Pedro T, Peiró R, Vilanova J, Olmos A, Gisbert C. *In vitro* multiplication of the *Vitis vinifera* L. cv. 'Monastrell. Electronic Journal of Biotechnology. 2017;**27**:80-83. DOI: 10.1016/j.ejbt.2017.03.006

[72] Goebel-Tourand I, Mauro MC, Sossountazov L, Miginiac E, Deloire A. Arrest of somatic embryo development in grapevine: Histological characterization and the effect of ABA, BAP and zeatin in stimulating plantlet development. Plant Cell Tissue and Organ Culture. 1993;**33**:91-103. DOI: 10.1007/BF01997603

[73] López-Perez AJ, Carreño J, Dabauza M. Somatic embryo germination and plant regeneration of three grapevine cvs: Effect of IAA, GA_3 and embryo morphology. Vitis. 2006;**45**:141-143

[74] Ben Amar A, Cobanov P, Boonrod K, Krczal G, Bouzid S, Ghorbel A, Reustle GM. Efficient procedure for grapevine embryogenic suspension establishment and plant regeneration: Role of conditioned medium for cell proliferation. Plant Cell Reports. 2007;**26**:1439-1447. DOI: 10.1007/s00299-007-0341-8

[75] Digiaro M, Martelli GP, Savino V. Proceedings of the Mediterranean network on grapevine closteroviruses 1992–1997 and the viruses and virus-like diseases of the grapevine a bibliographic report, 1985–1997. Bari: Centre international de Hautes Études Agronomiques Méditerranéennes; 1999. p. 280

[76] Gambino G, Bondaz J, Gribaudo I. Detection and elimination of viruses in callus, somatic embryos and regenerated plantlets of grapevine. European Journal of Plant Pathology. 2006;**114**:397-404. DOI: 10.1007/s10658-006-0004-6

[77] Goussard PG, Wiid J, Kasdorf GGF. The effectiveness of *in vitro* somatic embryogenesis in eliminating fanleaf virus and leafroll associated viruses from grapevines. South African Journal of Enology and Viticulture. 1991;**12**:77-81. DOI: 10.21548/12-2-2214

[78] Reed BM, Engelmann F, Dulloo ME, Engels JMM. Technical Guidelines for the Management of Field and In Vitro Germplasm Collections. Rome: International Plant Genetic Resources Institute; 2004. p. 116

[79] Gisbert C, Dabauza M, Correal E, Swennen R, Panis B. Cryopreservation of *Bituminaria bituminosa* varieties and hybrids. Cryobiology. 2015;**71**:279-285. DOI: 10.1016/j.cryobiol.2015.07.011

Chapter 2

Fingerprints of Anthocyanins and Flavonols in Wild Grapes (*Vitis vinifera* L. ssp. *sylvestris* (Gmelin) Hegi)

Eugenio Revilla, Rosa Arroyo-Garcia, Alberto Bellido,
David Carrasco, Anna Puig and Leonor Ruiz-Garcia

Additional information is available at the end of the chapter

http://dx.doi.org/10.5772/intechopen.70861

Abstract

Phenolic compounds are a group of natural products that play an important role in the quality of wines. Most phenolic compounds present in wine are derived from those contained in grapes and extracted from skins, seeds, and pulp during the initial steps of winemaking. Among them, anthocyanins and flavonols are involved in the color of red wines as pigments or copigments and also as precursors of polymeric pigments after reaction with other phenols. Biosynthesis of those phenolics in grapes is regulated by different genes; thus, each grape genotype presents a characteristic phenolic fingerprint, which is modulated by different environmental conditions. In this chapter, the anthocyanins and flavonols composition of different genotypes of wild grapes preserved at El Encin Germplasm Bank has been examined in detail. Wild grapevines are a remarkable genetic resource that may be used in breeding programs to improve the phenolic composition of cultivated grapes and, hence, the quality of red wines.

Keywords: wild grapes, anthocyanins, flavonols, fingerprint, HPLC

1. Introduction

Anthocyanins and flavonols are two families of phenolic compounds that play important roles in Enology. Free anthocyanins are the pigments responsible for the coloration of young red wines and take part in the reactions leading to the formation of stable polymeric pigments responsible for the coloration of aged red wines [1]. On the other hand, flavonols are involved in copigmentation of the flavylium form of anthocyanins in young red wines [2]. Moreover, flavonols present antioxidant properties that pose positive effects on human health [3]. The pathways involved in the biosynthesis of these molecules are well-known, and the core structural genes of those pathways, leading to the formation of primitive anthocyanins (delphinidin-3-*O*-glucoside and cyanidin-3-*O*-glucoside) and flavonol aglycones, like myricetin and quercetin, have been cloned and characterized [4, 5]. Moreover, several *O*-methyltransferases involved in the methylation of anthocyanins and flavonol glycosides have been identified [6, 7]; it has been demonstrated that the color exhibited by different grape cultivars may be associated with the *VvmybA1* and *VvmybA2* regulatory genes [8–10] that activate the expression of structural genes involved in the late steps of the anthocyanins biosynthetic pathway.

Anthocyanins are red pigments accumulated in skins during grape maturation (and also in pulp in teinturier cultivars), and their content has been related to several agroecological factors [11, 12], especially light and temperature, light being indispensable for anthocyanin biosynthesis and accumulation in the skins of berries and for phenylalanine ammonia lyase activity. Thus, their concentration is quite variable, even if the same cultivar or the same clone grown in a given location has been examined in several consecutive years [13, 14]. Nevertheless, the proportion of different anthocyanins, or anthocyanin fingerprint, is quite similar in the late stages of grape maturation of a cultivar grown in a given location from year to year [14]. On the other hand, the accumulation of flavonols (that are yellow pigments predominantly synthesized in grape skins [15]), is affected by shading treatments. The studies carried out in Shiraz grapes suggest that the branch of flavonoid biosynthetic pathway leading to flavonol biosynthesis is light-dependent, in contrast to anthocyanin and flavanol biosynthesis, which are little affected by shading treatments [16].

Cultivated grapevines are thought to be domesticated from genotypes of *Vitis vinifera* L. ssp. *sylvestris* (Gmelin) Hegi, which are present in small, isolated wild populations, located in riverbanks from the Western Himalayas to Western Europe [17, 18]. The sanitary status of those populations and their morphological and genetic characteristics have been recently studied [19–22]. Mature fruits of wild grapevines usually show high acidity, low pH, and a high intensity of color if compared with cultivated grapes [23, 24]; these features might be used to adapt Viticulture to the new climatic conditions, mitigating the potential effects of global warming on grape production.

The qualitative and quantitative anthocyanin composition of wild grape accessions preserved at El Encin Germplasm Bank has been examined by our research group after 2006 [25, 26], as well as their flavonol fingerprint [27–29]. The main objective of this study is to evaluate whether the anthocyanin and flavonol composition of wild grapes differs from that presented by cultivated grapes, and to determine whether some wild genotypes present some genetic characters of interest related to anthocyanins and flavonols accumulation during grape maturation. For this purpose, different female genotypes preserved at El Encin Germplasm Bank, that were collected in various natural populations located in different Spanish regions, were sampled in 2008 and have been fully examined for anthocyanins and flavonols content by HPLC.

2. Sampling of grapes

Samples of 25 genotypes of wild grapevines from different Spanish natural populations preserved at El Encin Grapevine Germplasm Bank (IMIDRA, Alcalá de Henares, Spain), grafted

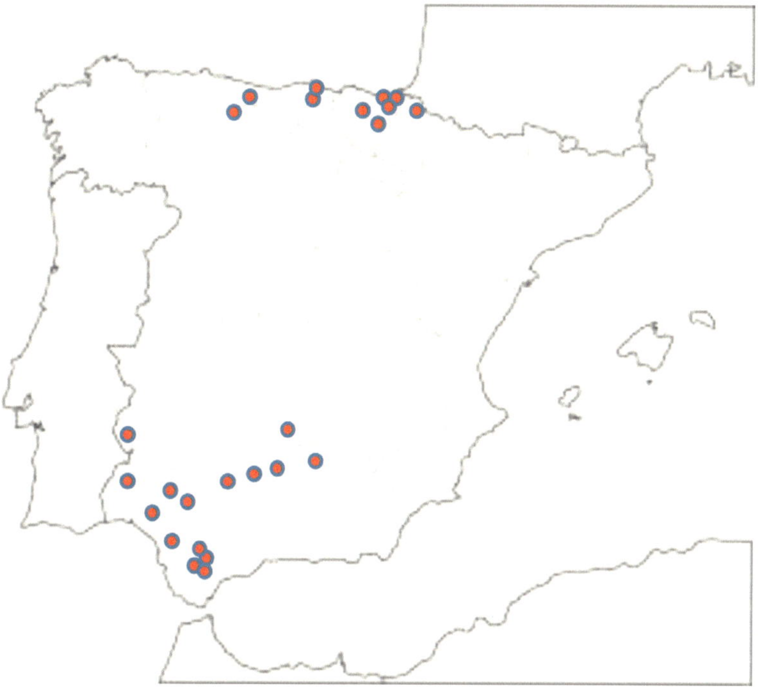

Figure 1. Location of natural populations of wild grapes where genotypes under study were originally collected.

on 110R and trained to cordon Royat, were collected in October 2008 at optimum stage of maturation (between 200 and 240 g of glucose + fructose by kg of must). Each sample consisted of four clusters, as only two plants of each genotype were available. Those genotypes, grown in El Encin, were originally collected from natural populations located in different Spanish regions (see **Figure 1**); 10 of them came from Northern Spain (Asturias, Cantabria, Castilla-León, Basque Country, and Navarra), the other 15 from Southern Spain (Andalousie, Castilla-La Mancha, and Extremadura). Every natural population was identified by two letters and by one or two numbers, and each genotype was identified with the population code and an additional number, as well as the suffix bis in some cases. Once in the laboratory, samples were stored at −20°C until sample preparation.

3. Sample preparation

Fifty berries were randomly selected and weighed once berries were separated from clusters, and grape skins were removed from pulps and seeds and stored at −20°C in methanol. Afterward, grape skins were grinded in a Kinematica PCU-2 blender for 1 minute. Then, they were sequentially extracted, using 25 mL of solvent for each extraction step: methanol for 16 hours at −25°C, 80% methanol for 4 hours at room temperature, 50% methanol for 4 hours at room temperature, deionized water for 16 hours at −25°C, and 75% acetone for 1 hour at room temperature [30]. At the end of each extraction step, the liquid was centrifugated at 3500 rpm for 20 minutes in a Rotofix 32A centrifuge, and the residue was submitted to extraction again. The volume of the combined liquid extracts was raised between 125 and 200 mL with methanol. Then, the extracts were stored at −20°C prior to analysis.

Flavonols were isolated prior to HPLC analysis to avoid interferences caused by anthocyanins, using solid-phase extraction on Oasis MCX cartridges (6 mL capacity) filled with 500 mg of an adsorbent containing a mixture of reverse-phase and cationic-exchanger materials (Waters Corp., Milford, MA), following a procedure described previously [31]. For this purpose, 3 mL grape skins extract was dried in a rotary evaporator (40°C) and resolved in 0.1 M hydrochloric acid (3 mL). Then, it passed through the MCX cartridges, previously conditioned with methanol (5 mL) and water (5 mL). After washing with 0.1 M hydrochloric acid (5 mL) and water (5 mL), the flavonol fraction was eluted with methanol (3 × 5 mL). This fraction also contained other neutral or acidic polyphenols. Fixed anthocyanins were removed using 2% ammonia in 80% methanol (3 × 5 mL). Finally, the cationic-exchanger material was regenerated with 0.52 M hydrochloric acid in 80% methanol (3 × 5 mL). Subsequent conditioning of the cartridge with methanol and water allows its reuse at least four or five more times. The eluate containing flavonols was dried in a rotary evaporator (30°C) and resolved in 1 mL of methanol.

4. Analytical procedures

The anthocyanin and flavonol fingerprints of skin extracts were obtained with HPLC-DAD [27], using a Waters Corp. liquid chromatograph consisting of a 600 quaternary pump, a 717 automatic injector, a TC2 controller for a column oven, a 996 photodiode array detector, and a Millennium 32 workstation. The separations were performed using a Waters Nova-Pak C18 steel cartridge (3.9 × 250 mm), filled with 5-μm particles, and furnished with a Waters Sentry Nova-Pack C18 guard cartridge (20 × 3.9 mm), both thermostated at 55°C. Water/acetonitrile (95:5) adjusted to pH 1.3 with trifluoroacetic acid (solvent A), and water/acetonitrile (50:50) adjusted to pH 1.3 with trifluoroacetic acid (solvent B) were used as mobile phases. Elution was performed at a 0.8 mL/min flow rate. For anthocyanins, a linear gradient from 15% B to 35% B in 20 min, from 35% B to 50% B in 10 min, 50% B for 6 min, from 50% B to 100% B in 5 min, 100% B for 5 min, 100% B to 15% B in 1 min was used. A linear gradient from 10% B to 35% B in 30 min, from 35% B to 50% B in 6 min, from 50% B to 100% B in 8 min, 100% B for 3 min, and from 100% B to 10% B in 1 min was used for flavonols. Samples (20 μL) were injected in triplicate. Spectra were recorded every second between 250 and 600 nm, with a bandwidth of 1.2 nm. Samples, standard solutions, and mobile phases were filtered before analysis through a 0.45-μm pore size membrane. The identity of the different anthocyanins and flavonols was elucidated by HPLC-MS, using an 1100 HPLC system (Agilent Technologies, Santa Clara, CA) with a PDA UV–Vis detector coupled to a QTOF mass spectrometer (AB SCiex, Framingham, MA). Chromatographic conditions were those used for the HPLC-DAD analysis. The MS analysis was carried out in the ESI+ mode, scanning from m/z 50 to 2000, with the following conditions: spray voltage, 5500 V; gas pressure, 80 psi; declustering potential, 50 V; focus potential, 210 V; CAD, 3 psi.

Total anthocyanins were determined in grape skins extracts, using the procedure described by Niketic-Aleksic and Hrzadina [32] using a BOECO S-22 UV–Vis spectrophotometer. Quantitative analysis of flavonols was carried out by HPLC, considering the surface of the different peaks, using standard solutions of quercetin-3-*O*-glucoside in the range of 20–100 mg/L.

5. Anthocyanin fingerprint of wild grapes

The HPLC analysis of anthocyanins extracted from wild grape skins permits the separation of 15 different anthocyanins. **Table 1** shows name, abbreviation, and number of peaks for each compound considered. The anthocyanin fingerprint of wild grapes revealed the presence of three groups of wild grapes genotypes, as it has been previously reported [25, 26]. **Figure 2** displays three typical chromatograms of those groups of genotypes.

The three groups of genotypes differ in different aspects linked to the pathways involved in anthocyanin biosynthesis [33] that are shown in **Figures 3** and **4**. First, the presence or absence of acylated anthocyanins, which implies important differences in the expression of genes involved in acyltransferase activity. Second, the prevalence of anthocyanins derived from delphinidin (Dp) or from cyanidin (Cy), which implies the differential expression of genes that control flavonoid-3′-hydroxylase and flavonoid-3′,5′-hydroxylase activities. Finally, the extent of methylation of Dp-3-gl and Cy-3-gl, due to the differential expression of genes controlling *O*-methyltransferase activity.

Anthocyanin	Abbreviation	Number of peaks
Delphinidin-3-*O*-glucoside	Dp-3-gl	1
Cyanidin-3-*O*-glucoside	Cy-3-gl	2
Petunidin-3-*O*-glucoside	Pt-3-gl	3
Peonidin-3-glucoside	Pn3-gl	4
Malvidin-3-glucoside	Mv-3-gl	5
Delphinidin-3-acetyl-glucoside	Dp-3-acgl	6
Cyanidin-3-acetylglucoside	Cy-3-acgl	7
Petunidin-3-acetylglucoside	Pt-3-acgl	8
Delphinidin-3-*p*-coumarylglucoside	Mv-3-cmgl	9
Peonidin-3-acetylglucoside	Pn-3-acgl	10
Malvidin-3-acetylglucoside	Mv-3-acgl	11
Petunidin-3-*p*-coumarylglucoside	Pt-3-cmgl	12
Malvidin-3-caffeoylglucoside	Mv-3-cfgl	13
Peonidin-3-*p*-coumarylglucoside	Pn-3-cmgl	14
Malvidin-3-*p*-coumarylglucoside	Mv-3-cmgl	15

Table 1. Name, abbreviation, and number of peaks for the different anthocyanins analyzed by HPLC.

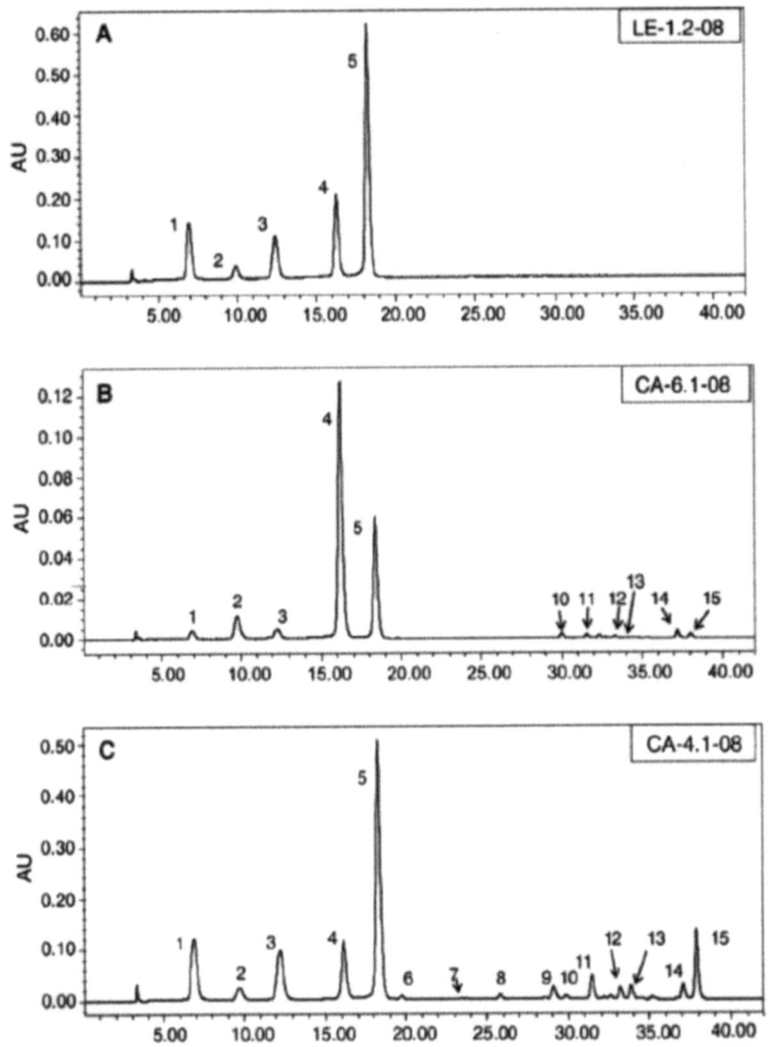

Figure 2. Chromatograms registered at 520 nm for three grape skins extracts representative of phenotypic groups A, B, and C. For key to peaks, see **Table 1**.

Genotypes of group A (tree samples) did not contain acylated anthocyanins (**Figure 2A**). This character is unusual in cultivated grapevines, occurring primarily in cv. Pinot Noir and its colored mutants [34, 35]. In these genotypes, genes encoding or regulating acyltransferase activity is neither presented nor expressed. To our knowledge, this type of anthocyanin fingerprint has not been described in grape cultivars usually considered of Spanish origin [35–37]. **Table 2** displays the percentages of several groups of anthocyanins presented in these genotypes. As can be observed, one genotype (BI-1.3bis) contained a remarkable amount of Cy-derived anthocyanins, over 50%, and the extent of methylation was very high (over 60%) in two genotypes (BI-1.3bis and LE-1.2). Similar trends were observed in several wild grapevine accessions that do not contain acylated anthocyanins in a previous report [26].

http://dx.doi.org/10.5772/intechopen.70861

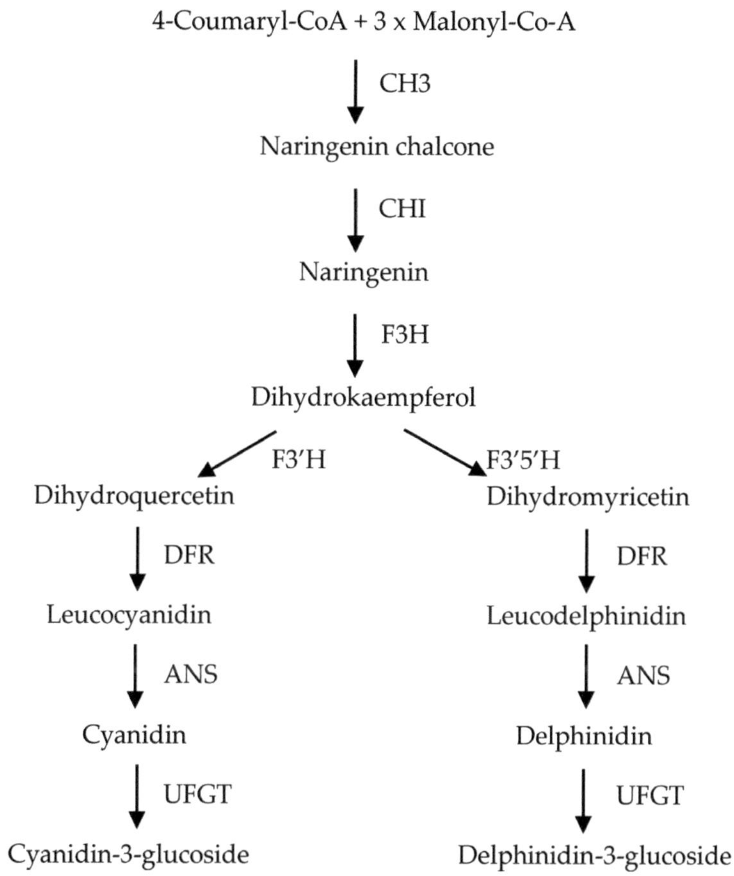

Figure 3. Biosynthesis of Dp-3-gl and Cy-3-gl. CH3, chalcone synthase; CHI, chalcone isomerase; F3H, flavanone-3-hydroxylase; F3′H, flavonoid-3′-hydroxylase; F3′5′H, flavonoid-3′,5′-hydroxylase; DFR, dihydroflavonol-4-reductase; ANS, anthocyanidin synthase; UFGT, UDP-Glc-flavonoid 3-O-glucosyltransferase. See full compound names in **Table 1**.

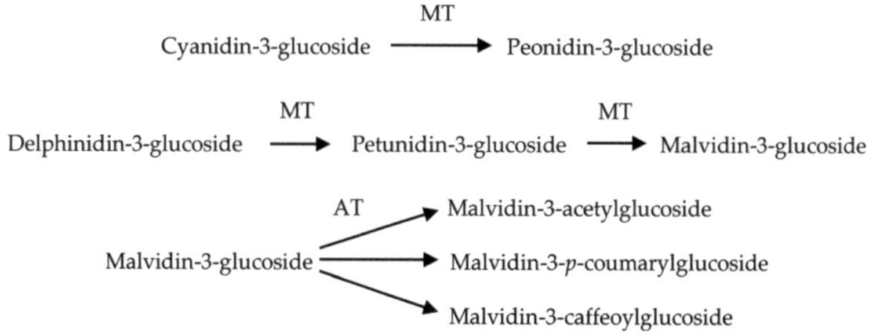

Figure 4. Biosynthesis of anthocyanins derived from Dp-3-gl and Cy-3-gl. MT, methyltransferase; AT, acyltransferase. See full compound names in **Table 1**.

Genotype	Dp-derived	Cy-derived	Methylated
BI-1.3bis	48.14	51.86	62.25
LE-1.2	80.61	19.39	82.14
SS-3.5bis	70.10	29.90	36.75

See full compound names in **Table 1**.

Table 2. Percentages of Dp-derived, Cy-derived, and methylated anthocyanins in genotypes of group A.

Genotypes of group B (six samples) contained acylated anthocyanins and a high proportion of Cy-derived anthocyanins (**Figure 2B**). This character is rare in cultivated grapevines, was observed only in 12 cultivars among 64 studied by Mattivi *et al.* [34], and it has been reported only in a cultivar considered of Spanish origin, cv. Brancellao [38]. Most cultivars of this type are gray or rosé cultivars, or even mutants of white cultivars. **Table 3** displays the percentages of several groups of anthocyanins presented in these genotypes. The percentage of Cy-derived anthocyanins ranged between 37 and 68%, and usually the percentage of methylated anthocyanins was up to 50%. Thus, Pn-3-gl usually was the major anthocyanin; the most remarkable exception was genotype SS-6.5bis. In this genotype, Cy-3-gl was the major anthocyanin (40.82%), as methylation was not very intense. Acylation was quite variable; it was too low in genotype CO-5.1, but considerably high in genotypes CA-13.3 and H-6.1. Moreover, most acylated anthocyanins were *p*-coumarylated derivatives; this character is quite common in red cultivars usually considered as Spanish, like Garnacha and Tempranillo [35].

Genotypes of group C (16 samples) contained acylated anthocyanins and a high proportion of delphinidin-derived anthocyanins (**Figure 2C**), as do most grapevine cultivars [33–37]; this implies that the expression of genes controlling flavonoid-3′,5′-hydroxylase is too high if compared with that of genes controlling flavonoid-3′-hydroxylase. The percentages of several groups of anthocyanins presented in these genotypes are displayed in **Table 4**. As can be noted, these genotypes also presented a high extent of methylation; the percentage of methylated anthocyanins was higher than 60%, except in four genotypes (BA-1.1, NA-1.4bis, SE-3.4,

Genotype	Dp-derived	Cy-derived	Methylated	Acylated	Acetylated	*p*-Coumarylated
CA-6.1	32.98	67.02	91.26	4.84	1.58	2.83
CA-13.3	54.35	45.65	80.84	14.77	4.42	9.97
CO-5.1	48.04	51.96	55.07	1.78	0.63	1.06
H-6.1	62.83	37.17	54.62	17.41	9.05	8.29
SS-6.5bis	42.70	57.30	31.17	4.46	2.09	2.38
VI-2.1bis	58.34	41.66	52.55	6.63	2.96	3.58

See full compound names in **Table 1**.

Table 3. Percentages of Dp-derived, Cy-derived, methylated, acylated, acetylated, and *p*-coumarylated anthocyanins in genotypes of group B.

Genotype	Dp-derived	Cy-derived	Methylated	Acylated	Acetylated	*p*-Coumarylated
BA-1.1	85.75	14.25	59.97	3.56	1.85	1.61
CA-4.1	82.69	17.31	71.97	14.70	4.76	9.01
CA-9.7	87.90	12.10	93.26	6.33	2.62	1.90
CA-11.3	83.92	16.08	65.99	34.26	27.74	6.06
CO-2.2	87.13	12.87	68.84	15.40	7.60	6.55
CO-3.7	81.95	18.05	73.27	16.32	8.00	7.08
CR-1.6	91.87	8.13	84.54	10.73	5.12	4.16
H-1.1	79.67	20.33	72.15	3.65	1.97	1.21
J-2.4	88.23	11.77	90.76	9.01	4.08	3.90
NA-1.4bis	73.82	26.18	43.96	2.82	1.26	1.56
O-1.5bis	80.48	19.52	62.43	6.64	3.26	3.15
S-1.3bis	88.95	11.05	66.03	16.43	7.42	8.10
S-1.9	90.97	9.03	65.99	8.84	4.37	4.11
SE-1.5	92.97	7.03	86.35	25.39	11.12	12.96
SE-3.4	84.65	15.35	58.70	16.42	6.52	9.59
SS-3.5	78.68	21.32	54.09	21.27	17.41	3.58

See full compound names in **Table 1**.

Table 4. Percentages of Dp-derived, Cy-derived, methylated, acylated, acetylated, and *p*-coumarylated anthocyanins in genotypes of group C.

and SS-3.5). In these late genotypes, Dp-3-gl was the major anthocyanin, but in the other 11 genotypes, the major genotype was Mv-3-gl. Sometimes, its content was higher than 90%.

The extent of acylation among genotypes included in group C was quite variable and not related to the extent of methylation. The percentage of acylated anthocyanins ranged from less than 3% (NA-1.4bis, 2.82%) to nearly 35% (CA-11.3, 34.26%). In two genotypes with a high extent of acylation (CA-11.3 and SS-3.5), acetylated anthocyanins were much more abundant than *p*-coumarylated anthocyanins. This character is well-documented in several French cultivars (e.g., Cabernet Sauvignon and Merlot), but is rare in Spanish cultivars. Most genotypes present less than 15% acylated anthocyanins, and percentages of acetylated and *p*-coumarylated anthocyanins were quite similar, as it has been observed in many grape cultivars considered of Spanish origin [35].

Data reported in **Table 5** point out that the total content of anthocyanins was quite variable, ranging from 273 to 3534 mg/kg, but, there is a remarkable difference among genotypes collected in populations located in Northern Spain and those from populations located in Southern Spain. As can be noted, genotypes from Northern Spain contained a higher amount of anthocyanins than those originated in Southern Spain ($p < 0.05$). As it is well-known, the accumulation of anthocyanins in grapes, that take place after veraison, is affected, at a great extent, by day-night thermal contrast [12], which can be considered neutral in our study, as all

	Range	Mean value	Standard deviation
Northern Spain	956–3078	2088	693
Southern Spain	273–3534	1697	797

Table 5. Range, mean value, and standard deviation for the content of total anthocyanins (mg/kg of grapes) in genotypes originated in northern Spain and southern Spain.

genotypes grew in the same environment. Thus, differences observed in anthocyanin content can be considered of genetic nature. The most probable explanation is that genotypes from Northern Spain have evolved in oceanic climate environments, where veraison takes place at the end of summer, and day-night thermal contrast is smaller than that in the Mediterranean climate environments in which evolved those genotypes collected from Southern Spain. Thus, it is probable that wild grapes in Northern Spain have evolved to accumulate enough anthocyanins capable of attracting birds and other animals to facilitate the dispersion of seeds, despite the limiting weather conditions for anthocyanin accumulation. Thus, when genotypes from Northern Spain grow in a warmer environment, like that of El Encin Germplasm Bank, the accumulation of anthocyanins may be very high.

6. Flavonol fingerprint of wild grapes

Six different flavonols were fully identified by HPLC-MS: a myricetin derivative (3-*O*-glucoside, My-3-gl), three quercetin derivatives (3-*O*-glucoside, Qu-3-gl; 3-*O*-glucuronide, Qu-3-gr; 3-*O*-rhamnoside, Qu-3-rh), a laricitrin derivative (3-*O*-glucoside, La-3-gl), and a syringetin derivative (3-*O*-glucoside, Sy-3-gl). All these flavonols have been identified in berries of several red grapevine cultivars [31, 38], and their presence in wild grapes should be expected. The flavonols tentatively identified were 3-*O*-galactosides of myricetin (My-3-gal) and quercetin (Qu-3-gal), which have been previously detected in red grape skins [38].

Among those flavonols, the most abundant were My-3-gl (trihydroxysusbtituted in B-ring, analogous to Dp-3-gl) and two quercetin derivatives (Qu-3-gl and Qu-3-gr, analogous to Cy-3-gl because they are dihydroxysubstituted in B-ring). Other myricetin derivatives, like laricitrin and syringetin derivatives (La-3-gl and Sy-3-gl), were minor components, and some of them were absent in several samples. **Figure 5** displays the chromatogram registered at 350 nm for an extract of genotype H-6.1, with three major peaks, corresponding to My-3-gl, Qu-3-gr, and Qu-3-gl.

Three phenotypic groups of wild grapes have been considered, taking into account the amounts of My-3-gl, Qu-3-gl, and Qu-3-gr. Group 1 includes eight genotypes, which did not contain My-3-gl (**Table 6**). This fact implies that, in these genotypes, dihydroxylation of dihydrokaempferol by flavonoid-3′,5′-dihydroxylase is blocked (**Figure 6**). In these genotypes, the major flavonol was Qu-3-gl or Qu-3-gr, and in some cases, contained very small amounts of other flavonol; anyway, La-3-gl and Sy-3-gl were absent (**Table 6**). Group 2 is formed by nine genotypes, which contain My-3-gl, but major flavonol was Qu-3-gl or Qu-3-gr (**Table 7**). These genotypes usually contained several minor flavonoids, including La-3-gl and Sy-3-gl,

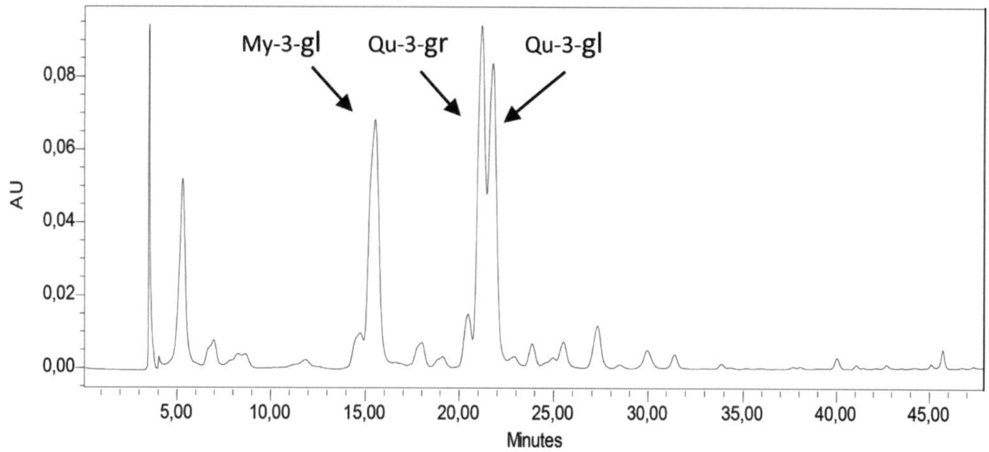

Figure 5. Chromatogram registered at 350 nm for a grape skin extract. My-3-gl, myricetin-3-*O*-glucoside; Qu-3-gr, quercetin-3-*O*-glucuronide; Qu-3-gl, quercetin-3-*O*-glucoside.

Genotype	My-3-gl	Qu-3-gr	Qu-3-gl	Other flavonols	Total
CA-6.1	nd	27.0	60.8	14.0	101.9
CA-13.3	nd	11.5	47.7	9.0	68.2
H-6.1	nd	54.2	73.6	17.3	145.1
NA-1.4bis	nd	18.0	10.7	0.0	28.7
S-1.3bis	nd	10.0	6.8	12.5	29.4
SS-3.5bis	nd	21.1	26.5	nd	47.6
SS-6.5bis	nd	26.1	61.5	nd	87.5
VI-2.1bis	nd	22.3	44.9	nd	67.2

nd: not detected. See full compound names in **Figure 5**.

Table 6. Content of flavonols (mg/kg of grapes) in genotypes that did not contain My-3-gl.

the exceptions being genotypes CA-11.3 and SE-3.4. In these two genotypes, methylation of My-3-gl by action of a *O*-methyltransferase is blocked. Finally, group 3 includes eight genotypes presenting My-3-gl as a major flavonol (**Table 8**). In most cases, these genotypes contained several minor flavonoids, including La-3-gl and Sy-3-gl, the exceptions being genotypes O-1.5bis and SS-3.5. Like in genotypes CA-11.3 and SE-3.4, methylation of My-3-gl by action of a methyltransferase is blocked.

Data reported in **Tables 6–8** point out that the total content of flavonols was quite variable, ranging from 29 to 324 mg/kg. Nevertheless, there is a remarkable difference among genotypes from populations located in Northern Spain and those from populations located in Southern Spain (**Table 9**). As can be noted, genotypes from Northern Spain contained a lower amount of flavonols than those originated in Southern Spain. The accumulation of flavonols in grapes

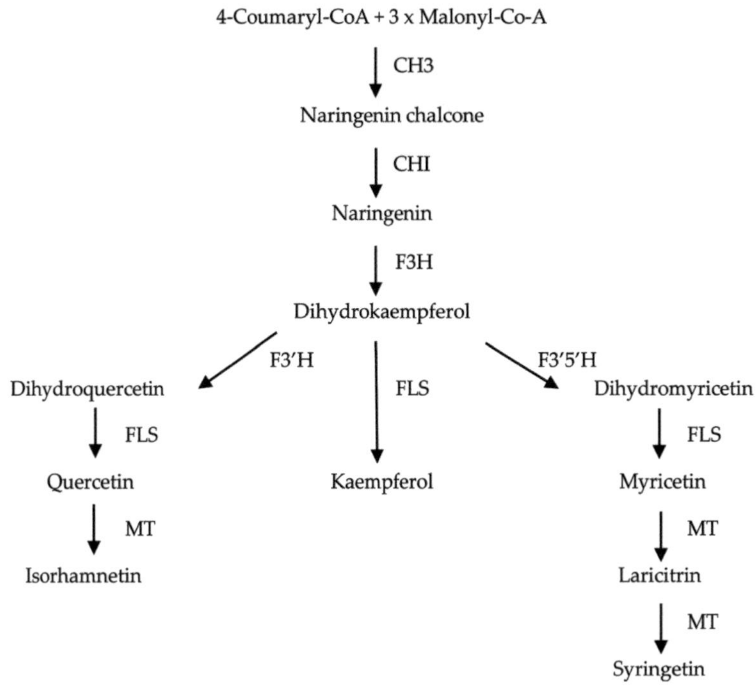

Figure 6. Biosynthesis of flavonol aglicones. CH3, chalcone synthase; CHI, chalcone isomerase; F3H, flavanone-3β-hydroxylase; FLS, flavonol synthase; F3′H, flavonoid-3′-hydroxylase; F3′5′H, flavonoid-3′,5′-hydroxylase; MT, methyltransferase.

Genotype	My-3-gl	Qu-3-gr	Qu-3-gl	Other flavonols	Total
BI-1.3bis	1.4	14.0	14.3	3.1	32.8
CA-9.7	2.9	13.8	9.4	13.7	39.8
CA-11.3	10.3	25.8	27.4	nd	63.4
CO-5.1	4.8	37.8	74.1	6.7	123.4
CR-1.6	6.3	31.4	20.3	25.2	83.2
J-2.4	8.1	14.4	7.8	14.2	44.5
LE-1.2	3.2	16.2	11.5	24.0	54.8
SE-1.5	80.8	106.1	71.4	46.2	304.4
SE-3.4	98.3	113.4	112.5	nd	324.1

nd: not detected. See full compound names in **Figure 5**.

Table 7. Content of flavonols (mg/kg of grapes) in genotypes that contained My-3-gl, being Qu-3-gl or Qu-3-gr the major flavonol.

increases with sun exposure, as demonstrated by other authors [16]. This factor can be considered as neutral in our study because all genotypes grown in the same environment; thus, differences observed in flavonols content can be considered of genetic nature. The most probable explanation is that genotypes originally collected in Northern Spain have evolved in

Genotype	My-3-gl	Qu-3-gr	Qu-3-gl	Other flavonols	Total
BA-1.1	84.8	47.0	42.0	9.5	183.3
CA-4.1	22.7	15.2	15.2	1.3	54.4
CO-2.2	48.7	14.6	22.8	14.0	100.1
CO-3.7	26.8	24.9	9.8	23.5	84.9
H-1.1	40.4	11.1	18.0	25.2	94.7
O-1.5bis	77.7	35.1	59.2	nd	172.1
S-1.9	58.0	30.1	28.3	12.2	128.6
SS-3.5	22.9	11.2	14.0	nd	48.1

nd: not detected. See full compound names in **Figure 5**.

Table 8. Content of flavonols (mg/kg of grapes) in genotypes in which My-3-gl was the major flavonol.

	Range	Mean value	Standard deviation
Northern Spain	28.7–172.1	69.7	47.3
Southern Spain	39.8–324.1	121.0	87.3

Table 9. Range, mean value, and standard deviation for the content of total flavonols (mg/kg of grapes) in genotypes originated in northern Spain and southern Spain.

oceanic climate environments, where sunlight exposure is lower than in the Mediterranean climate environments in which evolved genotypes from Southern Spain.

7. Comparison between anthocyanin fingerprint and flavonol fingerprint

The anthocyanin and flavonol fingerprints of wild grape genotypes are quite different, taking into account the pathways involved in their biosynthesis. Thus, in most genotypes B-ring trisubstituted anthocyanins (Dp-derived) predominate, but B-ring disubstituted flavonols (Qu derivatives) are more abundant than My derivatives (B-ring trisubstituted). Moreover, some genotypes do not present B-ring trisubstituted flavonols (**Table 7**), but they always present Dp-derived anthocyanins, sometimes in a high proportion (e.g., genotype S-1.3bis, see **Table 4**). On the other hand, some genotypes presenting a very low amount of Cy-derived anthocyanins (e.g., CR-1.6 and SE-1.5, see **Table 4**) contain a remarkable amount of Qu derivatives (**Table 7**). These data suggest that flavonol synthase activities linked to the formation of Qu and My are regulated in a different way than enzymatic activities linked to the formation of Dp-3-gl and Cy-3-gl from the corresponding dihydroflavonols. Other relevant biosynthetic difference among flavonols and anthocyanins is B-ring *O*-methylation. This reaction seems to be more intense for anthocyanins than for flavonols; thus, in most genotypes, methylated anthocyanins predominate. This fact suggests that primitive anthocyanins (Cy-3-gl and Dp-3-gl) are better substrates for *O*-methyltransferases (OMT) than quercetin and myricietin, as pointed out previously [34].

In general, phenylpropanoid biosynthesis and subsequent flavonoid production are tightly linked to primary metabolism through phenylalanine as a precursor of flavonoids. Catalyzing the first committed step into the flavonoid biosynthetic pathway, chalcone synthase (CHS) plays a pivotal role to provide a common chalcone precursor for the production of all intermediates and final products of the flavonoid biosynthetic pathway which are therefore biogenetically and structurally related (**Figures 3** and **6**). In most plants, including grapevine, flavanones are preferentially used as substrates for flavanone-3β-hydroxylase (F3H), which produces dihydroflavonols as an important branch point flavonoid and an essential substrate for all classes of downstream compounds (**Figures 3** and **6**). The biosynthesis of flavonol aglycones through flavonol synthase 1, FLS4 [39, 40]; as well as the biosynthesis of proanthocyanidins and anthocyanin precursors through dihydroflavonol 4-reductase (DFR) employs dihydroflavonols as substrates thereby directly competing for the same substrate (**Figures 3** and **6**). DFR reshuffles substrates away from flavonol biosynthesis and converts dihydroflavonols to leucoanthocyanidins, which are precursors for proanthocyanidin and anthocyanin biosynthesis [41]. While DFR is specific for the anthocyanin/proanthocyanidin pathway, flavonoid-3′-hydroxylase (F3′H) and flavonoid-3′,5′-hydroxylase (F3′5′H) gene products are necessary for the production of all subclasses, namely flavonols, anthocyanins, and proanthocyanidins. In general, hydroxylation of the B-ring of dihydroflavonols, flavanones, and flavones changes the color of the resulting anthocyanin-derived pigment and increases dramatically the chemodiversity of flavonols, proanthocyanidins, and anthocyanins [42].

The known biosynthetic pathway of flavonoids shares common enzymatic steps, whereas the activities of enzymes specific for anthocyanins or flavonols lead exclusively to the biosynthesis of the respective flavonoid by competing for common substrates (**Figures 3** and **6**). The accumulation of flavonol compounds in the berry is mediated by an increase of transcripts encoding FLS (*VvFLS4* or *VvFLS5*) under the regulation of the transcriptional factor *VvMYF* [40]. Later during veraison, the anthocyanins are synthesized through the flavonoid pathway in grapevine cultivars that harbor the wild-type VvmybA1 transcription factor for the expression of UFGT [43]. The encoded enzyme UFGT catalyzes the glycosylation of unstable anthocyanidin aglycones into pigmented anthocyanins (**Figure 3**). Two primitive anthocyanins (Cy-3-gl and Dp-3-gl) are synthesized in the cytosol of berry epidermal cells. The B-ring of Cy-3-gl is dihydroxylated at the 3′ and 4′ positions, whereas Dp-3-gl has a tri-hydroxylated B-ring because of an additional hydroxyl group at the 5′ position. The 3′ position of Cy-3-gl and Dp-3-gl and sequentially the 5′ position of Dp-3-gl can be methoxylated by *O*-methyltransferase (*VvOMT*), generating Pn-3-gl, Pt-3-gl, and Mv-3-gl, respectively [44]. Anthocyanins can be further modified by acyltransferases, which produce 3-*O*-acetyl-, 3-*O*-coumaroyl-, and 3-*O*-caffeoyl-monoglucosides by attaching acyl groups to the C6″ position of the glucose moiety [45].

Taking into consideration the regulations and biosynthesis pathway, we can suggest that the differences in the anthocyanin and flavonol fingerprints of wild grapes are putative due to the different expression level of the structural genes *F3′H* and *F3′5′H*, which are essential in the breach point for the final anthocyanin and flavonol compounds and the transcriptional factor involved in the pathway. Finally, the different level of methylation of Cy-3-gl and Dp-3-gl could be due to different expression of *OMT* genes and the gene encoding an anthocyanin acyltransferase, anthocyanin-3-*O*-glucoside-6″-*O*-acyltransferase (3), which is capable of producing the common acylated anthocyanins found in grape berries [46].

8. Conclusions

The anthocyanin fingerprint of wild grapes skins, considering the relative amount of 15 anthocyanins, showed a considerable variability, being possible to distinguish three phenotypic groups. Differences into those groups are related with the predominance of delphinidin- or cyanidin-derived anthocyanins and the expression of genes involved in acyltransferase activities. Moreover, it has been possible to separate 12 flavonol glycosides, eight of them were successfully identified. Major flavonols were Qu-3-gl, Qu-3-gr, and My-3-gl. The diversity and number of flavonols differed for each genotype. In most genotypes, Qu-3-gl or Qu-3-gr was the major flavonol, and My-3-gl was absent in some genotypes. Quantitative analysis of anthocyanins and flavonols revealed that genotypes collected in wild grapevine populations located in Northern Spain were richer in anthocyanins and poorer in flavonols than those collected in populations located in Southern Spain. This difference may be explained by the different expression level of the structural genes and transcriptional factors in the biosynthesis pathway in relation with the impact of climatic conditions on the evolution of wild grapes in different environments.

Acknowledgements

The authors acknowledge the financial support of Spain's Ministry of Economy and Competitiveness (Project RTA2014-0016-C003-01, MINECO-INIA-CCAA). AB holds an INIA fellowship for this project.

Author details

Eugenio Revilla[1]*, Rosa Arroyo-Garcia[2], Alberto Bellido[2], David Carrasco[2], Anna Puig[3] and Leonor Ruiz-Garcia[4]

*Address all correspondence to: eugenio.revilla@uam.es

1 Departament of Agricultural and Food Chemistry, Universidad Autonoma de Madrid, Madrid, Spain

2 Center for Plant Biotechnology and Genomics, CBGP-INIA, Pozuelo de Alarcón, Spain

3 Catalonian Grape and Wine Institute, INCAVI-IRTA, Vilafranca del Penedès, Spain

4 Murcia Institute for Agricultural and Food Research and Development, IMIDA, Murcia, Spain

References

[1] Fulcrand H, Duenas M, Salas E, Cheynier V. Phenolic reactions during winemaking and ageing. American Journal of Enology and Viticulture. 2006;**57**:289-297

[2] Trouillas P, Sancho-Garcia JC, De Freitas V, Gierschner J, Otyepka M, Dangles O. Stabilizing and modulating color by copigmentation: Insights from theory and experiment. Chemical Reviews. 2016;**116**:4937-4982. DOI: 10.1021/acs.chemrev.5b00507

[3] Burda S, Oleszek W. Antioxidant and antiradical activities of flavonoids. Journal of Agricultural and Food Chemistry. 2001;**49**:2774-2779. DOI: 10.1021/jf001413m

[4] Boss PK, Davies C, Robinson SP. Expression of anthocyanin biosynthesis genes in red and white grapes. Plant Molecular Biology. 1996;**32**:565-569. DOI: 10.1007/BF00019111

[5] Sparvoli F, Martin C, Scienza A, Gavazzi G, Tonelli C. Cloning and molecular analysis of structural genes involved in flavonoid and stilbene biosynthesis in grape (*Vitis vinifera* L.). Plant Molecular Biology. 1994;**24**:743-755. DOI: 10.1007/BF00029856

[6] Hugueney P, Provenzano S, Vemes C, Ferrandino A, Meudec E, Barelli G, Merdinoglu D, Cheynier V, Schubert A, Ageorges A. A novel cation-dependent O-methyltransferase involved in anthocyanin methylation in grapevine. Plant Physiology. 2009;**150**:2057-2070. DOI: 10.1104/pp.109.140376

[7] Lucker J, Martens S, Lund ST. Characterization of a *Vitis vinifera* cv. Cabernet Sauvignon 3′,5′-O-methyltransferase showing strong preference for anthocyanins and glycosilated flavonols. Phytochemistry. 2010;**71**:1474-1484. DOI: 10.1016/j.phytochem.2010.05.027

[8] This P, Lacomb T, Cadle-Davinson M, Owens CL. Wine grape (Vitis vinifera L.) color associates with alellic variation in the domestication gene VvmybA1. Theoretical and Applied Genetics. 2007;**114**:723-730. DOI: 10.1007/s00122-006-0472-2

[9] Lijavetzky D, Ruiz-Garcia L, Cabezas JA, De Andres MT, Bravo G, Ibanez A, Carreno J, Cabello F, Ibanez J, Martinez-Zapater JM. Molecular genetics of berry colour variation in table grape. Molecular Genetics and Genomics. 2006;**276**:427-435. DOI: 10.1007/s00438-006-0149-1

[10] Walker AR, Lee E, Bogs J, Mcdavid D, Thomas MR, Robinson SP. White grape arose through the mutation of two similar and adjacent regulatory genes. The Plant Journal. 2007;**49**:772-785. DOI: 10.1111/j.1365-313X.2006.02997.x

[11] Jackson DI, Lombard PB. Environmental and management practices affecting grape composition and wine quality. A review. American Journal of Enology and Viticulture 1993;**44**:409-430

[12] Teixeira A, Eiras-Dias J, Castellarin SD, Geros H. Berry phenolics of grapevine under challenging environments. International Journal of Molecular Sciences. 2013;**14**:18711-18739. DOI: 10.3390/ijms140918711

[13] Ryan JM, Revilla E. Anthocyanin composition of Cabernet Sauvignon and Tempranillo grapes at different stages of ripening. Journal of Agricultural and Food Chemistry. 2003;**51**:3372-3378. DOI: 10.1021/jf020849u

[14] Revilla E, García-Beneytez E, Cabello F. Anthocyanin composition of several clones of Tempranillo grapes and wines made with them. Australian Journal of Grape and Wine Research. 2009;**15**:70-78. DOI: 10.1111/j.1755-0238.2008.00037.x

[15] Price SF, Breen PJ, Valladao M, Watson BT. Cluster sun exposure and quercetin in Pinot noir grapes and wine. American Journal of Enology and Viticulture. 1995;**46**:187-194

[16] Downey MA, Harvey JS, Robinson SP. The effect of bunch shading on berry development and flavonol accumulation in Shiraz grapes. Austr. J. Grape Wine Res. 2004;**10**:55-73. DOI: 10.1111/j.1755-0238.2004.tb00008.x

[17] Levadoux L. Les populations sauvages et cultivées de Vitis vinifera L. Ann. Amelior. Plant. 1956;**6**:59-119

[18] Zohary D, Hopf M, Weiss E. Domestication of Plants in the Old World, 4th ed. Oxford: Oxford University Press. 2012. DOI: 10.1093/acprof:osobl/9780199549061.001.0001

[19] Arnold C, Guillet F, Gobat JM. Situation de la vigne sauvage Vitis vinifera ssp. sylvestris en Europe. Vitis. 1998;**37**:159-170

[20] Ocete R, Lopez MA, Perez MA, Tio R. Las Poblaciones Españolas de Vid Silvestre. Madrid: INIA; 1999

[21] Cunha J, Teixeira-Santos M, Carneiro LC, Fevereiro P, Eiras-Dias JE. Portuguese traditional grapevine cultivars and wild vines (Vitis vinífera L.) share morphological and genetic traits. Genetic Resources and Crop Evolution. 2009;**56**:975-989. DOI: 10.1007/s10722-009-9416-4

[22] De Andres MT, Benito A, Perez-Rivera G, Ocete R, Lopez MA, Gaforio L, Muñoz G, Cabello F, Martinez-Zapater JM, Arroyo-Garcia R. Genetic diversity of wild grapevine populations in Spain and their genetic relationships with cultivated grapevines. Molecular Ecology. 2012;**21**:800-816. DOI: 10.1111/j.1365-294X.2011.05395.x

[23] Ocete R, Arroyo-Garcia R, Morales ML, Cantos M, Gallardo A, Perez MA, Gomez I, Lopez MA. Characterization of Vitis vinifera L. subspecies sylvestris (Gmelin) in the Ebro river Basin (Spain). Vitis. 2011;**50**:11-16

[24] Ocete R, Cantos M, Lopez-Gallardo A, Perez-Troncoso A, Lara M, Failla O, Ferragut FJ, Liñan J. Caracterización y conservación del recurso fitogenético vid silvestre en Andalucía. Falcor: *sevilla*; 2007

[25] Revilla E, Carrasco D, Benito A, Arroyo-Garcia R. Anthocyanin composition of several wild grape accesions. American Journal of Enology and Viticulture. 2010;**61**:636-642. DOI: 10.5344/ajev.2010.09134

[26] Revilla E, Carrasco D, Carrasco V, Benito A, Arroyo-Garcia R. On the absence of acylated anthocyanins in some wild grapevine accessions. Vitis. 2012;**51**:161-165

[27] Revilla E, Bellido A, Yus J, Ortiz P, Carrasco D, Arroyo RA. Flavonols in skins of wild grapes (*Vitis vinifera* L., subsp. *sylvestris* (Gmelin) Hegi). Bio Web of Conferences. 2016;**7**: UNSP 01018. DOI: 10.1051/bioconf/20160701018

[28] Bakker J, Timberlake CF. The distribution of anthocyanins in grape skin extracts of Port wine cultivars as determined by high performance liquid chromatography. Journal of the Science of Food and Agriculture. 1985;**36**:1315-1324. DOI: 10.1002/jsfa.2740361217

[29] García-Beneytez E, Cabello F, Revilla E. Analysis of grape and wine anthocyanins by HPLC-MS. Journal of Agricultural and Food Chemistry. 2003;**51**:5622-5629. DOI: 10.1021/jf0302207

[30] Revilla E, Ryan JM, Martin-Ortega G. Comparison of several procedures used for the extraction of anthocyanins from red grapes. Journal of Agricultural and Food Chemistry. 1998;**46**:4592-4597. DOI: 10.1021/jf9804692

[31] Castillo-Muñoz N, Gomez-Alonso S, Garcia-Romero E, Hermosin-Gutierrez I. Flavonol profiles of *Vitis vinifera* red grapes and their single-cultivar wines. Journal of Agricultural and Food Chemistry. 2007;**55**:992-1002. DOI: 10.1021/jf062800k

[32] Niketic-Aleksic GK, Hrzadina G. Quantitative analysis of the anthocyanin content in grape juices and wines. Lebensmittel-Wissenschaft & Technologie. 1972;**5**:163-165

[33] Holton TA, Cornish EC. Genetics and biochemistry of anthocyanin biosynthesis. The Plant Cell. 1995;**7**:1071-1083. DOI: 10.1105/tpc.7.7.1071

[34] Mattivi F, Guzzon R, Vrhovsek U, Stefanini M, Velasco R. Metabolite profiling of grapes: flavonols and anthocyanins. Journal of Agricultural and Food Chemistry. 2006;**54**:7692-7702. DOI: 10.1021/jf061538c

[35] Garcia-Benitez E, Revilla E, Cabello F. Anthocyanin pattern of several red grape cultivars and wines made with them. European Food Research and Technology. 2002;**215**:32-37. DOI: 10.1007/s00217-002-0526-x

[36] Pomar F, Novo M, Masa A. Varietal differences among the anthocyanin profiles of 50 red table grapes cultivars studied by high performance liquid chromatography. Journal of Chromatography. A. 2005;**1094**:34-41. DOI: 10.1016/j.chroma.2005.07.096

[37] Gomez-Alonso S, Fernandez-Gonzalez M, Mena A, Martinez J, Garcia-Romero E. Anthocyanin profile of Spanish *Vitis vinífera* L. red grape varieties in danger of extinction. Australian Journal of Grape and Wine Research. 2007;**13**:150-156. DOI: 10.1111/j.1755-0238.2007.tb00245.x

[38] Figueirido-Gonzalez M, Martinez-Carballo E, Cancho-Grande B, Santiago JL, Martinez MC, Simal-Gandara J. Pattern recognition of three *Vitis vinifera* L. red grapes varieties based on anthocyanin and flavonol profiles, with correlations between their biosynthesis pathways. Food Chemistry. 2012;**130**:9-19. DOI: 10.1016/j.foodchem.2011.06.006

[39] Downey MO, Harvey JS, Robinson SP. Synthesis of flavonols and expression of flavonol synthase genes in the developing grape berries of Shiraz and Chardonnay (*Vitis vinifera* L.). Australian Journal of Grape and Wine Research. 2003;**10**:55-73. DOI: 10.1111/j.1755-0238.2003.tb00261.x

[40] Fujita A, Goto-Yamamoto N, Aramaki I, Hashizume K. Organ specific transcription of putative flavonol synthase genes of grapevine and effects of plant hormones and shading on flavonol biosynthesis in grape berry skins. Bioscience, Biotechnology, and Biochemistry. 2006;**70**:632-638. DOI: 10.1271/bbb.70.632

[41] Martens S, Teeri T, Forkmann G. Heterologous expression of dihydroflavonol 4-reductases from various plants. FEBS Letters. 2002;**531**:453-458. DOI: 10.1016/S0014-5793(02)03583-4

[42] Bogs J, Ebadi A, McDavid D, Robinson SP. Identification of the flavonoid hydroxylases from grapevine and their regulation during fruit development. Plant Physiology. 2006;**140**:279-291. DOI: 10.1104/pp.105.073262

[43] Kobayashi S, Goto-Yamamoto N, Hirochika H. Retrotransposon- induced mutations in grape skin color. Science. 2004;**304**:982. DOI: 10.1126/science.1095011

[44] Fournier-Level A, Hugueney P, Verriès C, This P, Ageorges A. Genetic mechanisms underlying the methylation level of anthocyanins in grape (*Vitis vinifera* L.). BMC Plant Biology. 2011;**11**:179-192. DOI: 10.1186/1471-2229-11-179

[45] Nakayama T, Suzuki H, Nishino T. Anthocyanin acyltransferases: specificities, mechanism, phylogenetics, and applications. Journal of Molecular Catalysis B: Enzymatic. 2003;**23**:117-132. DOI: 10.1016/S1381-1177(03)00078-X

[46] Rinaldo A, Cavallini E, Jia Y, Moss SMA, McDavid DAJ Hooper LC, Robinson SP, Tornielli GB, Zenoni S, Ford CM, Boss PK, Walker AR. A grapevine anthocyanin acyltransferase, transcriptionally regulated by VvMYBA, can produce most acylated anthocyanins present in grape skins. Plant Physiology. 2015;**169**:1897-1916. DOI: 10.1104/pp.15.01255

Chapter 3

Grapevine Trunk Diseases (GTDs): Impact on Table Grapes and Wine Vineyards in Chile

Ximena Besoain

Additional information is available at the end of the chapter

http://dx.doi.org/10.5772/intechopen.72607

Abstract

This chapter involves the description and information about the main grapevine trunk diseases (GTDs) affecting *Vitis vinifera* in Chile. There is a complete description of Esca-type disease, bot canker disease affecting more than 7-year-old plants, and Petri disease affecting young plants. Symptom descriptions of GTD diseases are done, and also the principal species as causal agents and the importance of them are analyzed. The dispersal of Botryosphaeriaceae spores is considered, and finally, the principal management measures can be taken in order to avoid or control these important diseases. The information is contrasted with different researches done in the most important countries where these diseases are relevant.

Keywords: bot canker, Esca-like disease, *Diplodia seriata*, *D. mutila*, Botryosphaeriaceae, *Phaeomoniella chlamydospora*, chlorotic leaf roll

1. Introduction

In Chile, the first notice of a disease caused by Botryosphaeriaceae occurred in 1986 [1], where *Botryosphaeria dothidea* was reported causing damage to a grapevine grown in a Spanish trellis for table grape production. Since the 1960s, worldwide studies in grapevine trunk diseases (GTDs) have expanded incrementally; four important types of diseases have been recognized: Esca disease, bot canker disease, *Eutypa* dieback, and *Phomopsis* dieback [2], and actually, the reported studies had acquired an exponential curve, but up to date, there is not a satisfactory field control of these important diseases. In the study done by Fontaine et al. [2], there is an important contribution to understanding these important diseases in many countries, but the

Chilean situation was not included; therefore, this chapter contributes to understand the actual status of GTDs in Chile.

Chile is the principal fruit exporter of fresh table grapes in the world. In addition, it is important as wine producer. An analysis of the main table grapes and wine producer regions is done, with special association to GTD predisposition. The economic importance of GTDs and potential damage that it causes is described. The principal species associated to Botryosphaeriaceae canker and a disease called Esca-like, caused mainly by *Phaeomoniella chlamydospora* in Chile, are descripted in detail, analyzing previous published information [3]. The symptomatology associated to bot canker and Esca disease causal agents is detailed, and a comparison of the situation for the main wine- and table grape-producing countries is done. Furthermore, the importance of these causal agents affecting another high-value species or plants associated to natural habitats, commonly interacting with vineyard's and table grape's orchards, is considered. The factors associated to GTDs will be described, including the analysis associated to climatic conditions. A spatial analysis will be done including different stress-affecting plants and the importance of the Botryosphaeriaceae family as endophytic pathogens. The chapter includes information about the tissues commonly colonized by Botryosphaeriaceae and by *Phaeomoniella*, considering analysis of their preference and age of the colonized tissue. Finally, the chapter includes a description of a complete integrated management and future trends for control of GTDs, with special emphasis in bot canker and Esca diseases. Up to date there is no official report of *Eutypa* disease affecting *Vitis vinifera* in Chile [3, 4], and *Phomopsis* dieback was reported, but the prevalence of its causal agent, *Phomopsis viticola*, is very low [3]. So, this chapter includes an analysis of bot canker and Esca-type disease.

The ability of GTD-associated species to be endophytes is specially a threat for the quarantine programs worldwide, which can be overlooked considering its latent phase. This is particularly important because most of the causal agents have the ability to rapidly express pathogenicity when their hosts are under stress [5]. This is particularly important considering that the Servicio Agrícola y Ganadero (SAG) of Chile requires a 2-year period of quarantine when propagation material is introduced from foreign countries, and not always these types of diseases will express in this period. In addition, it is also important to consider that under the climate change, there is an increase in drought periods in many places, with an increase in stress for many plant communities and an increase of GTD-type diseases.

2. Grapevine trunk diseases (GTDs) in Chile

2.1. Chilean table grape and wine industry

Considering world table grape production, Chile is the seventh main producer worldwide, but considering the export to other countries, Chile is the first exporter, doubling the exports of the second most important country, the United States. Actually, there are 52,234 ha for table grape fresh fruit production in Chile, with vineyards developed in seven different regions, from the North in the Atacama Region and to the South up to the Bío Bío Region. In the Atacama

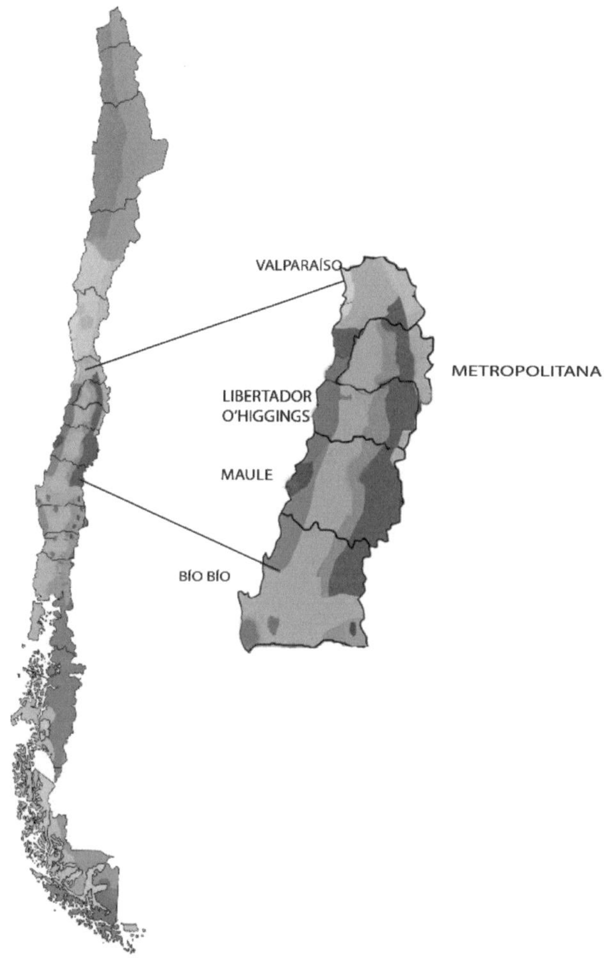

Figure 1. Map of Chile; in detail the five main producing regions with vineyards for wine.

Region, there are about 8515.9 ha, mainly grown for pisco (distilled liqueur) production. In relation to the wine industry, Chile is the sixth most important wine producer and the fourth most important wine exporter. Chile exported in 2016 about 9,100,000 hectoliters, corresponding to 8.74% of the total volume exported in the world [6]. The main wine-producing regions in Chile considering the cultivated area are Maule (37.94%), O'Higgins (32.81%), Bío Bío (10.65%), Metropolitana (9.2%), and Valparaíso (7.1%) (**Figure 1**) with a total of 141,918 ha in 2015 [7]. So, this delineates the significance of GTDs affecting vineyards and the importance of prevention, detection, diagnosis, and opportune control.

2.2. Main diseases and economic importance of GTDs in Chile

In Chile Petri disease [8] has been detected affecting young *Vitis* plants produced for the wine industry, mainly associated to *Phaeomoniella* and *Phaeoacremonium* genera. In vineyards more than 7-year-old, the main diseases affecting plants are *Botryosphaeria* dieback (or bot canker) and a disease similar to Esca found associated to Diatrypaceae fungi and with Basidiomycete fungi associated to wood spongy tissues [1, 3, 4, 8].

Pathogenicity of bot canker has been consistently proven in Chile [4] as well as necrotic strikes associated to *Phaeomoniella chlamydospora* [3, 9]. But there is an important disease called chlorotic leaf roll (*enrollamiento clorótico*) of the vines that is frequently associated to V-shaped necrosis and necrotic streaks in the xylem and also with spongy tissues mainly associated to Basidiomycetes colonization, which Koch's postulates have not been done. So, this type of disease and symptoms needed to be solved by the Chilean plant pathologists.

The economic importance of diseases affecting grapevines has been partially studied in Chile. Morales et al. [4] found that bot canker disease affects table grape vineyards and that the disease increases with the age of the plants. With an increasing incidence of the disease of 22% in an 11-year-old vineyard to 70% in the 22-year-old vineyard, and with maximum severity up to 22% of damage index in a 16-year-old table grape vineyard. In these table grape vineyards, the main species and detected prevalence (%) were *Diplodia seriata* (83.3%), *D. mutila* (8.3%), and *Spencermartinsia viticola* (8.3%). In a national survey done by Díaz et al. [3], with about 694 wood samples being analyzed obtained from 67 Chilean vineyards, from Copiapó (27° 18′S) to Los Angeles (37°42′S), a total of 1363 fungal isolates were obtained from diseased plants with different symptoms, and the most prevalent identified species were *Phaeomoniella chlamydospora* (in 85% of the samples); Botryosphaeriaceae (in 56% of the samples) with the main detected species *Diplodia seriata, D. mutila, Neofusicoccum parvum,* and *Spencermartinsia viticola*; and Diatrypaceae (in 4.8% of the samples, but *Eutypa lata* was not found). *Phomopsis viticola* and *Cylindrocarpon* sp. were only found in 0.4% of the analyzed samples. Finally, Valencia et al. [10] analyzing conidial dissemination in four vineyards from the Valparaíso Region found in the spore traps conidia from *D. seriata, S. viticola,* and *Neofusicoccum australe,* but the greater peaks found corresponded to *D. seriata.*

On the other hand, Torres et al. [11] reported the prevalence and damage observed in 10 vineyards cv. Cabernet Sauvignon during the 2010–2011 growing season. In the research done in 14 blocks of the O'Higgins (n = 8) and Maule (n = 6) regions, the average damages associated with bot canker were 36 and 48% for each region, respectively, with an average of production loss estimated at 5800 kg ha^{-1}. The most prevalent species detected were *Diplodia seriata* (69.2%), followed by *D. mutila* (12.8%), and *Neofusicoccum parvum* (10.2%).

GTDs are the main biotic factors affecting grapes wherever are grown and affecting vineyard productivity and longevity [12]. In France, there is a national crisis with 12% of the vineyards not economically viable, causing annual losses for about 1 billion euros [13]. In China, where *Botryosphaeria* dieback is a serious disease, Yan et al. [14] made a survey in 72 vineyards of 20 grape-growing regions; they found that dieback occurs in 18/20 provinces, and they confirmed the presence of *B. dothidea, D. seriata, L. theobromae,* and *N. parvum.* These species were spatially distributed: *L. theobromae* and *N. parvum* only occurring in subtropical monsoon climate regions, *Diplodia seriata* only occurring in temperate climatic regions, and *B. dothidea* occurring in both types of climates. Of the 25 cultivars most grown in China, none was resistant to the four (taxa) species.

Black dead arm (BDA) disease was first reported by Lehoczky [15] associated to *B. stevensii* (formerly named as *D. mutila*). This pathogen did not infect the green parts of the vines or the chesters. He reported that the phloem and xylem tissues were black discolored; therefore, he named the disease as "black dead arm disease" instead of the "dead arm disease" caused by *Phomopsis viticola*, previously described by Pine [16]. Actually, there are four major diseases in vineyards. And, their names are Esca, *Eutypa* dieback, *Botryosphaeria* canker, and *Phomopsis* dieback [2].

2.3. GTD causal agents

According to the 10th International Workshop on Grapevine Trunk Diseases (2017) taken place in Reims, France, in the introduction done by Dr. José Ramón Úrbez-Torres "Main achievements and future prospects in GTDs Pathogen Characterization and Identification," he analyzed that 133 species from 34 genera are associated to GTDs worldwide. Considering these information, the analysis of this type of diseases affecting grapevines is much complex to organize. Fortunately, in Chile the situation is more easy considering that there are only five species causing bot canker [3, 4, 8, 10, 17], two main species associated to Petri and Esca-like diseases [3, 9, 18], plus three species of Basidiomycetes [3]. *Phomopsis* dieback is very infrequently seen, and black foot also seems to be associated with problems at the plantation site of grafted material and with very low prevalence [3].

In table grape and wine vineyard research done in Chile, only five Botryosphaeriaceae species have been reported to affect these crops: *Diplodia seriata* De Not. (≡ *Botryosphaeria obtusa* (Schwein.) Shoemaker); *D. mutila* (Fr.) Mont. (= *Botryosphaeria stevensii* Shoemaker); *Neofusicoccum parvum* (Pennycook & Samuels) Crous, Slippers & A.J.L. Phillips (= *Botryosphaeria parva*) Pennycook & Samuels; *N. australe* (Slippers, Crous & M.J. Wingf.) Crous, Slippers & A.J.L. Phillips (= *Botryosphaeria australis* Slippers, Crous & M.J. Wingf.); and *Spencermartinsia viticola* (A.J.L. Phillips & J. Luque) A.J.L. Phillips, A. Alves & Crous (= *Botryosphaeria viticola* A.J.L. Phillips & J. Luque) [1, 3, 4, 8, 11, 17]. The names in parentheses are in accordance with the name and authors described by Phillips et al. [19].

In accordance with Díaz et al. [3] from a total of 694 analyzed samples obtained from different varieties grown for table grapes or wine industry, 56.3% of them were Botryosphaeriaceae species, and from 85% of these samples, the principal recovered species was *Phaeomoniella chlamydospora* (*Pa. chlamydospora*) (W. Gams, Crous, M.J. Wingf. & L. Mugnai) Crous & W. Gams, obtained from vascular streaks. But considering cankers with V-shaped form, 91% of the recovered species were Botryosphaeriaceae, and from these 68% of them was *Diplodia seriata*. From a soft-spongy canker (n = 314), the main species isolated in 96% of the samples was *Inocutis* sp. In pathogenicity tests carried out by Díaz et al. [3], *Inocutis* was able to cause damage but significantly less than that caused by Botryosphaeriaceae or *Phaeomoniella* species.

Considering Úrbez-Torres' [20] review of species affecting *Vitis vinifera*, he could recognize 21 different Botryosphaeriaceae species affecting this crop in California. In France the disease known as Esca complex is caused by *Pa. chlamydospora* and by *Phaeoacremonium* spp. plus the basidiomycete *Fomitiporia mediterranea* [2].

2.3.1. Reproductive structures

Taking into account *Botryosphaeria* dieback disease, all of the pathogens involved produce their asexual spores or conidia in pycnidia, and depending on the species, it is the shape and color of the conidia (**Figure 2A–C**). They are formed in conidiogenous cells, but some species have conidiophores [19]. In their sexual stage (if present), the ascospores are produced in a bitunicate asci, and this is inside an ascomata, most of them gregarious. In Chile, there is no report of sexual stage of Botryosphaeriaceae occurring in *Vitis*.

The *Phaeomoniella* genus produces yeast-like colonies that were olivaceous with sparse aerial mycelium and erect non-branched, smooth, and septate conidiophores [18]. The conidia are cellular, slightly pigmented, smooth, oblong to ellipsoid, and with small dimensions (3.0–4.2 μm large). Black globose pycnidia were produced on pine needles. White dense cirrus was produced on pine needles.

Figure 2. Grapevine trunk symptoms associated to bot canker and Esca-like disease symptoms in Chile. (A) V-shaped canker on the main branch of *Vitis*. (B) Main brunch associated with canker necrosis, yellow soft-spongy canker, and mottling in the trunk. (C) Dead spur attached to a diseased brunch with a typical grayish color. (D) Table grape vineyard with symptoms of distorted leaf and necrosis in the margin associated to chlorotic leaf roll.

2.4. Principal symptoms associated to GTDs

According to the research carried out by Díaz et al. [3, 4], Esca-like symptoms and their causal agents are associated to brown streaking in the xylem of the trunk and cordon of young grapevines (≤ 10 years old) (**Figure 2A–C**). Cordons and trunk from young grapevines showed this brown strike with the appearance of necrotic mottling in cross sections or dark brown necrotic streaking in longitudinal sections. They reproduce part of the symptoms associated to Esca in Europe [2] but not the characteristics foliar symptoms of this disease.

On the other hand, brown V-shaped cankers were consistently associated in table grape vineyards to four Botryosphaeriaceae species by Díaz et al. [3] and Morales et al. [4]. In a vast survey done along the main viticulture regions, Díaz et al. [3] found a total of 12 genera including *Diplodia, Dothiorella, Neofusicoccum, Spencermartinsia, Cryptovalsa, Eutypella, Cylindrocarpon, Phomopsis, Inocutis, Phaeoacremonium, Phaeomoniella,* and *Seimatosporium*. When the fungal isolates obtained from these genera were correlated to the principal symptoms observed in the field, these authors associated dark brown streaking symptoms to *Pa. chlamydospora* (85.2% of the samples), *D. mutila* (3.8% of samples), and *Inocutis* sp. (3.7% of samples). Symptoms of brown hard V-shaped canker (**Figure 3A**) were associated to *Diplodia seriata* (68.1% of samples), *Neofusicoccum parvum* (8.3% of samples), *Spencermartinsia viticola* (8.1% of samples), and *Pa. chlamydospora* (5.3%), and yellowish soft-spongy canker (**Figure 3B**) was associated to *Inocutis* sp. (96.2% of samples), *Diplodia seriata* (2.5% of samples), and *Pa. chlamydospora* (1.3% of samples). In vineyards, the presence of dead spurs that show a grayish color is frequently observed (**Figure 3C**).

The causal agent of leaf chlorotic roll symptom associated to decay plants of table grape and vineyard cultivars (**Figure 2D**) has not solved up to date. This symptom is frequently associated to cankers caused by Botryosphaeriaceae species, but up to date, no one has reproduced these symptoms in Chile. It consists in a leaf malformation and chlorosis of the leaf lamina, short internodes of the shoots or canes, death of spurs, and dieback commonly associated to cordon and/or trunk dieback [3, 4]. In Argentina, a similar situation was studied by Gatica et al. [21] with the disease called "hoja de malvón," where they found similar symptoms in association to mature vines with *Phellinus* sp. that prevailed over *Phaeoacremonium aleophilum, Pm. parasiticum,* and *Phaeomoniella chlamydospora*. In a previous study done by Gatica et al. [22], they made an in vitro pathogenicity test with *Phellinus* sp., *Botryodiplodia* sp., *Phaeoacremonium aleophilum,* and *Phaeomoniella chlamydospora,* where all these were pathogenic and cause decline and death of young plants, but the chlorotic leaf symptoms with reddish edges were produced by *Phellinus* sp. and by *P. chlamydospora*. Nevertheless, this situation needed to be studied reproducing the symptoms in mature plants.

2.5. Cultivar susceptibility

The first report of a *Botryosphaeria* species affecting the *Vitis* genus in Chile was done by Latorre et al. [1], where these authors report a severe attack of a canker disease affecting 2-year-old plants of cv. Flame Seedless grown in a Spanish trellis for fresh table grape production. Extensive cankers were found in the trunk, causing mild to moderate chlorosis and leaf necrosis and a poor seasonal growth. All that time, the authors classified the causal agent as

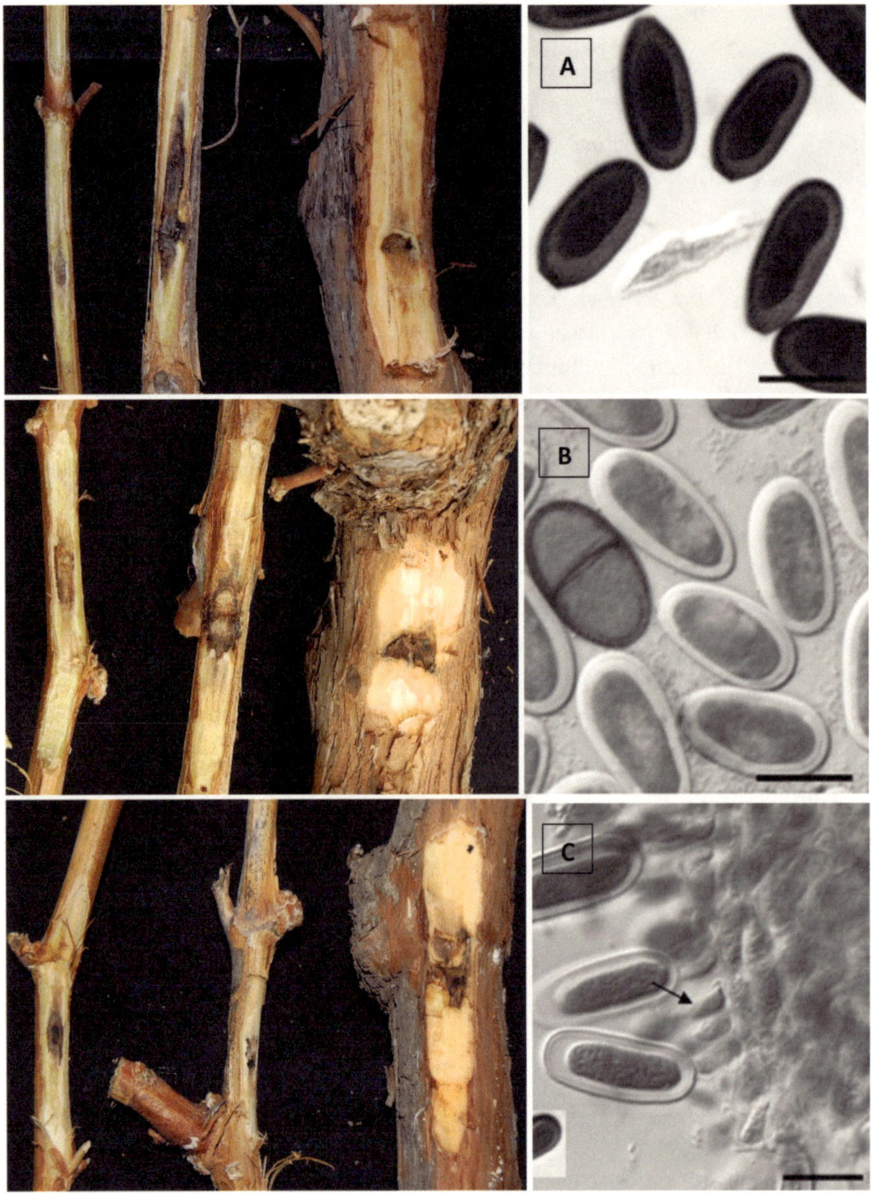

Figure 3. Pathogenicity tests done in shoots of 1-, 2-, and 5-year-old *Vitis* plants and the spores of different Botryosphaeriaceae. (A) Lessons and spores of *Diplodia seriata*. (B) Lessons and spores of *D. mutila*. (C) Lessons and spores of *Spencermartinsia* (= *Dothiorella*) *viticola*.

being *B. dothidea*, but with the increasing knowledge in the genera, the detected species correspond to *D. mutila*. Considering the most important table grape cultivars grown in Chile, in trials done by Morales et al. [4] with excised shoots of cvs. Thompson Seedless, Red Globe, and Flame Seedless, the most aggressive species was *D. mutila*. In a field trial done with 25-year-old plants of the cv. Flame Seedless, inoculating three different tissues, 1-, 2-, and

5-year-old shoots, the most aggressive species of the Botryosphaeriaceae tested was also *Diplodia mutila* (**Figure 3B**), and *Diplodia seriata* (**Figure 3C**) was less aggressive in all of the tested tissues considering the necrotic lesion. However, considering vascular streak length, *Diplodia seriata* was more aggressive than *D. mutila*. This is not in accordance with another pathogenicity tests reported by Úrbez-Torres and Gubler [23], where the most virulent species was *Lasiodiplodia theobromae*, followed by *Neofusicoccum luteum*, *N. parvum*, and *N. australe*, and they considered *Diplodia mutila* and *D. seriata* as weakly virulent. Maybe, the climatic conditions make the difference in the pathogenicity of the isolates or the age of the tissue done in the pathogenicity test. On the other hand, in pathogenicity field trials done in two regions, Sonoma and Colusa, employing adult plants of Chardonnay and Zinfandel, respectively, infecting nine different GTD causal agents, the most infectious were the four Botryosphaeriaceae species tested, while *Pa. chlamydospora* was in an intermediate position, and the less infective were *Togninia minima*, *Phaeoacremonium parasiticum*, *P. richardsiae*, and *Eutypa lata* [24].

In young tissue the most aggressive species are *Neofusicoccum* and *Lasiodiplodia*. Also, it is important to note that in Chile only five species of Botryosphaeriaceae have been detected affecting *Vitis vinifera* [1, 4, 8, 17]. Nevertheless, these species are capable to cause almost 50% of less yield in some regions [11]. Up to date, *Lasiodiplodia theobromae*, an important causal agent described in different countries, has not been detected or described affecting table grape vines or wine cultivars in Chile [3, 4, 10].

2.6. Botryosphaeriaceae main hosts

In order to understand how Botryosphaeriaceae species can arrive to *Vitis* plants, one way could be in the association of different species crops in a same orchard or in surrounding orchards. If we analyze the statement done by De Wet et al. [25], this authors classify different species of Botryosphaeriaceae associated to gymnosperms or angiosperms: *Diplodia* species occurred mainly in gymnosperms; *Dothiorella* species restricted to angiosperms; *Lasiodiplodia* occurred equally on gymnosperms or angiosperms, and *Fusicoccum* anamorphs occurred mostly on angiosperms with rare reports in gymnosperms. Nevertheless, in the case of *Vitis vinifera*, this species has been affected by all the different genera of the Botryosphaeriaceae family. In Chile, other important hosts for Botryosphaeriaceae are almond trees [26], loquat [27], blueberry [28], avocado trees [29], and araucaria trees [30], among others.

2.7. Plant physiology damage

In the trunk a depletion of starch reserves in woody tissues is not only associated with fungal colonization but also affects leaves (where no colonization is achieved) affecting photosynthetic rate. This can cause a decrease of plant development and vigor during the subsequent year. Also, lipid and amino acid are downregulated [2]. There is a response of the affected plants resulting in an accumulation of tyloses and gummosis, not only causing the blockage of the limiting fungal invasion but also causing disease. But, their progression in the wood is also inhibited by the formation of polyphenol-rich reaction zones and by the accumulation of pathogenesis-related (PR) proteins and the production of oxidative damage and reactive oxygen species (ROS) production.

It has been demonstrated that fresh pruning wounds are the main infection route for grape trunk disease pathogens, bot canker disease [3, 4], and Esca-like disease [3].

2.8. Endophytic species

Many authors highlight the fact that these fungi are common endophytes mainly in woody plants and that these fungi have the ability to cause serious diseases, particularly when plants are predisposing due to environmental stress [5, 31]. In general, these species have been recognized as natural opening colonizers. The first report of a forest pathogen isolated from stem and xylem healthy tissues was done by Petrini and Fisher [32]. Later, Johnson et al. [33] could recover *Fusicoccum, Neofusicoccum, Pseudofusicoccum,* and *Lasiodiplodia* from healthy mango plant parts.

Muruamendiaraz and Legorburu [34] could recover *D. seriata* from asymptomatic tissues of *Vitis* plants. This finding could agree with the endophytic character of this species. In the same work, *E. lata* was rarely found in asymptomatic vines. On the other hand, *Phaeomoniella chlamydospora* and *Phaeoacremonium* spp. were present in healthy rootstock propagation material as endophytes [35].

2.9. GTD spore's dispersion and optimal temperature

In Chile according to Valencia et al.'s [10] work, the dispersal of Botryosphaeriaceae conidia was studied in two vineyards in the Valparaíso Region of Chile, where semiarid Mediterranean conditions prevail. Due to the climate situation in this region characterized by the habitual rainy period mainly occurring in winter months, peaks of inoculum also occurred in the season associated to precipitation events equal or greater than 2 mm. The main species associated to the spore traps were *D. seriata, Spencermartinsia* (= *Dothiorella*) *viticola, Neofusicoccum,* and probably *N. australe.* In this work no association was observed between the volume of water used in the application of agrichemical products ($\leq$1000 L ha^{-1}) and peaks of Botryosphaeriaceae.

Considering the dispersion of Botryosphaeriaceae species, there are mainly two works done in other countries: Úrbez-Torres et al.'s [36] work done in California, United States, and Amponsah et al.'s [37] work done in Canterbury, New Zealand. In California, Úrbez-Torres et al. [36] found a strong regression between the spore release and precipitation and that more spores were trapped during the winter period. Amponsah et al. [37] work on the dispersal of spores of Botryosphaeriaceae that are responsible for dieback in grapevines. They work in one vineyard, with the Vaseline-coated slide trap. During all the trapping period, 59.8% of the total species trapped were *Neofusicoccum* spp., and 40.2% of the rest were *Diplodia* spp. The spores were detected during the entire period but most abundant during December, January, and February associated to high temperatures and rainfall, where routine trimming of mature canes provides wounds for infection. Under these conditions, the main collected species were *Neofusicoccum luteum, N. australe, N. parvum,* and *Diplodia mutila.* These species infect mainly trunks, canes, green shoots, and plant debris. In this region, rainfall occurs not only in the winter period, mainly June, July, and August but also in the rest of the year.

Taking into account Díaz and Latorre's [18] work, they reported that they could not find *Pa. chlamydospora* in pruning debris samples (n = 168), sap samples (n = 82), soil samples (n = 100), or weed samples (n = 293) that were collected during 2010–2012. Maybe, infection of this species is mainly associated to young plants.

Studies carried out by Úrbez-Torres et al. [36], evaluating the conidial germination of different Botryosphaeriaceae at different temperatures found that conidia did not germinate at 5°C with the exception of *Botryosphaeria dothidea* and *Neofusicoccum parvum*. On the other hand, only *B. dothidea, Diplodia seriata,* and *Lasiodiplodia theobromae* germinate at 40°C. Optimum germination at 25°C was for *B. dothidea* and *Dothiorella iberica*; 25–30°C for *Spencermartinsia viticola*; 30°C for *D. corticola, D. mutila, D. seriata,* and *Neofusicoccum parvum*; and 40°C for pigmented *L. theobromae*. This is not strange considering the observations done by Yan et al. [14] that *L. theobromae* was found in *Vitis* under subtropical monsoon climatic conditions.

2.10. Disease management

Considering the recent statement done by Gramaje et al. [38] about the establishment, progress, severity, and spread of GTDs have been associated with (i) drought, (ii) limited availability of effective fungicides, (iii) pressure to increase yields, (iv) lack of pruning wound protection, and (iv) poor low-quality propagation material. All the efforts should first be done in order to avoid these types of situation.

2.10.1. Petri disease

In order to avoid early infection of grapevines by *Phaeomoniella* or *Phaeoacremonium* species, the main treatments considered are during the production of plants in the nursery. The program should start with an adequate clean mother plants and use of fungicides in the hydration process when the new plants are produced. Details of this process can be followed in Auger et al. [39]. This is complemented with the heat treatment [40].

2.10.2. Bot canker control

The disease's integrated management of this disease should start with an efficient cleaning process of all the diseased or dead shoots, arms, and trunks present in a vineyard. This material should be immediately taken out of the vineyard and efficiently destroyed in order to avoid the formation of reproductive structures such as pycnidia or ascomata of Botryosphaeriaceae and also to avoid the colonization of diseased tissues associated to Basidiomycetes like *Inocutis* spp.

A cultural management interesting to note is that the actual pruning cordon system produces many little injuries that are difficult to protect, more over if a mechanical pruning is used. Something that should be evaluated, especially in regions where GTDs are highly prevalent, the use of double Guyot pruning system should be considered.

Botryosphaeriaceae can cause cankers and dieback in grapevines, but different species affect grapevines in different grape-growing regions and countries. There are regions or countries where the rainfall period is concentrated mainly in winter time, while there are some regions or countries where rainfall conditions occur around the whole year. According to the climatic specific conditions of each area, the management will be oriented to prevent the disease mainly in the winter period or during the whole year. In Chile, according to the work published by Díaz et al. [3] and Valencia et al. [10], the main risk period for bot canker is in winter, where pruning management takes place, so the main effort should be oriented to protect pruning wounds. Nevertheless, the wine-producing area has expanded to the southern regions of Chile, where it usually rains during the summer months, so at these locations the problem probably expands to another species and attack green shoots that currently occurs in New Zealand [37] or in subtropical regions of China [14].

One of the first works done in evaluating different fungicides under in vitro conditions was the work done by Amponsah et al. [41]. They evaluate mycelial and conidial inhibition and also if they were able to protect pruning wounds against infection. In this work the tested species differed in their response to the different fungicides, and considering all the evaluated species, *N. luteum* was the least sensitive. The most effective fungicides were flusilazole, thiophanate-methyl, and mancozeb. According to the in vitro mycelial inhibition, flusilazole was the most effective (0.002 ppm) but was less effective considering conidial germination inhibition. For the in vivo trial, the best treatments were carbendazim, mancozeb, iprodione, and flusilazole. Under field conditions the best treatment was based on flusilazole active ingredient.

In the work done in Chile by Torres et al. [42], they evaluate demethylation-inhibiting (DMI) fungicides against *D. mutila*, *D. seriata*, *N. australe*, and *N. parvum*. The results demonstrate that the fungicides tebuconazole, myclobutanil, prochloraz, and prochloraz plus epoxiconazole were effective under in vitro conditions in inhibit mycelial and conidial germination for all the evaluated species.

Díaz and Latorre [18] evaluate the efficacy of benomyl, pyraclostrobin, tebuconazole, and thiophanate-methyl, applied as paste or liquid, in pre- or postinoculation. The paste application provides a better control than the sprayed form. **Table 1** shows the actual registered fungicides to be used for GTD control in Chile.

In California, USA, Rolshausen et al. [24] probed four fungicides (previously registered for use in vineyards) and their effect in the control of nine different GTD causal agents. The fungicides were Topsin M (thiophanate-methyl), Biopaste (boron), Cabrio EG (pyraclostrobin), and Garrison

Commercial product	Active ingredient	Manufacturer	Pathogens controlled
Podexal	Pyraclostrobin	BASF	Plateado, *Phaeomoniella*, and *Botryosphaeria obtusa*
Podastik Max	Tebuconazole	Arysta LifeScience	Silver leaf, *Nectria*, *Neofusicoccum* spp., and black dead arm of *Vitis*
Pasta Poda TPN-50	Chlorothalonil	Anasac	Plateado, *Phomopsis*, and European canker
Pasta Poda Full	Tebuconazole + Kresoxim-methyl	Anasac	*Neofusicoccum parvum* and *Phaeomoniella* spp.

Table 1. Commercial products actually commercialized in Chile for GTD control in vineyards.

(cyproconazole and iodocarb). All of these fungicides were effective in reducing the lessons caused by this GTD species, but the most effective according to the authors was Topsin M.

In Portugal, Rego et al. [43] in a 3-year field trial conducted in 15-year-old vineyard cv. Aragonez evaluate the efficacy of six different treatments against *Botryosphaeria* canker and *Phomopsis* dieback, applying three sprays each year, and the most efficient treatment was the application of Bion plus Score (difenoconazole).

3. Conclusion and future trends

In Chile the most important diseases affecting vineyards for table grapes and wine production are bot canker and Esca-like disease. In addition, for young plants the main problem is Petri disease. There are several studies done in determining causal agents, epidemiology, and control strategies, but there still remaining work to be done, for example, the fulfillment of Koch's postulates in order to determine the etiology of chlorotic leaf roll disease. Also, the convenience of the use of a double Guyot pruning system instead of a cordon pruning system, in order to avoid so many exposed fresh cuts during the pruning period, should be analyzed.

Something that is important to note is that some management change occurred worldwide that makes a point of inflection in the increase of GTDs. In Europe some change was that sodium arsenite was excluded in 2001 for Esca control, but in Chile sodium arsenite was not used. In the management of vineyards, a change in the gray mold control was done, mainly associated with the discontinuous use of benzimidazoles because of the increase of resistant *Botrytis cinerea* isolates. It is important to note that in many trials done in different countries, benzimidazoles were highly efficient in controlling most of the GTD agents.

These considerations are something that should be approached in an interdisciplinary work done with different researchers from different countries.

Author details

Ximena Besoain

Address all correspondence to: ximena.besoain@pucv.cl

School of Agronomy, Pontifical Catholic University of Valparaíso (Escuela de Agronomía, Pontificia Universidad Católica de Valparaíso), Quillota, Chile

References

[1] Latorre BA, Besoain X, Flores V. *Botryosphaeria* canker of table grapes. Phytopathology. 1986;**76**(10):1112. (Abstract)

[2] Fontaine F, Pinto C, Vallet J, Clément C, Gomes AC, Spagnolo A. The effects of grapevine trunk diseases (GTDs) on vine physiology. European Journal of Plant Pathology. 2016;**144**: 707-721. DOI: 10.1007/s10658-015-0770-0

[3] Díaz GA, Auger J, Besoain X, Bordeu E, Latorre BA. Prevalence and pathogenicity of fungi associated with grapevine trunk diseases in Chilean vineyards. Ciencia e Investigación Agraria. 2013;**40**(2):327-339. DOI: 10.7764/rcia.v40i2.1101

[4] Morales A, Latorre BA, Piontelli E, Besoain X. Botryosphaeriaceae species affecting table grape vineyards in Chile and cultivar susceptibility. Ciencia e Investigación Agraria. 2012; **39**(3):445-458. DOI: 10.7764/rcia.v39i3.37

[5] Slippers B, Wingfield WJ. Botryosphaeriaceae as endophytes and latent pathogens of woody plants: Diversity, ecology and impact. Fungal Biology Reviews. 2007;**21**(2–3):90-106. DOI: 10.1016/j.fbr.2007.06.002

[6] OIV. Top Ten Wine-Exporting Countries. 2017. Available from: https://italianwinecentral.com/top-ten-wine-exporting-countries/ [Accessed: June 15, 2017]

[7] ODEPA. Vides: Superficie y producción. 2017. Available from: http://www.odepa.cl/vides-superficie-y-produccion-3/. [Accessed: June 30, 2017]

[8] Auger J, Esterio M, Ricke G, Pérez I. Black dead arm and canker of *Vitis vinifera* cv. Red Globe caused by *Botryosphaeria obtusa* in Chile. Plant Disease. 2004;**88**:1286. DOI: 10.1094/PDIS.2004.88.11.1286A

[9] Díaz GA, Esterio M, Auger J. Effects of *Phaeomoniella chlamydospora* and *Phaeoacremonium aleophilum* on grapevine rootstocks. Ciencia e Investigación Agraria. 2009;**36**(3):381-390. DOI: 10.4067/S0718-16202009000300005

[10] Valencia D, Torres C, Camps R, López E, Celis-Diez JL, Besoain X. Dissemination of Botryosphaeriaceae conidia in vineyards in the semiarid Mediterranean climate of the Valparaíso Region of Chile. Phytopathologia Mediterranea 2015;**54**(2):394-402. DOI: 10.14601/Phytopathol_Mediterr-16055

[11] Torres C, Camps R, Latorre BA, Besoain X. Prevalence, damage and potential yield loss of vineyards cv. Cabernet Sauvignon in two regions of Chile. In: 10th International Workshop on Grapevine Trunk Diseases. Book of Abstracts. Reims, July 4–7, 2017. p. 73 (Abstract)

[12] Bertsch C, Ramírez-Suero M, Magnin-Robert M, Larignon P, Chong J, Abou-Mansour E, Spagnolo A, Clément C, Fontaine F. Grapevine trunk diseases: Complex and still poorly understood. Plant Pathology. 2013;**62**:243-265. DOI: 10.1111/j.1365-3059.2012.02674.x

[13] Lorsch W. Fatal wood diseases affect 12 percent of French vineyards. 2014. Available from: http://www.wine-searcher.com/m/2014/10/fatal-wooddiseases-affect-12-percent-offrench-vineyards [Accessed: Jul 14, 2017]

[14] Yan JY, Xie Y, Zhang W, Wang Y, Liu JK, Hyde KD, Seem RC, Zhang GZ, Wang ZY, Yao SW, Bai XJ, Dissanayake AJ, Peng YL, Li XH. Species of Botryosphaeriaceae involved in grapevine dieback in China. Fungal Diversity. 2013;**61**:221-236. DOI: 10.1007/s13225-013-0251-8

[15] Lehoczky J. Black dead arm disease caused by *Botryosphaeria stevensii* infection. Acta Phytopathologica Academia Scientiarum Hungariae. 1974;**9**:319-327

[16] Pine TS. Etiology of the dead arm disease of grapevines. Phytopathology. 1958;**48**:192-197

[17] Besoain X, Torres C, Díaz GA, Latorre BA. First report of *Neofusicoccum australe* associated with *Botryosphaeria* canker of grapevine in Chile. Plant Disease. 2013;**97**:143-143. DOI: 10.1094/PDIS-07-12-0652-PDN

[18] Díaz GA, Latorre BA. Infection caused by *Phaeomoniella chlamydospora* associated with esca-like symptoms in grapevine in Chile. Plant Disease. 2014;**98**:351-360. DOI: 10.1094/PDIS-12-12-1180-RE

[19] Phillips AJL, Alves A, Abdollahzadeh J, Slippers B, Wingfield MJ, Groenewald JZ, Crous PW. The Botryosphaeriaceae: Genera and species known from culture. Studies in Mycology. 2013;**76**:51-167. DOI: 10.3114/sim0021

[20] Úrbez-Torres JR. The status of Botryosphaeriaceae species infecting grapevines. Phytopathologia Mediterranea. 2011;**50**(Supplement):S5-S45

[21] Gatica M, Césari C, Magnin S, Dupont J. *Phaeoacremonium* species and *Phaeomoniella chlamydospora* in vines showing "hoja de malvón" and young vine decline symptoms in Argentina. Phytopathologia Mediterranea. 2001;**40S**:317-324. DOI: 10.14601/Phytopathol_Mediterr-1609

[22] Gatica M, Dubos B, Larignon P. The "hoja de malvón" grape disease in Argentina. Phytopathologia Mediterranea. 2000;**39**:41-45. DOI: 10.14601/Phytopathol_Mediterr-1556

[23] Úrbez-Torres JR, Gubler WD. Pathogenicity of Botryosphaeriaceae species isolated from grapevine cankers in California. Plant Disease. 2009;**95**:584-592. DOI: 10.1094/ PDIS-93-6-0584

[24] Rolshausen PE, Úrbez-Torres JR, Rooney-Latham S, Eskalen A, Smith RJ, Gubler WD. Evaluation of pruning wound susceptibility and protection against fungi associated with grapevine trunk diseases. American Journal of Enology and Viticulture. 2010;**61**:113-119

[25] De Wet J, Slippers B, Preisig O, Wingfield B-D, Wingfield MJ. Phylogeny of the Botryosphaeriaceae reveals patterns of host association. Molecular Phylogenetics and Evolution. 2008;**46**:116-126. DOI: 10.1016/j.ympev.2007.08.016

[26] Besoain X, Briceño E, Piontelli E. *Fusicoccum* sp. as the cause of canker in almond trees and susceptibility of three cultivars. Fitopatología. 2000;**35**:176-182

[27] Gisbert AD, Besoain X, Llácer G, Badenes ML. Protección del cultivo II: Enfermedades. In: Agustí M, Reig C, Undurraga P, editors. El cultivo del níspero japonés. España: Pontificia Universidad Católica de Valparaíso y Universidad Politécnica de Valencia; 2006. pp. 227-250

[28] Espinoza JG, Briceño EX, Chávez ER, Úrbez-Torres JR, Latorre BA. *Neofusicoccum* spp. associated with stem canker and dieback of blueberry in Chile. Plant Disease. 2009;**93**: 1187-1194. DOI: 10.1094/PDIS-93-11-1187

[29] Auger J, Palma F, Pérez I, Esterio M. First report of *Neofusicoccum australe* (*Botryosphaeria australis*), as a branch dieback pathogen of avocado trees in Chile. Plant Disease. 2013;**97**: 842. DOI: 10.1094/PDIS-10-12-0980-PDN

[30] Besoain X, Camps R, Guajardo J. First report of *Diplodia mutila* causing gummy canker in *Araucaria araucana* in Chile. Plant Disease. 2017;**101**:1328. DOI: 10.1094/PDIS-12-16-1730-PDN

[31] Crous PW, Slippers B, Groenwald JZ, Wingfield MJ. Botryosphaeriaceae: Systematics, pathology, and genetics. Fungal Biology. 2017;**121**:305-306. DOI: 10.1016/j.funbio.2017.01.003

[32] Petrini O, Fisher PJ. A comparative study of fungal endophytes in xylem and whole stem of *Pinus sylvestris* and *Fagus sylvatica*. Transactions of the British Mycological Society. 1988;**91**:233-238. DOI: 10.1016/S0007-1536(88)80210-9

[33] Johnson GI, Mead AJ, Cooke AW, Dean JR. Mango stem end rot pathogens – Fruit infection by endophytic colonization of the inflorescence and pedicel. Annals of Applied Biology. 1992;**120**:225-234. DOI: 10.1111/j.1744-7348.1992.tb03420.x

[34] Muruamendiaraz A, Legorburu J. Suitable of an increment borer as a sampling device for grapevine trunk disease. Phytopathologia Mediterranea. 2009;**48**:145-149. DOI: 10.14601/Phytopathol_Mediterr-2884

[35] Haleen F, Crous PW, Petrini O. Fungi associated with healthy grapevine cuttings in nurseries, with special reference to pathogens involved in the decline of young vines. Australasian Plant Pathology. 2003;**32**(1):47-52. DOI: 10.1071/AP02062

[36] Úrbez-Torres JR, Bruez E, Hurtado J, Gubler WD. Effect of temperature on conidial germination of Botryosphaeriaceae species infecting grapevines. Plant Disease. 2010;**94**: 1476-1484. DOI: 10.1094/ PDIS-06-10-0423

[37] Amponsah NT, Jones EE, Ridgway HJ, Jasper MV. Rainwater dispersal of *Botryosphaeria* conidia from infected grapevines. New Zealand Plant Protection. 2009;**62**:228-233

[38] Gramaje D, Baumgastnerk K, Halleen F, Mostert L, Sosnowski MR, Úrbez-Torres JR, Armengol J. Fungal trunk diseases: A problem beyond grapevines? Plant Pathology. 2016; **65**:355-356. DOI: 10.1111/ppa.12486

[39] Auger J, Pérez I, Esterio M. Enfermedades causadas por hongos de la madera de la vid. Red agrícola. 2017. Available from: www.redagricola.com/manejo-control-enfermedades-causadas-hongos-la-madera-la vid/

[40] Elena G, Di Bella V, Armengol J, Luque J. Viability of Botryosphaeriaceae species pathogenic to grapevine after hot water treatment. Phytopathologia Mediterranea 2015;**54**:325-334. DOI: 10.14601/Phytopathol_Mediterr-15526

[41] Amponsah N, Jones E, Ridway HJ, Jaspers MV. Evaluation of fungicides for the management of *Botryosphaeria* dieback diseases of grapevines. Pest Management Science. 2012;**68**: 676-683. DOI: 10.1002/ps.2309

[42] Torres C, Latorre B, Undurraga P, Besoain X. Evaluation of DMI fungicides against species of Diplodia and Neofusicoccum associated with Botryosphaeria canker of grapevine. Ciencia e Investigación Agraria. 2013;**40**(1):131-138. DOI: 10.7764/rcia.v40i1.457

[43] Rego C, Reis P, Dias A, Correia L. Field evaluation of fungicides against Botryosphaeria canker and Phomopsis cane and leaf spot. Phytopathologia Mediterranea. 2014;**53**:581-582. DOI: 10.14601/Phytopathol_Mediterr-15167

Chapter 4

Viticulture in Warmer Climates: Mitigating Environmental Stress in Douro Region, Portugal

Manuel Oliveira

Additional information is available at the end of the chapter

http://dx.doi.org/10.5772/intechopen.71155

Abstract

There is evidence that temperatures all over the planet are increasing slowly but steadily. The projections for the Mediterranean basin indicate a strong warming and decreased precipitation, especially in summer, and they are not favorable for agriculture in general and, in particular, for viticulture, which is a very climate-sensitive system. The vineyards of Douro Demarcated Region, located in the northeastern Portugal, are grown in marginal conditions, and the expected daily higher temperatures and decreased rainfall are bound to affect the yield and wine quality. Touriga Nacional and Tinta Roriz are two of the most valuable varieties for high-end wines of Douro, and they were subjected to a number of experiments on mitigation techniques to counteract the worse effects of adverse weather conditions. Irrigation, canopy shading, water nebulization, and kaolin coating can mitigate the losses caused by increasing stressful conditions, but they have detrimental effects on some wine characteristics. They even might not be sufficient to protect the already producing vineyards from aggravated weather conditions.

Keywords: viticulture, temperature stress, water stress, irrigation, canopy management, cooling techniques, climate change

1. Introduction

Today there is a large consensus that the earth's average temperature has been increasing slowly but steadily. The planet's average surface temperature has risen about 1.1°C since the late nineteenth century, and most of the warming occurred in the past 35 years; 2016 was the warmest year on record and every month from January through September, except for June, which was the warmest on record for those respective months [1]. Both projections from general circulation and regional climate models show that Southern Europe and the Mediterranean basin will experience strong warming and decreased precipitation, especially in summer, and more frequent and longer extreme climate events such as heat waves and prolonged drought [2, 3].

Such scenarios are not favorable for agriculture for an economic activity dependent upon and deeply connected to climate and weather. The cultivation of grapevines (*Vitis vinifera*), the fruit in Europe that is primarily used for winemaking and is a climate-sensitive agricultural system that has been used as an indicator of both historic and contemporary climate change [4]. Despite adaptation of *Vitis vinifera* to a wide range of climatic conditions, individual varieties, in particular premium wine grapes, have narrow climate ranges making the system very susceptible to minor changes in climate [5, 6]. Climate, especially air temperature, together with soil properties are considered the most important factors in defining the suitability of an area for wine production, affecting physiological behavior of the grapevine, the phenological phases of vine, the formation and maturity of the berry and its chemical changes, and finally, the characteristics of wines produced in a particular area [7, 8].

Portugal is expected to face aggravated weather conditions in coming years because the territory has suffered a pronounced decrease in precipitation in the spring, in particular in March, since the early 1950s [9] while a warming trend has been observed revealing a significant increase in extreme heat events for both spring and summer seasons, and a decrease in extreme cold events in winter [10]. All Portuguese wine regions are expected to experience an upper shift in temperature and lower precipitation. The Douro Demarcated Region (DDR), located in the northeastern Portugal, is classified as a Denomination of Controlled Origin (DOC), the highest Portuguese wine classification, given its distinctive climatic, topographic, and soil characteristics. It is of particular concern given its economic preeminence in Portuguese wine industry. Its strong Mediterranean climate influence shows the typical variability in precipitation along with high evapotranspiration during summer, the factors that limit grapevine development, yield, and quality components [11]. In these regions, meteorological averages from the period 1950–2000 show that during active vine development from April to September, rainfall varies from 190 to 326 mm but as low as 50 to 85 mm during the ripening stage (from July to September) coupled with high temperatures and intense solar radiation that give rise to very stressful conditions, particularly in the eastern end of the region (Alto Douro) [12] where grapevines can be subjected to a difference between evapotranspiration and precipitation as high as 730–750 mm from bud break to harvest [13]. Under these conditions, vineyards of DDR are, in general, grown in marginal conditions for agriculture production [14] that contribute to large fluctuations in interannual production and low yields averaging 30 hl ha^{-1} of wine well below the legal limit of 55 hl ha^{-1} permitted for DOC wine [12]. The expected higher daily temperatures and decreased rainfall [15] are bound to affect yield and wine quality.

Similar trends were observed in other wine regions worldwide where average temperatures have increased by 1.7°C during the growing seasons since 1950 and are expected to rise by an average of 2°C by 2050 leaving some regions at or near their optimum growing-season temperatures [16]. There are forecasts of further warming in Greek wine regions, pushing some of them past suitability for viticulture [17]. Three wine regions of the Northeastern Spain (Alt Penedès, Priorat, and Segrià) have experienced, from 1952 to 2006, overall growing-season warming of

1.0 to 2.2°C, while precipitation during the bud break to veraison period declined significantly, indicating more intense soil moisture stress during this critical growth stage [10].

Facing the most likely scenarios of climate change, wine farmers must adapt strategies to continue producing quality wines and to preserve their typicity from each growing location [18]. As a perennial crop, grapevines require a few years to reach maturity and can remain economically productive for several decades; thus, adaptation strategies must account for both short- and long-term impacts of a shifting environment. Perennial practices, such as choice of the planting site, choice of grapevine and rootstock variety, and vineyard layout, must be decided prior to planting. For those vineyards already under production, winegrowers will have to adapt using suitable short-term mitigation strategies [19], such as irrigation, canopy management, and cooling techniques.

2. Weather conditions in Alto Douro

Alto Douro has a total area of 120,000 ha with an average elevation of 420 m, but the best prized vintages are produced at lower elevations where records for the period 1950–2000 show an annual precipitation ranging from 600 to 800 mm, but only 200 to 300 mm occur during the growing season from April to October, while annual temperatures ranged between 14 and 16°C and growing season oscillated from 19 to over 21°C [20]. Located in Alto Douro, close to Douro river (41° 08′ North, 7° 08′ West), some mitigation techniques have been experimented to protect *Vitis vinifera* cv Touriga Nacional and cv Tinta Roriz (Syn. Aragonêz, Tempranillo) from intense abiotic stresses that might prove useful under more extreme conditions eventually brought about by climate change.

Touriga Nacional is one of the most valuable premium varieties of DDR. It is considered one of the best red varieties to produce quality wines with a unique aroma profile that can fetch high market prices [21]. It has good adaptation to environmental stresses, withstanding high light intensities that allow for better adjustment to warm conditions, provided an adequate water supply [22]. Tinta Roriz can be cultivated in a variety of soils and climatic conditions, because it resists well to moderate drought. It has a thick-skinned berry with a high anthocyanin concentration that makes for deep-colored wines with moderate tannins and moderate acidity. It provides some of the best red wines of Portugal and Spain [23, 24]. Despite Touriga Nacional and Tinta Roriz good adaptability, they can suffer serious yield losses under unfavorable weather conditions. An on-site automatic weather station provided records for the period 2004 to 2016 (**Table 1**). Average annual temperature was 16.1 and 21°C for the growing season, whereas total precipitation was 594 mm annually with 284 mm for the growing season. Calculated reference evapotranspiration (ET_0) [25] averaged 1946.8 mm a year, mostly occurring during the growing season (1631.3 mm) creating an imbalance between ET_0 and rainfall of 1349.3 and 1321.4 mm annually and for the growing season, respectively. Temperature was higher and rainfall was lower than the average values recorded for Alto Douro during the period 1950–2000. The average temperature for the growing season is already at the upper limit that is considered adequate for quality wine grape production [26].

	Rad (Kw m^{-2})	T (°C)	R (mm)	ET_0 (mm)
Annual	1757.8	16.1	594.0	1946.2
Nov–Mar	346.8	9.2	309.9	315.6
Apr–Oct	1411.0	21.0	284.2	1631.3

Table 1. Total solar radiation (rad), air temperature (T), precipitation (R), and reference evapotranspiration (ET_0) averages for the period 2004–2016 at location in Alto Douro 41° 08′ North, 7° 08′ West (on-site meteorological station).

2.1. Mitigation techniques in Alto Douro

2.1.1. Irrigation

Historically, vineyards of DDR in general and Alto Douro in particular are water stressed during the growing season, given the high atmospheric demand and the scarcity of rainfall that limits the soil water (SW) availability. SW has a complex relationship with air temperature that, when it rises, increases vapor pressure deficit, and more water is evapotranspirated from the soil; eventually, water availability becomes critical. In [27], SW will be the limiting resource in vineyards of Douro and it will become relatively more important than temperature for maintaining grapevine production. A model of SW annual distribution in Douro valley shows a sinusoidal pattern (**Figure 1**) that is consistent with temperature and rainfall distributions [28].

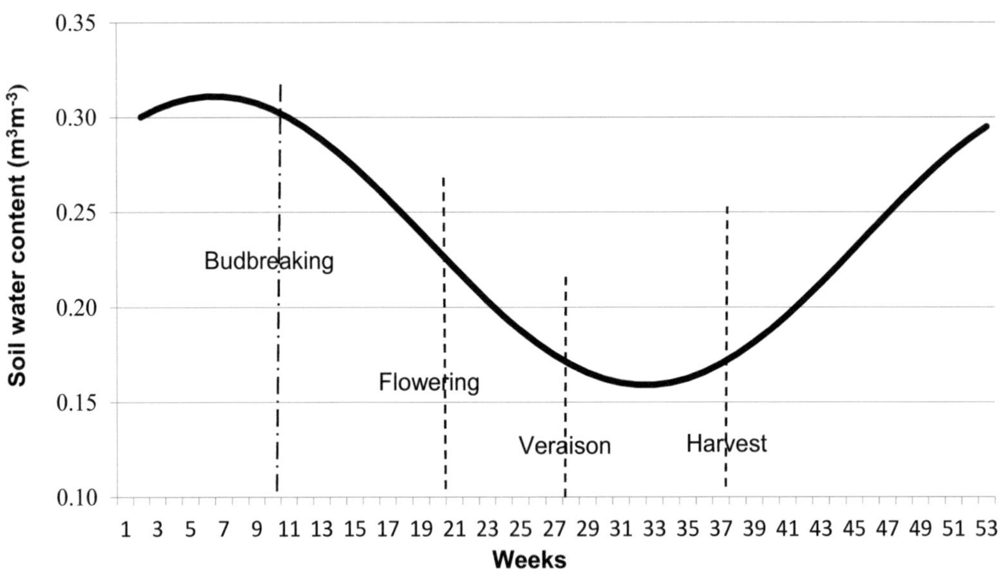

Figure 1. Time frequency (Fourier series) generated data of soil water content (m^3 m^{-3}) based on 364 historical observations from 1991 to 1997 and annual occurrence of main phenological stages of the grapevines: bud breaking, flowering, veraison, and harvest. (Adapted from Oliveira [28]).

The time frequency generated data in **Figure 1** shows that the minimum SW content is lower than the permanent wilting point (0.19), bringing the SW content of the important period from flowering to veraison to a critical level when irrigation will be most useful. As climate becomes warmer, grapevines in Douro show a clear tendency toward earlier phenological events [18, 29] a phenomenon also observed in Touriga Nacional and Tinta Roriz for the period 2004–2016, and in 2017, flowering occurred about 15–20 days earlier than usual across Alto Douro according to empirical observations. Earlier events could overlap with periods of higher SW content, but ripening might also occur in time of higher temperatures that are detrimental to quality [30, 31]. As rainfall is expected to diminish, SW will become scarcer and a short growing season might not bring any benefit. Under present conditions, rainfed vineyards in Alto Douro produce as little as 0.7 kg per plant and even less in drier years [32], not enough to cover production costs. Irrigation is then essential to ensure a stable production and minimize the risk of drought damage [33]. Experimental fields were set up in DDR to study the effects of irrigation in grapevines [32–35]. The main experimental field was a 2.5-hectare commercial vineyard (*Vitis vinifera* cv. Tinta Roriz) where six treatments comprising three adjacent grapevine rows each, 120 plants per treatment, laid out in a random design pattern and trained to a vertical shoot position (VSP). Each grapevine row had its own drip irrigation line and they were nonirrigated, irrigated from flowering to veraison at rates of 4 (total of 160 mm) and 8 mm a day (total of 320 mm), irrigated from veraison to commercial maturation at rates of 4 (total of 225 mm) and 8 mm a day (total of 450 mm), and irrigated from flowering to commercial maturation at a rate of 8 mm a day (total of 720 mm). Water use efficiency (WUE) was calculated as the ratio between a given parameter (*X*) and crop evapotranspiration (*Ec*) [36].

$$WUE = X/Ec \tag{1}$$

Ec was determined by soil water balance [37, 38] using SW data measured with TDR probes.

$$\left(h_i - h_{i-1}\right) = (P + I) - (Ec + R + D) \tag{2}$$

where *h* is the soil water storage at measurement time *i* and at preceding measurement time *i*-1, *P* is precipitation, *I* is irrigation, *Ec* is the water used by plant transpiration plus evaporation from soil, *R* is the surface run-off, and *D* is the soil drainage below the root zone (all units in mm). *R* and *D* can be safely ignored because low SW content and sparse precipitation during growing season, conditions similar to those found by other authors [39, 40].

At commercial harvest, three samples were collected from every treatment for laboratory analysis; each sample was a composite of berries collected from 10 randomly chosen grapevines (18 analyses repeated over 3 years).

Irrigation caused a significant increase in yield (**Table 2**) due to larger berries, rather than to a greater number of clusters or berries per cluster, and simultaneously, the compounds related to must quality (color, aroma, and soluble solids) were either diluted or their amount significantly reduced compared with musts from rainfed plants. On the other hand, there were no significant differences in pH, free anthocyanins, and total flavonols.

Irrigation (mm)	Ec (mm)	Y (g plant^{-1})	TSS (°Brix)	TA (g L^{-1})	TAc (g L^{-1})	MAc (g L^{-1})
0	193	721^a (3.7^a)	30^a (1.5^a)*	2.3^a(1.2^a)**	6.6^a(2.3^a)**	1.5^a(0.8^a)**
160	302	1002^b (3.3^b)	28^b (0.9^b)*	2.4^b(0.8^b)**	6.5^a(2.1^b)**	2.2^c(0.7^b)**
320	522	1003^b (1.9^c)	26^c (0.5^c)*	2.7^c (0.5^c)**	6.5^a(1.2^c)**	2.7^e(0.5^c)**
225	308	865^c (2.8^d)	27^c (0.9^b)*	2.4^b(0.8^b)**	6.9^b(2.2^a)**	2.3^d(0.7^b)**
450	548	889^c (1.6^e)	25^d (0.4^d)*	2.6^c (0.5^c)**	7.0^b(1.3^c)**	2.2^c(0.4^c)**
720	744	1224^d (1.6^e)	24^e (0.3^e)*	3.0^d(0.4^d)**	6.8^b(0.9^d)**	2.1^b(0.3^d)**

Different superscript letters on same column indicate a significant difference at $\alpha \leq 0.05$ (Tukey's HSD test). *Vitis vinifera* cv Tinta Roriz, Alto Douro (adapted from Refs. [32, 35]).

*WUE × 10.

**WUE × 100.

Table 2. Effect of irrigation on crop evapotranspiration (Ec), yield (Y), total soluble solids (TSS), titrable acidity (TA), tartaric acid (TAc), malic acid (Mac), and respective water use efficiency of each parameter (in parenthesis).

Titrable acidity increased with irrigation, but it is still too low for high-end wines with a preferred acidity of 6–7 mg L^{-1} that is equivalent of tartaric acid [41], which is a manifestation of the influence of high temperatures on degrading the main organic acids of the must, namely, malic and tartaric acids (**Table 2**).

Total soluble solids (TSS), polyphenol index (PI), total anthocyanins (TAn), and color index (CI) significantly decreased with irrigation (**Table 3**). TSS is very important for grape growers and winemakers because it determines how much sugar is available for conversion into alcohol. The relationship between irrigation and TSS is complex; more available water induces higher photosynthetic rate and accumulation of soluble solids but also augments yields and the berry size that can dilute the soluble solids, and in the end, their lower concentration produces a total volume of wine poorer in alcohol. PI is a phenolic parameter that measures the

Irrigation	TSS (°Brix)	PI	TAn (mg berry^{-1})	CI
0	30^a (1.5^a)*	233^a (13.4^a)*	13.8^a (7.9^a)**	0.25^a (12.0^a)**
160	28^b (0.9^b)*	203^b (7.2^b)*	11.9^b (4.1^b)**	0.21^b (6.9^b)**
320	26^c (0.5^c)*	196^b (3.9^d)*	12.3^b (2.4^c)**	0.18^c (3.5^d)**
225	27^c (0.9^b)*	176^c (6.1^c)*	10.6^c (3.8^b)**	0.16^d (5.3^c)**
450	25^d (0.4^c)*	174^c (3.2^d)*	9.2^d (1.7^d)**	0.15^d (2.8^d)**
720	24^e (0.3^d)*	138^d (1.8^e)*	9.1^d (1.2^d)**	0.13^e (1.5^e)**

Different superscript letters on same column indicate a significant difference at $\alpha \leq 0.05$ (Tukey's HSD test). *Vitis vinifera* cv Tinta Roriz, Alto Douro (adapted from Oliveira et al. [35]).

*WUE × 10.

**WUE × 100.

Table 3. Effect of irrigation treatments on total soluble solids (TSS), polyphenol index (PI), total anthocyanins (TAn), color intensity (CI), and respective water use efficiency of each parameter (in parenthesis).

polyphenolic content of must and wine, and it relates to color stability of red wines as it ages; irrigated plants produced musts with lower phenolic potential that contribute for producing red wines with lower aging potential irrespective of the irrigation schedule. Accumulation of anthocyanins (glucosides and aglycones) in grape skin is responsible for the dark color of red grapes, and their concentration is related to the berry size that increases with irrigation, resulting in musts with lower concentration of anthocyanins as the amount of available water is higher, especially if irrigation occurs close to harvest date. CI is directly related to TAn, and its values in must of irrigated plants follow the pattern of TAn.

It is then clear that irrigation rate and its schedule had a significant influence on yield and berry quality. Irrigated grapevines had better yields but, in general, the must characteristics revealed lower quality. The must quality of grapevines receiving water before veraison was better than from those irrigated after veraison. Excessive water stress before veraison causes poor fruit setting and reduced metabolic functions; berry growth is stunted and essential metabolites, such as organic acids, do not accumulate. Irrigation starting after veraison cannot make up for the losses already sustained. If water stress before veraison is moderate, berries still can attain size and composition adequate for winemaking but increasing the SW content after veraison might dilute soluble solids, acids, anthocyanins, and phenols, turning the must inadequate to high-quality wine.

For grape growers, profits are made up of both yield and quality and very low yield, no matter how the quality is, which is not profitable. The best balance between quality and yield and still attaining acceptable water use efficiency was reached with irrigation from flowering to veraison amounting to about 50% of the reference evapotranspiration of that period. When irrigation is halted at veraison, SW content declines slowly until harvest and probably there is enough water to sustain the vine metabolic functions and yield at higher levels than rainfed grapevines [32, 35]. In addition, it also permits a larger foliar area providing additional shade to partially offset rising temperatures.

Tinta Roriz had lower water use efficiency as more water is added (**Tables 2** and **3**), an important consideration in scenarios of scarcer water resources and competing uses. In dry weather conditions and sparse soil cover as vineyards, direct evaporation from soil becomes an increasingly larger proportion of evapotranspiration, and when more water is available, it results in higher crop water use that is not directly proportional to plant yield and to its metabolic functions; thus, WUE lowers in irrigated plants. Again, a compromise between yield, quality, and WUE can be achieved with deficit irrigation before veraison. Deficit irrigation at critical moments contribute to moderate water stress during the entire growing season to produce quality musts, acceptable yields, and still keep a judicious use of water.

2.1.2. Canopy management

High solar radiation intensity and elevated temperatures damage the grape berries. The fruit shrivels and can desiccate completely (raisining), making it inappropriate for winemaking. This is a phenomenon very common in Douro, also in other wine regions [42], given the actual climate conditions (**Table 1**), and it is bound to increase in the future. A north-south row orientation allows the fruits, on both sides of the canopy, reach a balance in photosynthetic efficiency, and their exposure to radiation [43] and shade propitiated by a denser canopy can minimize the losses. However, slope might not permit a north-south orientation and additional shade provided by a denser canopy actually is not enough to prevent significant yield reductions. Shading the plants can reduce their exposure to radiation and heat but creates an imbalance in carbon budget, reducing vine biomass, and affects negatively the canopy photosynthesis [44]. To avoid these negative effects and still protect a rainfed commercial vineyard of *Vitis vinifera* cv Touriga Nacional in Alto Douro, planted in east-west oriented rows and VSP trained, had the lower third of their south-facing side (from ground to 20 cm above the insertion point of grape clusters) partially covered with a vertical double layer of white plastic netting attached to iron spikes driven into the ground at 3-m intervals [45]. At about 30 randomly laid out rows with 70 vines, each was divided into three groups with equal number of rows. One group was shaded after fruit setting, another after veraison, and the third was nonshaded (control). The net was removed just prior to harvest. The net reduced the total solar radiation by 23% and photosynthetic active radiation by 27%.

From the end of flowering to commercial harvest, every week on clear sky days at noon, two well lit, adult leaves of each five randomly chosen plants on each group were used to measure stomatal conductance (porometry) and leaf water potential (pressure chamber) (300 readings repeated over 3 years). On the same days and time but only after veraison, the temperature (infrared thermometry) of one cluster on 10 randomly chosen plants per treatment was measured (180 readings repeated over 3 years). At harvest, the same procedure described for irrigation section was used to sample berries for laboratory analysis (nine analyses repeated over 3 years).

Shading did not influence significatively either water status (leaf water potential) of the plants or their stomatal conductance; thus, it is reasonable to assume that they did not affect the net photosynthesis per unit of leaf area. Shaded vines consistently had significant higher yields (2100 g plant^{-1}) than nonshaded vines (1500 g plant^{-1}) because the percentage of shriveled berries per cluster was about 6.5% against 14%, respectively (**Table 4**). The number of clusters, weight, and size of the berries showed no significant differences. Shading does not totally prevent berry shriveling because the phenomenon is also caused by other factors like water stress [46].

Shading did not alter significatively the concentrations of total soluble solids (24.3 °Brix), tartaric acid (4.3 g L^{-1}), malic acid (1.7 g L^{-1}), glucose (118.2 g L^{-1}), and fructose (110 g L^{-1}) in

Shading	Yield (g plant^{-1})	% Shriveled berries per cluster	Weight 200 berries (g)	Volume 200 berries (cm^3)
Nonshaded	1504^a (6000)*	14^a	270^a	136^a
After fruit setting	2172^b (8690)*	6^b	257^a	131^a
After veraison	2076^b (8300)*	7^b	256^a	131^a

Different superscript letters on same column indicate a significant difference at $\alpha \leq 0.05$ (Tukey's HSD test). *Vitis vinifera* cv Touriga Nacional, Alto Douro (adapted from Oliveira et al. [45]).

*In parenthesis calculated yield kg ha^{-1}.

Table 4. Effect of shading on yield, percentage of shriveled berries per cluster, weight, and volume of 200 berries for each shading treatment.

the must. Must pH was higher (3.9) in nonshaded vines against shaded (3.8). These results suggest that photosynthetic carbon acquisition was not significantly different for all groups of plants, and they are consistent with values reported for the same region and they are typical for hot climate vineyards [47, 48]. The shaded berries were approximately 1°C cooler than nonshaded ones, a difference probably due to less radiation reaching the clusters [49]. There is evidence that berry temperature affects its composition, but only differences in 3°C or above are reported as causing changes [50, 51]. On the other hand, the biosynthesis of anthocyanins is sensitive to light environment and it decreases with lower light intensity [52, 53], and in fact, shaded berries from Touriga Nacional had lower concentration of total anthocyanins and extractable anthocyanins than nonshaded berries (**Table 5**).

Lower concentrations of anthocyanins, responsible for the red color intensity of wines, can be detrimental to the must quality [54, 55] with loss in market value but that might be offset by an additional yield of 2300 kg ha^{-1} in favor of shaded grapevines.

According to the net maker, the net can last to 10–12 years. The grower did support the cost of the net, 3000 euros per hectare, plus an expense outlay of 150 euros per hectare for handling of the net. These added expenses look reasonable for vineyards already planted in an inconvenient layout that can sustain heavy losses under spells of extreme weather conditions.

2.1.3. Cooling techniques

The temperature of plant leaves are closely related to air temperature, and the rising temperatures reduce stomatal conductance [55], decrease photosynthesis rate significantly due to the increase in respiration [56], restrict carbon assimilation per unit leaf area, affecting negatively the vegetative growth and yield [57], and impair several physiological processes, notably sugar accumulation [58]. Berries after veraison are very sensitive to a combination of intense solar radiation and high temperature, reducing in size and in soluble solid concentration [59].

High temperatures and light intensity cause berries to ripe more slowly and contribute to berry sunburn and shriveling as cells of the mesocarp die [60]. The negative effects of high-temperature events on vine physiology are more severe in vines experiencing water stress [61].

Two mitigation techniques are actually being experimented in Alto Douro region with rainfed Touriga Nacional to reduce the effects of adverse temperature and radiation. Nontreated

Shading	**TAn (mg L^{-1})**	**EAn (mg L^{-1})**
Nonshaded	1534^a	825^a
After fruit setting	990^b	679^b
After veraison	1240^b	726^b

Different superscript letters on same column indicate a significant difference at $\alpha \leq 0.05$ (Tukey's HSD test). *Vitis vinifera* cv Touriga Nacional, Alto Douro (adapted from Oliveira et al. [45]).

Table 5. Effect of shading on total anthocyanins (TAn) and extractable anthocyanins (EAn) for each shading treatment.

grapevines (control) are compared with plants treated from the end of flowering for 2 weeks before the forecast harvest date: (1): coated with a suspension of kaolin [$Al_4Si_4O_{10}(OH)_8$] at a rate of 12.5 kg ha^{-1} for every 20–25 days and (2): nebulized with water from misters located above the canopy for 30 seconds every 30 minutes as long as air temperature is higher than 32°C (The aperture of electro valves is automatically triggered by temperature records of the weather station.)

The sampling procedures described on above sections (irrigation, canopy management) were followed from the end of flowering to commercial harvest, to measure canopy temperature (average temperature of both sides of the canopy), cluster temperature, stomatal conductance, and berry sampling for laboratory analysis.

Infrared thermometry shows that nebulization keeps the canopies cooler at least 1°C than coated and control canopies (**Table 6**). Despite the reflective characteristics of kaolin, temperature of coated canopy is not significantly different from canopy temperature of control plants. Reduced stomatal opening, conducive to lower stomatal conductance, higher resistance of leaf boundary layer, and low photosynthetic photon flux density during part of the day are pointed out as explanations for these results [62, 63]. Stomatal conductance (g_s, cm s^{-1}) of nebulized plants (3.11) is significantly higher than of nontreated plants (2.14) (**Table 6**). Stomatal conductance is negatively correlated with leaf temperature [55], and as temperature raises, the plant reduces the opening of its stomata to conserve water and this mechanism reduces transpiration, a mechanism that rises the leaf temperature.

Evaporative cooling and kaolin coating had the same effect on cluster temperature, measured after veraison (infrared thermometry). Clusters of treated plants show a reduced temperature in relation to control plants, which are at least 1.3°C warmer (**Table 6**). Here, the scattering of radiation caused by kaolin is not counter acted by reduced transpiration and cluster temperature was maintained lower than air temperature.

Excessive temperature and radiation that cause fruit sunburn, a major factor in loss of production in grapevines of Alto Douro region, were effectively avoided by nebulization and kaolin coating (**Table 7**). Treated plants have 5–7% of shriveled berries per cluster compared with 12–13% in control vines. Reduced berry loss contributed to higher yield per plant.

Total soluble solids are 26.1°Brix for nebulized vines, and it is significantly higher than kaolin coated and control, 24.6 and 24.7°Brix, respectively (**Table 7**). Nebulized canopies show lower

	Temperature (°C)		**g_s (cm s^{-1})**
	Canopy	**Cluster**	
Control	30.3^a	32.1^a	2.14^a
Nebulization	28.4^b	31.1^b	3.11^b
Kaolin coating	30.1^a	31.5^b	2.09^a

Different superscript letters on same column indicate a significant difference at $\alpha \leq 0.05$ (Tukey's HSD test). *Vitis vinifera* cv Touriga Nacional, Alto Douro.

Table 6. Effect of nebulization and kaolin coating on average canopy temperature and stomatal conductance (g_s) from flowering to harvest and for cluster temperature from veraison to harvest.

	Yield		Sunburnt berries	TSS
	g plant^{-1}	Kg ha^{-1}	%	°Brix
Control	868^{a}	3470	12^{a}	24.7^{a}
Nebulization	983^{b}	3930	7^{b}	26.1^{b}
Kaolin coating	1012^{b}	4050	5^{b}	24.6^{a}

Different superscript letters on same column indicate a significant difference at $\alpha \le 0.05$ (Tukey's HSD test). *Vitis vinifera* cv Touriga Nacional, Alto Douro.

Table 7. Effect of nebulization and kaolin coating on yield, percentage of sunburnt berries, and total soluble solids (TSS).

temperatures and, consequently, higher stomatal conductance (**Table 6**) that might permit an increased photosynthetic rate as observed by other authors [55, 56]. In turn, increased photosynthetic activity contributes to larger berry sugar content, evaluated as total soluble solids, as seen in **Table 7** and corroborated in other works [58, 59].

However, other characteristics of the must (pH, titrable acidity, tartaric acid, malic acid, total tannins, total polyphenols, and total anthocianins) do not show significant differences among treated and control plants (data not shown).

The cost of installation (2500 euros ha^{-1}) and running the water nebulization system (100 euros ha^{-1}) might be impeditive of its adoption. Infection by fungus and insect repellency are factors associated to water spray and kaolin coating, respectively, that must be considered but they have not been assessed at the time of this report.

3. Conclusions

Meteorological records and grapevine phenological occurrences in DDR are in agreement with forecasts of higher temperatures, lower precipitations, and advances of phenological stages. The environment of Douro is expected to become more stressful in coming years, and adaptations are imperative to save the viticulture as an economic activity. For the vineyards already in commercial production, efficient irrigation associated with shading and cooling techniques can reduce significatively water and temperature stress on grapevines and they can maintain acceptable yields with marketable quality wine. However, these techniques have some drawbacks in must quality and might not suffice to a large shift in temperature and precipitation given the actual conditions already at the limit of tolerance of the best grape varieties.

Acknowledgements

Financial support was provided by Quinta do Vale Meão (Olazabal & Sons, LDA), COTESI: Companhia de Texteis, S.A., and UTAD. My special thanks go to Francisco Olazabal, Pedro Barbosa, André Guedes, and Frederico Lobão.

Author details

Manuel Oliveira

*Address all correspondence to: mto@utad.pt

CITAB – UTAD Department of Agronomy, Vila Real, Portugal

References

[1] NASA. Global Climate Change. Vital Signs of the Planet [Internet]. 2017. Available from: https://climate.nasa.gov/evidence/ [Accessed: May 8, 2017]

[2] Lionello P, Abrantes F, Gacic M, Planton S, Trigo R, Ulbrich U. The climate of the Mediterranean region: Research progress and climate change impacts. Regional Environmental Change. 2014;**14**:1679-1684. DOI: 10.1007/s10113-014-0666-0

[3] Alo CA, Anagnostou EN. A sensitivity study of the impact of dynamic vegetation on simulated future climate change over southern Europe and the Mediterranean. International Journal of Climatology. 2017;**37**:2037-2050. DOI: 10.1002/joc.4833

[4] Mosedale JR, Abernethy KE, Smart RE, Wilson RJ, Maclean IMD. Climate change impacts and adaptive strategies: Lessons from the grapevine. Global Change Biology. 2016;**22**:3814-3828. DOI: 10.1111/gcb.13406

[5] Jones GV, Webb LB. Climate change, viticulture, and wine: Challenges and opportunities. Journal of Wine Research. 2010;**21**:103-106. DOI: 10.1080/09571264.2010.530091

[6] Hannah L, Roehrdanz P, Ikegami M, Shepard A, Shaw M, Tabo G, Zhi L, Marquet P, Hijmans R. Climate change, wine, and conservation. Proceedings of the National Academy of Sciences of the United States of America. 2013;**110**:6907-6912. DOI: 10.1073/pnas.1210127110

[7] Jackson R. Wine Science, Principles and Applications. London: Academic Press; 2014. p. 776. 9780123814685

[8] Moral FJ, Rebollo FJ, Paniagua LL, García A, Salazar EM. Application of climatic indices to analyse viticultural suitability in Extremadura, south-western Spain. Theoretical and Applied Climatology. 2016;**123**:277-289. DOI: 10.1007/s00704-014-1363-0

[9] Gallego MC, Trigo RM, Vaquero JM, Brunet M, García JA, Sigró J, Valente MA. Trends in frequency indices of daily precipitation over the Iberian Peninsula during the last century. Journal of Geophysical Research. 2011;**116**:D02109. DOI: 10.1029/ 2010JD014255

[10] Ramos MC, Jones GV, Martínez-Casasnovas JA. Structure and trends in climate parameters affecting winegrape production in northeast Spain. Climate Research. 2008;**38**:1-15. DOI: org/10.3354/cr00759

[11] Sotés V. Bases écophysiologiques et choix techniques dans la gestion de l'eau dans les vignobles méditerranéens: Expérience de l'Espagne. Le Progrès agricole et viticole. 2001;**18**:389-396 ISSN 0369-8173

[12] Jones GV, Alves F. Impacts of climate change on wine production: A global overview and regional assessment in the Douro Valley of Portugal. International Journal of Global Warming. 2012;**4**:383-406. DOI: 10.1504/IJGW.2012.049448

[13] Malheiro A, Ferreira T, Mathews R. Transpiration of vineyards of Douro Region. In: Proceedings of I Congreso Internacional sobre Clima y Viticultura (Conclivit II); 10-14 April 2007; Zaragoza. Zaragoza: Diputación General de Aragón; 2007. p. 400-406

[14] Gouveia C, Liberato MLR, Da Camara CC, Trigo RM, Ramos AM. Modelling past and future wine production in the Portuguese Douro Valley. Climate Research. 2011;**48**:349-362. DOI: 10.3354/cr01006

[15] Fraga H, Malheiro AC, Moutinho-Pereira J, Jones GV, Alves F, Pinto JG, Santos JÁ. Very high resolution bioclimatic zoning of Portuguese wine regions: present and future scenarios. Regional Environmental Change. 2014;**14**:295-306. DOI: 10.1007/s10113-013-0490-y

[16] Jones GV, White MA, Cooper OR, Storchmann K. Climate change and global wine quality. Climatic Change. 2005;**73**:319-343. DOI: 10.1007/s10584-005-4704-2

[17] Koufos G, Mavromatis T, Koundouras S, Fyllas NM, Jones GV. Viticulture–climate relationships in Greece: The impacts of recent climate trends on harvest date variation. International Journal of Climatology. 2014;**34**:1445-1459. DOI: 10.1002/joc.3775

[18] van Leeuwen C, Darriet P. The impact of climate change on viticulture and wine quality. The Journal of Wine Economics 2016;**11**:150-167. DOI: 10.1017/jwe.2015.21

[19] Marx W, Haunschild R, Bornmann L. Climate change and viticulture – A quantitative analysis of a highly dynamic research field. Vitis. 2017;**56**:35-43. DOI: 10.5073/vitis.2017.56.35-43

[20] Jones GV. A Climate Assessment for the Douro Wine Region: An Examination of the Past, Present, and Future Climate Conditions for Wine Production [Internet]. 2012. Available from: http://www.advid.pt/imagens/outros/1368437718350.pdf [Accessed: Oct 15, 2016]

[21] De Pinho PG, Falqué E, Castro M, Silva E, Machado B, Ferreira AC. Further insights into the floral character of Touriga Nacional wines. Journal of Food Science. 2007;**72**:S396-S401. DOI: 10.1111/j.1750-3841.2007.00405.x

[22] Zarrouk O, Garcia-Tejero I, Pinto C, Genebra T, Sabir F, Prista C, David TS, Loureiro-Dias MC, Chave MM. Aquaporins isoforms in cv. Touriga Nacional grapevine under water stress and recovery – Regulation of expression in leaves and roots. Agricultural Water Management. 2016;**164**:167-175. DOI: 10.1016/j.agwat.2015.08.013

[23] Borges PM. Estudo comparativo das castas tintas nobres do Dão: Touriga Nacional, Jaen, Tinta Roriz e Alfrocheiro [thesis]. Viseu: Instituto Politecnico de Viseu; 2012. Available from: http://repositorio.ipv.pt/bitstream/10400.19/1773/1/BORGES%2C%20Pedro%20

Manuel%20Marques_Estudo%20comparativo%20das%20castas%20tintas%20nobres%20do%20D%C3%A3o.pdf Accessed: Feb 5, 2016]

[24] Ramos MC, Jones GV, Yuste J. Phenology and grape ripening characteristics of cv Tempranillo within the Ribera del Duero designation of origin (Spain): Influence of soil and plot characteristics. European Journal of Agronomy. 2015;**70**:57-70. DOI: org/10.1016/j.eja.2015.07.009

[25] Allen RG, Pereira L, Raes D, Smith M. Crop Evapotranspiration. Guidelines for Computing Crop Water Requirements. FAO Irrigation and Drainage paper No. 56 [Internet]. 1998. Available from: http://www.fao.org/docrep/x0490e/x0490e00.htm [Accessed: Sep 18, 2015]

[26] Jones GV. In: Macqueen RW, Meinert LD, editors. Climate and Terroir: Impacts of Climate Variability and Change on Wine. Fine Wine and Terroir – The Geoscience Perspective. St. John's, Newfoundland: Geological Association of Canada; 2006. p. 203-216

[27] Cunha M, Richter C. The impact of climate change on the winegrape vineyards of the Portuguese Douro region. Climatic Change. 2016;**138**:239-251. DOI: 10.1007/s10584-016-1719-9

[28] Oliveira M. Modeling water content of a vineyard soil in the Douro region, Portugal. Plant and Soil. 2001;**233**:213-221. DOI: 10.1023/A:1010361420899

[29] Real AC, Borges J, Cabral JS, Jones GV. Partitioning the grapevine growing season in the Douro Valley of Portugal: Accumulated heat better than calendar dates. International Journal of Biometeorology. 2015;**59**:1045-1059. DOI: 10.1007/s00484-014-0918-1

[30] Webb L, Whetton P, Barlow EWR. Climate change and wine grape quality in Australia. Climate Research. 2008;**36**:99-111. DOI: 10.3354/cr00740

[31] Santos JA, Grätsch SD, Karremann MK, Jones GV, Pinto JG. Ensemble projections for wine production in the Douro Valley of Portugal. Climatic Change. 2013;**117**:211-225. DOI: 10.1007/s10584-012-0538-x

[32] Oliveira M, Sousa T. Organic acids and sugars in musts of irrigated grapevines in Northeast Portugal. Journal of Wine Research. 2009;**20**:1-13. DOI: 10.1080/09571260902978485

[33] Resco P, Iglesias A, Bardají I, Sotés V. Exploring adaptation choices for grapevine regions in Spain. Regional Environmental Change. 2016;**16**:979-993. DOI: 10.1007/s10113-015-0811-4

[34] Oliveira M. Effect of soil water availability upon the production of the variety Touriga Franca in the demarcated region of Douro (Cima-Corgo). Ciência e Técnica Vitivinícola. 1995;**12**:59-71

[35] Oliveira M, de Freitas V, Sousa T. Water use efficiency and must quality of irrigated grapevines of north-eastern Portugal. Archiv für Acker- und Pflanzenbau und Bodenkunde 2012;**58**:871-886. DOI: 10.1080/03650340.2011.551875

[36] Howell TA. Enhancing water use efficiency in irrigated agriculture. Agronomy Journal. 2001;**93**:281-289. DOI: 10.2134/agronj2001.932281x

[37] Roygard JFK, Alley MM, Khosla R. No-till corn yields and water balance in the mid-Atlantic coastal plain. Agronomy Journal. 2002;**94**:612-623. DOI: 10.2134/agronj2002.6120

[38] Lane PNJ, Morris J, Ningnan Z, Guangyi Z, Guoyi Z, Daping X. Water balance of tropical eucalypt plantations in south-eastern China. Agr Forest Meteorol. 2004;**124**:253-267. DOI: org/10.1016/j.agrformet.2004.01.015

[39] Wythers KR, Lauenroth WK, Paruelo JM. Bare-soil evaporation under semiarid field conditions. Soil Science Society of America Journal. 1999;**63**:1341-1349. DOI: 10.2136/sssaj1999.6351341x

[40] Fares A, Alva AK. Soil water components based on capacitance probes in a sandy soil. Soil Science Society of America Journal. 2000;**64**:311-318. DOI: 10.2136/sssaj2000.641311x

[41] Esteban MA, Villanueva MJ, Lissarrague JR. Effect of irrigation on changes in berry composition of Tempranillo during maturation. Sugars, organic acids, and mineral elements. American Journal of Enology and Viticulture. 1999;**50**:418-434

[42] Greer DH, Weedon MM. Interactions between light and growing season temperatures on, growth and development and gas exchange of Semillon (*Vitis vinifera* L.) vines grown in an irrigated vineyard. Plant Physiology and Biochemistry. 2012;**54**:59-69. DOI: 10.1016/j.plaphy.2012.02.010

[43] Cuevas E, Baeza P, Lissarrague JR. Variation in stomatal behaviour and gas exchange between midmorning and mid-afternoon of north–south oriented grapevines (*Vitis vinifera* L. cv. Tempranillo) at different levels of soil water availability. Scientia Horticulturae. 2006;**108**:173-180. DOI: 10.1016/j.scienta.2006.01.027

[44] Morandi B, Zibordi M, Losciale P, Manfrini L, Pierpaoli E, Grappadelli LC. Shading decreases the growth rate of young apple fruit by reducing their phloem import. Scientia Horticulturae. 2011;**127**:347-352. DOI: org/10.1016/j.scienta.2010.11.002

[45] Oliveira M, Teles J, Barbosa P, Olazabal F, Queiroz J. Shading of the fruit zone to reduce grape yield and quality losses caused by sunburn. Journal International des Sciences de la Vigne et du Vin. 2014;**48**:1-9. DOI: 10.20870/oeno-one.2014.48.3.1579

[46] Bondada B, Shutthanandan J. Understanding differential responses of grapevine (*Vitis vinifera* L.) leaf and fruit to water stress and recovery following re-watering. American Journal of Plant Sciences. 2012;**3**:1232-1240. DOI: 10.4236/ajps.2012.39149

[47] López MI, Sánchez MT, Díaz A, Ramírez P, Morales J. Influence of a deficit irrigation regime during ripening on berry composition in grapevines (*Vitis vinifera* L.) grown in semi-arid areas. International Journal of Food Sciences and Nutrition. 2007;**58**:491-507. DOI: 10.1080/09637480701311801

[48] Guerra J, Abade E. Caracterização Enológica de Castas Autóctones da Região do Douro [Internet]. 2009. Available from: http://www.drapn.minagricultura.pt/drapn/conteudos/

fil_trab/Trabalho%20Castas%20Brancas%20e%20Tinas%20do%20Douro.pdf [Accessed: Jul 21, 2016]

[49] Bergqvist J, Dokoozlian N, Ebisuda N. Sunlight exposure and temperature effects on berry growth and composition of cabernet sauvignon and Grenache in the Central San Joaquin Valley of California. American Journal of Enology and Viticulture. 2001;**52**:1-7. DOI: 10.1186/s12870-016-0850-0

[50] Pieri P, Fermaud M. Effects of defoliation on temperature and wetness of grapevine berries. Acta Horticulturae (ISHS). 2005;**689**:109-116. DOI: 10.17660/ActaHortic.2005.689.9

[51] Tarara JM, Lee J, Spayd SE, Scagel CF. Berry temperature and solar radiation alter acylation, proportion, and concentration of anthocyanin in merlot grapes. American Journal of Enology and Viticulture. 2008;**59**:235-247

[52] Pollastrini M, Di Stefano V, Ferretti M, Agati G, Grifoni D, Zipoli G, Orlandini S, Bussotti F. Influence of different light intensity regimes on leaf features of *Vitis vinifera* L. in ultraviolet radiation filtered condition. Environmental and Experimental Botany. 2011;**73**:108-115. DOI: org/10.1016/j.envexpbot.2010.10.027

[53] Koyama K, Ikeda H, Poudel PR, Goto-Yamamoto N. Light quality affects flavonoid biosynthesis in young berries of cabernet sauvignon grape. Phytochemistry. 2012;**78**:54-64. DOI: 10.1016/j.phytochem.2012.02.026

[54] Bonfante A, Alfieri SM, Albrizio R, Basile A, De Mascellis R, Gambuti A, Giorio P, Langella G, Manna P, Monaco E, Moio L, Terribile F. Evaluation of the effects of future climate change on grape quality through a physically based model application: A case study for the Aglianico grapevine in Campania region, Italy. Agricultural Systems. 2017;**152**:100-109. DOI: 10.1016/j.agsy.2016.12.009

[55] Gonzalez-Dugo V, Zarco-Tejada P, Berni JAJ, Suárez L, Goldhamer D, Fereres E.Almond tree canopy temperature reveals intra-crown variability that is water stress-dependent. Agricultural and Forest Meteorology. 2012;**154-155**:156-165. DOI: 10.1016/j.agrformet.2011.11.004

[56] Goldammer T. Grape grower's Handbook. A Guide to Viticulture for Wine Production. NY: Apex Publishers; 2015. p. 728 978-0-9675212-7-5

[57] Zsófi ZS, Tóth E, Rusjan D, Bálo B. Terroir aspects of grape quality in a cool climate wine region: Relationship between water deficit, vegetative growth and berry sugar concentration. Scientia Horticulturae-Amsterdam. 2011;**127**:494-499. DOI: 10.1016/j.scienta.2010.11.014

[58] Greer DH, Weston C. Heat stress affects flowering, berry growth, sugar accumulation and photosynthesis of *Vitis vinifera* cv. Semillon grapevines grown in a controlled environment. Functional Plant Biology. 2010;**37**:206-214. DOI: 10.1071/FP09209

[59] Hulands S, Greer DH, Harper JDI. The interactive effects of temperature and light intensity on *Vitis vinifera* cv: "Semillon" grapevines. I. Berry growth and development.

European Journal of Horticultural Science. 2014;**78**:249-257. Available from: http://www.pubhort.org/ejhs/2013/file_4119074.pdf [Accessed: Apr 24, 2016]

[60] Greer DH, Weedon MM. The impact of high temperatures on *Vitis vinifera* cv. Semillon grapevine performance and berry ripening. Frontiers. Plant Science. 2013;**4**:1-9. DOI: 10.3389/fpls.2013.00491

[61] Edwards EJ, Smithson L, Graham DC, Clingeleffer PR. Grapevine canopy response to a high-temperature event during deficit irrigation. Australian Journal of Grape and Wine Research. 2011;**17**:153-161. DOI: 10.1111/j.1755-0238.2011.00125.x

[62] Cantore V, Pace B, Albrizio R. Kaolin-based particle film technology affects tomato physiology, yield and quality. Environmental and Experimental Botany. 2009;**66**:279-288. DOI: 10.1016/j.envexpbot.2009.03.008

[63] Brillante L, Belfiore N, Gaiotti F, Lovat L, Sansone L, Poni S, Tomasi D. Comparing kaolin and pinolene to improve sustainable grapevine production during drought. PLoS One. 2016;**11**:e0156631. DOI: 10.1371/journal.pone.0156631

Chapter 5

Water Balance Indices for Tropical Wine Grapes

Antônio Heriberto de Castro Teixeira,
Jorge Tonietto and Janice F. Leivas

Additional information is available at the end of the chapter

http://dx.doi.org/10.5772/intechopen.72751

Abstract

Over the last few decades, the Brazilian semiarid region has appeared as one of the main tropical wine production areas in the country. The aim of this research was the elaboration and application of water balance indices to upscale them in the wine grape growing regions of the Petrolina and Juazeiro counties in the states of Pernambuco (PE) and Bahia (BA), respectively, simulating different pruning dates along the year. Previous energy balance measurements were used for relating the crop coefficient (Kc) with the accumulated degree-days (DD_{ac}). The model was applied to upscale the water balance indices during the growing seasons (GS). It was concluded that if irrigation water is available, the best pruning periods are for GS from May to July because of better natural thermal and moisture conditions. Much care should be taken for pruning done in other periods of the year, with regard to the effect of increasing thermal conditions on wine quality. The classifications and delimitations done, joined with other environmental characteristics, are important for a rational planning of the commercial tropical wine production expansion, mainly in the actual situations of climate and land use changes together with rising water competition along the years in the Brazilian semiarid region.

Keywords: evapotranspiration, crop coefficient, vineyard adaptation, water resources, *Vitis vinifera*

1. Introduction

The influence of climatic variables during the vineyard growing seasons on wine quality is well known because they influence the grapevine growth and then the berry composition. For wine grape crops, plant phenology, wine quality, and yield are very dependent on climate at regional, local and microclimatic scales [1–4]. For any of these spatial scales, considerations of grape site selection, cultural practices, and water management are important, which are very important issues for potential adaptations to different climate scenarios; however, large-scale climate has been the focus for assessing the climate change impacts [5–8].

The optimal vineyard response to air temperature (T_a) ranges between 20 and 35°C [9]. According to Fraga et al. [3], a 10°C basal temperature (T_b) is needed for the growing season onset. However, high thermal conditions contribute to increase sugar content in grapes, resulting in large alcohol concentration and low acidity in wines, rising pH [2, 10–12]. There are also important secondary thermal effects, such as increase of pest and disease risks, like downy and powdery mildew, especially under rainy conditions [4, 13, 14].

Rising thermal conditions ($T_a > 30°C$) should increase suspended solid concentrations, but high Brix levels may be attributed to large evapotranspiration (ET) rates [15]. According to Webb et al. [16], high T_a during the harvest period may reduce berry quality due to increasing ET. These issues make the rational water management an important issue for controlling water deficiencies and excesses, and the need of vineyard water requirements (WR) quantification [12, 17, 18].

Taking into account the water balance, on the one hand, warming conditions can directly affect the vineyard WR, which, together with low precipitation (P) amounts, promote high levels of aridity and water demand. On the other hand, high soil moisture throughout the growing season may cause excessive vigor, increased risks of pest and diseases, and other problems related to wine quality and to the balance of its chemical components [4, 9, 19–21]. All these thermohydrological effects during the vineyard growing seasons on wine quality and production show the importance of water accounting to delimitate areas and seasons with suitable climatic aptitude for winemaking processes.

A large number of climate models have been used worldwide to classify winemaking regions by using different methodologies. For aptitude delimitation, aiming at grape and wine production, one can apply bioclimatic indices based on the thermohydrological requirements. The multicriteria climatic classification (MCC) system proposed by Tonietto and Carbonneau [22] has been used under temperate climate conditions in Europe [23] and in South America [24, 25]. However, the method has worked well considering a single, 6-month growing season per year under temperate climate conditions.

Over the last years, the Brazilian semiarid region has appeared among the main tropical wine-producing areas in the country, typically growing under irrigation conditions and trained mainly in vertical shoot-positioning systems. With proper irrigation and cultural management practices, the farmers can produce grapes and carry out winemaking at any time of the year, allowing a potential average of between two and three vineyard-growing cycles per year, in accordance with and depending on each variety. The T_a rising with a consequent increase in ET rates and aridity in this region will affect both the wine grape quality and vineyard water requirements.

The coupled effect of increasing water consumption and decreasing precipitation, together with land use change, makes it important to elaborate and upscale indices for subsidizing winemaking adaptations and water productivity improvements. Vineyard water variables have been quantified in this region by point measurements [26, 27] but to upscale these punctual results, tools such as remote sensing and geographic information system (GIS) can be used. For the vineyard climatic suitability determination, one can apply water balance indices by using GIS and long-term weather data [4, 18, 21, 28]. For the vineyard water balance indices used in this chapter, distinctions are important between reference (ET_0) and actual (ET) evapotranspiration. ET_0 is the water flux from a reference surface, not a shortage of water, which may be considered as a hypothetical grass surface with specific characteristics, while ET is the real water flux occurring from the surfaces in a specific situation involving all environmental conditions [29].

This chapter aims to elaborate on and apply water balance indices to be scaled up by using a GIS in the wine grape growing regions of Petrolina and Juazeiro counties, located respectively in the semiarid regions of Pernambuco (PE) and Bahia (BA) states, simulating different pruning dates along the year. These indices were delimitated and analyzed, generating criteria for a rational expansion of irrigated and rain-fed vineyards with higher probability of success for tropical wine elaboration. The results generate criteria for a rational expansion of irrigated and rain-fed vineyards with higher probability of success for the Brazilian tropical wine elaboration, under the actual scenario of rising water competition by irrigated agriculture and nonagricultural sectors.

2. Materials and methods

2.1. Study region and data set

A net of agrometeorological stations was used throughout interpolation processes in a GIS environment. The stations are spread in the Petrolina (PE) and Juazeiro (BA) counties, with seven of them inside irrigated farms and the other seven in the natural vegetation, called "Caatinga." The gridded weather data well characterize the horizontal thermohydrological contrast between these mixed agroecosystems (**Figure 1**).

According to Teixeira [19], in the Brazilian Northeast semiarid region, disturbed currents from the South, North, East, and West influence the climatology. Excluding the places of high altitude, all areas present long-term annual T_a higher than 24°C, with the average maximum of 33°C and the average minimum of 19°C. The warmest months are October and November when the Sun is close to the zenith position with low cloud cover, and the coldest ones are June and July at winter solstice in the Southern hemisphere. The thermal homogeneity strongly contrasts with the spatial and temporal heterogeneity of the rainfall regime, with the rainy period from November to April (90% of the annual total), the period January to April representing 68% of the annual rainfall. The weather variables in the current study involved a 10-year period (2003–2012). Monthly data were used to calculate the reference evapotranspiration (ET0) by the Penman-Monteith method [29].

2.2. Modeling vineyard water balance indices

Previous Bowen ratio energy balance data from Teixeira et al. [26] for the cv. *Syrah* in the Brazilian semiarid region were used. **Figure 2** shows the details of the field experiment carried out in Petrolina (PE), Northeast of Brazil.

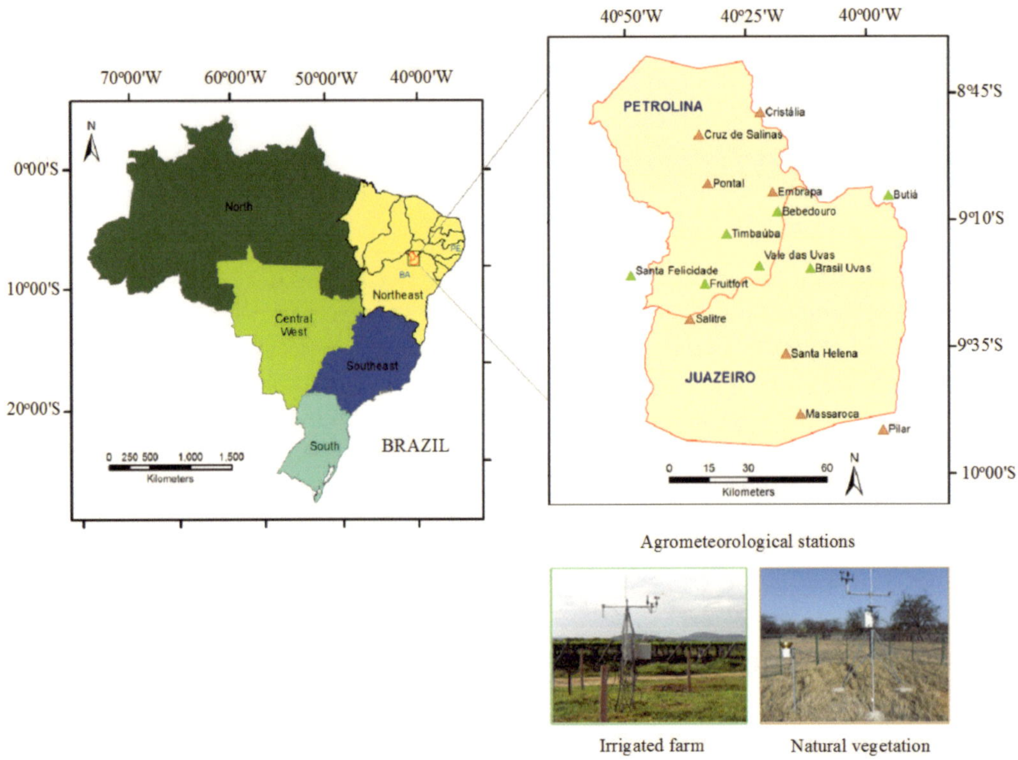

Figure 1. Petrolina and Juazeiro counties, respectively in Pernambuco (PE) and Bahia (BA) states, in the Brazilian Northeast, together with the agrometeorological stations used for the interpolation processes.

The energy balance equation for the wine grape can be expressed by means of bulk energy and heat fluxes:

$$R_n - \lambda E - H - G = 0 \quad (1)$$

where R_n is the net radiation, λE is the latent heat flux, H is the sensible heat flux, and G is the soil heat flux.

The vineyard λE was obtained by a partitioning parameter:

$$\lambda E = \frac{R_n - G}{1 + \beta} \quad (2)$$

where β is the Bowen ratio:

$$\beta = \gamma\left(\frac{\Delta T}{\Delta e}\right) \quad (3)$$

and γ (kPa °C^{-1}) is the psychrometric constant, ΔT(°C) the temperature gradient measured by dry thermocouples, and Δe (kPa) is the water vapor pressure gradient measured by the difference between dry and wet thermocouples over the height interval above the vineyard canopy surface.

Figure 2. Measurements of gradients of air temperature, vapor pressure and wind speed; incident and reflected shortwave; net radiation; acquisition data system; and soil moisture in Bowen ratio system of wine grape.

Actual evapotranspiration (ET) was derived from the latent heat of vaporization (λ), density of water, and λE. The field experiment was close (3 km) to Bebedouro station (**Figure 1**), which ET0 data allowed the acquirement of the crop coefficient (Kc) along the crop stages [29]:

$$Kc = \frac{ET}{ET0} \quad (4)$$

Considering a base temperature (T_b) of 10°C, Kc was related with the accumulated degree days DD_{ac} [30]:

$$Kc = aDD_{ac}^{\,2} + bDD_{ac} + c \quad (5)$$

where $a = -2 \times 10^{-7}$, $b = 4 \times 10^{-4}$ and $c = 0.54$ are the regression coefficients ($R^2 > 0.70$).

Further, Kc was used to obtain the ET under potential conditions, which in turn considered the vineyard water requirements (WR), using the cv. *Syrah* as a reference wine grape in the study region. WR for a growing season (GS) was acquired by simulating different pruning date and considering a 4-month mean GS duration under the Brazilian semiarid conditions:

$$WR_{GS} = Kc_{GS} ET0_{GS} \quad (6)$$

Five Kc values were taken into account, being DD_{ac} zero at the start of a GS, while the other DD_{ac} values were calculated, along the GS, with the average T_a for the subsequent months. The five-Kc averaged values and the total ET0 for a GS ($ET0_{GS}$) were considered for acquiring WR_{GS}.

Another indicator, the water balance difference (WBd), was applied to quantify the magnitude of excess or deficiencies of water in the vineyards on large scales for a GS, where P_{GS} is the total growing season precipitation:

$$WBd_{GS} = P_{GS} - WR_{GS} \quad (7)$$

Neglecting the water storage in the root zones in the vineyard water balance, difficult to consider in large-scale analyses, positive WBd values are a quantification of vineyard water excess, while the negative ones are related to the vineyard water deficiencies.

The WR_{GS} together with P_{GS} values allowed the development and application of the water balance ratio (WBr) indicator:

$$WBr_{GS} = \frac{P_{GS}}{WR_{GS}} \tag{8}$$

WBr takes into account the thermohydrological conditions, and it is a measure of the water availability in the vineyard root zone. When it is around 1.00, imply the feasibility for rain-fed wine grape, while those much higher should indicate moisture excess problems, independently of the absence or not of irrigation. Low WBr values mean possibility of natural water deficiencies and the degree of irrigation needs according to the pruning dates.

3. Results and discussion

Figure 3 shows the T_{GS} maps, for different wine grape pruning dates along the year, considering a mean 4-month GS, and the 10-year period from 2003 to 2012.

The coldest and the hottest growing seasons (GS) are those for pruning dates between April and July and from September to December, respectively. Considering the standard deviation (SD) values, there are low thermal spatial variations, due to the proximity of the counties to the equator. The lowest air temperatures for a growing season (T_{GS}) occur at the winter

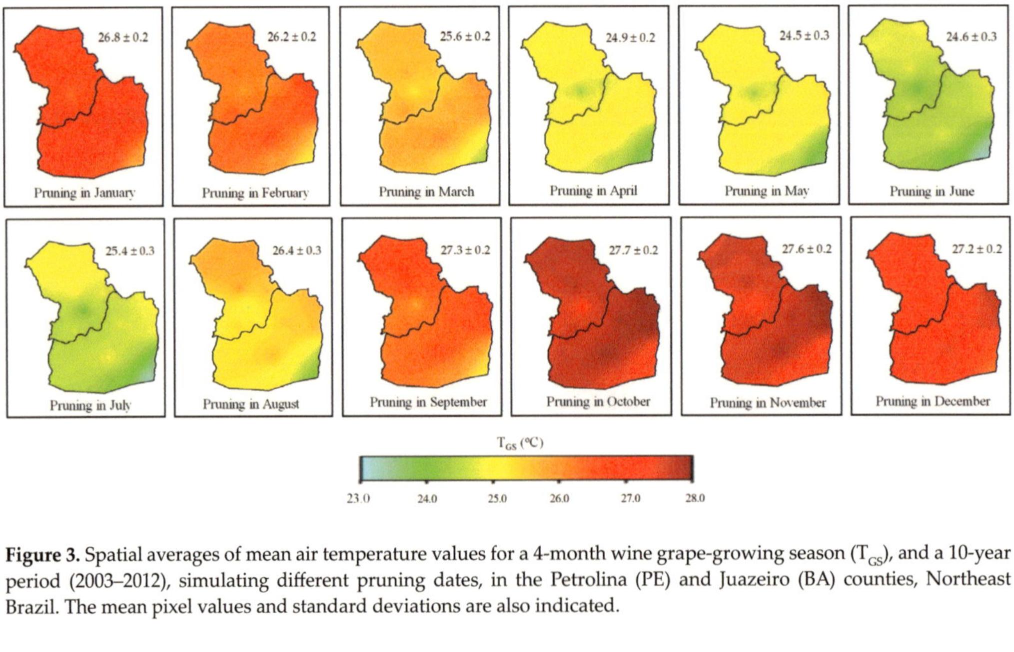

Figure 3. Spatial averages of mean air temperature values for a 4-month wine grape-growing season (T_{GS}), and a 10-year period (2003–2012), simulating different pruning dates, in the Petrolina (PE) and Juazeiro (BA) counties, Northeast Brazil. The mean pixel values and standard deviations are also indicated.

solstice time in the Southern hemisphere, while the highest ones are when the Sun is around the zenith position over the Brazilian tropical wine grape growing region. For pruning done during the coldest periods, several pixels present T_{GS} values lower than 24°C, while one can see all the areas with T_{GS} higher than 26°C, for pruning in the hottest months.

There are no thermal limitations for wine grape crop in the Brazilian semiarid region, with pruning dates in the middle of the year. On the one hand, for all pruning periods, T_{GS} pixels are below 30°C, which conditions around or above this value should increase suspended solid concentrations [15]. On the other hand, T_{GS} values are not below the threshold of 10°C, which could introduce a dormancy stage in temperate climates [3]. The T_{GS} values in the study area are between 23°C and 28°C, inside the optimum thermal range pointed by Gouveia et al. [9].

However, when the pruning is done from September to December, many areas with T_{GS} above 27°C often occur and could affect negatively the wine quality. These latter conditions will contribute to high sugar content in grapes but wines with increasing levels of alcohol, low acidity, and large pH values. These effects together will promote a wine unbalance with instability for the phenolic and aromatic composition [2, 3, 10–12].

Figure 4 shows the P_{GS} maps, for different wine grape pruning dates along the year, considering a mean 4-month GS, and the 10-year period from 2003 to 2012.

The pruning dates with the highest P_{GS} are those from December to February, with several pixel values larger than 300 mm GS^{-1} in the Petrolina County. During this period, the largest moisture spatial variation is also verified according to the SD values. The lowest P_{GS} are for pruning between May and July. High pixel values occur in the northwestern side of Petrolina (PE), while the lowest ones occur in the southwestern area of Juazeiro (BA).

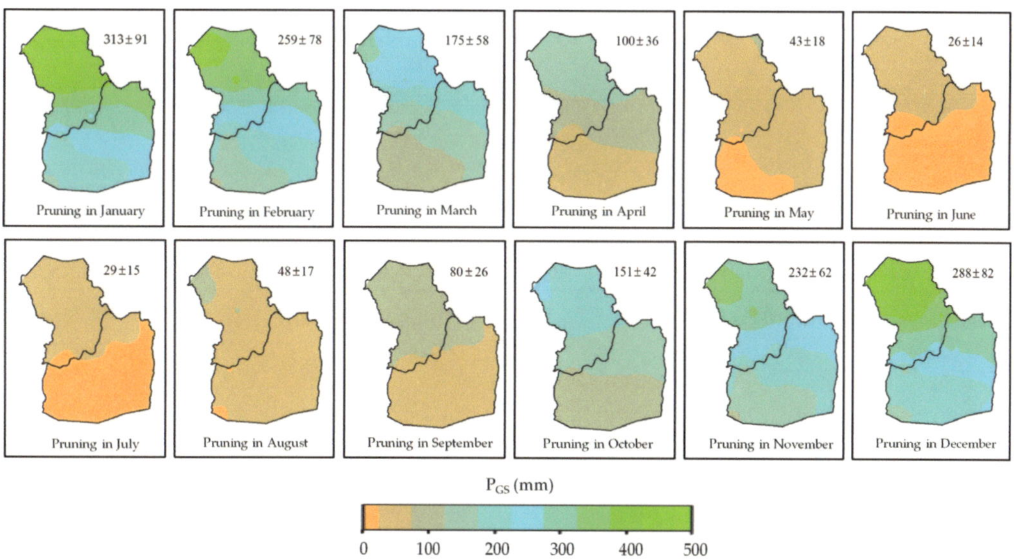

Figure 4. Spatial averages of total precipitation for a 4-month wine grape-growing season (P_{GS}), and a 10-year period (2003–2012), simulating different pruning dates, in the Petrolina (PE) and Juazeiro (BA) counties, Northeast Brazil. The mean pixel values and standard deviations are also indicated.

Taking into account all pruning dates, the rainfall amounts in Petrolina (PE) are 61% larger than in Juazeiro (BA). Thus, in the first county, there are more possibilities of matching the vineyard water requirements with rainfall together with supplementary irrigations, whenever irrigation water is available. However, as a first guess, risks of pest and diseases and other problems related to wine quality and to the balance between its chemical components are higher for pruning done from December to February [4, 9, 19–21].

Considering the cv. *Syrah* as reference for wine grapes in the growing regions of Petrolina (PE) and Juazeiro (BA), and the long-term weather conditions (2003–2012), the WR_{GS} spatial values for a 4-month mean GS are presented in **Figure 5**.

The pruning dates with the highest WR_{GS} are from August to October, with average pixel values larger than 420 mm GS^{-1}, when, according to the SD values, there are also the highest spatial variations. Pruning done from March to May will promote the lowest water consumptions, with mean WR_{GS} below 350 mm GS^{-1}, and the smallest SD values, below 25 mm in March and April. Large WR_{GS} occur in the northwestern side of Petrolina (PE), what might correspond to good grape yield and wine quality if water is available together with techniques to avoid natural water excesses in the root zones [4, 12]. However, special attention should be given under water scarcity conditions, when there is ample room for water productivity improvements in situations of lower atmospheric demands [21].

Taking into account all pruning dates along a year, the water demands in the Petrolina (PE) county are 10% larger, when comparing with those for Juazeiro (BA) one. Daily average WR values in the study region were between 2.7 and 3.6 mm day^{-1}, being similar to the ET rates

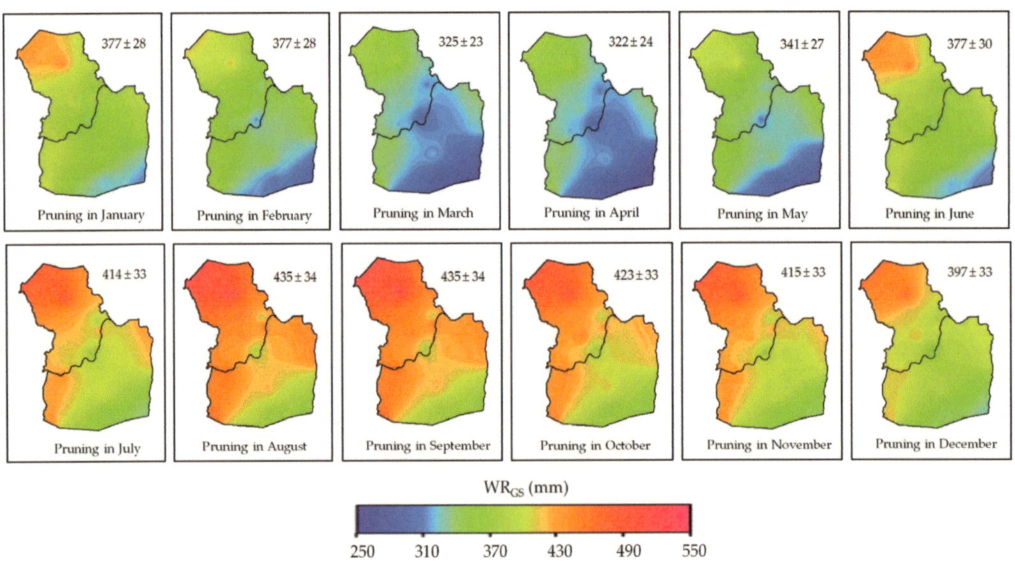

Figure 5. Spatial averages of the water requirements for a 4-month wine grape-growing season (WR_{GS}), and 10-year period (2003–2012), simulating different pruning dates, in the Petrolina (PE) and Juazeiro (BA) counties, Northeast Brazil. The mean pixel values and standard deviations are also indicated.

found throughout field experiments in different wine grapes growing regions of the world [27, 30, 31], bringing confidence to the upscaling techniques applied in the current case study.

Table 1 resumes the averages and SD values of the vineyard water balance indices for each 4-month pruning date per county, for the wine grape, cv. *Syrah* considering the period of weather data from 2003 to 2012 in the growing region of Petrolina – Pet (PE) e Juazeiro – Jua (BA), in the semiarid region of Northeast Brazil.

No significant differences arise among the T_{GS} mean values from Petrolina (PE) and those from Juazeiro (BA) with average for all pruning periods of 26°C GS^{-1} for both counties; however, the second one presents larger spatial thermal variation, according to the SD values.

In case of P_{GS}, the values for Juazeiro (BA) are lower those for Petrolina (PE), indicating higher possibility of rainfall water use by the vineyards in the second county. As the vineyard thermal conditions between them did not differ so much, the WR_{GS} values for Juazeiro (BA) were, in average, 90% of those for Petrolina (PE). These differences could be attributed to the effect of relative humidity (RH) in the ET0 calculations, as lower P_{GS} reduce RH in Juazeiro (BA) when comparing with Petrolina (PE).

Keeping in mind that the wine quality depends on both thermal and water conditions, these conditions were analyzed throughout the mean pixel values and standard deviations (SD)

Pruning date	**T_{GS} (°C)**		**P_{GS} (mm GS^{-1})**		**WR_{GS} (mm GS^{-1})**	
	Pet	**Jua**	**Pet**	**Jua**	**Pet**	**Jua**
January	26.7 ± 0.1	26.8 ± 0.3	396 ± 65	253 ± 52	398 ± 23	362 ± 20
February	26.2 ± 0.1	26.3 ± 0.3	329 ± 57	209 ± 47	372 ± 22	337 ± 18
March	25.6 ± 0.1	25.6 ± 0.3	224 ± 47	139 ± 35	342 ± 19	313 ± 18
April	25.0 ± 0.1	24.9 ± 0.3	130 ± 28	78 ± 22	340 ± 19	310 ± 19
May	24.6 ± 0.2	24.4 ± 0.3	58 ± 14	32 ± 11	361 ± 22	327 ± 20
June	24.8 ± 0.2	24.5 ± 0.3	40 ± 10	17 ± 7	399 ± 25	360 ± 22
July	25.5 ± 0.2	25.3 ± 0.3	44 ± 8	18 ± 7	439 ± 28	396 ± 24
August	26.5 ± 0.2	26.3 ± 0.3	65 ± 11	37 ± 7	461 ± 29	416 ± 24
September	27.3 ± 0.2	27.2 ± 0.3	105 ± 17	62 ± 12	461 ± 29	417 ± 23
October	27.7 ± 0.1	27.7 ± 0.3	191 ± 28	121 ± 21	448 ± 27	404 ± 22
November	27.6 ± 0.1	27.6 ± 0.3	292 ± 40	189 ± 33	441 ± 26	396 ± 22
December	27.2 ± 0.1	27.3 ± 0.3	365 ± 57	233 ± 45	420 ± 24	368 ± 21
Mean	26.1 ± 0.1	26.2 ± 0.3	187 ± 32	116 ± 25	407 ± 24	367 ± 21

Air temperature (T_{GS}); Precipitation (P_{GS}); and Water requirements (WR_{GS}).

Table 1. Mean values and standard deviations (SD) of the vineyard water balance indices for the wine grape, cv. *Syrah*, considering a 4-month average growing season (GS) and a 10-year period (2003–2012), in the Petrolina—Pet (PE) and Juazeiro—Jua (BA) counties, Northeast Brazil.

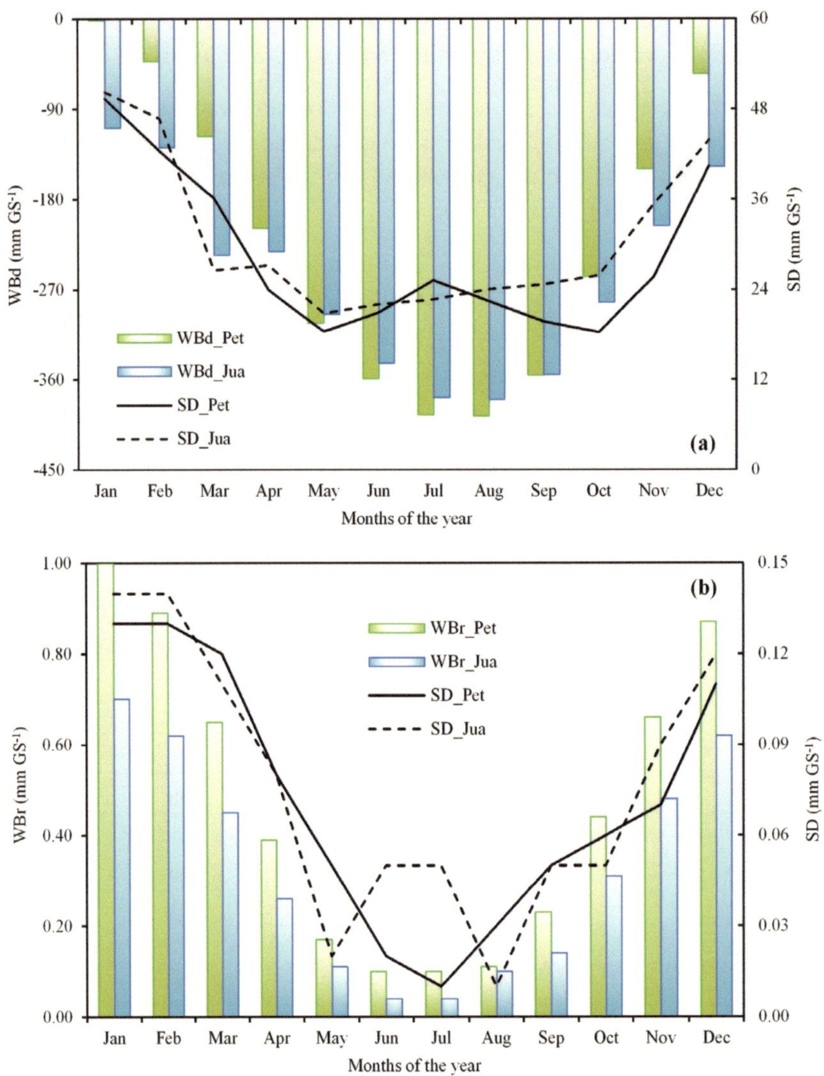

Figure 6. Mean pixel values and standard deviations (SD) for the water balance indices, considering a 10-year period (2003–2012) and an average 4-month wine grape-growing season, cv. *Syrah* according to the simulated pruning dates, in the Petrolina—Pet (PE) and Juazeiro—Jua (BA) counties, Northeast Brazil: (a) water balance difference (WBd) and (b) water balance ratio (WBr).

of WBd and WBr for a 4-month average growing season and according to the pruning dates considering the 10-year period of weather data (2003–2012) (**Figure 6**).

According to **Figure 6a**, there are no positive WBd mean values, meaning that considering the whole area and the average conditions; in general, there is absence of vineyard water excesses for any pruning dates. Disregarding the water storage in the root zones in the vineyard water balance, the pruning periods with the highest water deficiencies (the most negative WBd) are from June to September, when both counties presented average WBd pixel values lower than −340 mm GS^{-1} and also the lowest SD, around 23 mm GS^{-1}. The less negative WBd values obtained for pruning dates are from December to February, when the average was above −150 mm GS^{-1}. These last thermohydrological conditions indicated the feasibility of rain-fed wine grape with supplementary irrigation. Natural water deficiency in Petrolina is 87% of that for Juazeiro, with better chances of success for rain-fed wine grapes, once rainfall-water storage techniques are applied.

As, in average, P_{GS} in Petrolina is 61% higher than that for Juazeiro, but WR_{GS} is only 11% larger (**Table 1**), the differences regarding rainfall amounts will affect more the water balance than the different evapotranspiration rates between the counties. Similarly to the WBd index, pruning dates from December to February present the highest $WBr_{,}$ with averages ranging from 0.60 to 1.00, meaning that rainfall amounts met from 60 to 100% of the vineyard water demands during these pruning periods; however, SD values for both of them are around 0.10 (**Figure 6b**).

Although rainy conditions having the beneficial aspect of natural water availability, increasing soil moisture may reduce the ripening capacity of grapes, and the difficulty of water stress management is unfavorable for the organoleptic wine quality. In this sense, care should be also taken for improving drainage for both irrigated and rain-fed vineyards during the periods of high WBr.

The natural climate dryness conditions occur when the pruning is done from May to August. Under these circumstances, the WBr values are around 0.10 with almost no spatial variation, favoring more the irrigated vineyards. These conditions avoid plant diseases, root respiration problems, and direct damage to the berries promoted by excess of precipitation, favoring the quality of must and wine [3, 4, 9, 12, 20, 21].

4. Conclusions

Water balance indices are successfully developed and applied, allowing the large-scale analyses of the thermohydrological conditions for wine grape production under the semiarid conditions of the Brazilian Northeast, considering different pruning dates along the year.

On the one hand, under irrigation conditions, the best wine grape pruning dates in the Brazilian Northeast are from May to August, with the thermohydrological conditions favoring a better tropical wine quality. On the other hand, the most problematic pruning periods for irrigated crops are from December to February because the joint effects of higher air temperatures and precipitations. Considering the possibility of rainfed crops, this last period should be considered in situations with the possibility of supplementary irrigation applying rainfall water storage techniques.

The spatial delimitations carried out in the current research, joined with other environmental characteristics, are important for the success of the commercial tropical wine production expansion, considering also the sustainability of the activity in the Brazilian semiarid region, where the land use and climate changes are happening together with water competition during the last decades.

Acknowledgements

The authors acknowledge the National Council for Scientific and Technological Development – CNPq (Processe 404229/2013-1), for the financial support to a project. The research deals with agricultural water productivity in hydrological basins with land use changes in Brazil.

Author details

Antônio Heriberto de Castro Teixeira[1]*, Jorge Tonietto[2] and Janice F. Leivas[3]

*Address all correspondence to: heriberto.teixeira@embrapa.br

1 Embrapa Coastal Tablelands, Aracaju-SE, Brazil

2 Embrapa Grape and Wine, Bento Gonçalves, Rio Grande do Sul, Brazil

3 Embrapa Satellite Monitoring, Campinas, São Paulo, Brazil

References

[1] Jones GV, Webb LB. Climate change, viticulture, and wine: Challenges and opportunities. Journal of Wine Research. 2010;**21**:03-106. DOI: 10.1080/09571264.2010.530091

[2] Fraga H, Malheiro AC, Moutinho-Pereira J, Santos JA. An overview of climate change impacts on European viticulture. Food and Energy Security. 2012;**1**:94-110. DOI: 10.1002/fes3.14

[3] Fraga H, Malheiro AC, Moutinho-Pereira J, Santos JÁ. Climate factors driving wine production in Portuguese Minho region. Agricultural and Forest Meteorology. 2014;**185**:26-36. DOI: 10.1016/j.agrformet.2013.11.003

[4] Lorenzo MN, Ramos AM, Brands S. Present and future climate conditions for winegrowing in Spain. Regional Environmental Change. 2016;**16**:617-627. DOI: 10.1007/s10113-015-0883-1

[5] Malheiro AC, Santos JA, Fraga H, Pinto JG. Climate change scenarios applied to viticultural zoning in Europe. Climate Research. 2010;**43**:163-177. DOI: 10.3354/cr00918

[6] Neethling E, Petitjean T, Quénol H, Barbeau G. Assessing local climate vulnerability and winegrowers' adaptive processes in the context of climate change. Mitigation Adaptation Strategy for Global Change. 2017;**22**:777-803. DOI: 10.1007/s11027-015-9698-0

[7] Dunn MR, Lindesay JA, Howden M. Spatial and temporal scales of future climate information for climate change adaptation in viticulture: A case study of user needs in the Australian winegrape sector. Australian Journal of Grape and Wine Research. 2015;**21**:226-239. DOI: 10.1111/ajgw.12138

[8] Tóth JP, Végvári Z. Future of winegrape growing regions in Europe. Australian Journal of Grape and Wine Research. 2016;**22**:64-72. DOI: 10.1111/ajgw.12303, 10.1111/ajgw.12138

[9] Golveia C, Liberato MLR, DaCamara CC, Trigo RM, Ramos AM. Modelling past and future wine production in the Portuguese Douro Valley. Climate Research. 2011;**48**:349-362. DOI: 10.3354/cr01006

[10] Tarara JM, Lee J, Spayd SE, Scajel CF. Berry temperature and solar radiation alter acylation, proportion, and concentration of anthocyanin in merlot grapes. American Journal of Enology and Viticulture. 2008;**59**:235-247. DOI: 10.5344/ajev.2016.15106. ISSN 0002-9254

[11] Orduña RM. Climate change associated effects on grape and wine quality and production. Food Research International. 2010;**43**:1844-1845. DOI: 10.1016/j.foodres.2010.05.001

[12] Maxwell JT, Ficklin DL, Harley GL, Jones GV. Projecting future winegrape yields using a combination of *Vitis vinifera* L. growth rings and soil moisture simulations, northern California, UAS. Australian Journal of Grape and Wine Research. 2016;**22**:73-80. DOI: 10.1111/ajgw.12138

[13] Centeno A, Baeza P, Lissarrague JR. Relationship between soil and plant water status in wine grapes under various water deficits regimes. HortTechnology. 2010;**20**:585-593. ISSN: 1063-0918

[14] Caffarra A, Rinaldi MF, Eccel E, Rossi V, Pertot I. Modelling the impact of climate change on the interaction between grapevine and its pests and pathogens: European grapevine moth and powdery mildew. Agriculture Ecosystems and Environments. 2012;**148**:89-101. DOI: 10.1016/j.agee.2011.11.017

[15] Keller M. Managing grapevines to optimize fruit development in a challenging environment: A climate change primer for viticulturists. Australian Journal of Grape and Wine Research. 2010;**16**:56-69. DOI: 10.1111/j.1755-0238.2009.00077.x

[16] Webb LB, Whetton PH, Bhend J, Darbyshire R, Briggs PR, Barlow EWR. Earlier winegrape ripening driven by climatic warming and drying and management practices. Nature Climate Change. 2012;**2**:259-264. DOI: 10.1038/nclimate1417

[17] Anderson K, Findlay C, Fuentes S, Tyerman S. Garmaute Climate Change Review: Viticulture, Wine and Climate Change. Adelaide: University of Adelaide; 2008

[18] Teixeira AH de C, Hernandez FBT, Lopes HL. Up scaling table grape water requirements in the low-middle São Francisco River basin. Acta Horticulturae. 2014;**1038**:655-662. DOI: 10.17660/ActaHortic.2014.1038.83

[19] Teixeira AH de C. Water Productivity Assessments from Field to Large Scale: A Case Study in the Brazilian Semi-Arid Region. Saarbrücken: LAP Lambert Academic Publishing; 2009. 226 p. ISBN: 978-3838327785

[20] Santos JA, Grätsch SD, Karremann MK, Jones GV, Pinto JG. Ensemble projections for wine production in the Douro Valley of Portugal. Climate Change. 2013;**117**:211-225. DOI: 10.1007/s10584-012-0538-x

[21] Teixeira AH de C, Hernandez FBT, Scherer-Warren M, Andtrade RG, Victoria D de C, Bolfe EL, Thenkabail PS, Franco RAM. Water productivity studies from earth observation data: Characterization, modeling, and mapping water use and water productivity. In: Prasad ST, editor. Remote Sensing of Water Resources, Disasters, and Urban Studies. Vol. III. 1st ed. Boca Raton: Taylor and Francis; 2015. pp. 101-126. ISBN: 9781482217919

[22] Tonietto J, Carbonneau A. A multicriteria climatic classification system for grape-growing regions worldwide. Agricultural and Forest Meteorology. 2004;**124**:81-97. DOI: 10.1016/j.agrformet.2003.06.001

[23] Blanco-Ward D, García Queijeiro JM, Jones GV. Spatial climate variability and viticulture in the Miño River Valley of Spain. Vitis. 2007;**46**:63-70. ISSN: 0042-7500

[24] Ferrer M, Pedocchi R, Michelazzo M, Gonzalez-Neves G, Carbonneau A. Delimitación y descripción de regiones vitícolas del Uruguay en base al método de clasificación climática multicriterio utilizando índices bioclimáticos adaptados a las condiciones del cultivo. Agrociencia. 2007;**11**:47-56. ISSN: 1405-3195

[25] Montes C, Perez-Quezada JF, Peña-Neira A, Tonietto J. Climatic potential for viticulture in Central Chile. Australian Journal of Grape and Wine Research. 2012;**18**:20-28. DOI: 10.1111/j.1755-0238.2011.00165.x

[26] Teixeira AH de C, Bastiaanssen WGM, Bassoi LH. Crop water parameters of irrigated wine and table grapes to support water productivity analysis in Sao Francisco River basin, Brazil. Agricultural Water Management. 2007;**94**:31-42. DOI: 10.1016/j.agwat.2007.08.001

[27] Azevedo PV, Soares JM, Silva V de P, Silva BB, Nascimento T. Evapotranspiration of "superior" grapevines under intermitente irrigation. Agricultural Water Management. 2008;**95**:301-308. DOI: 10.1016/j.agwat.2007.10.011

[28] Teixeira AH de C, Scherer-Warren M, Hernandez FBT, Andrade RG, Leivas JF. Large-scale water productivity assessments with MODIS images in a changing semi-arid environment: A Brazilian case study. Remote Sensing. 2013;**5**:5783-5804. DOI: 10.3390/rs5115783

[29] Allen RG, Pereira LS, Raes D, Smith M. Crop Evapotranspiration: Guidelines for Computing Crop Water Requirements. Rome: Food and Agriculture Organization of the United Nations; 1998. 300 p

[30] Williams LE, Ayars JE. Grapevine water use and the crop coefficient are linear functions of the shaded area measured beneath the canopy. Agricultural and Forest Meteorology. 2005;**132**:201-211. DOI: 10.1016/j.agrformet.2005.07.010

[31] Ortega-Farias S, Carrasco M, Olioso A, Acevedo C, Poblete C. Latent heat flux over Cabernet Sauvignon vineyard using the Shuttleworth and Wallace model. Irrigation Science. 2007;**25**:161-170. DOI: 10.1007/s00271-006-0047-7

Chapter 6

Convenience of Applying of Viticulture Technique as a Function of the Water Status of the Vine-Stock

Ester Gamero, Francisco Espinosa, Daniel Moreno,
David Uriarte, María Henar Prieto,
Inmaculada Garrido and María Esperanza Valdés

Additional information is available at the end of the chapter

http://dx.doi.org/10.5772/intechopen.72799

Abstract

This study determined the effect that the irrigation dosage and cluster thinning showed over the antioxidant activities and total phenols in grapes, and phenolic profile and chromatic characteristics of cv. Tempranillo wines. The experiment was carried out over two consecutive seasons at an experimental vineyard in Extremadura (Spain). The plants were subjected to two post-*véraison* irrigation treatments, one at 100% (R100) and the other at 25% (R25) of crop evapotranspiration (ETc). Two crop loads (C and T) were additionally established for each irrigation treatment by cluster thinning at *véraison*. The winemaking process involved separate use of the grapes from each of the four resulting treatment groups (R100C, R100T, R25C and R25T) and followed a common protocol. In grapes, the effect of thinning enhanced when combined with water deficit, resulting in increases phenylpropanoids and flavonoids at the harvest, while leaving the polyphenol oxidase´s activity unaffected. In wines, the higher post-*véraison* water stress of the R25 treatments resulted in higher values of dimeric flavonol and flavanol concentrations. The wine hue was affected by water status. The cluster thinning caused additional increases in the concentrations of monomeric flavanols and anthocyanins, as well as resulting in stronger wine colour intensity.

Keywords: irrigation, crop load level, antioxidant activity, phenolic composition, wine, grapes

1. Introduction

Because of the high dependence of fruit quality on various environmental and endogenous factors [1, 2], several agronomic techniques have been used to improve the quality of the berries. Appropriate viticultural practices that enhance ripening, by creating favourable mesoclimates, or by achieving adequate but not excessive vine vigour, will improve wine quality [3]. Soil water availability is a critical factor for vine performance and wine composition. Reports show that severe water stress can influence various physiological and metabolic plant processes, including growth, photosynthesis, and respiration and might be detrimental to fruit quality because of a poor canopy development and reduced leaf assimilation rate thus leading to an inadequate vine capacity to ripen the crop [4–6], particularly under high yield levels [7]. These, in turn, can affect the production, composition and characteristics of the must [8] and, in consequence, the quality of the resulting wine [9]. However, controversy remains as to whether there is any direct effect on berry metabolism other than inhibition of growth [10]. In semi-arid regions, irrigation is often applied as a technique to increase yield and in occasions to achieve a proper supply-demand vine balance. Regulated deficit irrigation (RDI) consists of applying short episodes of water restriction, typically starting after bloom (anthesis), whereby irrigation water is supplied at amounts below those lost to vineyard or crop evapotranspiration (ETc) [11]. Several studies have reported that water stress during the growing season has some beneficial effects (increasing anthocyanin and polyphenol concentrations and soluble solids content) on grape berry quality [12–15]. Water stress applied during the period from fruit set to *véraison*, heavily reduces fruit size [16] and in the late season water restriction may reduce fruit cell enlargement and water accumulation [17]. The timing and intensity of water deficits influence the extent of alterations occurring in berry metabolism and therefore in wine colour and flavour [18–20]. Irrigation is now a commonly used technique in vine-growing in Spain, where cv. Tempranillo is the most widespread cultivar for red wine production. Previous studies have been conducted to analyse the effect of irrigation in phenolic profile of grapes [9, 21–23]. In another hand, it is widely accepted that very high crop yields delay ripening and reduce fruit and wine quality [24]. This leads to increases in the required substrate levels for the synthesis and accumulation in thinned vine berries of, amongst others, phenolic substances [25–29]. Water restriction in Tempranillo grapes led to alterations in the levels of peroxidation and peroxidase activity [30].

In another hand, it is widely accepted that very high crop yields delay ripening and reduce fruit and wine quality [24]. The crop level is a determining factor in berry quality and one of the goals of modern viticulture is to establish field practices which are able to limit vineyard yield and improve grapevine composition [31, 32]. The cluster thinning is a common practice designed to control yield and ripening under adverse conditions of climate. This practice influences the sink-source ratio, restricting part of the yield without lowering leaf surface area. Thinning allows the plant to concentrate its activity on the remaining clusters, making it an effective method of regulating production and modifying the composition of the berries [31, 33, 34]. Several authors have found that cluster thinning enhances berry ripening, affecting the contents of sugars, acids, polyphenols and aromas of the harvested grape and therefore positively affecting the quality of the wine [34, 35]. Cluster thinning is in cases applied to regulate the yield levels and to help ripening the crop under poor climatic conditions or excessive crop demand. However, results presented in literature have reported contrasting results, with cluster thinning leading to better fruit quality in some cases [36, 37], but with no clear effect in some others [33, 38]. Studies on cluster thinning have also shown the importance of the time when it is performed. When thinning is performed at *véraison*, the effect on berry weight and cluster size is lower than when performed immediately after fruit set, since it is at *véraison* that potential berry size and the number of berries per cluster are determined, as well as some compounds, such as the organic acids [39]. In addition, the result of cluster thinning can depend to a large extent on the climatological conditions during the season, with contrasting and even opposite results being obtained depending on the varieties and yields [36, 37].

The overall effect of irrigation might change according to other cultural practices particularly those affecting the crop level [24, 40]. Vines with higher crop level seem to benefit more of a higher amount of irrigation both in terms of yield [41] and of fruit composition [42]. Numerous studies exist in the effects of water deficit and control of yield by eliminating fruits (thinning) on berry composition, but less research has been developed to the combination of both techniques in its composition of wines. Determination of such relationships between composition and chromatic characteristics allow greater understanding of how interactions between cultural practices and chemical composition are reflected in the perception of colour. This type of study is particularly important in vineyards with resource limitations, as would be the case in deficit irrigated grapevines with high crop loads. Such studies are also relevant in relation to red grapevine cultivars, since the repercussions of these practices on the phenolic and chromatic characteristics of the wines obtained need to be analysed. Up to new, few studies have analysed how the combined practice of these techniques affects wine phenolic composition, and the results are not conclusive as to how the water status of the vine impacts on the effect of cluster thinning or vice-versa [18, 19, 39, 43, 44].

The phenolic compounds from the grape are responsible for some of the main organoleptic characteristics of the resulting wine, including its colour, astringency and aroma. In the red wine making process, the phenolic compounds of the skin and seeds are transferred to the must during the fermentation and maceration process [45]. Phenolic compounds can be divided into flavonoid (anthocyanins, flavan-3-ols and flavonols) and non-flavonoid (stilbenes and *hydroxybenzoic and hydroxycinnamic* acids). Given the importance of colour in the quality of red wines, phenolic compound concentration can be used as a parameter for the evaluation of grape, must and wine quality [46]. Besides the variety, other factors such as location, climate, soil type, berry maturity and the vine-growing practices affect the phenolic composition of the grape berries [47]. Improved vine management techniques can therefore be used to modify phenolic content and enhance wine quality [48]. The question of grape berry size and its effect on wine composition has produced inconclusive results, and the relationship between grape berry size and the phenolic composition of wine remains unclear [49]. Wine colour is linked to the accumulation of anthocyanins in the grape berries and particularly in the skin. However, it is not only the anthocyanin concentration and profile that is responsible for wine colour: copigmentation can account for 30–50% of colour in young wines [50]. Copigmentation in wine results from molecular interactions between anthocyanic pigments and other organic molecules, called cofactors, which form molecular associations or complexes. For copigmentation in young wines to increase, there must be sufficient quantities of substances that can act as copigments. The most common cofactors are compounds such as phenolic acids, flavonoids, and in particular, flavonol and flavone derivatives [51]. Good quality grapes should have high concentrations of both stable pigments and substances that can act as cofactors.

In plants, increased synthesis of phenols is a common response to stress [52] and the same environmental conditions that cause oxidative stress are associated with the induction of phenylpropanoid metabolism. Phenylpropanoids and flavonoids are involved in the protection against oxidative stress [53]. The level of lipid peroxidation and how the total antioxidant capacity and the phenolic compound content evolve can both be used as indicators of stress and the response to it. One response to stressors during ripening is an increase in the amount of phenolic compounds. These include flavonoids and phenylpropanoid glycosides, compounds that are part of the cellular antioxidant system, and which are involved in the elimination of ROS [52–55].

ROS have been implicated in plant growth and stress responses. The production and detoxification of ROS are both highly regulated processes, and ROS levels are kept under tight control. The antioxidant system plays an important part in ROS homeostasis [56, 57]. It includes such enzymes as peroxidase (POX), catalase (CAT) and superoxide dismutase (SOD). Stresses and physiological processes in plants both induce the development of a complex network of oxidant/antioxidant reactions that modulate the resulting oxidative response [58, 59]. Thus, a key role in the response to stresses is played by the activities involved in ROS production and elimination, and they are also involved in such physiological processes as ripening [60]. In this regard, while the participation of peroxidases, catalases and superoxide dismutase is crucial for the control of the production or elimination of ROS, their activities are affected by water stress and defoliation [30].

Tempranillo is the most common variety of red grape in Spain comprising a total of 205,975 ha (20% of Spain's total vineyard area), though most of these are cultivated under rainfed conditions with typically low yields (≤ 6500 kg ha^{-1}). This cultivar is originally from northern, cool regions of Spain and today is the most widely cultivated for production of red wines all over Spain. It is reputedly to be sensitive to water stress and prone to early leaf senescence [61].

This chapter analyses and studies the effects of the combination of water deficit and thinning on the constitution and composition of the Tempranillo grape, namely on the lipid peroxidation, oxidant and antioxidant enzymes activities and their interactions on phenolic profile and chromatic characteristic from the wines. Also, relations between wine chromatic parameters and phenolic substances have been investigated.

2. Materials and methods

2.1. Vineyard site and experimental design

The experiment was conducted in a *Vitis vinifera* L. cv. Tempranillo vineyard in Finca La Orden (owned by the Regional Government of Extremadura) in Extremadura, Spain, over the 2007 and 2008 vintages. The vineyard was planted in 2001 on Richter 110 rootstock at a spacing of 2.5 by 1.2 m (3333 vines ha^{-1}). Row orientation was north-south and vines were trained to a bilateral cordon system and vertical trellis.

The experimental design was a complete randomised block with 16 experimental plots, 4 replicates by 4 treatments (irrigation and cluster thinning). Experimental plots consisted of

48 vines across 6 rows. The experiment comprised 768 vines in total. The irrigation regimes, applied during *véraison* and maturation, were as follows:

- Full irrigation (R100), corresponding to 100% of crop evapotranspiration (ETc).
- Deficit irrigation (R25) corresponding to 25% of ETc

ETc was calculated using the equation ETc = ETo × Kc, where ETc was estimated as the product of reference evapotranspiration (ETo), measured by a weather station at the site, and crop coefficient (Kc), following the methodology of Allen et al. [62]. Irrigation was initiated when stem water potential (Ψs) reached −0.5 MPa and stopped after harvest, in mid-September. Irrigation was applied with pressure-compensated emitters supplying 4 L h^{-1} and spaced 120 cm apart. Two cluster load levels were established for each irrigation regime:

- Control treatment (C) without cluster thinning (7–9 clusters m^{-2} of planting area)
- Cluster-thinning treatments (T), in which the load was adjusted to 4–5 clusters m^{-2} of planting area by removing clusters at *véraison*.

Table 1 represents the ETo, ETc and Kc, as well as the irrigation amount applied in the R100 and R25 treatments for each of the study years. Also shown is the water status of the plant during the growing period in the two campaigns (stem water potential during the flowering-*véraison* and *véraison*-harvest periods) [18].

2.2. Physico-chemical composition of grapes and wine chromatic characteristics

Table 2 represents the physico-chemical characteristics of the grapes at harvest and the chromatic characteristics of the wines. The winemaking protocol and analytic methodology can be consulted in Gamero et al. [19].

2.3. Biochemical assay and enzymatic activities of the grapes

Phenols, flavonoids and phenylpropanoid glycosides were assayed colourimetrically. First, grapes were homogenised with methanol, chloroform and 1% NaCl (1:1:0.5). The homogenate was filtered and centrifuged at 3200 × g for 10 min. Total phenols (expressed as μg caffeic acid g^{-1} FW) were determined at 765 nm with Folin:Ciocalteu reagent according to the method of singleton et al. [63]. Total flavonoids (expressed as μg rutin g^{-1} FW) were determined according to the method of Kim et al. [64], calculating the content on the basis of the rutin standard curve. Phenylpropanoids (expressed as μg verbascoside g^{-1} FW) were determined at 525 nm based on estimating an *O*-dihydroxycinnamic derivative using the Arnow reagent as described in Gálvez et al. [65], calculating the content on the basis of the 3,4-dihydroxyphenylalanine standard curve.

Lipid peroxidation was determined by measuring malondialdehyde (MDA) formation using the thiobarbituric acid (TBA) method as described by Madhava Rao and Sresty [66]. The MDA

Year	Seasonal measurements				Irrigation (mm)		Ψs flowering-*véraison* (MPa)		Ψs *véraison*-harvest (MPa)	
	Rainfall (mm)	ETo (mm)	ETc (mm)	Kc	R100[1]	R25	R100	R25	R100	R25
2007	182	975	508	0.52	172	62	−0.54	−0.58	−0.56	−0.77
							−0.81–0.51			
2008	129	1026	755	0.74	611	153	−0.52	−0.76	−0.49	−0.89

Effective seasonal rain, reference evapotranspiration (ETo), crop evapotranspiration (ETc), crop coefficient (Kc) from April to September and mean values in each season of stem water potential (Ψs). Irrigation: Volumes of water applied in the 2007 and 2008 treatments. In 2007 and 2008, the ETc values were obtained by means of a lysimeter installed in the vineyard.[1]R100 = Full irrigation (100% ETc); R25 = Water deficit (25% ETc)

Table 1. Water balance variables in Tempranillo vineyard, Extremadura (Spain).

concentration (expressed as nmol MDA g^{-1} FW) was calculated using an extinction coefficient of $\varepsilon = 155\ mM^{-1}\ cm^{-1}$.

Enzymatic activities were determined on a crude extract of the grapes. The grapes (2 g mL^{-1}) were homogenised at 4°C in 50 mM phosphate buffer, pH 6.0. The homogenate was filtered and centrifuged at 39,000 × g for 30 min at 4°C. The pellet was discarded, and the supernatant filtered and collected for the enzyme assays. The protein content was determined by the method of Bradford [67].

Peroxidase (EC 1.11.1.7) activity, POX, was measured at 590 nm ($\varepsilon = 47.6\ mM^{-1}\ cm^{-1}$) [68], and the coniferyl alcohol (CA) peroxidase activity, CA-POX, was determined by measuring the decrease in absorbance at 265 nm of a reaction medium consisting of the enzyme extract and 0.1 mM CA in 25 mM acetate buffer pH 5.0 ($\varepsilon = 7.5\ mM^{-1}\ cm^{-1}$). A unit of CA-POX is defined as the amount of enzyme required to cause the oxidation of 1 nmol CA per minute at 25°C, pH 5.0. The superoxide dismutase (SOD) (EC 1.15.1.1) activity was determined from the absorbance at 560 nm according to [69]. A unit of SOD is defined as the amount of enzyme required to cause 50% inhibition of NBT reduction. Polyphenol oxidase (PPO) (EC 1.14.18.1) activity was determined from the absorbance at 390 nm [70]. A unit of PPO is defined as the amount of enzyme required to cause a decrease in absorption of 0.001 units min^{-1}.

2.4. Individual phenolic composition of the grapes analysed by HPLC-DAD-FLD

Low molecular weight phenols were analysed by HPLC-DAD, following the method described in Gómez-Alonso [71], though with some slight modifications to enhance the resolution. After filtering (0.25 m diameter Chromafil filters, Düren, Germany), 10 μl of wine were directly injected into a LC-Agilent 1200 (Agilent Technologies, Palo Alto, CA, USA) chromatographic system. Separation was performed in an Ace® 5 C18 250 × 4.6 mm (Advanced Chromatography Technologies, Aberdeen, Scotland) column used as stationary phase.

Parameters	Treatment	Irrigation dosage[1,3]		Significance level[2]		
		R100	R25	R	T	R × T
Yield (kg ha^{-1})	C	21403.75[a]	18749.58[a]	ns	**	ns
	T	14987.03[b]	14704.16[b]			
Berry weight (g)	C	2.17[a]	2.00[b]	ns	**	ns
	T	2.03[b]	1.90[b]			
Total soluble solids (°Brix)	C	22.75[ab]	21.93[b]	ns	***	*
	T	23.87[ab]	24.37[a]			
Titratable acidity (g tartaric acid L^{-1})	C	6.43[a]	5.51[ab]	***	***	ns
	T	5.12[ab]	4.32[b]			
Maturity index[4]	C	3.60[b]	4.08[b]	*	**	*
	T	4.66[ab]	5.81[a]			
Colour intensity (AU)	C	6.69[b]	6.71[b]	ns	***	***
	T	9.51[a]	9.70[a]			
Colour hue	C	0.65[a]	0.62[b]	*	ns	ns
	T	0.67[a]	0.60[b]			
% red	C	52.81[b]	54.00[a]	***	ns	ns
	T	52.06[b]	54.88[a]			
% blue	C	13.00[a]	12.31[b]	*	ns	ns
	T	13.20[a]	12.08[b]			
% yellow	C	34.18[a]	33.67[b]	*	ns	ns
	T	34.73[a]	33.02[b]			

[1]R100C = Full irrigation (R100) without cluster thinning (C); R100A = Full irrigation with cluster thinning; R25C = Water deficit without cluster thinning and R25A = Water deficit with cluster thinning

[2]Significance level: ns = not significant; * = $p \leq 0.05$; ** = $p \leq 0.01$; *** = $p \leq 0.001$

[3]Values followed by different letters indicate significant differences between them ($p \leq 0.5$)

[4]Maturity index = Total soluble solids/titratable acidity

Table 2. Mean values of 2007 and 2008 seasons combined for yield, berry composition and wine chromatic characteristics of cv. Tempranillo wines, subjected to different irrigation dosage levels with cluster-thinning treatment (T) and without cluster-thinning treatment (C) in Extremadura (Spain).

The various phenolic compounds were identified by order of elution and the retention times of their respective standards. A total of 42 phenolic compounds were identified, distributed between simple and conjugated anthocyanins and monomeric and dimeric flavonols and flavanols. The anthocyanins were identified in the monoglucoside, acetyl-glucoside and *p*-coumaroyl-glucoside forms of delphinidin, cyanidin, petunidin, peonidin and malvidin. The quantified flavanols were (+)-catechin, (−)-epicatechin and the procyanidins (B1, B2 and B3). The analysed flavonols were myricetin, quercetin, kaempherol and isorhamnetin in free form and their respective 3-glucosides. Also quantified were the quercetin and kaempherol rutinosides, quercetin-3-glucuronide and quercetin-3-galactoside.

The concentration was calculated in mg L^{-1} of malvidin-3-glucoside, myricetin-3-glucoside and catechin for anthocyanins, flavonols and flavanols, respectively.

2.5. Wine colour composition

Wine colour was evaluated using the indices proposed by Glories [72]: colour intensity (CI) was calculated by adding the absorbance readings at 420, 520 and 620 nm, while hue (CT) was determined as the ratio of absorbance readings at 420 and 520 nm, and percentages of red, blue and yellow (red %, blue % and yellow %). The wines were centrifuged for 3 min at 1100 × g before spectrophotometric analysis. Measurements were taken using a Shimadzu UV-1700 spectrophotometer (Shimadzu, Japan).

2.6. Statistical analysis

Data were analysed using a general linear model, with water status (two levels), cluster thinning (two levels) and their interactions as factors. Differences between treatments were obtained using Tukey's test for significant differences ($p \leq 0.05$).

A principal component analysis (PCA) was carried out to observe the distribution of the wines of the different treatments as a function of their phenolic composition. A correlation analysis was performed to determine the relationships between phenolic compounds and agronomic factors and Tukey's test was used to evaluate significant correlations ($p < 0.05$).

The partial least squares (PLS) approach was used to analyse the relationships between wine colour and HPLC-quantified phenolic compounds. Collinearity between X variables was discarded as significant correlations were not found between them. The statistical analyses were performed with the XLSTAT-Pro (Addinsoft 2009, París, France) statistical software package.

3. Results

3.1. Effect of treatments on the phenolic content and oxidant/antioxidant activities in grapes

Table 3 shows the different values obtained for oxidant and antioxidant activities, and phenolic content in grapes at harvest. With thinning, all these phenolic components were found in higher or equal concentrations, but not lower. The greater value of lipid peroxidation corresponds to CT (rainfed and thinned). The SOD activity was always lower than in the unthinned treatments. It stands out that the lowest values of SOD activity corresponded to CT. Regarding the activity of PPO, an enzyme which catalyses the formation of *o*-quinones, at harvest, the PPO activity levels were similar in all treatments, although showing a minimum value in CT. The lowest POX values correspond to the thinning treatments (CT and R100T). The activity levels of CA-POX, an enzyme which is involved in processes of lignification, were slightly higher with thinning than without thinning.

Treatment*	Lipid peroxidation nmol MDA g^{-1} FW	Superoxide dismutase (SOD) U mg^{-1} prot	Peroxidase (POX) U mg^{-1} prot	Conyferilalcohol peroxidase (CA-POX) U mg^{-1} prot	Polyphenoloxidase (PPO) U mg^{-1} prot	Total Phenols µg caffeic acid g^{-1} FW	Total Phenylpropanoid glycosides µg verbascoside g^{-1} FW	Total flavonoids µg rutín g^{-1} FW
C	94.6 ± 20.5^{b}	89.3 ± 21.7^{b}	4.9 ± 1.2^{b}	53.6 ± 16.5^{b}	1415.5 ± 615.4^{b}	988.6 ± 142.2^{a}	1689.9 ± 290.5^{b}	1122.3 ± 211.0^{b}
CT	130.1 ± 23.0^{a}	71.6 ± 18.0^{b}	4.2 ± 1.5^{b}	66.2 ± 19.3^{ab}	900.3 ± 189.0^{a}	1108.9 ± 158.0^{a}	2013.5 ± 188.3^{b}	1589.7 ± 132.4^{a}
R100C	88.3 ± 18.4^{b}	158.7 ± 22.8^{a}	5.2 ± 1.2^{b}	78.2 ± 20.8^{a}	1510–3 ± 312.5^{b}	1045.6 ± 100.1^{a}	1865.0 ± 366.0^{ab}	1325.6 ± 126.1^{ab}
R100T	91.4 ± 24.3^{b}	81.5 ± 20.1^{b}	2.6 ± 1.0^{a}	79.0 ± 22.1^{a}	1389.0 ± 340.0^{b}	966.0 ± 144.5^{a}	1924.4 ± 171.3^{b}	1159.0 ± 206.6^{b}

Data were the means ± SD of at least 10 replicates obtained from 5 different experiments. The values were subjected to the Tukey test (different letters within each column indicate significant difference, $p \leq 0.05$)[1]C = Rainfed without cluster thinning; R100C = Full irrigation without cluster thinning; CT = Rainfed with cluster thinning; R100T = Full irrigation with cluster thinning

Table 3. Average values (2007 and 2008 vintage) of lipid peroxidation, antioxidant activities, total phenols, phenylpropanoid glycosides and flavonoids on grapes cv. Tempranillo at harvest.

3.2. Effect of treatments on the phenolic profile of wines

Table 4A and **B** shows the mean concentration values over the 2 year study period of anthocyanin and non-anthocyanin flavonoid substances (flavanols and flavonols) of the R100C, R100T, R25C and R25T wines, as well as the results of the statistical analysis.

Table 4A shows the amount of monoglucosides (G), acetates (A) and coumaroyl derivatives (C) and the amounts of delphinidin (Dp), cyanidin (Cy), petunidin (Pt), peonidin (Pn) and malvidin (Mv) compounds.

In the case of the flavanols (**Table 4B**), as well as the mean concentration values of (+)-catechin (Cat) and (−)-epicatechin (Ep), also shown are the global mean values of these monomers (Mon) and, as well as the individual mean of (−)-epicatechin-(4ß-8)-(+)-catechin (PB1), (−)-epicatechin-(4ß-8)-(−)-epicatechin (PB2) and (+)-catechin-(4ß-8)-(+)-catechin (PB3) values, the combined mean values of these dimers (Dim). The flavonols are grouped together by the flavonol molecule present, such that myricetin (My), quercetin (Qc), kaempherol (Kp) and isorhamnetin (Ir) glucosides represent the total values, respectively.

Results of the general linear model analysis only found statistical significance of the irrigation and thinning interaction in Cy, PB1, PB3, Dim and total flavonoid substances. A statistical analysis was then performed of the overall effect of irrigation dosage and its individual effect on each crop load, as well as of the overall effect of cluster thinning and its effect on each water status.

3.2.1. *Effect of irrigation*

As can be seen in **Table 4A**, the different irrigation dosages had almost no effect on wine anthocyanin concentration values. Slightly lower concentrations of these substances were found in the wine from R25 treatment compared to the wine from R100 treatment, but the differences were non-significant ($p > 0.05$). This trend was also observed when studying separately the effect of irrigation dosage on the different crop loads (R100T vs. R25T and R100C vs. R25C).

However, the opposite trend was observed for the rest of the flavonoid families (flavanols and flavonols). The R25 wines contained significantly higher concentrations of Cat, the flavonol compounds My, Qc, Kp and Ir, Mon and total flavanols (**Table 4B**). An analysis of the effect of irrigation dosage on the different crop loads revealed significant differences only in the wines from the cluster-thinned treatments (R25T vs. R100T).

3.2.2. *Effect of cluster thinning*

In terms of wine anthocyanin composition, the effect of thinning was greater than of irrigation, with the A-treatment wines having significantly higher values of all the substances of this family, except for the acetyl-acetate derivatives. The biggest differences were found in the G (18%), Mv (15%) and Dp (29%). **Table 4A** also shows the separate analysis made of the effect of thinning on the R25 and R100 wines, with a greater and more significant response in R100. These results confirm the importance of vine water status in the effect of cluster thinning on

Compound	Load	Water status[2]		Significance level[1]						
				R	T	Irrigation		Thinning		R and T
		R100	R25			in C	in T	in R25	in R100	
Σ Glucosides	C	90.21^{ab}	83.35^{b}	ns	*	ns	ns	ns	**	*
	T	111.08^{a}	100.19^{a}							
Σ Acetates	C	11.96^{a}	10.26^{a}	ns	ns	ns	ns	ns	ns	ns
	T	11.28^{a}	11.2^{a}							
Σ Cumarates	C	10.18^{b}	11.65^{b}	ns	**	ns	ns	ns	ns	*
	T	14.45^{a}	14.32^{ab}							
Σ Malvidins	C	82.88^{a}	79.59^{a}	ns	**	ns	ns	ns	**	ns
	T	100.35^{a}	92.06^{a}							
Σ Petunidins	C	15.04^{ab}	13.23^{b}	ns	***	ns	ns	ns	**	*
	T	18.73^{a}	17.58^{ab}							
Σ Delphinids	C	7.97^{b}	8.11^{ab}	ns	***	ns	ns	ns	***	**
	T	11.80^{a}	11.06^{ab}							
Σ Peonidins	C	3.32^{a}	2.62^{a}	ns	*	ns	ns	ns	ns	ns
	T	3.93^{a}	3.49^{a}							
Σ Cyanidins	C	3.15^{a}	1.70^{b}	Significant interaction						*
	T	2.01^{b}	1.53^{b}							
Total Anthocyanins	C	112.36^{a}	105.25^{a}	ns	**	ns	ns	ns	*	ns
	T	136.82^{a}	125.67^{a}							

[1]ANOVA test significance level. R: irrigation; A: cluster thinning; R and A: combination of both treatments. ns = not significant; * = $p \leq 0.05$; ** = $p \leq 0.01$; *** = $p \leq 0.001$

[2]For the same compound, values followed by different letters indicate significant differences between treatments ($p \leq 0.5$)

Table 4A. Average values (2007 and 2008 vintage) of anthocyanins (mg L^{-1}) in Tempranillo wines from vines subjected to different irrigation dosages and with different crop level loads, cluster thinned (T) and non-thinned (C), in Extremadura (Spain).

the concentration of anthocyanin substances present in the wines. In addition, when compared to the C-treatment wines, the T-treatment wines had lower concentrations ($p < 0.001$) of Cat, and higher concentrations of My and Qc (**Table 4B**).

3.2.3. *Effect of treatments*

Table 4A and **B** also shows the results of the combined statistical analysis of the four treatments. In concordance with the results described above, the highest and lowest values for anthocyanin substances were found in the 100T and 25C wines, respectively. Of these substances, Cy (the sum total of cyanidin compounds) is a special case as both the lower irrigation

Flavanols										
Compound	Load	Water status[2]		Significance level[1]						
				R	T	Irrigation		Thinning		R and T
		R100	R25			in C	in T	in R25	in R100	
(+)-catechin	C	25.67^{b}	31.89^{a}	**	***	**ns**	***	*	***	***
	T	16.35^{c}	23.08^{b}							
(−)-epicatechin	C	4.35^{a}	4.01^{a}	**ns**	**ns**	**ns**	*	**ns**	**ns**	**ns**
	T	5.42^{a}	4.31^{a}							
Procyanidin 1 (PB1)	C	34.54^{c}	49.44^{b}	**Significant interaction**						***
	T	40.95^{ab}	71.18^{a}							
Procyanidin 2 (PB2)	C	6.64^{b}	6.00^{b}	**ns**	***	**ns**	**ns**	**	***	**
	T	6.00^{b}	8.24^{a}							
Procyanidin 3 (PB3)	C	17.08^{bc}	16.14^{c}	**Significant interaction**						***
	T	21.96^{a}	18.27^{b}							
Monomers	C	30.03^{b}	35.91^{a}	***	***	**ns**	**	*	**	***
	T	21.77^{c}	27.38^{b}							
Dimers	C	58.26^{c}	71.59^{b}	**Significant interaction**						***
	T	71.04^{b}	97.69^{a}							
Total Flavanols	C	88.29^{c}	107.49^{b}	***	**	***	***	**	**ns**	***
	T	92.81^{c}	125.07^{a}							
Σ Myricetins	C	8.06^{c}	9.64^{b}	***	***	**ns**	*	***	***	***
	A	11.46^{b}	14.06^{a}							
Σ Quercetins	C	6.52^{c}	8.20^{b}	***	***	**ns**	**	**	**	***
	A	7.84^{bc}	10.5^{a}							
Σ Kaempherols	C	1.40^{c}	1.74^{b}	*	**ns**	***	**ns**	***	***	***
	A	1.55^{bc}	2.17^{a}							
Σ Isorhamnetins	C	0.96^{c}	1.28^{b}	***	***	*	*	**	**	***
	A	1.24^{b}	1.62^{a}							
Total Flavonols	C	16.94^{c}	20.85^{b}	***	***	**ns**	**	***	***	***
	A	22.09^{b}	28.37^{a}							

[1]ANOVA test significance level. R: irrigation; A: cluster thinning; R and A: combination of both treatments. ns = not significant; * = $p \leq 0.05$; ** = $p \leq 0.01$; *** = $p \leq 0.001$

[2]For the same compound, values followed by different letters indicate significant differences between treatments ($p \leq 0.5$)

Table 4B. Average values (2007 and 2008 vintage) of flavanols, flavonols, (mg L^{-1}) in cv. Tempranillo wines from vines subjected to different irrigation dosages and with different crop load levels, cluster thinned (T) and non-thinned (C) in Extremadura (Spain).

http://dx.doi.org/10.5772/intechopen.72799

dosage as well as cluster thinning resulted in lower concentration values, with the consequent concentration value for the R100C wines being significantly higher than for the other wines.

Values of the PB1 and PB3 dimers increased as irrigation dosage fell, with the 25T and 100C wines having the minimum and maximum values, respectively. Finally, and given the sensitivity of flavonols to both irrigation dosage and crop load, the R25T wines had significantly higher values than the R100C.

3.2.4. *Correlation between agronomic parameters and grape berry ripeness indicators and concentration of wine phenolic substances*

To determine the degree of correlation between the agronomic parameters (*véraison*-harvest Ψs and yield values of the different treatments), berry weight and berry ripeness indicators (TSS—total soluble solids and MI—maturity index) and the values of the different phenolic families of the vines, the mean values for the 2 years of the study were subjected to a Pearson's correlation and a PCA.

No significant correlation was found between *véraison*-harvest Ψs and wine phenolic values. However, significant negative correlations were found between yield and concentration

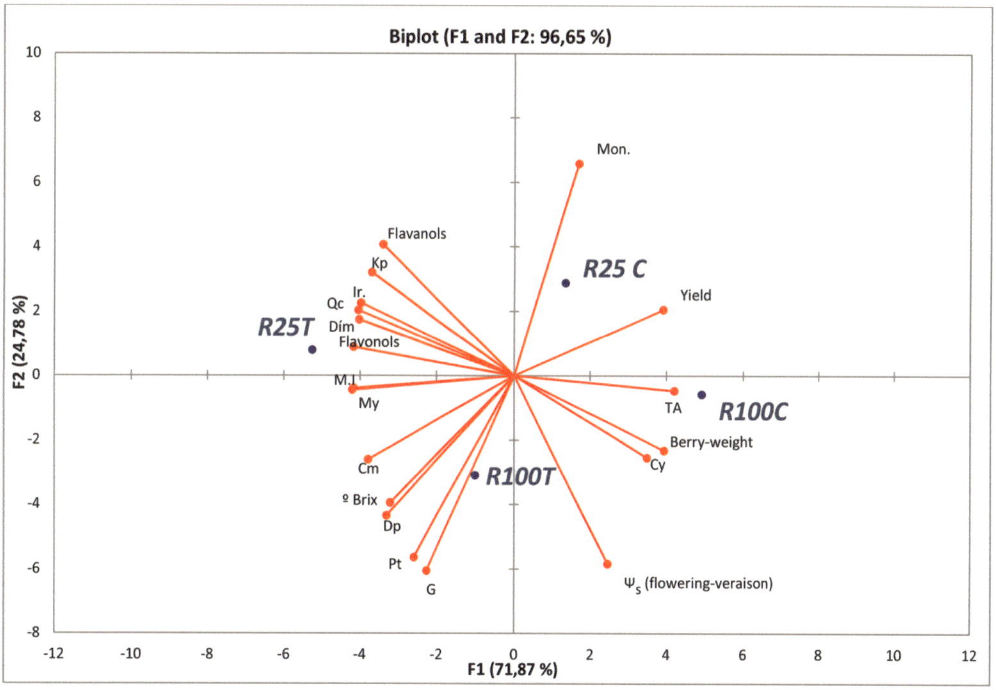

Figure 1. Principal component analysis (PCA) between the different parameters studied (stem water potential Ψs, yield, ° brix and maturity index (MI) and wine phenolic compounds. G: ∑ Glycosides, Cm: ∑ Coumarates, Pt: ∑ Petunidins, Dp: ∑ Delphinids, Cy: ∑ Cyanidins, My: ∑ Myricetins, Qc: ∑ Quercetins; Kp: ∑ Kaempherols; Ir: ∑ Isorhamnetin; Mon:∑ Monomers; Dim: ∑ Dimers, TA: total acidity.

of C (anthocyanin coumarates, r = −0.997) and between Qc and Ih (r = −0.961 and −0.984, respectively) and berry weight. This latter parameter was also positively correlated with Cy (r = 0.955). Conversely, berry acidity was negatively correlated with dimer values (r = −0.956) and all the flavonols except Kp. Finally, berry ripeness as reflected by the MI was positively correlated with dimers and My (r = 0.955 and 0.997 respectively).

Figure 1 represents the projection of the variables and samples in the plane defined by the first two principal components. The two principal components explain 96.65% of total variance. Principal component 1, which explains 71.87% of the variance, allows to distinguish the T wines from the C wines. This component is characterised by titratable acidity (TA), yield and berry weight on the positive side of the axis and by practically all the phenolic substances except G and Pt on the negative side, such that as we shift from left to right the wines are poorer in phenolic substances. The second principal component explains 24.78% of the variance and allows to distinguish the R100 from the R25 wines and is characterised by high monomer concentrations on the positive side of the axis and by *véraison*-harvest Ψs and G values on the negative side.

3.2.5. Correlation between wine phenolic composition and chromatic parameters

A PLS regression was performed to determine correlations between the chromatic parameters of the different wines, Colour intensity and Colour hue (CI, CT), percentage of red, blue and

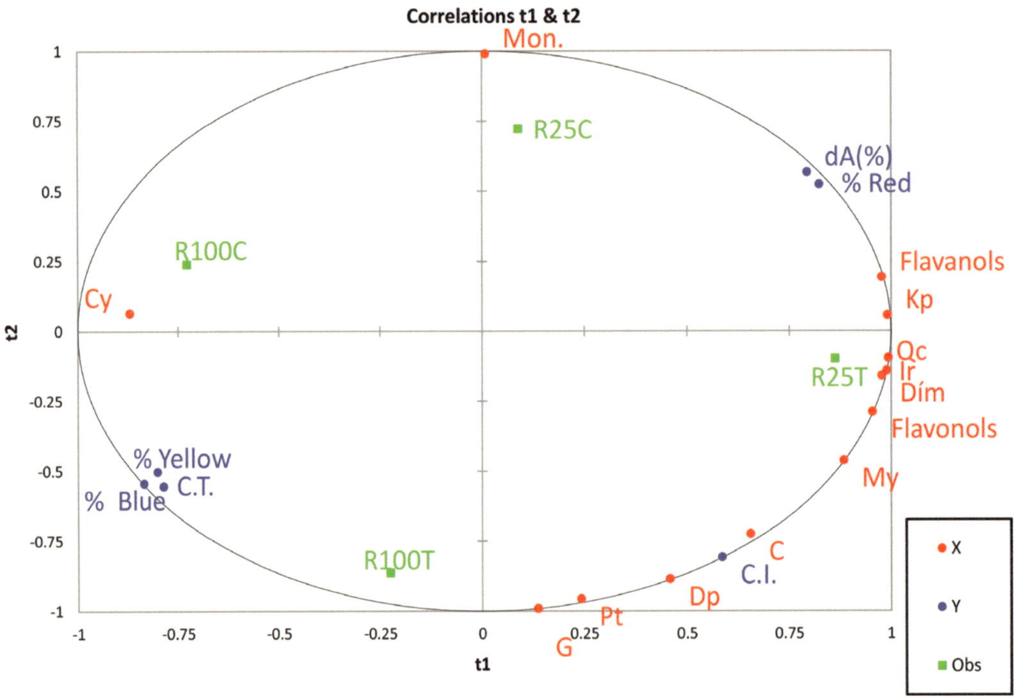

Figure 2. Partial least squares (PSA) regression between the colorimetric parameters and wine phenolic compounds. G: ∑ Glycosides, Cm: ∑ Coumarates, Pt: ∑ Petunidins, Dp: ∑ Delphinids, Cy: ∑ Cyanidins, My: ∑ Myricetins, Qc: ∑ Quercetins; Kp: ∑ Kaempherols; Ir: ∑ Isorhamnetin; Mon:∑ Monomers; Dim: ∑ Dimers.

yellow and dA%, and the phenolic compounds with significant inter-treatment differences. The chromatic data were taken as Y variables and phenolic substance concentrations as X variables. The data were also standardised to approximately the same scale to better reveal the correlations between the colour and chemical variables. Two new variables (axes) were obtained through this technique which explains together 97% of total variance.

Figure 2 shows how CI was high and positively correlated with the presence of G, C, Dp, Pt and My, with this correlation being higher and more intense in the R100T and R25A wines. Conversely, %dA and %red were positively correlated with Kp (sum total of kaempherol glucosides) and the overall amount of flavonol, with this correlation being higher in the R25T wines. These results allow to make wine colour predictions based on wine phenolic composition. Accordingly, colour intensity can be predicted by the amount of certain phenolic compounds in the wines: G and C, Dp and Pt, and My. In the same way, the %dA and %red of a wine can be predicted by the concentration of Kp and flavonols.

4. Discussion

The results obtained for phenols, flavonoids and PPGs in grapes at harvest are consistent with those obtained by Tardáguila et al. [73] who describe increased overall amounts of phenols and anthocyanins in grapes from thinned cv. Tempranillo vines. Likewise, Diago et al. [27] also describe an increase in these compounds caused by thinning in this variety, although the treatment was much less effective in the Garnacha variety. Esteban et al. [74] and Ojeda et al. [75] describe similar results in the content of total phenols, flavonoids, anthocyanins and flavonols as a result of water stress. Again, the combination of water deficit and thinning is the treatment with the greatest incidence on the evolution and content of these phenolic compounds. The SOD activity was always lower than in the unthinned treatments, an aspect which could have been the cause of the observed greater level of lipid peroxidation in the water deficit and thinned treatment. The nonspecific peroxidase activity was very low, especially in the thinned treatments. On the contrary, in the harvest the CA-POX activity's levels were slightly higher with thinning than without thinning. This enzyme might be involved in the production of H_2O_2 during the formation of lignins [76]. According to our previous results [30], it does not appear to be affected by either the stage of ripening or the different treatments (thinning and water deficit).

Most studies on the effect of irrigation on grape berry and wine anthocyanin composition have found a decrease in the concentration of different anthocyanin derivatives and in their overall concentration as irrigation dosage is increased [77, 78]. Until a few years ago, these results were explained as being secondary effects of irrigation, including modification of berry size and consequently of skin/pulp ratio, which were responsible for the decrease in the concentration of these substances in the berry [75]. In this respect, sufficiently large increases in R100 berry size were not obtained in our study (**Table 2**) to generate, by dilution effect, a decrease in the concentration of these substances in the wines. However, recent studies have found that these substances are synthesised in the biosynthetic phenylpropanoid pathway, and it has been shown that it is the pre-*véraison* deficit which regulates the expression of genes involved in this biosynthesis [79]. Accordingly, the lack of response to irrigation in the G, and

T and C forms, as well as in the Mv, Pt, Df and Pn derived compounds can be explained based on the different irrigation dosages applied in the post-*véraison* period.

In a similar study to ours conducted in Requena (Spain) with Tempranillo vines, Intrigliolo et al. [9] found increased anthocyanin concentrations in wines from the most irrigated vines. They explained their results as the consequence of the greater water requirements and the inability of cv. Tempranillo, typically grown in colder regions, to adapt to edapho-climatic conditions. This could also be the case in our study, as the highest values were observed in the R100 wines. This result might also suggest a greater synchrony between sugar and anthocyanin accumulation in the berry of the higher irrigation treatments [18].

The significant differences that were found in the flavanol- and flavonol-type substances are not a decisive factor for these substances. Studies have shown that while water deficit has a moderate effect on flavonol synthesis, the effect of the irrigation application period is almost negligible [23, 80]. This would explain the significant increase in My, Qc, Kp and Ir in the R25 wines compared to the R100 wines. The synthesis of these substances is light dependent, so the lower leaf area observed in the R25 treatment may also have contributed to this increase. In this respect, Matthews et al. [81] reported that the decrease in berry weight was higher, and consequently the increase in flavonoid concentration was higher, when water stress was imposed to pre-*véraison*.

Given the higher content of anthocyanin substance in the A-treatment musts [18], the formation of copigments must have taken place in them in a greater proportion during fermentation, which would in turn have resulted in greater flavonol extraction. This may be one of the reasons why, when the effect of irrigation was analysed separately on the two load levels, significant differences were only found in My, Qc and Ir values in the R100T versus R25T comparison. This result is very important, as copigments stabilise the colour of young red wines [50].

The results shown in **Table 4A** and **B** concurs with those from previous studies which show that cluster thinning is a useful tool to increase the concentration of different flavonoid compounds in grapes and wines [29, 37, 82]. Oenologically speaking, and in full concurrence with the results observed by Peña-Neira et al. (29), the most important and notable effects of cluster thinning are those related to the increase in phenolic compound values via the phenylpropanoid pathway.

Among the anthocyanin substances, the increase in the coumarate forms is particularly important, as these are more stable than the monoglucosides [83, 84] and offer greater stability to wine colour. Also notable is the low significance of cluster thinning on the R25 wines. In a similar study performed with the Syrah grape [82], cluster thinning led to significant increases in the coumarate forms of malvidin and peonidin. In our study, the maximum anthocyanin content in the R25 berries may have been generated at lower TSS values than those at which the berries were harvested (24.37°Brix), so that anthocyanin material may have been lost in these berries.

The increase in flavonols is especially interesting, since these contribute to astringency and also play an important role in anthocyanin copigmentation and in the formation of more complex pigments during the ageing process [85, 86]. In the copigmentation phenomenon which contributes to the stabilisation of the colour of red wines [50, 87], Baranac et al. [88] reported that flavonol substances are amongst the best copigments, especially quercetin [89].

The increased concentrations of catechin and procyanidin B1 are also important given their relation to colour and body stability through these copigmentation and polymerisation reactions with other flavonols.

Similar to the results from other studies [19, 26, 39], wines from the cluster-thinned treatments had higher CI values. Colour hue variation between wines as a function of vine water status has also been described in other studies, though with contradictory results [21]. In our study, the greater presence of cyanidin due to the later ripening of the R100C grapes may have been responsible for the higher percentage of blue hues ($p < 0.05$) of these wines.

Many studies have used statistical techniques to find correlations between phenolic compounds and colour parameters during the maturation and ageing processes of red wine [90]. In one interesting work, Monagas et al. [91] showed that chromatic attributes of red wines could be predicted by their phenolic profile using polynomial regression techniques. The substances which provided the best fitting model in that study were the anthocyanin compounds, and in the specific case of cv. Tempranillo wines the pyruvic adducts. According to the PLS regression which we preformed, the CI of Tempranillo wines can also be predicted as a function of anthocyanin compounds, specifically C, G, and Df and Pt compounds. Along the same line, our work also confirms the findings of Escudero-Gilete et al. [92]. They studied the correlations between colour and anthocyanin composition of wines made from a blend of cv. Tempranillo and cv. Graciano grapes and found that the petunidin compounds (monoglucoside, acetyl-acetate and acetyl-coumarate) were the ones most closely related to the chromaticity of these wines. Similarly, the negative correlation between the percentages of yellow and blue and the flavanols which were also found by the PLS regression concur with the findings of the previously cited work of Monagas et al. [91]. From all of the above, it can be concluded that the thinning treatment generated, most notable, an increase in wine colour intensity and that vine water stress modified, most notably, wine hue as a result of differences in flavonol concentrations.

5. Conclusion

This work contributes to our understanding of how vine-growing practices can affect the compounds responsible for the sensory attributes of wines. It is shown how modification of water status and crop load through deficit irrigation and cluster-thinning techniques can induce variations in the phenolic composition and consequently the chromatic characteristics of a wine. It can be concluded from the results that post-*véraison* water stress increased the concentration values of non-anthocyanin flavonoids, though the extent of the increase varied depending on the compound under consideration and, in the case of flavanols, on the crop load. Consequently, special care should be taken to adapt and control irrigation dosages in regions of very dry and hot summers where post-*véraison* irrigation is an absolute necessity. The results of our study also show an increase in the phenolic composition of wines as a result of the cluster thinning of irrigated vines and, above all, show how the extent and significance of this technique depends in the case of anthocyanins and flavanols on vine water status. Furthermore, our work demonstrates how these practices can affect wine colour. With the increase in anthocyanin content, thinning augmented wine colour intensity, while the different levels of vine water status affected colour hue by modifying the concentration of flavonols which participate in the reactions that help to stabilise wine colour.

Acknowledgements

This research was supported by funds from INIA, FEDER and Junta de Extremadura, Projects RTA2005-00038-C06 and RTA2008-00037-C04. The authors gratefully acknowledge the technical assistance of Dra. Mar Vilanova (Misión Biológica de Galicia, CSIC) in statistical analysis.

Author details

Ester Gamero[1], Francisco Espinosa[2*], Daniel Moreno[1], David Uriarte[3], María Henar Prieto[3], Inmaculada Garrido[2] and María Esperanza Valdés[1]

*Address all correspondence to: espinosa@unex.es

1 Technological Institute of Food and Agriculture, INTAEX Junta of Extremadura, Badajoz, Spain

2 U. Plant Physiology, Facultad de Ciencias, University of Extremadura, Badajoz, Spain

3 Institute of Agricultural Research Finca La Orden-Valdesequera, CICYTEX Junta of Extremadura, Badajoz, Spain

References

[1] Jackson DI, Lombard PB. Environmental and management practices affecting grape composition and wine quality. A review. American Journal of Enology and Viticulture. 1993;**44**:409-430

[2] Coombe BG, McCarthy MG. Dynamics of grape berry growth and physiology of ripening. Australian Journal of Grape and Wine Research. 2000;**6**:131-135. DOI: 10.1111/j.1755-0238.2000.tb00171.x

[3] Kounduras S, Marinos V, Gkoulioti A, Kotseridis Y, Van Leeuwen C. Influence of vineyard location and vine water status on fruit maturation of non-irrigated cv Agiorgitiko (*Vitis vinifera* L.). Effects on wine phenolic and aroma components. Journal of Agriculture and Food Chemistry. 2006;**54**:5077-5086. DOI: 10.1021/jf0605446

[4] Hardie W, Considine J. Response of grapes to water-deficit stress in particular stages of development. American Journal of Enology and Viticulture. 1976;**27**:55-61

[5] Osakabe Y, Osakabe K, Shinozaki K, Tran LSP. Response of plants to water stress. Frontiers in Plant Sciences. 2014;**5**:1-7. DOI: 10.3389/fpls.2014.00086

[6] Procházková D, Jan S, Abd-Allah EF, Ahmad P. Water stress in grapevine (*Vitis vinifera* L.). In: Ahmad P, editor. Water Stress and Crop Plants: A Sustainable Approach. Chichester: Wiley; 2016

[7] Freeman BM, Kiewer WM. Effect of irrigation, crop level and potassium fertilization on Carignane vines. II. Grape and wine quality. American Journal of Enology and Viticulture. 1983;**34**:197-207

[8] Wang ZP, Deloire A, Carbonneau A, Federspiel B, Lopez F. Study of sugar phloem unloading in ripening grape berries under water stress conditions. Journal International Science de la Vigne et du Vin. 2003;**37**:213-222

[9] Intrigliolo DS, Castel JR. Response of grapevine cv. 'Tempranillo' to timing and amount of irrigation: Water relations, vine growth, yield and berry and wine composition. Irrigation Science. 2010;**28**(2):113-125. DOI: 10.1007/s00271-009-0164-1

[10] Castellarin SD, Matthews MA, Di Gaspero G, Gambetta GA. Water deficits accelerate ripening and induce changes in gene expression regulating flavonoid biosynthesis in grape berries. Planta. 2007;**227**:101-112. DOI: 10.1007/s00425-007-0598-8

[11] Keller M, Smithyman RP, Mills LJ. Interactive effects of deficit irrigation and crop load on Cabernet Sauvignon in an arid climate. American Journal of Enology and Viticulture. 2008;**59**(3):221-234

[12] Bravdo B, Hepner Y, Loinger C, Cohen S, Tabacman H. Effect of irrigation and crop level on growth, yield and wine quality of Cabernet Sauvignon. American Journal of Enology and Viticulture. 1985;**36**:132-139

[13] Williams LE, Matthews MA. Grapevine. In: Stewart BA, Nielsen DR, editors. Irrigation of Agricultural Crops, Agronomy Monograph. Vol. 30. Madison: ASA-CSSA-SSSA; 1990. pp. 1019-1055

[14] Ginestar C, Eastham J, Gray S, Iland P. Use of sap-flow sensors to shedule vineyard irrigation. II. Effects of post-véraison water deficits on composition of Shiraz grapes. American Journal of Enology and Viticulture. 1998;**49**(4):421-428

[15] Roby G, Matthews MA. Relative proportion of seed, skin and flesh in ripe berries from Cabernet Sauvignon grapevines grown in a vineyard either well irrigated on under water deficit. Australian Journal of Grape and Wine Research. 2004;**10**:74-82. DOI: 10.1111/j.1755-0238.2004.tb00009.x

[16] McCarthy MG. The effect of transient water deficit on berry development of cv. Shiraz (*Vitis vinifera* L.). Australian Journal of Grape and Wine Research. 1997;**3**(3):102-108

[17] Smart RE, Coombe BG. Water relations of grapevines. In: Kozlowski TT, editor. Water Deficits and Plant Growth, Vol. VII: Additional Woody Crop Plants. 1983. pp. 137-196

[18] Gamero E, Moreno D, Talaverano I, Prieto MH, Guerra MT, Valdés ME. Effects of irrigation and cluster thinning on Tempranillo grape and wine composition. South African Journal of Enology and Viticulture. 2014;**35**(2):196-204

[19] Gamero E, Moreno D, Vilanova M, Uriarte D, Prieto MH, Valdes ME. Effect of bunch thinning and water stress on chemical and sensory characteristics of Tempranillo wines. Australian Journal of Grape and Wine Research. 2014;**20**(3):394-400. DOI: 10.1111/ajgw.12088

[20] Talaverano I, Valdes E, Moreno D, Gamero E, Mancha L, Vilanova M. The combined effect of water status and crop level on Tempranillo wine volatiles. Journal of the Science of Food and Agriculture. 2017;**95**(5):1533-1542. DOI: 10.1002/jsfa.7898

[21] Casassa L, Keller M, Harbertson J. Regulated deficit irrigation alters anthocyanins, tannins and sensory properties of Cabernet Sauvignon grapes and wines. Molecules. 2015;**20**:7820e7844. DOI: 10.3390/molecules20057820

[22] Zarrouk O, Francisco R, Pinto-Marijuan M, Brossa R, Santos RR, Pinheiro C, Chaves MM. Impact of irrigation regime on berry development and flavonoids composition in Aragonez (Syn. Tempranillo) grapevine. Agricultural Water Management. 2012;**114**:18-29. DOI: 10.1016/j.agwat.2012.06.018

[23] Kennedy JA, Matthews MA, Waterhouse AL. Effect of maturity and vine water status on grape skin and wine flavonoids. American Journal of Enology and Viticulture. 2002;**53**:268-274

[24] Bravdo B, Hepner Y, Loinger C, Cohen S, Tabacman H. Effect of crop level on growth, yield and wine quality of a high yielding Carignane vineyard. American Journal of Enology and Viticulture. 1984;**35**:247-252

[25] Petrie PR, Clingeleffer PR. Crop thinning (hand versus mechanical), grape maturity and anthocyanin concentration: Outcomes from irrigated Cabernet Sauvignon (*Vitis vinifera* L.) in a warm climate. Australian Journal of Grape and Wine Research. 2006;**12**(1):21-29. DOI: 10.1111/j.1755-0238.2006.tb00040.x

[26] Diago MP, Tardágila LJ. Estudio y desarrollo del deshojado precoz como técnica para el control del rendimiento productivo de la vid (*Vitis vinifera* L.). Efectos sobre el desarrollo vegetativo, los componentes de la producción, así como sobre la composición y la calidad de la uva y del vino, 2010

[27] Fanzone M, Zamora F, Jofré V, Assof M, Peña-Neira A. Phenolic composition of Malbec grape skins and seeds from Valle de Uco (Mendoza, Argentina) during ripening. Effect of cluster thinning. Journal of Agricultural and Food Chemistry. 2011;**59**(11):6120-6136. DOI: 10.1021/Jf200073k

[28] Uriarte D, Intrigliolo DS, Mancha L, Valdes E, Gamero E, Prieto MH. Combined effects of irrigation regimes and crop load on 'Tempranillo' grape composition. Agricultural Water Management. 2016;**165**:97-107. DOI: 10.1016/j.agwat.2015.11.016

[29] Peña-Neira A, Cáceres A, Pastenes C. Low molecular weight phenolic and anthocyanin composition of grape skins from cv. Syrah (*Vitis vinifera* L.) in the Maipo Valley (Chile): Effect of clusters thinning and vineyard yield. Food Science and Technology International. 2007;**13**(2):153-158. DOI: 10.1177/1082013207077920

[30] Garrido I, Llerena JL, Valdés ME, Mancha LA, Uriarte D, Prieto MH, Espinosa F. Effects of defoliation and water restriction on total phenols and antioxidant activities in grapes during ripening. Journal International des Sciences de la vigne et du vin. 2014;**48**:31-42

[31] Gatti M, Bernizzoni F, Civardi S, Poni S. Effects of cluster thinning and preflowering leaf removal on growth and grape composition in cv. Sangiovese. American Journal of Enology and Viticulture. 2012;**63**(3):325-332

[32] Palliotti A, Poni S. Traditional and innovative summer pruning techniques for vineyard management. Advances in Horticultural Science. 2011;**25**(3):151-163. DOI: 10.13128/ahs-12764

[33] Keller ML, Mills L, Wample R, Spayd S. Cluster thinning effects on three deficit-irrigated *Vitis Vinifera* cultivars. American Journal of Enology and Viticulture. 2005;**56**:91-103

[34] Sun Q, Sacks GL, Lerch SD, Vanden Heuvel JE. Impact of shoot and cluster thinning on yield, fruit composition, and wine quality of Corot noir. American Journal of Enology and Viticulture. 2012;**63**(1):49-56

[35] Ferrer M, Gonzalez Neves G, Camussi G, Echeverria G. Effect de l'éclaircissage des grappes sur les paramètres productivs, vegetatifs et oenologiques de la vigne (*Vitis vinifera* L.) cv. Tannat. Progrès Agricole et Viticole. 2009;**126**(1):12-18

[36] Prajitna A, Dami IE, Steiner TE, Ferree DC, Scheerens C, Schwartz SJ. Influence of cluster thinning on phenolic composition, resveratrol, and antioxidant capacity in Chambourcin wine. American Journal of Enology and Viticulture. 2007;**58**(3):346-350

[37] Guidoni S, Allara P, Schubert A. Effect of cluster thinning on berry skin anthocyanin composition of *Vitis vinifera* cv. Nebbiolo. American Journal of Enology and Viticulture. 2002;**53**:224-226

[38] Ough CS, Nagaoka R. Effect of cluster thinning and vineyard yields on grape and wine composition and wine quality of Cabernet Sauvignon. American Journal of Enology and Viticulture. 1984;**35**:30-34

[39] Valdés ME, Moreno D, Gamero E, Uriarte D, Prieto MH, Manzano R, Intrigliolo DS. Effects of cluster thinning and irrigation amount on water relations, growth, yield and fruit and wine composition of Tempranillo grapes in Extremadura (Spain). Journal International des Sciences de la Vigne et du Vin. 2009;**43**(2):67-76

[40] Poni S, Lakso AN, Turner JR, Melious RE. Interactions of crop level and late season water stress on growth and physiology of field-grown Concord grapevines. American Journal of Enology and Viticulture. 1994;**45**:252-258

[41] Poni S, Lakso AN, Turner JR, Melious RE. The effects of pre- and post-véraison water stress on growth and physiology of potted Pinot Noir grapevines at varying crop levels. Vitis. 1993;**32**:207-214

[42] Hepner Y, Bravdo B. Effect of crop level and drip irrigation scheduling on the potassium status of Cabernet Sauvignon and Carignane vines and its influence on must and wine composition and quality. American Journal of Enology and Viticulture. 1985;**36**:140-147

[43] Bravdo B, Hepner Y. Irrigation management and fertigation to optimize grape composition and vine performance. In: Symposium on Grapevine Canopy and Vigor Management, XXII IHC 206; 1986. pp. 49-68

[44] Santesteban LG, Miranda C, Royo JB. Regulated deficit irrigation effects on growth, yield, grape quality and individual anthocyanin composition in *Vitis vinifera* cv. Tempranillo. Agricultural Water Management. 2011;**98**:1171-1179. DOI: 10.1016/j.agwat.2011.02.011

[45] González-Neves G, Charamelo D, Balado J, Barreiro L, Bochicchio R, Gatto G, Gil G, Tessore A, Carbonneau A, Moutounet M. Phenolic potential of Tannat, Cabernet-Sauvignon and Merlot grapes and their correspondence with wine composition. Analytica Chimica Acta. 2004;**513**(1):191-196. DOI: 10.1016/j.aca.2003.11.042

[46] Gollücke AP, Catharino RR, De Souza JC, Eberlin MN, Tavares DQ. Evolution of major phenolic components and radical scavenging activity of grape juices through concentration process and storage. Food Chemistry. 2009;**112**(4):868-873. DOI: 10.1016/j.foodchem.2008.06.060

[47] Downey MO, Dokoozlian NK, Krstic MP. Cultural practice and environmental impacts on the flavonoid composition of grapes and wine: A review of recent research. American Journal of Enology and Viticulture. 2006;**57**(3):257-268

[48] Shange L, Conradie WJ. Effects of soil parent material and climate on the performance of *Vitis vinifera* L. cv. Sauvignon blanc and Cabernet Sauvignon—Part I. Soil analysis, soil water status, root system characteristics, plant water status, cane mass and yield. South African Journal of Enology and Viticulture. 2012;**33**:161-173

[49] Holt HE, Francis IL, Field J, Herderich MJ, Iland PG. Relationships between berry size, berry phenolic composition and wine quality scores for Cabernet Sauvignon (*Vitis vinifera* L.) from different pruning treatments and different vintages. Australian Journal of Grape and Wine Research. 2008;**14**(3):191-202. DOI: 10.1111/j.1755-0238.2008.00019.x

[50] Boulton R. The copigmentation of anthocyanins and its role in the colour of red wine: A critical review. American Journal of Enology and Viticulture. 2001;**52**(2):67-87

[51] Rustioni L, Bedgood DR, Failla O, Prenzler PD, Robarbs K. Copigmentation and anti-copigmenation in grape extracts studied by spectrophotometry and post-column-reaction HPLC. Food Chemistry. 2012;**132**:2194-2201. DOI: 10.1016/j.foodchem.2011.12.058

[52] Dixon RA, Paiva NL. Stress-induced phenylpropanoid metabolism. Plant Cell. 1995;**7**:1085-1097. DOI: 10.1105/tpc.7.7.1085

[53] Grace SC. Phenolics as antioxidants. In: Smirnoff N, Grace SC, editors. Antioxidants and Reactive Oxygen Species in Plants. Oxford: Blackwell; 2005. pp. 141-168

[54] Fortes AM, Agudelo-Romero P, Silva MS, Ali K, Sousa L, Maltese F, Choi YH, Grimplet J, Martinez-Zapater JM, Verpoorte R, et al. Transcript and metabolite analysis in Trincadeira cultivar reveals novel information regarding the dynamics of grape ripening. BMC Plant Biology. 2011;**11**:149. DOI: 10.1186/1471-2229-11-149

[55] Grace SC, Logan BA. Energy dissipation and radical scavenging by the plant phenylpropanoid pathway. Philosophical Transactions of the Royal Society of London. Series B, Biological Sciences. 2000;**355**:1499-1510. DOI: 10.1098/rstb.2000.0710

[56] Apel K, Hirt H. Reactive oxygen species: Metabolism, oxidative stress, and signal transduction. Annual Review of Plant Physiology and Molecular Biology. 2004;**55**:373-399. DOI: 10.1146/annurev.arplant.55.031903.141701

[57] Miller G, Suzuki N, Ciftci-Yilmaz S, Mittler R. Reactive oxygen species homeostasis and signalling during drought and salinity stresses. Plant Cell and Environment. 2010;**33**: 453-467. DOI: 10.1111/j.1365-3040.2009.02041.x

[58] Kawano T. Roles of the reactive oxygen species generating peroxidase reactions in plant defense and growth induction. Plant Cell Reports. 2003;**21**:829-837. DOI: 10.1007/s00299-003-0591-z

[59] Mittler R, Vanderauwera S, Gollery M, van Breusegem F. Reactive oxygen gene network of plants. Trends in Plant Science. 2004;**9**:490-498. DOI: 10.1016/j.tplants.2004.08.009

[60] Jiménez A, Romojaro F, Gómez JM, Llanos MR, Sevilla F. Antioxidant systems and their relationship with the response of pepper fruits to storage at 20°C. Journal of the Science of Food and Agriculture. 2003;**51**:6293-6299. DOI: 10.1021/jf030052i

[61] Gomez del Campo M, Ruiz C, Sotes V, Lissarrague JR. Water consumption, leaf area development, and production in four grapevine varieties. Acta Horticulturae. 2000;**526**:193-200. DOI: 10.17660/ActaHortic.2000.526.18

[62] Allen RG, Pereira LS, Raes D, Smith M. Crop Evapotranspiration. Guidelines for Computing Crop Water Requirements. FAO Irrigation and Drainage Paper 56; Rome; 1998

[63] Singleton VL, Salgues M, Zaya J, Troudsale E. Caftaric acid disappearance and conversion to products of enzymatic oxidation in grape must and wine. American Journal of Enology and Viticulture. 1985;**36**:50-56

[64] Kim D, Jeong SW, Leo CY. Antioxidant capacity of phenolic phytochemicals from various cultivars of plums. Food Chemistry. 2003;**81**:321-326. DOI: 10.1016/S0308-8146(02)00423-5

[65] Gálvez M, Martín-Cordero C, Houghton PJ, Ayuso MJ. Antioxidant activity of methanol extracts obtained from *Plantago* species. Journal of Agricultural and Food Chemistry. 2008;**53**:1927-1933. DOI: 10.1021/jf048076s

[66] Madhava Rao KV, Sresty TVS. Antioxidative parameters in the seedling of pigempea (*Cajanus cajan* L. Milspaugh) in response to Zn and Ni stresses. Plant Science. 2000;**157**:113-128. DOI: 10.1016/S0168-9452(00)00273-9

[67] Bradford MMA. Rapid and sensitive method for quantification of microgram quantities of protein utilizing the principle of protein-dye binding. Analytical Biochemistry. 1976;**72**:248-259

[68] Ngo TT, Lenhoff HM. A sensitive and versatile chromogenic assay for peroxidase and peroxidase-coupled reactions. Analytical Biochemistry. 1980;**195**:389-397. DOI: 10.1016/0003-2697(80)90475-3

[69] Beauchamp C, Fridovich I. Superoxide dismutase: Improved assays and an assay applicable to acrylamide gels. Analytical Biochemistry. 1971;**44**:276-287. DOI: 10.1016/0003-2697(71)90370-8

[70] Thipyapong P, Hunt MD, Steffens JC. Systemic wound induction of potato (*Solanum tuberosum*) polyphenol oxidase. Phytochemistry. 1995;**40**:673-676. DOI: 10.1016/0031-9422(95)00359-F

[71] Gómez-Alonso S, García-Romero E, Hermosin-Gutiérrez I. HPLC analysis of diverse grape and wine phenolics using direct injection and multidetection by DAD and fluorescence. Journal of Food Composition and Analysis. 2007;**20**:618-626. DOI: 10.1016/j.jfca.2007.03.002

[72] Glories Y. La couleur des vins rouges. Mesure, origine et interpretation. Connaissance de la Vigne et du Vin. 1984;**18**:253-271

[73] Tardáguila J, Blanco JA, Poni S, Diago MP. Mechanical yield regulation in wine grapes: Comparison of early defoliation and crop thinning. Australian Journal of Grape and Wine Research. 2012;**18**:344-352. DOI: 10.1111/j.1755-0238.2012.00197.x

[74] Esteban MA, Villanueva MJ, Lisarrague JR. Effect of irrigation on changes in the anthocyanin composition of the skin of cv. Tempranillo (*Vitis vinifera* L.) grape berries during ripening. Journal of the Science of Food and Agriculture. 2001;**81**:409-420. DOI: 10.1002/1097-0010(200103)81:4<409::AID-JSFA830>3.0.CO;2-H

[75] Ojeda H, Andary C, Kraeva E, Carbonneau A, Deloire A. Influence of pre- and postvéraison water deficit on synthesis and concentration of skin phenolic compounds during berry growth of *Vitis vinifera* cv. Shiraz. American Journal of Enology and Viticulture. 2002;**53**:261-267

[76] Kärkönen A, Warinowski T, Teeri TH, Simola LK, Fry SC. On the mechanism of apoplastic H_2O_2 production during lignin formation and elicitation in culture spruce cells-peroxidases after elicitation. Planta. 2009;**203**:553-567. DOI: 10.1007/s00425-009-0968-5

[77] Koundouras S, Hatzidimitriou E, Karamolegkou M, Dimopoulou E, Kallithraka S, Tsialtas JT, Kotseridi Y. Irrigation and rootstock effects on the phenolic concentration and aroma potential of *Vitis vinifera* L. cv. Cabernet Sauvignon grapes. Journal of Agricultural and Food Chemistry. 2009;**57**(17):7805-7813. DOI: 10.1021/jf901063a

[78] Salón JL, Chirivella C, Castel JR. Response of cv. Bobal to timing of deficit irrigation in Requena, Spain: Water relations, yield, and wine quality. American Journal of Enology and Viticulture. 2005;**56**(1):1-8

[79] Casassa LF, Harbertson JF. Extraction, evolution, and sensory impact of phenolic compounds during red wine maceration. Annual Review of Food Science and Technology. 2014;**5**:83-109. DOI: 10.1146/annurev-food-030713-092438

[80] Grimplet J, Wheatley MD, Jouira HB, Deluc LG, Cramer GR, Cushman JC. Proteomic and selected metabolite analysis of grape berry tissues tissues under well-watered and water-deficit stress conditions. Proteomics. 2009;**9**:2503-2528. DOI: 10.1002/pmic.200800158

[81] Matthews MA, Ishii R, Anderson MM, O'Mahony M. Dependence of wine sensory attributes on vine water status. Journal of the Science of Food and Agriculture. 1990;**51**:321-335. DOI: 10.1002/jsfa.2740510305

[82] Gil M, Esteruelas M, González E, Kontoudakis N, Jiménez J, Fort F, Zamora F. Effect of two different treatments for reducing grape yield in *Vitis vinifera* cv Syrah on wine composition and quality: Berry thinning versus cluster thinning. Journal of Agricultural and Food Chemistry. 2013;**61**(20):4968-4978. DOI: 10.1021/jf400722z

[83] Heller W, Forkmann G. Biosynthesis of Flavonoids. The Flavonoids, Advances in Research Since. London: Chapman and Hall; 1996

[84] Degenhardt A, Hofmann S, Knapp H, Winterhalter P. Preparative isolation of anthocyanins by high-speed countercurrent chromatography and application of the color activity concept to red wine. Journal of Agricultural and Food Chemistry. 2000;**48**:5812-5818. DOI: 10.1021/jf0007481

[85] Gomez-Miguez M, Heredia FJ. Effect of the maceration technique on the relationships between anthocyanin composition and objective colour of Syrah wines. Journal of Agricultural and Food Chemistry. 2004;**52**(16):5117-5123. DOI: 10.1021/jf049570z

[86] Liang Z, Yang Y, Cheng L, Zhong GY. Polyphenolic composition and content in the ripe berries of wild Vitis species. Food Chemistry. 2012;**132**(2):730-738. DOI: 10.1016/j.foodchem.2011.11.009

[87] Spayd S, Tarara J, Mee D, Ferguson J. Separation of sunlight and temperature effects on the composition of *Vitis vinifera* L. cv. Merlot berries. American Journal of Enology and Viticulture. 2002;**53**(3):171-182

[88] Baranac JM, Petranovic NA, Dimitric-Markovic JM. Spectrophotometricstudy of anthocyanin copigmentation reactions. Journal of Agriculture and Food Chemistry. 1996;**44**:1333-1336. DOI: 10.1021/jf950420l

[89] Bakowska A, Kucharska AZ, Oszmianski J. The effects of heating, UV irradiation, and storage on stability of the anthocyanin–polyphenol copigment complex. Food Chemistry. 2003;**81**(3):349-355. DOI: 10.1016/S0308-8146(02)00429-6

[90] Gómez-Cordobés C, ML G-SJ, Junquera B, Estrella Y. Correlation between flavonoids and colour in red wines aged in wood. American Journal of Enology and Viticulture. 1995;**46**(3):295-298

[91] Monagas M, Hernández-Ledesma B, Gómez-Cordovés C, Bartolomé B. Commercial dietary ingredients from *Vitis vinifera* L. leaves and grape skins: Antioxidant and chemical characterization. Journal of Agricultural and Food Chemistry. 2006;**54**(2):319-327. DOI: 10.1021/jf051807j

[92] Escudero-Gilete ML, González-Miret ML, Heredia FJ. Implications of blending wines on the relationships between the colour and the anthocyanic composition. Food Research International. 2010;**43**:745-752. DOI: 10.1016/j.foodres.2009.11.004

Section 2

Wine Production and Characterization

Chapter 7

The Evolution of Polyphenols from Grapes to Wines

Violeta-Carolina Niculescu, Nadia Paun and
Roxana-Elena Ionete

Additional information is available at the end of the chapter

http://dx.doi.org/10.5772/intechopen.72800

Abstract

Polyphenols play an important role in the quality of wines, due to their contribution to the wine sensory properties: color, astringency and bitterness. They act as antioxidants, having positive role in human health. They can be divided into non-flavonoid (hydroxybenzoic and hydroxycinnamic acids and stilbenes) and flavonoid compounds (anthocyanins, flavan-3-ols and flavonols). Anthocyanins are responsible for the color of red grapes and wines, hydroxycinnamic and hydroxybenzoic acids act as copigments, stilbenes as antioxidants and the flavan-3-ols are mainly responsible for the astringency, bitterness and structure of wines, being involved also in the color stabilization during aging. This chapter will focus on the chemical structures of the main polyphenols, their identification and quantification in grapes and wines by advanced analytical techniques, highlighting also the maceration and aging impact on the polyphenols evolution. The factors influencing the phenolic accumulation in grapes are also reviewed, emphasizing as well the relationship between phenolic content in grapes versus wine. Polyphenolic changes during the wine making process are highlighted along with the main polyphenol extraction methods and analysis techniques. This research will contribute to the improvement in the knowledge of polyphenols: their presence in grapes, the relationship with wine quality and the influence of the external factors on their evolution.

Keywords: grapes, liquid chromatography, polyphenols, chemistry, wines

1. Introduction

First, a simple question can be addressed: what are phenolic compounds? For a simple question, a simple answer is that they are compounds that have one or more hydroxyl groups attached directly to an aromatic ring. Phenol (**Figure 1**) is the basic structure, the aromatic ring being benzene.

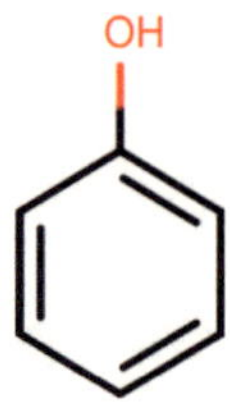

Figure 1. Phenol chemical structure.

Due to the aromatic ring, the hydrogen of the hydroxyl group is labile, which makes phenols weak acids. Polyphenols are compounds that have more than one phenolic hydroxyl group attached to one or more aromatic rings. The term is somewhat misleading since it tends to make people think of polymers of individual phenol molecules.

The general term 'phenolic' covers a very large and diverse group of chemical compounds, and they can be classified in many ways. Phenolic compounds are well connected with several functions in plants like protection against invading pathogens or from UV radiation, pigmentation, or attraction of pollinators and seed dispersers. These compounds were classified into groups based on the number of carbons in the molecule (**Table 1**) [1].

An alternative classification has been used by Swain and Bate-Smith [2]. They grouped the phenols in 'common' and 'less common' categories. Also, phenols were grouped into three families [3]:

1. Widely distributed phenols—ubiquitous to all plants, or of importance in a specific plant
2. Phenols that are less widely distributed—limited number of compounds known
3. Phenolic constituents present as polymers

Structure	Class
C6	Simple phenolics
C6-C1	Phenolics acids and related compounds
C6-C2	Acetophenones and phenylacetic acids
C6-C3	Cinnamic acids, cinnamyl aldehydes, cinnamyl alcohols
C6-C3	Coumarins, isocoumarins, chromones
C15	Chalcones, aurones, dihydrochalcones
C15	Flavans
C15	Flavones
C15	Flavanones
C15	Flavanonols
C15	Anthocyanidins
C15	Anthocyanins
C30	Biflavonyls

Structure	Class
C6-C1-C6, C6-C2-C6	Benzophenones, xanthones, stilbenes
C6, C10, C14	Quinines
C18	Betacyanins
Lignans, neolignans	Dimmers or oligomers
Lignins	Polymers
Tannins	Oligomers or polymers
Phlobaphenes	Polymers

Table 1. Classification of phenolic compounds.

Phenolic compounds play an important role on the sensory characteristics of both grapes and wine because they are responsible for some sensory properties: aroma, color, flavor, bitterness and astringency [4]. The chemical structure enables them to act as antioxidants, scavenging and neutralizing free radicals, and wine polyphenols have been extensively studied in relation to their positive role in human health [5].

2. Grape and wine polyphenols: a chemical perspective

During the grape ripening phase, the physiological and biochemical changes determine the quality of the fruit, the relative proportions of the grape components and wines, being influenced considerably by site (altitude, geological features, soil type, sunlight exposure), climate (cool climate or warm climate), time of harvesting (fungi and bacteria present in grapes), and viticultural practice. The studies about the formation and development of grapes showed that environmental stress factors like heat, drought and light intensity drive to changes in the development of grapes, their chemical composition and phenolic metabolism. The first period of grapes' growth consists mostly of cell division and expansion, followed by a rapid growth phase during which the berry is formed and the seed embryos are produced. In this period, several compounds accumulate in the berry, especially the tartaric and malic acids, conferring the acidity of the future wine. During the first growth period, other compounds accumulate: hydroxycinnamic acids—in the flesh and skin of the grapes; tannins and catechins—in the skin and seed tissues of the grapes; minerals, amino acids and amino acids. The most important changes in grapes' composition occur during the second growth phase (the ripening phase). Grapes switch from small, hard and acidic berries, to larger, softer, sweeter, less acidic, flavored and colored one. The majority of the solutes accumulated during the first growing period remain at harvest. During the second period, the malic acid is metabolized and used as an energy source, its proportion decreasing toward the tartaric acid concentration, which remains almost unchanged. Tannins also decrease considerably after *véraison*.

Winemaking techniques involve the extraction of juice from ripe grapes and fermentation with yeast, changes in polyphenolic composition occurring due to the participation of these compounds in various reactions such as copigmentation, cycloaddition, polymerization and oxidation. The reactions begin after grapes crushing, followed by fermentation and aging, contributing to the sensory properties of wines, mainly color and astringency.

The understanding of the relationship between the quality of a particular wine and its phenolic composition is, at the moment, one of the major challenges in oenology research. For example, the anthocyanin fingerprints of varietal wines are proposed as an analytical tool for authenticity certification [6].

Polyphenolics must be also considered from the taxonomical point of view, knowing that the patterns of some classes of flavonoids are under strict genetic control and that their distribution varies considerably among different grape cultivars [7]. There are several factors with impact on the nonvolatile wine phenolic compounds, including the '*terroir*,' the grape variety and its degree of maturation before harvesting, or the winemaking process with its specific conditions of fermentation or aging [8]. Certain technological procedures such as addition of sulfur dioxide (SO_2) and/or ascorbic acid ($C_6H_8O_6$) prior to crushing the grapes, maceration, yeast strain utilization and alcoholic fermentation, oxidation or adsorption, can also influence the levels of phenolic during the winemaking process [9].

The investigation of phenolic composition in grapes and wine may provide some specific biomarkers that allow us to better assess the chemical evolution of grapes during growth and maturation periods but also to improve the knowledge on wine authentication by developing and implementing new/improved control methods. Phenolic compound identification and quantification fit authentication purpose, protecting consumers against fraud [10].

There is a great chemical diversity in the nonvolatile phenolic composition of grapes and wines, due not only to the varieties of grapes, but also to the fact that they exist in both the free and conjugated forms, as they may be bound to one or more sugar molecules (glucose, galactose, sucrose and mannose) [11]. From the flavonoid compounds, the main phenolics involved in the red color of grapes and wines are the anthocyanins, while the flavan-3-ols are most linked with the astringency, bitterness and structure of wines, but also with the stabilization of color during aging. The nonflavonoid compounds act as copigments (hydroxycinnamic and hydroxybenzoic acids) and antioxidants (stilbenes).

2.1. Phenolic acids

They include two main groups, the benzoic acids, containing seven carbon atoms (C6-C1), and the cinnamic acids, with nine carbon atoms (C6-C3), and they exist mainly as hydroxybenzoic and hydroxycinnamic acids, in either the free or the conjugated form. Various types of hydroxybenzoic acids (HBA) have been identified in both grapes and wines, among them parahydroxybenzoic, protocatechuic, vanillic, gallic and syringic acids (**Figure 2**) [12]. Gallic acid is considered the most important phenolic acid, being the precursor of all hydrolyzable

http://dx.doi.org/10.5772/intechopen.72800

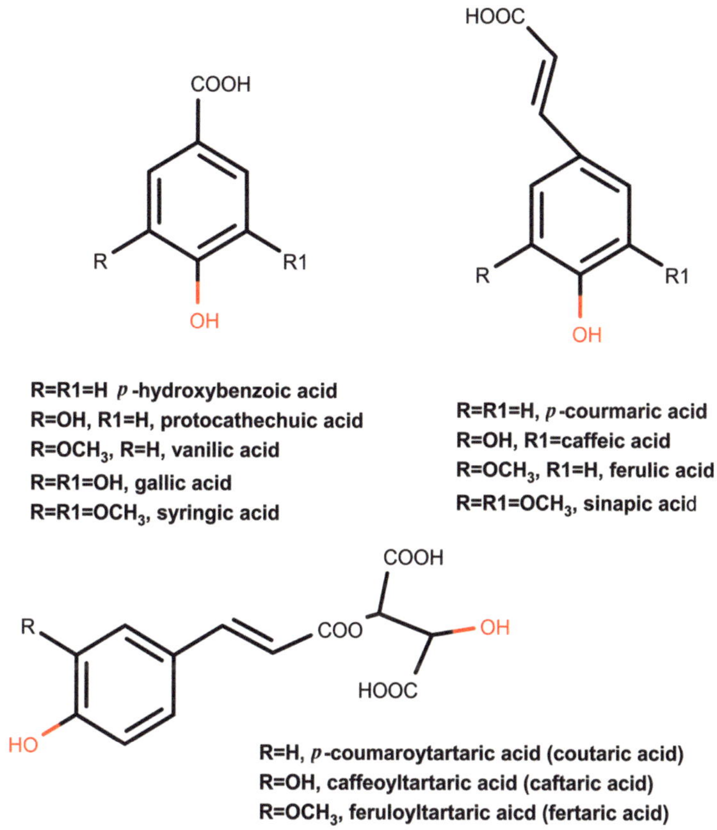

Figure 2. Chemical structures of hydroxybenzoic, hydroxycinnamic and hydroxycinnamoyltartaric acids.

tannins. Hydroxycinnamic acids (HCA) are found in both grapes and wine, the most referenced compounds among them being *p*-coumaric, caffeic, ferulic and sinapic acids (**Figure 2**), which are associated with the browning of wine [13].

Cinnamic acids exist as *cis* or *trans* isomeric forms, which are convertible either enzymatically or through the action of light. Other hydroxycinnamic acids detected in grapes and wines are *p*-coutaric acid, fertaric acid and caftaric acid (**Figure 2**). Caftaric and fertaric acids exist in their *trans* form, mainly localized in the grape pulp and, during the grape pressing, being quickly released into the juice, while a negligible fraction of the *cis* isomer has been found for *p*-coutaric acid. Contrary to this, the *trans* and *cis* isomers of *p*-coutaric acid are less extractable since they are mostly localized in the grape skin, being partially responsible for the astringent properties of both grapes and wines [14]. Although the white wines have a lower concentration of phenolic compounds compared to the red ones, they contain, in turn, a high quantity of caftaric acid. Hydroxycinnamic acids and their tartaric esters constitute the main class of nonflavonoid phenolics in red wines and the main class of phenolic compounds in white wines [14]. During the fermentation process, these esters are partially hydrolyzed, resulting free hydroxycinnamic acids, transformed then into ethyl coumarate and ethyl caffeate [15].

2.2. Flavonoids

They are a chemical class with a basic structure of 15 carbon atoms, including two aromatic rings bound through a three-carbon chain (C6-C3-C6), which is responsible for the chemical. Flavonoids are grouped into several classes, differing from the oxidation degree of the central pyran ring, except of chalcones, including flavanols, flavonols, flavononols, flavones, flavanones, flavanes, anthocyanidins and anthocyanins, chalcones and dihydrochalcones (**Figure 3I**) [12].

In grapes, the highest flavonol concentrations were found at flowering, decreasing as the grapes maturate. Some of the flavonoids present in both grapes and wine are represented in **Figure 3II**. In the flavonoid class, the chemical diversity *versus* complexity is related to the high variety of aglycones and glycosides, as well as to the occurrence of condensation reactions, three distinct categories being usually accepted: glycosylated flavonoids, flavonoid aglycones and anthocyanidin glycosides (anthocyanins) [16].

Flavanols (**Figure 3I**) are benzopyrans with a saturated carbon chain between C2 and C3, a hydroxyl function in C3, and no carbonyl group in C4. The most abundant flavan-3-ols in nature are catechin and its enantiomer epicatechin (**Figure 3II**), and they can be found in the grape skin, seeds and stems, as well as in wine. Some catechin derivatives, such as gallocatechin, epigallocatechin, epicatechin gallate and epigallocatechin gallate, were identified in grapes and wine [17].

Flavonols (**Figure 3I**) have a double bond between C2 and C3 atoms and a hydroxyl group in C3, being often named 3-hydroxyflavones. Approximately 90% of the flavonols are hydroxylated in C3, C5 and C7, being designated as 3,5,7-trihydroxylated derivatives [18]. In grapes, there were identified: quercetin (3′,4′-diOH), kaempferol (4′-OH), myricetin (3′,4′,5′-triOH), isorhamnetin (3′-MeO analog of quercetin), laricitrin (3′-MeO analog of myricetin) and syringetin (3′,5′-MeO analog of myricetin) (**Figure 3II**). In *Vitis vinifera* grapes, there was detected the simultaneous presence of these aglycones [19]. The most abundant condensed flavonoids are the *O*-glucosides, the *O*-sulfates and the derivatives containing acylated sugars and aliphatic or aromatic acid groups in their structure [20]. White wine contains only quercetin, kaempferol and isorhamnetin [20]. Depending on the nature of sugar moiety bound at position C-3, there are three different complete series for flavonol 3-*O*-glycosides in red grapes. The 3-*O*-glucosides are the main derivative of the flavonol aglycones, namely kaempferol, quercetin, isorhamnetin, myricetin, laricitrin and syringetin, while the minor ones are the 3-*O*-galactoside derivatives.

Flavononols (**Figure 3I**) are characterized by the presence of a hydroxyl group in the C3 position and the absence of a double bond in the heterocyclic ring, being also named 3-hydroxyflavonones or dihydroflavonols (e.g., taxifolin, astilbin and dihydromyricetin 3-*O*-rhamnoside) [16].

Flavones (**Figure 3I**) have a double bond between carbons C2 and C3, and the hydroxyl group is absent in the C3 position. Although they are widely encountered in plants, as aglycones or glycosides, they are not present in grapes in significant amounts, except for luteolin (**Figure 3II**) [8].

Flavanones (**Figure 3I**) have a saturated carbon chain between atoms C2 and C3, being often named dihydroflavones (e.g., eriodictyol, a flavanone that has been extracted from grapes) [21].

http://dx.doi.org/10.5772/intechopen.72800

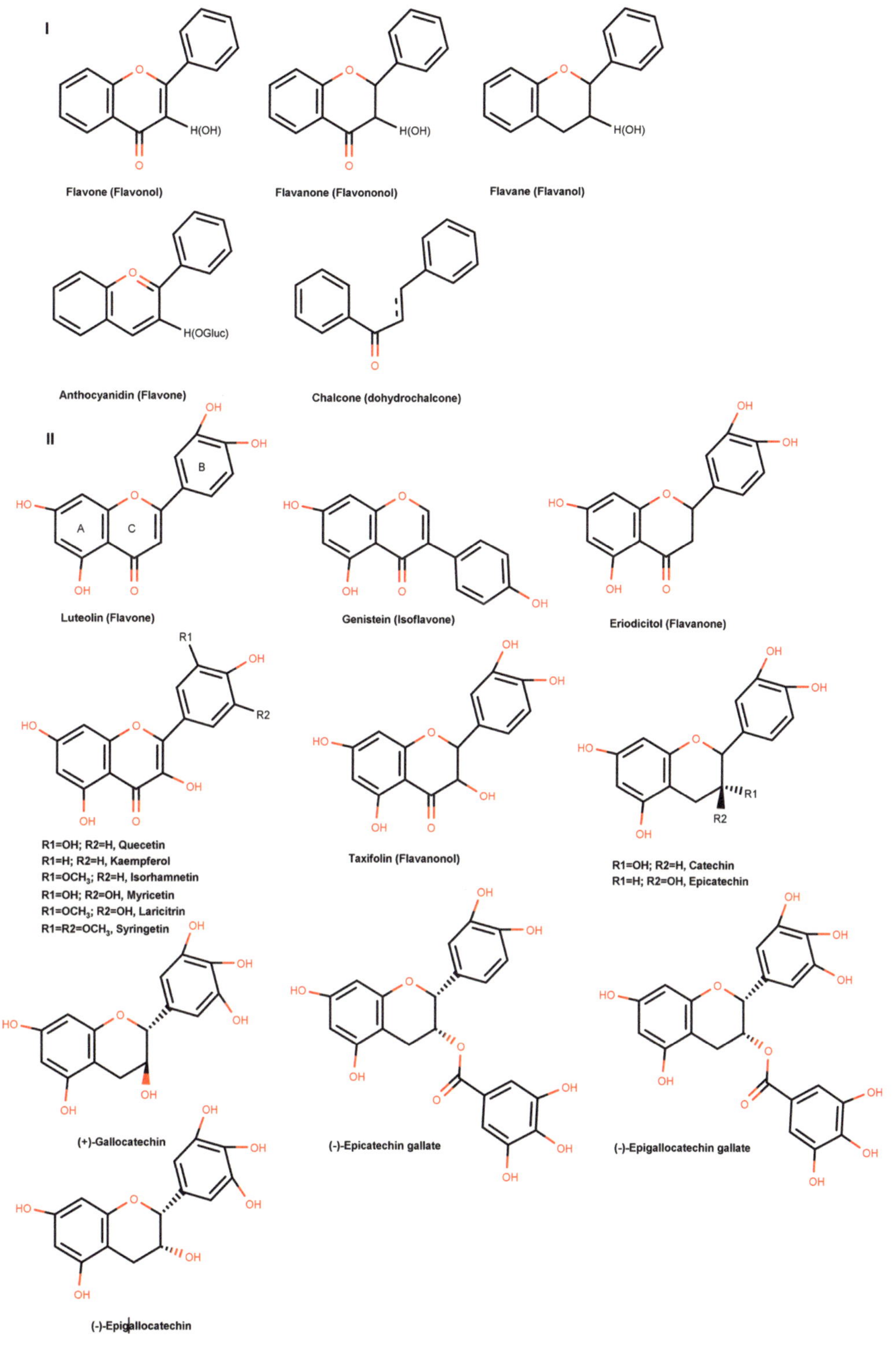

Figure 3. General chemical structure of flavonoid subgroups.

Flavanes (**Figure 3I**) contain a saturated carbon chain between C2 and C3, and no carbonyl group in C4 [22].

Chalcones have two aromatic rings linked by a carbonylic α,β-insaturated system (**Figure 3I**). Dihydrochalcones are obtained from chalcones through a reduction process [23].

Different families of anthocyanin derivatives have been reported in grapes, wine and wine-like solutions. Anthocyanidins and anthocyanins (**Figure 3I**) have as nucleus the 2-phenyl-benzopyrylium (the flavylium) cation. Anthocyanins are anthocyanidin glycosides. They are responsible for the color of grapes and wines. In red grapes and wines, six anthocyanidins have been detected: cyanidin (orange red), peonidin (red), delphinidin (bluish red), pelargonidin (orange), petunidin and malvidin (bluish red) [24], the last one being the most representative compound in *V. vinifera* [25, 26]. Also, the 3-*O*-monoglucosides of the mentioned six anthocyanidins have already been detected in grapes [24, 25, 27]. The anthocyanins are mainly found in the skin of grapes, while the other flavonoids occur in both skin and seeds. The amount of anthocyanins and other flavonoids extracted during vinification depends on the duration of the process, the temperature or the extent of disruption of the grapes [16]. In the grape vines, anthocyanins accumulate in the leaves during senescence and they determine the coloration of the grape skin in red and rose cultivars. The anthocyanidin composition of grapes is affected by factors such as the grape-growing origin (*terroir*) and varieties, the degree of maturity and the weather conditions. The profiles of anthocyanins for each grape variety are relatively stable, while absolute concentrations can vary widely between different vintages due to environmental and agronomical factors [11].

Stilbenes have two aromatic rings linked by an ethene bridge, resveratrol (3,5,4′-trihydroxystilbene, **Figure 4**) being the most important in grapes and wine. Also, it was identified in vine leaf and in the grapes skin, and its concentration is known to decrease significantly upon grape maturation [28]. Due to the presence of resveratrol in the grapes skin, the wine-processing methods determine its concentration in the final product. Wines from grapes with longer maturation periods have an increased content of resveratrol, its concentration being higher in red wines compared to white wines [28]. The stilbenes in wines can contribute to the correlation of their profiles with winemaking procedures and with the grape wine varieties.

Tannins are polyphenols with astringent properties, able to cause protein precipitation. They are usually divided in two classes: the hydrolyzable and the nonhydrolyzable or condensed tannins [22]. Most of the natural tannins present in grapes and wine are the condensed type. In young wines, tannins occur mainly in the form of dimers or trimers. However, their concentration decreases with aging due to oxidation and precipitation processes [29]. Hydrolyzable tannins can be degraded through pH changes as well as by enzymatic or nonenzymatic hydrolysis into smaller fragments, mainly sugars and phenolic acids. Their basic unit is gallic acid (**Figure 2**) and its derivatives (e.g., ellagic acid). These acids are usually esterified with D-glucose, resulting 500–2800 molecular weight species. Aging in oak barrels promotes the extraction of low molecular weight phenolic compounds, mainly from ellagitannins, into wine. For example, ellagic acid and myricetin are probably the major phenolic compounds present in muscadine grapes (*Vitis rotundifolia)* [30]. Condensed tannins

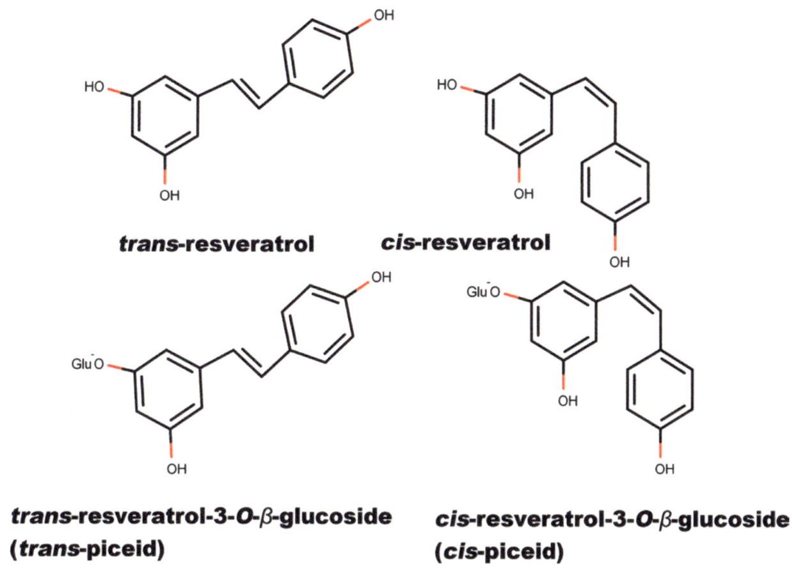

Figure 4. Chemical structures of resveratrol and resveratrol glycoside.

(proanthocyanidins) are polymeric compounds that are transformed into anthocyanidins, and they can be found in residual amounts of the solid grape components (e.g., seeds, skin) or in the pulp [27]. They are transferred into the must during winemaking operations such as crushing, maceration and fermentation [27]. Proanthocyanidins generally occur as oligomers and polymers of flavan-3-ols. The proportion of the polymer units and the average number of proanthocyanidins have been highlighted by LC–MS/MS and thiolytic degradation [31]. *V. vinifera* grapes contain mainly proanthocyanidins as oligomers and polymers of (+)-catechin and (−)-epicatechin linked through C4/C8 bonds [32]. The procyanidins and prodelphinidins, leading to cyanidin and delphinidin, are the most abundant condensed tannins in grapes and wine [32]. The relationship between monomeric, oligomeric and flavan-3-ol composition has been intensively studied as a prerequisite for improving wine quality, being showed that flavan-3-ols and their monomers ((+)-catechin and (−)-epicatechin) as predominant phenolic compounds extracted from grape seeds have a highly positive correlation with antioxidant activity. In *V. vinifera*, oligomers with a maximum degree of polymerization of 16 have been identified. The identification and quantification of proanthocyanidins is an important subject, since these compounds are mainly responsible for the sensory characteristics of wine and they play an important role in the wine aging process.

3. Polyphenolic changes in the winemaking and polyphenol's extraction techniques

The total extractable phenolic content in grapes is encountered in the pulp (about 10% or less), in the seeds (60–70%) and in the skin (28–35%). In the seeds, the phenolic content may range between 5% and 8%, by weight [33]. There are a great number of methods for polyphenol extraction from grapes. **Table 2** summarizes the most applied extraction techniques.

The main phenolic compounds in musts from white grapes are hydroxycinnamic tartaric acid esters (HCTA). Catechins and proanthocyanidins are found mainly in the skins of white grapes. They can sustain oxidation; as a consequence, the grapes must often be protected by limiting the contact with oxygen and inhibiting polyphenoloxidase (PPO) by SO_2 addition. Due to the oxidation, the phenolic profile of the musts is characterized from lower levels of HCTA and flavanols, and high 2-S-glutationyl caffeyl tartaric acid (GRP). Red wines are produced by the must fermentation in the presence of the grape skins and seeds. During the process, phenolic compounds such as anthocyanins are subjected to various reactions, such as enzymatic oxidation, electrophilic substitution, degradation of anthocyanins, cyclo-accession of the carbonyl compounds to anthocyanins and formation of vitisins (A, B and C), hydrolysis of proanthocyanidins and formation of carbocation in C4 position which attacks the positions C6 and C8 of the proanthocyanidins and anthocyanins, attack of C4 carbocation of anthocyanin to C6 or C8 of proanthocyanidins, attack of acetaldehyde to C6 and C8 of proanthocyanidins and formation of structures with flavanol linked to anthocyanin by ethyl bridge [33].

Polyphenol extraction from grapes can be achieved by using alcoholic or hydroalcoholic solutions either containing, or not containing, mineral acids, or using acetone/water solutions. The use of a mineral acid (usually methanol containing HCl 1% by volume) allows extraction of all phenolic compounds. The use of nonalcoholic acid containing solutions does not guarantee satisfying extraction of HCTA. Grapes' freezing after the harvest, followed by heating at room temperature before performing extraction, allows PPO to be in touch with and to oxidize the substrates [34].

A satisfying extraction method for all polyphenols is the use of a model wine solution containing: tartrate buffer pH 3.2 (with 12% ethanol, v/v) and sodium metabisulfite 2 g/L. This can be prepared by adding 5 g tartaric acid, 22.2 mL 1 M NaOH solution and 2 g of $Na_2S_2O_5$ to 500 mL

Sample preparation	Extraction	Clean-up isolation	Instrumental analysis
Freeze-drying Air drying Centrifugation Filtration	Supercritical fluid extraction (SFE)/solid phase extraction (SPE)/solid phase micro-extraction (SPME) Soxhlet Solid-liquid Liquid-liquid Accelerated solvent extraction/ pressurized liquid extraction (ASE/PLE) Microwave Matrix solid phase dispersion (MSPD) Sonication	Column chromatography/acidity-based fragmentation: SPE Thin-layer chromatography (TLC), MSPD High-speed countercurrent chromatography (HSCCC)	Liquid Chromatography (LC): UV, FLUD, ECD, NMR, MS, MS^n CE Gas Chromatography (GC): FID, ECD, MS, MS^n

Table 2. Polyphenol extraction methods.

ultrapure water, stirring continuously, then adding 125 mL ethanol and adjusting the final volume to 1 L with ultrapure water. Often, extracts have to be concentrated by solid phase extraction (SPE) before performing analysis by high-performance liquid chromatography (HPLC) or spectrophotometry. As soon as the grapes were harvested, the seeds and skins of 50 berries must be separated from the others parts of the berry, immersed in two solutions of 50 mL tartrate buffer solutions containing 1 g/L SO_2 and immediately frozen. Before analysis, the samples must be heated at room temperature and kept for 4 h; the extracts must be then homogenized and centrifuged; the supernatant transferred in a 100-mL volumetric flask, pellets being resuspended twice in 15–20 mL tartrate buffer; after centrifugation, the liquid phases can be collected in the same volumetric flask; the final volume must be adjusted to 100 mL with tartrate buffer [33].

When using acetone/water solutions for the extraction, skins from 300 berries are rinsed with ultrapure water, after separation from the flashes, immersed in 100 mL of aqueous solution containing 100 mg/L of SO_2 and stored at 4°C. In order to perform the extraction, the sample must be laid into a 500-mL Erlenmeyer flask, over which 200 mL of acetone is added. Next, the flask is covered and placed for 24 h on a shaker, at 20°C and 100 rpm. After this procedure, the acetone is removed at 40°C, under vacuum; the solution is centrifuged, and the supernatant is recovered and diluted with ultrapure water, at 100 mL, and finally stored at −20°C until analysis [35].

Another method for extracting polyphenols from seeds and skins is by immersion of the fine powder (previously obtained by grinding the grape skins and seeds in liquid nitrogen) in an acetone/water 7:3 (v/v) solution containing 0.1% ascorbic acid added to prevent oxidation [36].

Other methods use methanol for extraction: 250 mg skins must be stored at −80°C, then grounded with mortar and pestle in liquid nitrogen and soaked in 30 mL methanol at 4°C; the extract is then concentrated to almost dryness and dissolved in 1 mL methanol [37].

Ethanol/water/chloroform solutions can be also used for phenolic compound extraction; the skins and seeds are freeze-dried until obtaining 1 g (dry weight) and 0.5 g (dry weight), respectively, and subjected to an extraction procedure using 20 mL ethanol/water/chloroform 1:1:2 (v/v/v) for 2 min. The 50% upper layer of ethanol containing the phenolic compounds is separated by the chloroform lower layer which contains undesirable compounds like chlorophylls and lipids; the ethanol is removed under vacuum; the resulted aqueous solution is then filtered through a glass membrane, diluted with water up to 100 mL and stored at −20°C until analysis [38].

Another method proposes that 1 g of freeze-dried grape seeds and skins is to be extracted, using a blender, with 5 × 20 mL with ethanol/water 1:1 (v/v). Subsequently, the obtained extracts are adjusted to a final volume of 100 mL and pH 4.0 and then purified and fractionated on a 250 × 5 mm column, type Toyopearl Gel HW-40(s), using as eluent the methanol [39].

4. Advanced techniques for polyphenol analysis

Monomer phenols in grapes and wines are usually analyzed by high-performance liquid chromatography (HPLC) using a reverse phase C18 column (usually 250 × 4 mm, 5 μm) operating close to or at room temperature.

For example, in order to obtain the anthocyanin monomer profile of grapes, a Cabernet Sauvignon grape skin extract was used, the chromatogram being displayed in **Figure 5**. A column type C18 (250 × 4 mm, 5 μm), at 45°C, using a binary solvent mixture of formic acid/water 10:90 (v/v) and formic acid/methanol/water 10:50:40 (v/v/v) and detection at 520 nm wavelength were used for the analysis. For compound elution from the column, a gradient program was applied. The sample volume injected is 20 μL [33].

Peaks of some typical anthocyanins for *V. vinifera* varieties, namely 3-glucosides, 3-glucoside acetates, 3-glucoside para-coumarates and malvidin 3-glucoside caffeate, are highlighted in the **Figure 5**. Even if different separation column is used, the sequence for compound elution from a reverse phase column is due to their polarity and is always the same. The values of single anthocyanin concentration can be expressed as area percentage on the total area amount of the peaks of all identified anthocyanins or in mg/kg grapes of malvidin-3-glucoside (compound used as external standard).

HPLC with UV-Vis detection makes possible the analysis of some anthocyanin derivatives in wines, such as vitisins, several pyranoanthocyanins and flavanol-ethyl-anthocyanin derivatives. Due to the lack of commercially available standards, identification and quantification of these compounds is difficult. Usually, the column is a reverse-phase polystyrene divinylbenzene (250 × 4.6 mm, 5 μm) operating at 30°C, elution being performed by a binary mobile phase mixture composed of aqueous H_3PO_4 1.5% (w/w) and aqueous H_3PO_4/acetonitrile 20:80 (v/v) with a gradient program, at 520 nm, the compounds being quantified as malvidin-3-glucoside. Analysis can be performed by direct injection of the sample or, to increase the sensibility, after concentration on a C18 cartridge [40].

HCTA and flavonols can be simultaneously detected in a single run using a C18 (250 × 4 mm, 5 μm) column operating at 40°C, performing elution with a binary solvent mixture composed of H_3PO_4 10^{-3} M and methanol, and an elution gradient program. **Figure 6** shows a typical chromatogram for flavonols, respectively, recorded in the analysis of a Cabernet Sauvignon grape skin extract. Detection wavelength was 360 nm for flavonols (sample volume injected 25 μL). Identification of compounds is performed on the elution order from the column and by recording UV-Vis spectra on the basis of maximum wavelengths [33].

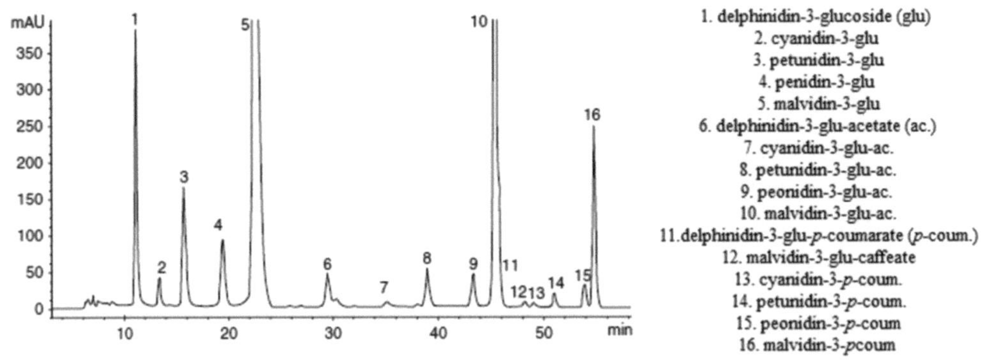

Figure 5. HPLC anthocyanin profile of Cabernet Sauvignon grape skin extract.

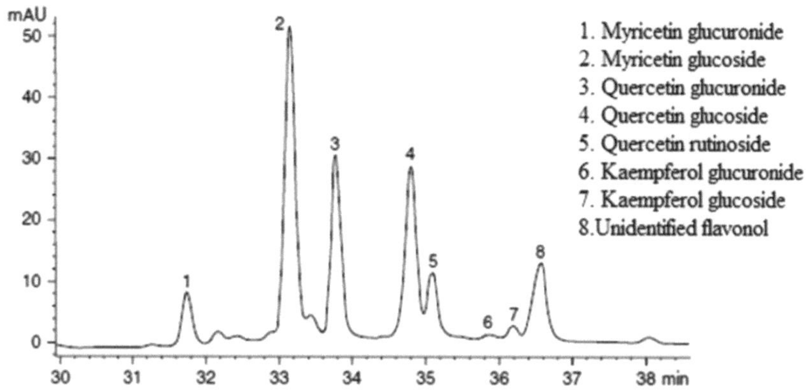

Figure 6. HPLC profile of flavonols of a Cabernet Sauvignon grape skin extract recorded at 360 nm [33].

Due to the lack of commercially available standards for HCTA, chlorogenic acid or free hydroxycinnamic acids are used as external standards for quantitative analysis. Quercetin and myricetin glucoside and other flavonols glycoside (e.g., rutin) are standards of flavonols commercially available.

For the HCTA and flavonols in wines, analysis can be performed on the acidified sample prepared by the addition of 0.5 mL 1 M H_3PO_4 to 4.5 mL of wine and filtration on 0.45 μm membrane. The HPLC conditions used are the same reported for skin extract.

The typical HPLC chromatograms of grape skins and red wines for the identification of *trans*-resveratrol in grape skins and red wines are presented in **Figures 7** and **8**, respectively. *Trans*-resveratrol was identified by the comparison with a commercial standard of *trans*-resveratrol. An Aquasil C18 (250 × 4.6 mm, 5 μm) analytical column was used for separation. Chromatographic conditions included a mixed mobile phase of water:acetonitrile:acetic acid

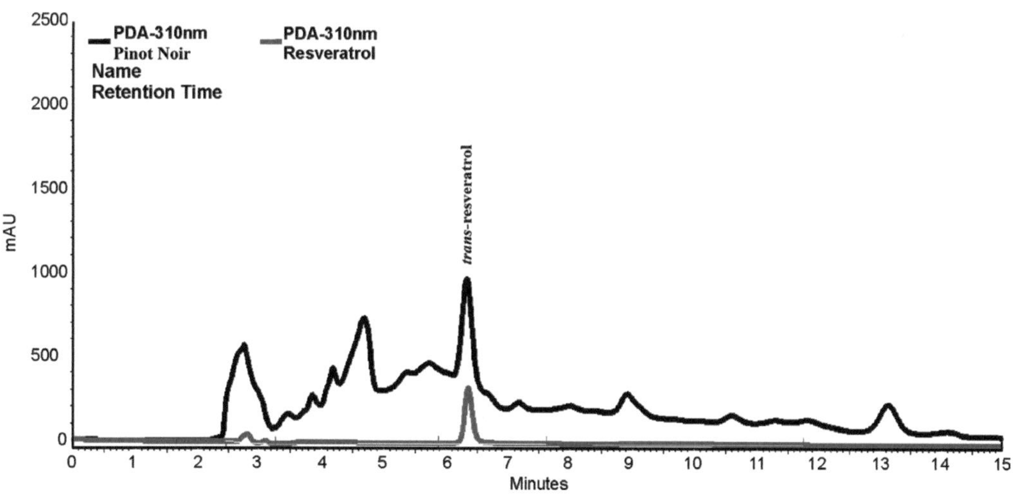

Figure 7. HPLC chromatogram of grape skin (*V. vinifera*, Pinot Noir) extract and standard [28].

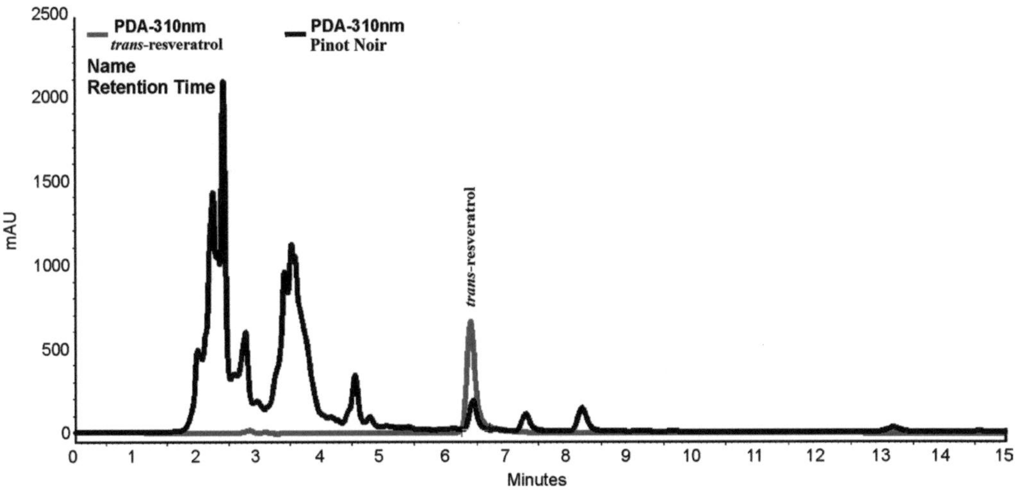

Figure 8. HPLC chromatogram of red wine (*V. vinifera*, Pinot Noir) and standard [28].

of 70:29.9:0.1, flowing through the system at the rate of 1 mL/min; the injection volume was 20 μL; the detection was set at 310 nm [28].

In order to evaluate the difference in phenolic composition of grape skins, 35 samples of five red grape varieties (*V. vinifera L.*), namely Pinot Noir, Merlot, Cabernet Sauvignon, Feteasca Neagra and Mamaia cultivated in Southeastern Romania, were investigated by reversed-phase high-performance liquid chromatography (RP-HPLC) using a system with: a diode array detector set at 280 nm; the mobile phase—water with 0.1% formic acid and acetonitrile with 0.1% formic acid; and an Accuacore PFP (100 × 2.1 mm, 2.6 μm) column, operated at 30°C. The concentrations of phenolic compounds in the extracts were calculated as mg/L using external calibration curves generated from standards, which were obtained for each phenolic compound (**Figure 9** and **Table 3**) [41].

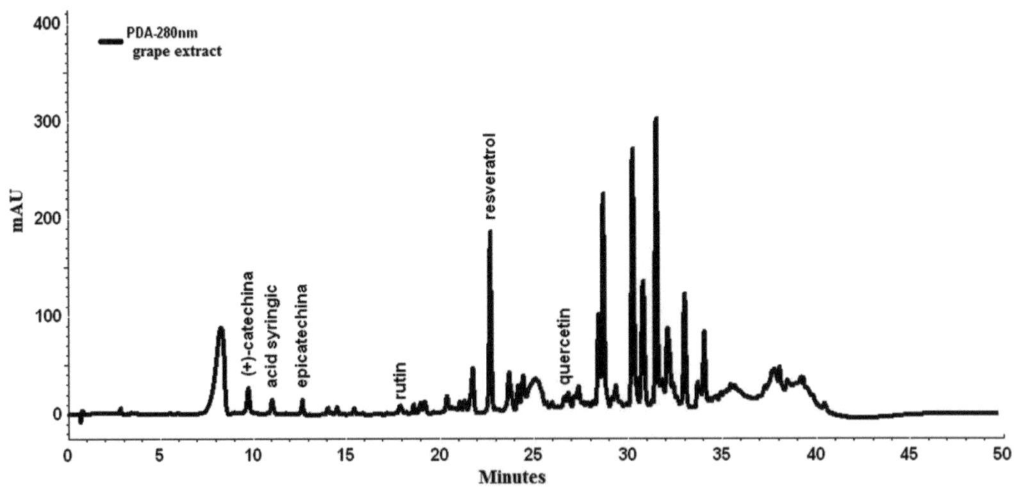

Figure 9. HPLC chromatogram of Cabernet Sauvignon grape skin extract [41].

Grape cultivar		Phenolic compound in grape skins				
		Syringic acid	Catechin	Epicatechin	Rutin	Trans-resveratrol
Cabernet Sauvignon	Average	6.1	44.2	20.3	30.0	70.6
Merlot	Average	11.5	52.6	41.5	63.5	151.6
Feteasca Neagra	Average	1.1	24.4	18.3	15.2	112.7
Mamaia	Average	3.4	14.4	13.6	18.3	18.1
Pinot Noir	Average	Not detected	28.1	38.1	88.2	203.6

Table 3. Phenolic compound in grape skins.

The phenolic profile in grape skins of red grape varieties grown in Southeastern Romania in 2013 season, under the same agronomic conditions and microclimate, showed significant differences. Each variety had a different ripening trend. The ripening influenced the phenolic composition of red grape skins, higher amounts of phenolic compounds being found on the last sampling week for Cabernet Sauvignon, Feteasca Neagra and Merlot grape cultivars, while phenolic maturity was reached for Mamaia and Pinot Noir. This observation must be taken into consideration in winemaking process, in order to obtain wines with high phenolic content [41].

Low molecular weight and volatile phenols are usually identified and quantified by Gas Chromatography coupled with Mass Spectrometry (GC-MS). For analysis and structural characterization of more polar compounds such as polyphenols, Liquid Chromatography Mass Spectrometry (LC-MS) and Multiple Mass Spectrometry (MS/MS and MS^n) techniques are used [42]. These methods are the most effective techniques to characterize individual polyphenols in grape extracts and wine, due to the soft ionization conditions and minor sample purification required. Opposed to the LC methods coupled with spectrophotometric detection which require hydrolysis or thiolysis for compounds identification, the LC-MS is a more reliable and advanced technique to be applied in the study of compound structure linked with the color-changing of red wines during aging formed by reactions of anthocyanins with other compounds. This technique also allows the characterization of proanthocyanidins (procyanidins, prodelphinidins) also called condensed tannins. Moreover, for the particular case of studying compounds, the MS/MS approach is a very powerful tool that allows characterization of aglycone and sugar moiety.

In order to analyze the grape flavonols, the extraction of berries with acidified methanol (0.01% 12 N HCl) must be performed. The extract is then filtered, the solvent removed under vacuum and the residue dissolved in 0.1 M citric acid buffer at pH 3.5. Polyphenolics are fractionated on the basis of their affinity to a C18 cartridge and then on a Sephadex LH-20 cartridge, the fractions being then eluted by ethyl acetate and methanol. To perform analysis of the isolate, ethyl acetate is evaporated and the residue is redissolved in the pH 3.5 buffer. The isolates can be analyzed by Liquid-Chromatography-Electrospray-Ionization-Mass Spectrometry. By connecting the LC/ESI-MS system to the probe of the mass spectrometer via the UV cell outlet, LC-UV chromatograms and spectra can also be recorded. Compounds such as vanillin and phenolic acid present in wine can be detected by LC/ESI-MS analysis [43].

To achieve structural and semiquantitative information on the anthocyanins from grapes, ESI-MS/MS direct-injection analysis of extract can be performed. An example of an ESI-MS spectrum is reported in **Figure 10**, the peaks being identified and quantified in **Table 4**.

Quantification by ESI/MS of anthocyanins present in grape extract can be done by constructing a calibration curve using two commercially available standard compounds, the Mv-3-*O*-glucoside (M+ species at m/z 493) for monoglucosides and the Mv-3,5-*O*-diglucoside (M+ species at m/z 655) for diglucosides [44].

Operating in the positive ionization mode, the ESI is also effective for analysis of flavan-3-ols. But since most phenolic acids in wine are not detectable in this mode, the negative ionization mode is preferable to work with (see an example in **Figure 11**) [43].

By operating with a cone voltage of 60 V, these compounds show high formation of the $[M\text{–}H]^-$ ion. The sample under investigation can be prepared by the liquid-liquid extraction of 50 mL wine, using diethyl ether (3 × 5 mL) and ethyl acetate (3 × 15 mL), after previous concentrating to 15 mL, at 30°C, under vacuum, to eliminate the ethanol. Next, the organic phases are combined and the resulting solution dried over Na_2SO_4. The solvent is removed under vacuum, then the residue is dissolved in 2 mL of methanol/water (1:1) and the solution is filtrated prior to analysis [45]. The compounds identified by LC-ESI-MS are reported in **Table 5** [43].

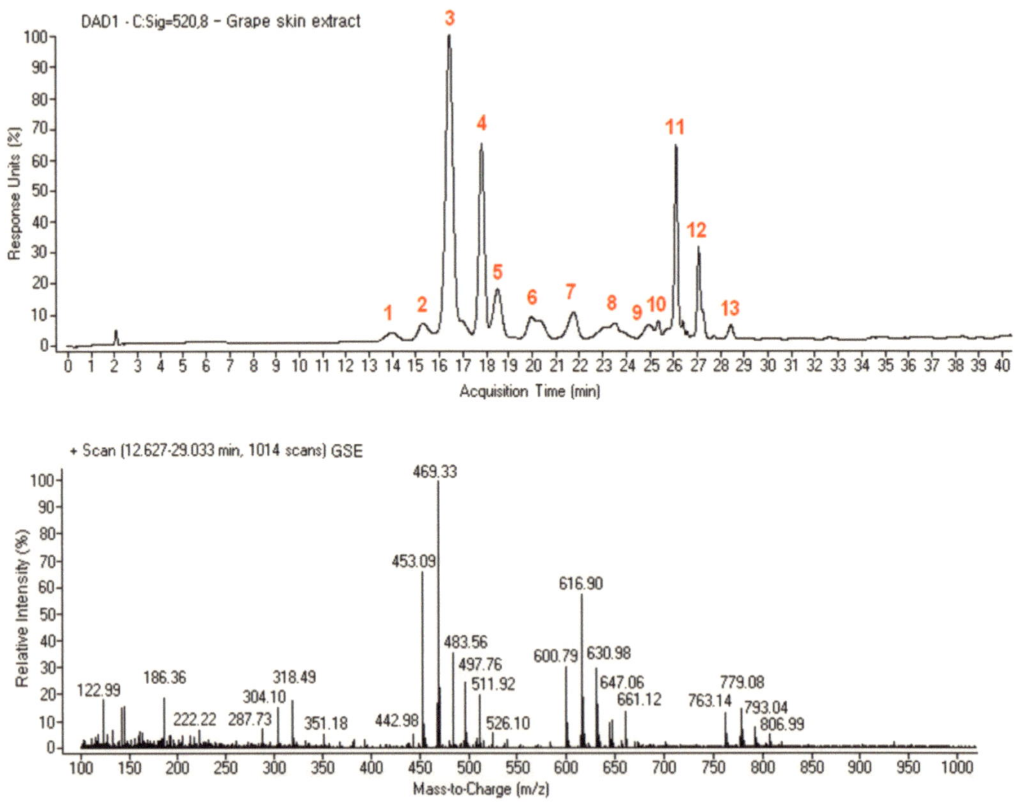

Figure 10. LC-ESI-MS anthocyanin profile of Othello grape skin extract.

http://dx.doi.org/10.5772/intechopen.72800

Peak number	RT (min)	m/z	Mass assignment (M^+ sau $[M + H]^+$)	Anthocyanins content (%)
1	13.9	633	Malvidin-Delphinidin (oligomeric form)	1.87
2	15.1	617	Malvidin-Cyanidin (oligomeric form)	1.87
3	16.2	469/303	Delphinidin-3-*O*-glucoside	40.64
4	17.6	453/287	Cyanidin-3-*O*-glucoside	17.82
5	18.2	453/287	Cyanidin-3-*O*-galactoside	5.75
6	19.7	467/303	Delphinidin-3-*O*-glucoside	5.44
7	21.5	511/303	Delphinidin-3-(6-*O*-acetylglucoside)	4.25
8	23.2	779/287	Malvidin-Cyanidin + glucoside	4.75
		495	Malvidin-3-*O*-glucoside	
9	24.7	763	Cyanidin-3-glucoside-ethyl-catechin	1.16
10	25.1	763	Cyanidin-3-glucoside-ethyl-catechin	0.53
		617	Malvidin-Cyanidin (oligomeric form)	
11	25.8	807	Malvidin-3-glucoside-8-vinyl(epi)catechin	9.71
		617	Malvidin-Cyanidin (oligomeric form)	
12	26.8	601	Malvidin-3-(6-*O*-acetylglucoside) pyruvate	6.01
13	28.1	645	Malvidin-Petunidin (oligomeric form)	0.16

Table 4. Peak assignments, retention time, mass special data, and percentage of grape skin anthocyanins detected by LC and electrospray ionization-MS.

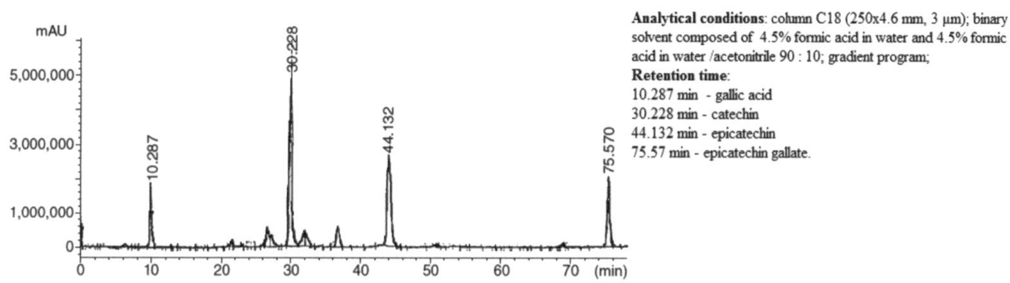

Figure 11. LC-ESI-MS of flavan-3-ols from a wine sample analysis performed in negative ionization mode.

MW	Compound	Main ions observed (m/z)	
		Fragm. 60 V	Fragm. 120 V
154	Protocatechuic acid	153 (109)	109 (153)
138	Protocatechuic aldehyde	137	137 (108)
138	*p*-Hydroxibenzoic acid	137 (93)	93 (137)
122	*p*-Hydroxibenzoic aldehyde	121	121 (92)
168	Vanillic acid	167	108 (167, 123, 91)
152	Vanillin	151 (136)	136 (151)

MW	Compound	Main ions observed (m/z)	
		Fragm. 60 V	Fragm. 120 V
194	Ferulic acid	193 (134, 149)	134 (193)
182	Syringic aldehyde	181 (166)	181 (166, 151)
164	*p*-Coumaric acid	147 (103)	147 (103)
180	Caffeic acid	179 (135)	135 (179)
178	Esculetin	177	177 (133)

Table 5. Some wine polyphenols determined by negative ionization mode.

5. Factors influencing phenolic content and composition of wine

Wine phenolic profile depends on the grapes and on the vinification technique, although other variables such as cultivar, viticulture practices can also affect it. The variety, growing season, environmental and climatic conditions, soil type and even maturity influence the concentration of phenolic compounds within grapes. The qualitative changes in the phenolic composition of the final wine, compared to the composition of the corresponding grapes, are mainly due to the production of new derivatives (such as flavenes, tyrosol and free phenolic acids—gallic, paracoumaric, caffeic, and so on). The environment in which both the must and the wine are produced can contribute to the generation of new phenolics [21]. Since wine composition is in constant evolution, the aging in bottles also seems to contribute to changes in the flavonol content, through the interaction of flavonols with other constituents [3]. It was found that the effect of vintage was significantly more pronounced on the anthocyanins than on the flavonols [46].

The total phenolic content in grapes is clearly affected by several factors: the cultivar, the year of production, the geographic origin of grapes, soil chemistry and the degree of maturation, many studies focusing on defining the effects of growing conditions on grape and wine phenolic composition and of the impact of light and temperature conditions on berry flavonoid composition [35]. The results revealed a combined effect of solar radiation and temperature on the compositional profile of flavonoids. It was observed that moderate exposure and temperature were favorable for the accumulation of anthocyanins, while an enhancement of skin tannins was noted in the grape berries from bunches exposed to sunlight versus those from shaded bunches or dense canopies. In fact, maturation is associated with polymerization of phenols, which leads to a marked decrease of astringency [16].

The phenolic composition, the relative proportions of anthocyanins and flavonols, changes mainly in the first steps of vinification and continues during storage, the applied techniques (maceration, fermentation, clarification, aging, etc.) influencing significantly both the concentration and the composition of phenolics and, therefore, also the color intensity and hue of red wines. Extraction of the beneficial components from the grape skins can occur before fermentation, during it, or after it. Fermentation can be delayed, if the must and the skins are kept cool enough, and maceration will occur in an aqueous medium. A key decision is when to separate wine and skins after fermentation has finished. After the wine has been separated from the skins, a pulp containing a mix of juice, skins and pips remains. This mixture is introduced in a press to squeeze out the remaining juice. The applied force and the type of the press determine the quality of the wine. This juice may be finished off separately from the rest of the wine or blended back into it. By pressing too hard, high concentrations of bitter compounds from the skins and seeds can be extracted, affecting the wine quality.

The phenolic composition of the wine also changes along the wine aging process, reflecting in the color and astringency degree of the final product. The relative anthocyanin content decreases upon aging although this chemical modification is associated with a very clear change in color, this characteristic being often used as a quality standard for aged wines. One of the main factors responsible for anthocyanin loss is the storage temperature [24].

The stability of the phenolics present in the wine is different from the one in situ (nonharvested grapes), several chemical changes, beginning in the grapes, and reaching completion only after the processing period. In general, the chemical composition of the final product is much more complex than the one of the raw material, due to the formation of a variety of new compounds.

Although procyanidins are related to grape composition, in wine, they were found to evolve during wine aging. In fact, they show a remarkable stability and resistance to several winemaking processes, such as sulfite bleaching. Oxygenated wines display characteristic color changes, along with a significant increase in the concentration of pyranoanthocyanins and related pigments [29]. In summary, one can say that a diversity of products can be obtained from condensation reactions between anthocyanins and tannins.

6. Conclusions

A main conclusion can be displayed: the resulted wine is a very complex medium influenced by winemaking techniques applied to different batches of grapes. The anthocyanin content of grapes at harvest is of vital importance in achieving quality wines. Tannins are the largest contributor, their content and composition in grapes probably determining the final wine quality. Presently, tannins are not routinely measured, the total phenolic content being determined and linked to the tannin content of grapes and wine, although it is not a specific assay. The wine phenolics can be used as a fingerprint for their differentiation, according to the geographical origin, vine variety and vintage. Phenolics compounds can be used as markers to discriminate the origin of wine (e.g., geographical, vine variety and vintage). For selection of winemaking practices aimed to improve the wine quality, the knowledge of intrinsic physicochemical characteristics of the grape cultivar employed in the vinification is required, but also the influence of different technological procedure to the final product, the wine. Since there is an increasing demand for producing and protecting high-quality wines, extensive studies on the factors influencing the chemical composition and biological effects of wine are still a necessity.

The use of advanced analytical techniques leads to the identification of some structures derived from tannins and anthocyanins in wine and determined how they are formed, the diversity of methods and experimental procedures reflecting the complexity of phenolics in grapes and wine. Improvement is still necessary, since each species is present in very small amounts and too many unidentified compounds still remain, especially with the polymeric fraction. Also, various extraction and analysis techniques have been developed as an alternative to conventional procedures, offering advantages with respect to analysis time, solvent consumption, extraction yields and reproducibility, each method providing specific advantages. Currently, the best approach to properly characterize the phenolic composition of the grapes and wines is using some assay combinations.

Author details

Violeta-Carolina Niculescu*, Nadia Paun and Roxana-Elena Ionete

*Address all correspondence to: violeta.niculescu@icsi.ro

National Research and Development Institute for Cryogenics and Isotopic Technologies—ICSI Ramnicu Valcea, Ramnicu Valcea, Romania

References

[1] Harborne JB, Simmonds NW. The natural distribution of the phenolic aglycones. In: Harbone JB, editor. Biochemistry of Phenolic Compounds. 1st ed. New York: Academic Press; 1964

[2] Swain T, Bate-Smith EC. Flavonoid compounds. In: Florkin M, Mason HS, editors. Comparative Biochemistry. Vol. III. New York: Academic Press; 1962. pp. 755-809

[3] Ribereau-Gayon P. Plant Phenolics. 1st ed. Edinburgh: Olyver and Boyd; 1972. p. 254

[4] Boulton RB. The copigmentation of anthocyanins and its role in the colour of red wine: A critical review. American Journal of Enology and Viticulture. 2001;**52**:67-87

[5] Scalbert A, Manach C, Morand C, Remesy C, Jimenez L. Dietary polyphenols and prevention of diseases. Critical Reviews in Food Science and Nutrition. 2005;**45**:287-306. DOI: 10.1080/1040869059096

[6] Kontoudakis M, Esteruelas M, Fort F, Canals JM, Freitas V, Zamora F. Influence of the heterogeneity of grape phenolicmaturity on wine composition and quality. Food Chemistry. 2011;**124**(3):767-774. DOI: 10.1016/j.foodchem.2010.06.093

[7] Revilla E, García-Beneytez E, Cabello F, Martin-Ortega G, Ryan JM. Value of high-performance liquid chromatographic analysis of anthocyanins in the differentiation of red grape cultivars and red wines made from them. Journal of Chromatography A. 2001;**915**(1-2):53-60. DOI: 10.1016/S0021-9673(01)00635-5

[8] Fang F, Li JM, Zhang P, Tang K, Wang W, Pan Q-H, Huang W-D. Effects of grape variety, harvest date, fermentation vessel and wine ageing on flavonoid concentration in red wines. Food Research International. 2008;**41**(1):53-60. DOI: 10.1016/j.foodres.2007.09.004

[9] Saucier C. How do wine polyphenols evolve during wine ageing? Cerevisia. 2010;**35**(1): 11-15. DOI: 10.1016/j.cervis.2010.05.002

[10] Anastasiadi M, Zira A, Magiatis P, Haroutounian SA, Skaltsounis AL, Mikros E. 1H NMR-based metabonomics for the classification of Greek wines according to variety, region, and vintage. Comparison with HPLC data. Journal of Agricultural and Food Chemistry. 2009;**57**(23):11067-11074. DOI: 10.1021/jf902137e

[11] Cheynier V, Schneider R, Salmon J, Fulcrand H. Chemistry of wine. In: Mander L, Liu HW, editors. Comprehensive Natural Products II. Oxford: Elsevier; 2010. pp. 1119-1172

[12] Baderschneider B, Winterhalter P. Isolation and characterization of novel benzoates, cinnamates, flavonoids, and lignans from Riesling wine and screening for antioxidant activity. Journal of Agricultural and Food Chemistry. 2001;**49**(6):2788-2798. DOI: 10.1021/jf010396d

[13] Kallithraka S, Salachaa MI, Tzouroua I. Changes in phenolic composition and antioxidant activity of white wine during bottle storage: Accelerated browning test versus bottle storage. Food Chemistry. 2009;**113**(2):500-505. DOI: 10.1016/j.foodchem.2008.07.083

[14] Vanzo A, Cecotti R, Vrhovsek U, Torres AM, Mattivi F, Passamonti S. The fate of trans-caftaric acid administered into the rat stomach. Journal of Agricultural and Food Chemistry. 2007;**55**(4):1604-1611. DOI: 10.1021/jf0626819

[15] Pozo-Bayon MA, Hernandez MT, Martin-Alvarez PJ, Polo MC. Study of low molecular weight phenolic compounds during the ageing of sparkling wines manufactured with red and white grape varieties. Journal of Agricultural and Food Chemistry. 2003; **51**(7):2089-2095. DOI: 10.1021/jf021017z

[16] Zoecklein BW, Fugelsang KC, Gump BH, Nury FS. Wine Analysis and Production. 1st ed. New York: Springer Science; 1995. p. 621

[17] Mattivi F, Guzzon R, Vrhovsek U, Stefanini M, Velasco R. Metabolite profiling of grape: Flavonols and anthocyanins. Journal of Agricultural and Food Chemistry. 2006; **54**(20):7692-7702. DOI: 10.1021/jf061538c

[18] Castillo-Munoz N, Gomez-Alonso S, Garcia-Romero E, Hermosin-Gutierrez I. Flavonol profiles of *Vitis vinifera* red grapes and their single-cultivar wines. Journal of Agricultural and Food Chemistry. 2007;**55**(3):992-1002. DOI: 10.1021/jf062800k

[19] Hsu YL, Liang HL, Hung CH, Kuo PL. Syringetin, a flavonoid derivative in grape and wine, induces human osteoblast differentiation through bonemorphogenetic protein-2/ extracellular signal-regulated kinase 1/2 pathway. Molecular Nutrition & Food Research. 2009;**53**(11):1452-1461. DOI: 10.1002/mnfr.200800483

[20] Munoz S, Mestres M, Busto O, Guasch J. Determination of some flavan-3-ols and anthocyanins in red grape seed and skin extracts by HPLC–DAD: Validation study and response comparison of different standards. Analytica Chimica Acta. 2008;**628**(1): 104-110. DOI: 10.1016/j.aca.2008.08.045

[21] Cieslik E, Gręda A, Adamus W. Contents of polyphenols in fruit and vegetables. Food Chemistry. 2006;**94**(1):135-142. DOI: 10.1016/j.foodchem.2004.11.015

[22] Macheix JJ, Fleuriet A. Phenolics in fruits and fruit products: Progress and prospects. In: Scalbert A, editor. Polyphenolic Phenomena. Paris: INRA; 1993. pp. 157-164

[23] Rentzsch M, Wilkens A, Winterhalter P. Non-flavonoid phenolic compounds. In: Moreno-Arribas MA, Polo C, editors. Wine Chemistry and Biochemistry. New York: Springer Science; 2008. pp. 509-527

[24] He F, Mu L, Yan G-L, Liang N-N, Pan Q-H, Wang J, Reeves MJ, Duan C-Q. Biosynthesis of anthocyanins and their regulation in coloured grapes. Molecules. 2010;**15**(12):9057-9091. DOI: 10.3390/molecules15129057

[25] Castillo-Munoz N, Winterhalter P, Weber F, Gomez MV, Gomez-Alonso S, Garcia-Romero E, Hermosin-Gutierrez I. Structure elucidation of peonidin 3,7-O-β-diglucoside isolated from Garnacha Tintorera (*Vitis vinifera* L.) grapes. Journal of Agricultural and Food Chemistry. 2010;**58**(20):11105-11111. DOI: 10.1021/jf102578u

[26] Paun N, Niculescu V. Isolation, purification and biological properties of anthocyanins. Progress of Cryogenics & Isotopes Separation. 2011;**14**(1):101-110

[27] Koponen JM, Happonen AM, Mattila PH, Törrönen AR. Contents of anthocyanins and ellagitannins in selected foods consumed in Finland. Journal of Agricultural and Food Chemistry. 2007;**55**(4):1612-1619. DOI: 10.1021/jf062897a

[28] Geana EI, Dinca OR, Ionete RE, Artem V, Niculescu VC. Monitoring *trans*-resveratrol in grape berry skins during ripening and in corresponding wines by HPLC. Food Technology and Biotechnology. 2015;**53**(1):73-80. DOI: 10.17113/b.53.01.15.3762

[29] Cheynier V, Duenas-Paton M, Salas E, Maury C, Souquet JM, Sarni-Manchado P. Structure and properties of wine pigments and tannins. American Journal of Enology and Viticulture. 2006;**57**(3):298-305

[30] Pastrana-Bonilla E, Akoh CC, Dsellappan S, Krewer G. Phenolic content and antioxidant capacity of muscadine grapes. Journal of Agricultural and Food Chemistry. 2003; **51**(18):5497-5503. DOI: 10.1021/jf030113c

[31] Gu L, Kelm MA, Hammerstone JF, Beecher G, Holden J, Haytowitz D, Prior RL. Screening of foods containing proanthocyanidins and their structural characterization using LC–MS/MS and thiolytic degradation tannins in grape seeds during maturation. Journal of Agricultural and Food Chemistry. 2003;**51**(25):7513-7521. DOI: 10.1021/jf034815d

[32] Zhao Q, Duan C-Q, Wang J. Anthocyanins profile of grape berries of *Vitis amurensis*, its hybrids and their wines. International Journal of Molecular Sciences. 2010;**11**(5): 2212-2228. DOI: 10.3390/ijms11052212

[33] Di Stefano R, Flamini R. High performance liquid chromatography analysis of grape and wine polyphenols. In: Flamini R, editor. Hyphenated Techniques in Grape and Wine Chemistry. West Sussex: John Wiley & Sons; 2008. pp. 33-79

[34] Paun N, Niculescu V, Tamaian R. Ultrasound assisted extraction of phenolic compounds from grapes. Bulletin of University of Agricultural Sciences and Veterinary Medicine Cluj-Napoca. Animal Science and Biotechnologies. 2010;**67**(1-2):5377

[35] Kennedy JA, Mathews MA, Waterhouse AL. Effect of maturity and vine water status on grape skin and wine flavonoids. American Journal of Enology and Viticulture. 2002;**53**(4):268-274

[36] Downey MO, Harvey JS, Robinson SP. Analysis of tannins in seeds and skins of Shiraz grapes throughout berry development. Australian Journal of Grape and Wine Research. 2003;**9**(1):15-27. DOI: 10.1111/j.1755-0238.2003.tb00228.x

[37] Mori K, Saito H, Goto-Yamamoto N, Kitayama M, Kobayashi S, Sugaya S, Gemma H, Hashizume K. Effect of abscissic acid treatment and night temperature on anthocyanin composition in Pinot noir grapes. Vitis. 2005;**44**(4):161-165

[38] De Freitas VAP, Glories Y, Monique A. Developmental changes of procyanidins in grape of red *Vitis vinifera* varieties and their composition in respective wines. American Journal of Enology and Viticulture. 2000;**51**(4):397-403

[39] Mateus N, Maqes S, Gonçalves AC, Machado JM, De Freitas VAP. Proanthocyanidin composition of red *Vitis vinifera* varieties from the Douro valley during ripening: Influence of cultivation altitude. American Journal of Enology and Viticulture. 2001;**52**(2):115-121

[40] Peng Z, Iland PG, Oberholster A, Sefton MA, Waters EJ. Analysis of pigmented polymers in red wine by reverse phase HPLC. Australian Journal of Grape and Wine Research. 2002;**8**(1):70-75. DOI: 10.1111/j.1755-0238.2002.tb00213.x

[41] Geana EI, Ionete RE, Niculescu V, Artem V, Ranca A. Changes in polyphenolic content of berry skins from different red grapes cultivars during ripening. Progress of Cryogenics and Isotopes Separation. 2016;**19**(2):95-104

[42] Flamini R, Dalla Vedova A, Cancian D, Panighel A, De Rosso M. GC/MS-positive ion chemical ionization and MS/MS study of volatile benzene compounds in five different woods used in barrel-making. Journal of Mass Spectrometry. 2007;**42**(5):641-646. DOI: 10.1002/jms.1193

[43] Perez-Magarino S, Revilla I, Gonzalez-SanJose ML, Beltran S. Various applications of liquid chromatography-mass spectrometry to the analysis of phenolics compounds. Journal of Chromatography A. 1999;**847**(1-2):75-81

[44] Pati S, Losito I, Gambacorta G, La Notte E, Palmisano F, Zambonin PG. Simultaneous separation and identification of oligomeric procyanidins and anthocyanin-derived pigments in raw red wine by HPLC-UV-ESI-MSn. Journal of Mass Spectrometry. 2006; **41**(7):861-871. DOI: 10.1002/jms.1044

[45] Flamini R, Traldi P. Mass Spectrometry in Grape and Wine Chemistry. New Jersey: John Wiley & Sons; 2010. p. 348

[46] Gris EF, Mattivi F, Ferreira EA, Vrhovsek U, Filho DW, Pedrosa RC, Bordignon-Luiz MT. Phenolic profile and effect of regular consumption of Brazilian red wines on in vivo antioxidant activity. Journal of Food Composition and Analysis. 2013;**31**:31-40 http://dx.doi.org/10.1016/j.jfca.2013.03.002

Chapter 8

Current State and Perspective in the Models Applicable to Oenology

Anca Şipoş

Additional information is available at the end of the chapter

http://dx.doi.org/10.5772/intechopen.71711

Abstract

Modeling, simulation, and control of the alcoholic fermentation of grape juice into wine are still not a totally resolved problem. A model that makes it possible to predict alcoholic fermentation development would be a valuable instrument to its technical and economical implications. Considering the field of bioprocess used in food production at the industrial scale, the chapter will be centered on models applicable to oenology. On the first part, the chapter proposes the different approaches that have been taken: "knowledge-based" models, non-physiological mathematical descriptions, behavior prediction models, and empirical models. The second part will deal with a nonlinear model for white wine alcoholic fermentation process which, besides the detailed kinetic model, involves equations corresponding to the physiological phases of yeast cells, the inhibitory effect of ethanol, heat transfer equations, and the dependence of kinetic parameters on temperature.

Keywords: modeling, simulation, advanced control of technological process

1. Introduction

If the twentieth century belonged to electronics and computer science, the twenty-first century would belong to biology and biotechnologies [1]. *Bioprocesses* are the foundation of life and especially of human health. Therefore, on an international scale, there are studies being conducted in order to find ways of improving the food safety quality.

A new science was born at the turn of two centuries of technical and scientific revolution: *bioinformatics*. The methods and the concepts of computer science began to hold the interest of many biologists, especially to the molecular biology specialists [2]. These methods are essential for various problems such as the analysis of evolutionary processes, the complex molecular structures shaping, and the simulating of some biological aspects. Researchers are currently overwhelmed by the enormous data quantity that comes from the multiple projects belonging to genomics, transcriptomics, proteomics, and metabolomics. The necessity of using computerized methods in manipulating and assembling these extremely complex high-level data is obvious. But the great challenge lies not only in the genetic sequencing and cartography but especially in understanding what do transcriptomics, proteomics, and metabolomics mean, when associated with certain life conditions and with a certain hereditary program at a given time. The identification of the structure and function of the proteins that were biosynthesized by the organisms, the complex mechanisms that allow the development of life, is therefore required. In order to achieve this, theoretical informatics—through some of its domains, which have reached a certain level of maturity, formal languages, data structures, and artificial intelligence—offer viable theoretical models.

At the same time, the beginning of our century is righteously marked by *biology,* determined and largely conditioned by the revolution within the biological sciences [1]. Therefore, modern biology begins to determine the main directions of some interdisciplinary developments, constantly calling on the discoveries of chemistry, physics, mathematics, and engineering. A new, promising branch of science was born at a specific meeting point: *bioengineering*. Bioengineering especially has developed in connection with biotransformation processes (biosynthesis-biodegradation), in order to obtain antibiotics, enzymes, vitamins, amino acids, organic acids, bicarbonates, biopolymers, as a result of the cooperation among microbiologist, biochemist, chemical, food industry and mechanical engineer, operator and computer scientist, in a domain which is called *Microbial Engineering* and *Biochemical Engineering* [3]. Food safety and its quality is a primary field in European and global policy and legislation of the twenty-first century because it concerns the required conditions for a healthy population. The key issues for improving the biochemical and micro-biochemical safety and the quality of food reside in knowing, understanding, and leading the *bioprocess* used in food production as well as possible [4]. The volume of bioprocesses in the technical development and implementation of the processes of obtaining new food products and a significant number of non-food items based on microbiological processes and enzyme technology have grown significantly in the last years.

It is very difficult to monitor, design, and control a bioprocess [5]. For example, in the last years, an impressive number of kinetic models were developed only for fermentation processes and the various phenomena that influence fermentation kinetics were taken into account. That is why these models needed the estimation of a large number of sizes, which are often very hard to identify [6–8]. A series of artificial intelligence-based control and optimization models and techniques were also developed (fuzzy and neuro-fuzzy techniques) [9–12]. These models allowed a more realistic definition of some model sizes, which carry a high degree of uncertainty.

The mathematical model of a biotechnological process is generally represented using differential equation systems, which are obtained by writing, for each component, the mass and energy balance equations. In these equations, the components or the concentrations of the components involved in the reactions that take place in the bioreactor appear. The dynamics of the temporal variation of each component is expressed by two types of phenomena. First, there are chemical and biochemical reactions that transform some components into some others. Second, there is the mass transfer, caused by the liquid or gas interchanges between the reactor and the environment or with other reactors. In this context, the mathematical model of a bioprocess is a mathematical description of two types of phenomena, in which chemical and biochemical reactions generate the *kinetics* of the bioprocess, while the mass transfer phenomena define the *transport dynamics*.

Another important aspect in connection to the control of the bioprocesses is the lack of transducers and on-line biosensors, specific to biochemical and micro-biochemical sizes, that characterize these kinds of processes [13, 14]. Attempts of obtaining these biosensors are made [11, 15, 16], but the research is far from being finished.

At the same time, the general model of a bioprocessor can provide a series of structural properties which might be useful for the resolution of some identification, state estimation (state observers development), or bioprocesses leading problems. The need for the state observers is imposed by the absence of reliable and cheap biosensors, capable of making online measurements of the biochemical and biological variables, used in implementing some convenient methods of monitoring and leading biotechnological processes [1]. The acquisition of biomass, sublayer, and the metabolism products are made through lab analysis; this method makes the leading of the bioreactors difficult (as for the direct adjustment of these sizes). The lab analysis requires taking a sample from the contents of the bioreactor, and this represents a higher risk of contaminating the culture. Also, lab models for determining the number of microorganisms, the concentration of the sublayer, as well as the concentrations of metabolism products are quite imprecise, thus generating uncertainties in appreciating the evolution of the abovementioned sizes. These problems are strongly amplified in the industrial environment, especially because of the lack of adequate equipment and sufficient staff in order to realize quality measurements in the lab. Normally, under these conditions, there are maximum three-four reliable analyses during the unfolding of a biotechnological process.

The estimation of sizes of the biotechnological process is considered to be a way of avoiding the various drawbacks connected to the acquisition of data from the bioreactors. The *state estimator* (also known as software sensor or observer in specialized works) is an algorithm used for determining some measurements of the process, which are not measurable in real time, based on other accessible measurements as related to their acquisition.

An essential problem, specific to industrial scale food industry bioprocesses, is the fact that the technological background results are very variable [1]. In many cases, the variability of raw materials used in the industrial processes leads to a non-reproducibility of the charges. Taking into consideration these aspects, the use of the technical operator's experience for leading the process is recommended, and it is possible by systems based on *system expert* type of knowledge.

Also, heat transfer aspects led to the development of models and automatic control techniques regarding the optimization of the thermal regime of the bioprocessors [17].

All these research and achievements are proper to the examined bioprocesses and usually have a low degree of reproducibility. A generalization has not been achieved at the moment; it is very difficult to achieve and this can only be made to a certain extent.

Another aspect is the improvement of the technological performances concerning the energy consumption. As a result, optimizing the energy consumption through the introduction of advanced leading systems leads to a conservation of fossil fuels. It can be affirmed that, indirectly, a bio-economy is being advocated for. In *Bio-economy versus biodiversity,* Hall [18] says "The bio-economy agenda is especially attractive to fossil fuel companies who want to be seen pursuing an exit- from-oil strategy; and to biotechnology companies desperately in need of a Trojan horse to provide safe passage for risky and unpopular new technologies."

After a brief introduction regarding the modeling of bioprocesses, on the first part, the chapter proposes the different approaches that have been taken: "knowledge-based" models, non-physiological mathematical descriptions, behavior prediction models, and empirical models. The second part will deal with a nonlinear model for white wine alcoholic fermentation process which, besides the detailed kinetic model, involves equations corresponding to the physiological phases of yeast cells, the inhibitory effect of ethanol, heat transfer equations, and the dependence of kinetic parameters on temperature.

2. Current stage in fermentation processes modeling

2.1. The biotechnological processes modeling

A *biotechnological process* implies a development (growth) of a microorganism population (culture medium), *biomass,* in a bioreactor (vessel) by the consumption of some nutrients (carbon, nitrogen, oxygen, vitamins, etc.) that represent the *substrate,* if the physical and chemical conditions (temperature, pH, aeration, etc.) are favorable. Customarily, the microorganisms' growth takes place in a liquid medium (aqua). It is obvious that in a bioreactor many biochemical and biological reactions take place simultaneously. Each elementary reaction is, usually, catalyzed by a protein (enzyme) and can form a specific *product* or a *metabolite.* The aim of such a culture can be

- biomass producing (bacteria, yeasts, etc.);

- producing a principal component (amino acid, medicines, marsh gas, etc.);

- biological decontamination (biological consumption of the pollutant substrates by the biomass).

Because the microbial mechanism of growth that involves alive organisms is very complex, a detailed modeling is not possible or is very complicated. Usually, the bioreactor assumes a perfect stirred and is described by a number of macro-scope variables, such as biomass, substrate, product, oxygen concentrations, pH, temperature, etc. Function of the fermentation type could also be defined by another concentration.

The mathematical model of a bioreactor depends on its operating mode. In this way, in practice, for these types of bioreactors three operating modes could be defined:

- *Batch mode*—A batch bioreactor is a reactor with a cyclical operating, without feed, and exit flows. The entire quantity of substrate and nutrients, with a small quantity of biomass, is introduced in the bioreactor at the beginning of the fermentation. During the fermentation, the bioreactor will not be fed with any substrate and the process will be completed after the substrate has been sufficiently consumed. The result is that the entire quantity of biomass is collected, from which then the desire product will be extracted. The possibilities of process control in this case are very limited and could be resumed with some physical variables: temperature, pH, energetically consumption, length of fermentation, and so on.

- *Fed-batch mode*—The fed-batch bioreactors type or with semi-continuous operating is a reactor with cyclical working, with a continuous or intermittent feeding and without an exit. In the vessel, both a small quantity of substrate and a biomass are initially introduced. Then, during the fermentation process, and in the function of the microorganisms' necessities, the reactor will be fed with a controlled flow of substrate. Therefore, the possibilities of process control of this kind of bioreactors are more diverse than that of the batch type. Aside from the physical variables, the biological variables: substrate concentration, biomass concentration, and so on, can also be controlled.
- *Continuous mode*—This mode is more efficient from the economical point of view and the bioreactors are stirred, with continuous operation, eventually with separation and recycling, as well as the gas-liquid bioreactors, used especially in the industry for obtaining a great volume of biomass (i.e., unicellular proteins) or for the biological treatment of residual water. In such a reactor, the biomass is evacuated with a flow equal with that of the substrate feeding.

The mathematical model of a biotechnological process is formed by an equation system in which the components or the concentrations of components implied in reactions that take place in bioreactor appear, equations which are obtained by writing, for each component, the mass and energetic balances. The dynamic or variation in time of the quantity of each component is expressed by two phenomena. Primarily, there are chemical, biochemical, and biological reactions which transform certain components in others. Secondarily, mass transfer due to the liquid and gas interchanges between the reactor and the environment or with other bioreactors exists. Within this context, the mathematical model of a bioprocess represents a mathematical description of two types of phenomena, where the chemical, biochemical, and biological reactions cause the bioprocess *kinetics* while the mass transfer phenomena define the *dynamics of transport*.

If the process is passed off in a reactor with perfect stirring, which means a reactor in which the culture medium composition is supposed to be homogeneous, the dynamics of the two types of phenomena can be represented by a unitary description, by differential equations system which involve the reaction components concentrations, as well the pH, temperature, and so on, variables which can be organized like a state vector. There are also bioreactors without perfect stirred and in which a set of concentration gradients, temperature, pressure, and so on, appear. Such types of bioreactors are tubular, bioreactors with fix layers, and so on. These reactors are described by mathematical models which contain partial derivate equations, where, in the excepted the temporal variables, at least a spatial variable will appear.

The modeling of a biotechnological system can be done progressive. It begins with the information that could be easily obtained, namely the reaction of the components involved in the bioprocess, the reactions which take place in the bioreactor and the liquid or/and gas interconnections of the system, elements which define the *reaction schema* and the *system architecture*. With these information, the *general structure of the model* is settled out. The general model of the bioprocess can wise up a set of structural properties useful for solving some issues concerning the state estimation or the control of the bioprocesses.

2.2. The reaction kinetics modeling

2.2.1. The kinetic of microorganisms growing

The kinetic of microorganisms' growth (**Figure 1**) is necessary from the point of view of knowing the process phases (**Table 1**), the duration of these phases, the factors which influence them in order to choose then the type of vessel, and the process control mode for the bioreactor.

2.2.2. The types of kinetic models

The phenomena that take place in a bioreactor are complex and coupled. The complexity is given by the heterogenic medium three phases (solid, liquid, and gas), in which the biosynthesis process takes place; the three phases medium is in a dynamic evolution and has a nonlinear character. The processes are coupled among them, an operational variable (the feed flow with substrate, the pressure, the temperature, the stirred, etc.) or the way in which the oxygen reaches at the microorganisms, are important both individually and for the whole microorganisms, for the biomass process control [19].

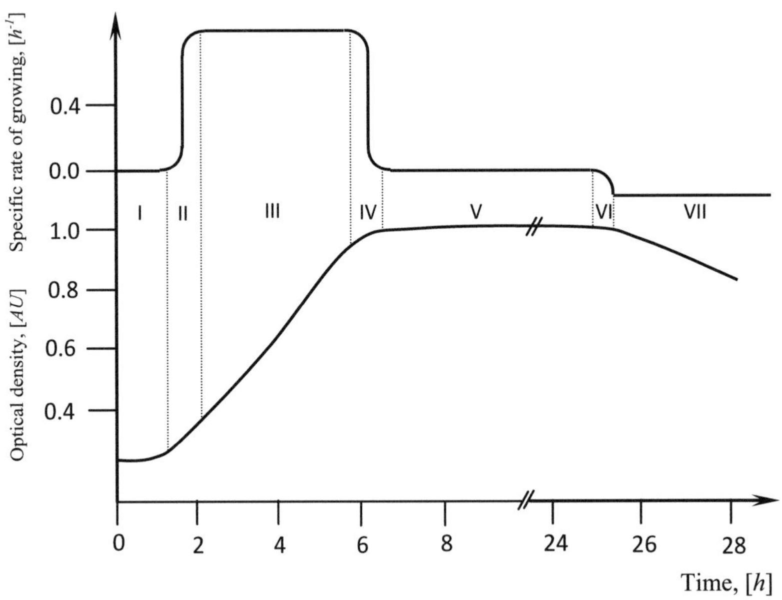

Figure 1. The microorganisms number evolution versus time.

Phase	Characterization	Observations
I. Latent (adaptation)	The regeneration processes of the hyphes or the germination of the spores, the function of the inoculums type take place.	Generally, this phase has a reduced practical importance. For this reason, in order to eliminate this phase, more generation of cells are cultivated on the respective medium aiming to accustom the cells with it.
II. Beginning of growth	The volume of the cells grows fast.	
III. Exponential growth	The specific growth rate μ is constant and the biomass growth is exponential.	This phase presents a practical importance when we have the obtaining of biomass in view. The metabolical activity from the exponential phase is in fact the primary metabolite; it corresponds with trot phase.
IV. Deceleration growing	The specific growth rate μ comes down.	It appears when the feed with an essential nutrient becomes insufficient or an element necessary for growth has been run out or intermediary substances have been accumulated in the medium.
V. Stationary	The microbial cells reach a maximum concentration, the proportion between alive and death cells number remains constant. The quantity of limiting nutrient influences this quantity of biomass, named total production.	The secondary metabolism is typical for the stationary phase; it corresponds to idiophase.
VI. Decay	The cells die, the autolysis takes place, and the quantity of biomass comes down.	At one point, it is possible that an easy growth of biomass takes place, due to the alive cells' consumption of the nutrients. The nutrients have been delivered by the lysated cells. For a valorous culture, the decline phase must be eliminated.

Table 1. The microbial population growing phases (adapted from [19]).

In **Figure 2**, the main phenomena are presented, the interactions and variables which influence the kinetic behavior of the population cells. The two systems, the biological system (represented by the microbial population) and the physical and chemical system (external medium), are in close correlation; the cells consume nutrients and transform the substrate in reaction products. The cells generate heat which warms the medium, and the medium temperature influences the cell behavior.

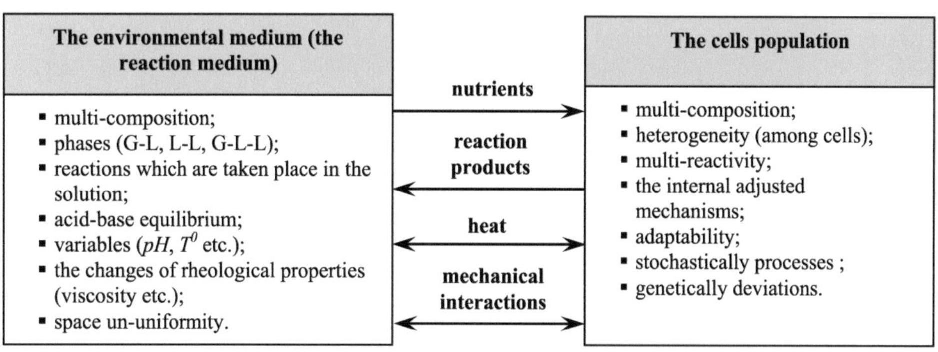

Figure 2. The main phenomena, interactions, and variables which influence the kinetic of the microbial populations.

The mechanical interactions are due to the hydrostatic pressure and to the flow effects of the medium on the cells and also to the medium viscosity changes, caused by the cells accumulation and/or by the cellular metabolites.

The medium is a multi-composition system that must contain the necessary nutrients for the growth of the microorganisms (carbon, nitrogen sources, mineral salts, vitamins, growing factors, oxygen, etc.) and in which different products of cellular metabolism (primary and secondary metabolites) are accumulated while the cells grow on.

In solution reactions that could modify the structure of the final products (i.e., the penicillin hydrolyses) could take place.

Often, the cells consume or produce components that could influence the medium acidity and the interrelation between cell needful and acid-base equilibrium determinates the medium pH, which, in turn, influences the cells' activity and the transport processes.

During the cellular reactions, the medium temperature, pH, ionic strength, and rheological properties can change in time.

A complex model of a bioreactor is multi-phases; it consists of a medium with solid particles among which liquid and gas particles (at the surface being the microbial culture) are dispersed or from a liquid medium in which gas bulls are dispersed. An example of three phases model, with two physical phases, gassy and liquid phases and a biotical one—the microorganisms culture, is presented in **Figure 3**, model which could be considered a model of a fermentation process.

Because of the great volume of bioreactors, of the higher viscosity and of the non-Newtonian nature of medium, in most cases, the technological conditions from vessel can vary from point to point.

Every individual of microbial population is a complex component of the system, frequently non-homogeneous, even at the level of a single cell. Many independent biochemical reactions take

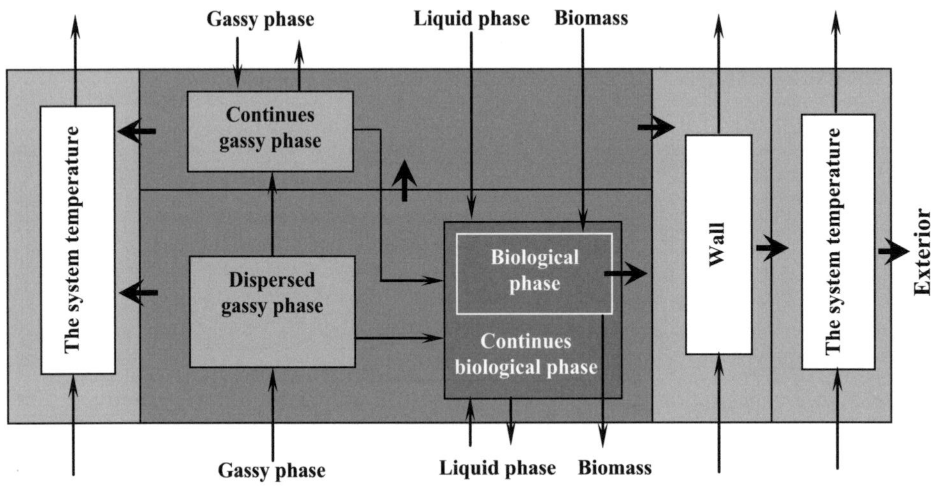

Figure 3. A bioreactor model with three-structured phases.

place simultaneously in every cell, which involves an internal complex control; this control allows the cell to adapt its activity and even to the biochemical type reaction in function of the external medium.

The dissimilarity among the cells is given by the variation in time and space of the cells' age (some cells are just born while others are dying or dividing); the cells with different ages are often characterized by various types of activities and metabolical functions.

In the cultivation for longer period, many spontaneous mutations of some individuals of cells population may appear.

Analyzing all these aspects, it becomes obvious that, in practice, a kinetic model that may include all the phenomena and interactions of the system cannot be formulated, a first simplification concerning the medium being necessary. It is considered that even a single component of medium, which is in quite great quantities, can influence the microbial growth of the medium. Sometimes, it could be necessary to include other components of medium such as the products with inhibitory role, which are accumulated in medium, in order to obtain a suitable description of the kinetics of the cells population.

2.2.3. *The unstructured kinetics models*

2.2.3.1. *Models based on Monod equation*

From the kinetics point of view, in order to construct a model it is necessary to study the rates and mechanisms of the physical, biochemical, and microbiological processes, in which microorganisms are involved (growing, cellular cycle, the components produced by reaction, the medium effects, and the biological interactions).

The *specific growth rate*, μ, represents the variation in time of the microbial cells concentration in synthesis medium:

$$\frac{1}{X} \cdot \frac{dX}{dt} = \mu \tag{1}$$

or in integrating form:

$$X = X_0 \cdot e^{\mu \cdot t} \tag{2}$$

where X represents the microbial cells concentration in biosynthesis medium [mol/m^3];

X_0 is the steady-state operation point of microbial concentration [mol/m^3];

t is the development time [s];

μ is the specific growth rate in exponential phase, [s^{-1}];

The specific growth rate depends, among other elements, on the limiting substrate concentration (i.e., the glucose concentration), S, taking place after the following equation:

$$\mu = \frac{\mu_{max} \cdot S}{K_S + S} \tag{3}$$

with the name Monod equation where S is the limiting substrate concentration (glucose concentration); μ_{max} is the maximum specific growth rate, achieving when $S>>K_s$ and the concentration of all the nutrients is unchanged; K_s is a constant which has a concentration dimension and represents the value of limited nutrient concentration at which $\mu = \frac{1}{2} \cdot \mu_{max}$.

The values of the two variables depend on the microorganism, on the work conditions (pH and temperature), and on the substrate complexity. The typical curves of biomass evolution and substrate consumption in time are presented in **Figure 4**, for a batch fermentation system. The value of substrate concentration S was pointed out for which $\mu = \frac{1}{2} \cdot \mu_{max}$.

K_s can be considered a measure of the microorganisms affinity against substrate:

- a small value of K_s indicates a great affinity, the microorganisms can grow in conditions of very small substrate concentrations (small dilution at continuous processes);

- a great K_s indicates a small affinity for the substrate, the microorganisms growth takes place slowly, even if the substrate concentrations are great.

The specific forming rate of the metabolism product:

$$q_p = \frac{r_p}{X} \quad (4)$$

where

$$r_p = k_1 \cdot X + k_2 \cdot r_x \quad (5)$$

is the forming rate of metabolism product $[mol/(m^3 \cdot s)]$; X is the biomass concentration; r_x is the growing rate of cellular mass $[mol/(m^3 \cdot s)]$; k_1 is the proportionality factor between the forming rate of product and biomass concentration $[mol\,product/(mol\,biomass \cdot s)]$; k_2 is the proportionality factor between the forming rate of the product and the growing rate of the cellular mass $[mol\,product/mol\,biomass]$.

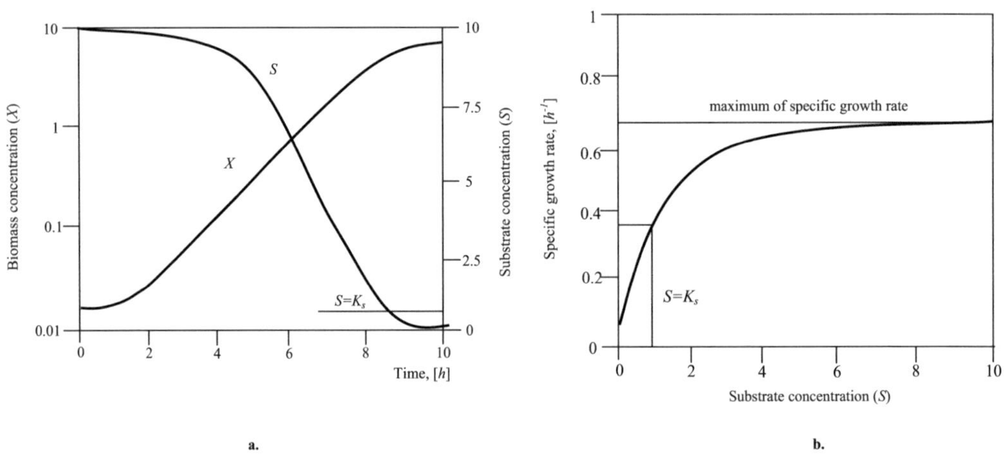

Figure 4. The evolution of biomass and substrate consume versus time, for a fermentation in batch system (a); for the same batch system-specific growth rate μ as a function of limiting substrate concentration, S (b).

Introducing Eq. (5) in Eq. (4), the specific rate of metabolism product can be determined:

$$q_p = k_1 + k_2 \cdot \mu \tag{6}$$

where $\mu = r_x/X$.

The forming rate of the product doesn't depend on the biomass X but on the growing rate of the cellular mass r_x at the obtained primary metabolites (ethanol, acetic acid, gluconic acid, butyric alcohol, acetone, etc.). Therefore, k_1= 0 and Eq. (5) become

$$r_p = k_2 \cdot r_x \tag{7}$$

In case of obtaining secondary metabolites, r_p depends only by biomass X:

$$r_p = k_1 \cdot X \tag{8}$$

If the development of microorganisms is limited by the concentration of a biosynthesis medium component, the microorganisms quantity that is formed is proportionally with the used quantity of substrate. Therefore, the *efficiency coefficient for biomass producing* can be defined:

$$Y'_{xs} = \frac{dX}{dS} = \frac{mols\ of\ biomass\ produced}{mols\ of\ substrate\ consumed\ for\ biomass\ producing} \tag{9}$$

and the *efficiency coefficient for metabolite producing*:

$$Y'_{ps} = \frac{dP}{dS} = \frac{mols\ of\ metabolism\ produced}{mols\ of\ substrate\ consumed\ for\ metabolism\ producing} \tag{10}$$

In *aerobic* conditions, at the biomass forming, the rate consume of substrate is

$$-r_s = \frac{r_x}{Y'_{xs}} + \frac{r_p}{Y'_{ps}} + m_s \cdot X \tag{11}$$

where m_s is a forming coefficient $[mol\,substrate/(mol\,biomass \cdot s)]$, specific for microorganism.

With the growth in the *anaerobic* conditions, the energy necessary to cell is obtained by the substrate transformation in reaction products; therefore, the metabolism products forming is a consequence of biomass growing:

$$-r_s = \frac{r_x}{Y'_{xs}} + m_s \cdot X \tag{12}$$

For *S>0*, the forming rate of metabolites in anaerobe conditions is

$$r_p = \frac{Y'_{ps}}{Y'_{xs}} \cdot r_s + m_p \cdot X \tag{13}$$

where m_p is a coefficient that describes the product forming during the growth $[mol\,substrate/(mol\,biomass \cdot s)]$. This coefficient can be determined from m_s on the basis that the reaction of the stoichiometric coefficients substrate consumed formed product. From Eqs. (5) and (3), the k_1 and k_2 coefficients could be obtained in anaerobic conditions.

The total efficiencies of biomass and metabolic products are used in practice.

During the *aerobic* processes, considering Eqs. (5) and (11), the result is

$$Y_{xs} = \frac{r_x}{-r_s} = \frac{r_x}{\frac{r_x}{Y'_{xs}} + \frac{r_p}{Y'_{ps}} + m_s \cdot X} = \frac{\mu \cdot X}{\frac{\mu \cdot X}{Y'_{xs}} + m_s \cdot X} = \frac{\mu}{\frac{\mu}{Y'_{xs}} + \frac{k_1 + k_2 \cdot \mu}{Y'_{ps}} + m_s} \tag{14}$$

At the specific growth very high rates, Y_{xs}=Y'_{xs}

$$Y_{ps} = \frac{r_p}{-r_s} = \frac{r_p}{\frac{r_x}{Y'_{xs}} + \frac{r_p}{Y'_{ps}} + m_s \cdot X} = \frac{k_1 \cdot X + k_2 \cdot r_x}{\frac{\mu \cdot X}{Y'_{xs}} + m_s \cdot X} = \frac{k_1 + k_2 \cdot \mu}{\frac{\mu}{Y'_{xs}} + \frac{k_1 + k_2 \cdot \mu}{Y'_{ps}} + m_s} \tag{15}$$

For the *anaerobic* processes:

$$Y_{xs} = \frac{r_x}{-r_s} = \frac{r_x}{\frac{r_x}{Y'_{xs}} + m_s \cdot X} = \frac{Y'_{xs}}{1 + \frac{m_s}{\mu} \cdot Y'_{xs}} \tag{16}$$

$$Y_{ps} = \frac{r_p}{-r_s} = \frac{\frac{Y'_{ps}}{Y'_{xs}} \cdot r_x + m_p \cdot X}{\frac{r_x}{Y'_{xs}} + m_s \cdot X} = \frac{Y'_{ps} + \frac{m_p}{\mu} \cdot Y'_{xs}}{1 + \frac{m_s}{\mu} \cdot Y'_{xs}} \tag{17}$$

There is a simplified modality to express the efficiencies with the following equations:

$$Y_{xs} = \frac{X - X_0}{S_0 - S} \quad \text{and} \quad Y_{ps} = \frac{P - P_0}{S_0 - S} \tag{18}$$

2.2.4. Alternative variants of models are based on Monod equation.

It is expected that the specific growth rate of biomass, expressed by Monod equation, does not match for all the fermentation processes. Many authors have tried to improve the Monod model and some examples will be presented in the subsequent part.

2.2.4.1. Teissier model

$$\mu = \mu_{\max} \cdot \left(1 - e^{-\frac{S}{K_S}}\right) \tag{19}$$

A disadvantage of the Monod and Teissier models is that these models do not take into consideration the inhibitory factor of the substrate when it is in excess. Andrews model abolishes this disadvantage by adding S^2/K_i at the denominator of biomass specific growth rate expression.

2.2.4.2. Andrews model

Therefore, a great quantity of substrate inhibits the cells' growth (i.e., glucose in quantities of >800 mol/m^3) and the following equation can be used:

$$\mu = \frac{\mu_{\max} \cdot S}{K_s + S + \frac{S^2}{K_i}} \tag{20}$$

where K_i is the inhibitory factor and $\mu_0 = \mu_{\max}\left(1 + \sqrt{\frac{K_S}{K_i}}\right)$.

Contois model, which takes into consideration the inhibitory effect of the biomass on the specific growth rate, is often used [1].

Contois model

$$\mu = \mu_{\max} \frac{S}{S + K_S X} \tag{21}$$

Because of the greater concentration of biomass, the biotical phase can be a substantial part from the total volume of the bioreactor and thus it is difficult to assimilate the substances by the biomass. Anyway, it is heavy to imagine in which mode the cells' concentration can inhibit its own growth and, probably, the ability of Contois kinetics to be in concordance with the experimental data is explained by the toxic effect of some metabolic products.

Another model that takes into consideration the inhibitory effect of the biomass on the specific growth rate is *Luong model*:

$$\mu = \frac{\mu_{\max} \cdot S}{K_s + S} \cdot \left(1 - \frac{S}{S_{\max}}\right) \tag{22}$$

where K_i is the inhibitory constant and $S_{\max}$ is the substrate concentration at which the microorganisms are in a stationary growth phase.

A similar variant of the specific growth rate of Monod biomass in which the substrate S appears at power n (empirical) is *Moser model*:

$$\mu = \frac{\mu_{\max} \cdot S^n}{K_S + S^n} \tag{23}$$

On the case of bioreactors with two substrates (S_1 and S_2), the models with Eqs. (3) and (19)–(23) are combined. For example:

Monod-Monod model

$$\mu = \mu_{\max} \cdot \frac{S_1}{K_1 + S_1} \cdot \frac{S_2}{K_2 + S_2} \tag{24}$$

Monod-Andrews model

$$\mu = \mu_{\max} \cdot \frac{S_1}{K_1 + S_1} \cdot \frac{S_2}{K_2 + S_2 + \frac{S_2^2}{K_i}} \tag{25}$$

When a metabolic product inhibits the growth, the specific growth rate has the following expression:

Jarusalimsky model

$$\mu = \mu_{max} \cdot \frac{S}{K_s + S} \cdot \frac{K_p}{K_p + P} \tag{26}$$

Levenspield model

$$\mu = \mu_{max} \cdot \frac{S}{K_s + S} \cdot \left(1 - \frac{P}{P_{max}}\right) \tag{27}$$

where K_p represents the concentration of the metabolic product P at which $\mu = \frac{1}{2} \cdot \mu_{max}$ and P_{max} is the maximum concentration of the metabolic product with inhibitory action.

The variables of the biotechnological process can also be described in function of the variables that represent the control variables of the bioreactor (temperature, pH, etc.). The control algorithms of pH and temperature are often complex because of frequent changes of their optimal values, during the bioprocess period [1].

The influence of the temperature on the maximum specific growth rate of biomass is important: decreasing or increasing with one grade the temperature of the optimal value caused the beginning of the proteins' denaturing process, which is undesirable process. At the smaller temperatures then that at which the denaturing of the proteins appears, the maximum specific growth rate of biomass can be modeled using the Arrhenius equation:

$$\mu_{max} = A \cdot e^{-\frac{Ea}{R \cdot T^0}} \tag{28}$$

where Ea is activation energy [kJ/mol]; A—pre-exponential factor; R—universal gas constant and T^0 is temperature [K].

Considering that the proteins were denaturized at a temperature of a chemical reversible reaction having the free energy ΔG_d [kJ/mol] and that the denaturized proteins are inactive, Roels [20] has proposed a mathematical equation for the maximum specific growth rate of biomass, equation which is relatively alike with Hougen-Watson equation for catalyses activity in heterogeneous chemical reactions:

$$\mu_{max} = \frac{A \cdot e^{-\frac{Ea}{R \cdot T^0}}}{1 + B \cdot e^{-\frac{\Delta G_d}{R \cdot T^0}}} \tag{29}$$

where B is a constant.

Topiwala and Sinclair model (for temperature)

$$\mu_{max} = \begin{cases} a_1 \cdot e^{-\frac{Ea_1}{RT^0}} - a_2 \cdot e^{-\frac{Ea_2}{RT^0}} - b, & if\ T_1^0 \leq T^0 \leq T_2^0 \\ 0 & , if\ T^0 < T_1^0\ or\ T^0 > T_2^0 \end{cases} \tag{30}$$

where Ea_1 and Ea_2 are activation energies;

$R$is universal gas constant;

a_1, a_2, and $b$are constants.

Eq. (30) shows that the specific growth rate of biomass is continually growing until a maximum $T^0{}_2$, value, at which the microorganisms enter in idiophase and then in autolysis.

The pH influence on the cellular activity is determined by the enzymes sensibility at the pH changes. Enzymes are active only in a particular interval of pH and therefore the enzymes total activity of cells is a complex function by medium pH.

Rozzi model (for pH)

$$\mu_{\max} = a \cdot pH^2 + b \cdot pH + c \tag{31}$$

For complex dependences $\mu = f(S_1, \ldots, S_{ms}, P_1, \ldots, P_{mp}, X, pH, T^0, \ldots)$, the multiplicative principle is often used:

$$\mu = \mu_{\max} \cdot f_1(pH) \cdot f_2(T) \cdot f_3(X) \cdot \prod_{j=1}^{ms} \varphi_j(S_j) \cdot \prod_{j=1}^{mp} \psi_j(P_j) \tag{32}$$

where $f_1(.)$, $f_2(.)$, $\varphi_j(.)$, j=1,…, ms and $\Psi_j(.)$, j=1,…,mp are penalization functions.

In **Table 2**, the most used mathematical models in biotechnological processes simulation, models used in process and optimal control of fermentative industrial processes are presented.

Nr. crt.	$\frac{dX}{dt}$	$\frac{dP}{dt}$	$\frac{dS}{dt}$	Model
1.	$\mu_{\max} \cdot \left(\frac{S}{K_s+S}\right) \cdot X$	$q_{p\max} \cdot \left(\frac{S}{K_{sp}+S}\right) \cdot X$	$-\left(\frac{1}{Y_{xs}} \cdot \frac{dX}{dt}\right) - \left(\frac{1}{Y_{ps}} \cdot \frac{dP}{dt}\right)$	**Monod**
Inhibitory effect of the substrate				
2.	$\mu_{\max} \cdot \left(\frac{S^n}{K_s+S^n}\right) \cdot X$	$q_{p\max} \cdot \left(\frac{S^n}{K_{sp}+S^n}\right) \cdot X$	$-\left(\frac{1}{Y_{xs}} \cdot \frac{dX}{dt}\right) - \left(\frac{1}{Y_{ps}} \cdot \frac{dP}{dt}\right)$	**Moser**
3.	$\mu_{\max} \cdot \left(1 - \exp\left(-\frac{S}{K_s}\right)\right) \cdot X$	$q_{p\max} \cdot \left(1 - \exp\left(-\frac{S}{K_{sp}}\right)\right) \cdot X$	$-\left(\frac{1}{Y_{xs}} \cdot \frac{dX}{dt}\right) - \left(\frac{1}{Y_{ps}} \cdot \frac{dP}{dt}\right)$	**Teissier**
Inhibitory effect of the substrate and of one of the metabolism products				
4.	$\mu_{\max} \cdot \left(\frac{S}{1+\frac{K_s}{S}+\frac{S}{K_{Xi}}}\right) \cdot X$	$q_{p\max} \cdot \left(\frac{S}{1+\frac{K_{sp}}{S}+\frac{S}{K_{Pi}}}\right) \cdot X$	$-\left(\frac{1}{Y_{xs}} \cdot \frac{dX}{dt}\right) - \left(\frac{1}{Y_{ps}} \cdot \frac{dP}{dt}\right)$	**Andrews and Noack**
5.	$\mu_{\max} \cdot \left(\frac{S}{K_s+S}\right) \cdot \exp\left(-\frac{S}{K_{Si}}\right) \cdot X$	$q_{p\max} \cdot \left(\frac{S}{K_{sp}+S}\right) \cdot \exp\left(-\frac{S}{K_{Pi}}\right) \cdot X$	$-\left(\frac{1}{Y_{xs}} \cdot \frac{dX}{dt}\right) - \left(\frac{1}{Y_{ps}} \cdot \frac{dP}{dt}\right)$	**Aiba**
6.	$\mu_{\max} \cdot \left(\frac{S}{K_s+S}\right) \cdot \left(1 - \frac{S}{S_{\max}}\right) \cdot n_x$	$q_{p\max} \cdot \left(\frac{S}{K_{sp}+S}\right) \cdot \left(1 - \frac{S}{S_{\max}}\right) \cdot n_x$	$-\left(\frac{1}{Y_{xs}} \cdot \frac{dX}{dt}\right) - \left(\frac{1}{Y_{ps}} \cdot \frac{dP}{dt}\right)$	**Loung**
7.	$\mu_{\max} \cdot \left(\frac{S}{K_s+S}\right) \cdot \left(1 - \frac{P}{P_{\max}}\right) \cdot n_x$	$q_{p\max} \cdot \left(\frac{S}{K_{sp}+S}\right) \cdot \left(1 - \frac{P}{P_{\max}}\right) \cdot n_x$	$-\left(\frac{1}{Y_{xs}} \cdot \frac{dX}{dt}\right) - \left(\frac{1}{Y_{ps}} \cdot \frac{dP}{dt}\right)$	**Levenspiel**
8.	$\mu_{\max} \cdot \left(\frac{S}{K_s+S}\right) \cdot \exp(-K_p \cdot P) \cdot X$	$q_{p\max} \cdot \left(\frac{S}{K_{sp}+S}\right) \cdot \exp(-K_{pp} \cdot P) \cdot X$	$-\left(\frac{1}{Y_{xs}} \cdot \frac{dX}{dt}\right) - \left(\frac{1}{Y_{ps}} \cdot \frac{dP}{dt}\right)$	**Aiba**
9.	$\mu_{\max} \cdot \left(\frac{S}{K_s+S}\right) \cdot \left(\frac{K_{Pi}}{K_{Pi}+P}\right) \cdot X$	$q_{p\max} \cdot \left(\frac{S}{K_{sp}+S}\right) \cdot \left(\frac{K_{ppi}}{K_{ppi}+P}\right) \cdot X$	$-\left(\frac{1}{Y_{xs}} \cdot \frac{dX}{dt}\right) - \left(\frac{1}{Y_{ps}} \cdot \frac{dP}{dt}\right)$	**Jerusalimsky**
10.	$\mu_{\max} \cdot \left(1 - \frac{P}{P_{\max}}\right) \cdot X$	$q_{p\max} \cdot \left(1 - \frac{P}{P_{\max}}\right) \cdot X$	$-\left(\frac{1}{Y_{xs}} \cdot \frac{dX}{dt}\right) - \left(\frac{1}{Y_{ps}} \cdot \frac{dP}{dt}\right)$	**Ghose and Tyagi**
11.	$\mu_{\max} \cdot \left(\frac{S}{K_s+S}\right) \cdot (1 - K_p \cdot P) \cdot X$	$q_{p\max} \cdot \left(\frac{S}{K_{sp}+S}\right) \cdot (1 - K_{pp} \cdot P) \cdot X$	$-\left(\frac{1}{Y_{xs}} \cdot \frac{dX}{dt}\right) - \left(\frac{1}{Y_{ps}} \cdot \frac{dP}{dt}\right)$	**Hinshelwood**

Table 2. The empirical models most used in fermentation industrial processes simulation.

2.2.5. A batch bioreactor modeling

A batch bioreactor is a closed system that is fed at the beginning of the process with sterile medium (with S_0 initial concentration) and pure culture of the microorganisms with X_0 initial concentration. The microorganisms will grow up by multiplication, and eventually they will produce metabolisms products by substrate consummation. Inside the bioreactor, the conditions for multiplication will be established: pH, temperature, oxygen feeding (at aerobic processes), and stirring (at processes with liquid substrate).

Suppose that the growth rate of the biomass depends only by the cells' mass, the mass balance equations could be written for

biomass: $\frac{dX}{dt} = r_x$;

substrate: $\frac{dS}{dt} = r_s$;

metabolism product: $\frac{dP}{dt} = r_p$.

One of the simplest models that respect the previous equation is those of Malthus model:

$$r_x = \mu \cdot X \tag{33}$$

where μ is constant; the equation corresponds to exponential growing phase.

The specific growth rate μ can be described by Monod model:

$$\mu = \frac{\mu_{\max} \cdot S}{K_s + S} \tag{34}$$

with the application of correction factors at the inhibitory process produce by the substrate or metabolism product.

With the forming rate of biomass r_x (33), the consuming rate of the substrate r_s and the forming rate of the metabolite r_p, which have been presented at the kinetics model, could be written by the following equations:

$$\text{biomass}: \frac{dX}{dt} = \mu \cdot X \tag{35}$$

$$\text{substrate}: \frac{dS}{dt} = -\frac{\mu \cdot X}{Y'_{xs}} - \frac{k_1 \cdot X + k_2 \cdot \mu \cdot X}{Y'_{ps}} - m_s \cdot X \tag{36}$$

$$\text{metabolism product}: \frac{dP}{dt} = k_1 \cdot X + k_2 \cdot \mu \cdot X \tag{37}$$

The biomass forming

It takes place in the *aerobic* conditions and in Eq. (36) the factor of metabolism products formed will not appear:

$$\frac{dS}{dt} = -\frac{\mu \cdot X}{Y'_{xs}} - m_s \cdot X \tag{38}$$

If the substrate is in excess (in exponential growing phase), this means $S \gg K_s$ and the process issues with maximum growth rate of biomass, the model is more simple. The total efficiency equation of the biomass, at maximum growth rates and constant forming efficiency of the biomass, becomes

$$\frac{dS}{dt} = -\frac{\mu \cdot X}{Y'_{xs}} \tag{39}$$

From Eq. (18) results:

$$S = S_0 - \frac{X - X_0}{Y'_{xs}} \tag{40}$$

By combining the above equation with Eq. (35), the following could be obtained:

$$\mu_{\max} \cdot dt = \frac{K_S + S_0 - \frac{X - X_0}{Y'_{xs}}}{\left(S_0 - \frac{X - X_0}{Y'_{xs}}\right) \cdot X} \cdot dX \tag{41}$$

and by integrating it between $t=0$, $X=X_0$ and t and X the named Monod integrated equation will be obtained, by which the biomass variation against time can be determined:

$$\mu_{\max} \cdot t = \ln\left(\frac{X}{X_0}\right) + \frac{K_s}{S_0 + \frac{X_0}{Y'_{xs}}} \cdot \left[\ln\left(\frac{X}{X_0}\right) - \ln\left(\frac{S_0 - \frac{X - X_0}{Y'_{xs}}}{S_0}\right)\right] \tag{42}$$

2.2.5.1. The metabolism products forming

2.2.5.1.1. Aerobic processes

In order to model this type of the processes, Eqs. (35)–(37) will be used. If μ will be considered constant and if the total efficiencies of the biomass and product will be introduced, the variation of the substrate quantity will be given by Eq. (36) in which the Y'_{xs} efficiency has been introduced, which means Eq. (39):

$$\frac{dS}{dt} = -\frac{\mu \cdot X}{Y'_{xs}} \tag{43}$$

and the equation that expresses the reaction product quantity variation will be obtained from the total efficiency of the product equation, Y'_{ps}, and Eq. (43):

$$\frac{dP}{dt} = \frac{Y'_{ps}}{Y'_{xs}} \cdot \mu \cdot X \tag{44}$$

2.2.5.1.2. Anaerobic processes

The equations that characterize the cultivation model in the aerobiosis conditions of the microorganisms, where the products of the reaction (usually primary metabolites) are formed always, are the mass balance equations, as follows:

$$\text{biomass}: \frac{dX}{dt} = \mu \cdot X \tag{45}$$

$$\text{substrate}: \frac{dS}{dt} = -\frac{\mu \cdot X}{Y'_{xs}} - m_s \cdot X \tag{46}$$

$$\text{metabolism product}: \frac{dP}{dt} = \frac{Y'_{ps}}{Y'_{xs}} \cdot \mu \cdot X + m_p \cdot X \tag{47}$$

The biomass and metabolism production will be completed when the substrate will be wasted.

In case the specific growth rate remains constant, simplifications can be made. Keeping tabs of total biomass and metabolism products efficiencies, Eqs. (46) and (47) become

$$\frac{dS}{dt} = -\frac{\mu \cdot X}{Y_{xs}} \tag{48}$$

$$\frac{dP}{dt} = \frac{Y_{ps}}{Y_{xs}} \cdot \mu \cdot X \tag{49}$$

which are identical with the equations that describe the aerobic fermentation. Unfortunately, during the anaerobic fermentation the biomass growth rate is inhibited by the metabolism products and the model becomes more complex. An example is the inhibition of the biomass growth in an alcoholic fermentation by the concentration of formed ethanol:

$$\mu = \frac{\mu_{max} \cdot S}{K_s + S} \cdot \left(1 - \frac{P}{P_{max}}\right) \tag{50}$$

where P_{max} represents the maximum concentration of ethylic alcohol at which the biomass production is stopped.

2.2.5.1.3. Energetically model for the batch bioreactor

The balance energy for reaction medium and bioreactor's jacket can been written as

$$\frac{\Delta Hr \cdot \frac{dS}{dt}}{\rho \cdot c_p} - \frac{K_T \cdot A_T}{V \cdot \rho \cdot c_p}\left(T^0 - T^0_{ag}\right) = \frac{dT^0}{dt} \tag{51}$$

$$\frac{F_{ag}}{V_{ag}}\left(T^0_{agi} - T^0_{ag}\right) + \frac{K_T A_T}{V_{ag} \cdot \rho_{ag} \cdot c_{pag}}\left(T^0 - T^0_{ag}\right) = \frac{dT^0_{ag}}{dt} \tag{52}$$

where ΔH_r is the reaction heat of the fermentation [*J/mol of the produced alcohol*]; K_T is the heat transfer coefficient [W/m^2.K] ; A_T is the heat transfer area [m^2]; ρ, ρ_{ag} is the density of the mass of the reaction, respectively, of the cooling agent [kg/m^3]; c_p, c_{pag} is the heat capacity of the mass of the reaction, respectively, of the cooling agent [J/kg.K]; V, $V_{ag}$is the fermentation medium volume, volume of the jacket [m^3]; F_{ag} is the flow of the cooling agent [m^3/h]; T^0 and $T^0{}_{ag}$are the temperature of the fermentation medium and temperature of the cooling agent in the jacket [K].

3. Study case—modeling of a white wine alcoholic fermentation process

Wine making is a complex ecological and biochemical process involving many interactions such as grape variety, microbiota, and several technological operations. The process' variables are often controlled empirically and traditionally. There are some factors that strongly affect the alcoholic fermentation. The most important ones are fermentation temperature, grape juice composition, anaerobic conditionsdue to CO_2 production, low media pH, sulfur dioxide concentration level, selected yeasts inoculation, and interaction with other microorganisms [21]. The models developed for these cases consequently have various domains of applications but none of them include the whole oenological aspects of the process. The majority of the models are of "knowledge-based" models type and they take into consideration a great number of phenomena that have an important effect on the kinetics of the process fermentation [22].

This part of the chapter proposes a complex nonlinear wine fermentation model based on previous researches by the author [23–25].

3.1. Experimental conditions

To evaluate the total fermentation yield losses under different operating conditions, four experiments were carried out and based on the data obtained within these experiments, a mathematical model was proposed. The strain and the culture medium, the equipment and the experimental conditions together with the measurements of the fermentation parameters were presented by Sipos and coworkers [23–25]. For the experiments, the *Saccharomyces cerevisiae* YEPD wine yeast was used, being seeded on a culture medium with the following composition: 5 g/L KH_2PO_4, 2 g/L $(NH_4)_2SO_4$, 0.4 g/L $(MgSO_4)\cdot 7H_2O$, 1 g/L yeast extract, 50 g/L glucose, and Mauzac must (sterilized through flash pasteurization). The sugar content of the grape must was supplemented with sucrose up to 180 g/L and 40 mL/h tiaminol was added. The SO_2 content reached 50 mg/L and the pH was adjusted at 3.8 mg/L H_3PO_4. Both the fermentation medium and the bioreactor were autoclaved for 20 min at 393 K. A New-Brunswick continuously stirred bioreactor equipped with *pH* and temperature sensors was used.

The following operating conditions were as follows: working volume, 8 L; temperature, 291 and 301 K; stirring speed, 150 rpm; pH, 3.8; influent glucose concentrations, 180 and 210 g/L; without aeration, the necessary oxygen was dissolved in must.

3.2. The mathematical model

The mathematical model of the alcoholic fermentation process was determined on the basis of the approach of the zone modeling principle, taking into consideration the evolution of the viable biomass ($X_v(t)$). Based on the analysis of the phenomenological aspects of the alcoholic fermentation process, the evolution of $X_v(t)$ was divided into three parts (**Figure 5**) as follows:

- latent phase (1);
- growing phase (2);
- decay phase (3).

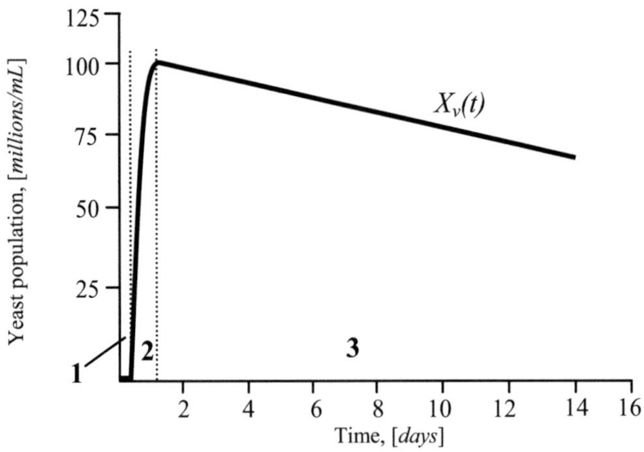

Figure 5. Evolution of the viable biomass concentration $X_v(t)$.

Table 3 presents the equations of the model. The parameters are adjusted through the nonlinear programming method, which compares the model predictions with experimental data and minimizes the errors.

Tables 4 and **5** present the list of the variables and parameters of the mathematical model.

Current phase	Model equations
Kinetic model	
Latent phase [23, 24]	$t_{lat} = \frac{a}{T^0} + b$
Exponential growing phase [1]	- biomass: $\frac{dX}{dt} = \mu_{\max} \cdot \left(\frac{S}{K_S+S}\right) \cdot e^{-K_p \cdot P} \cdot X;\ \mu_{\max} = A_1 \cdot e^{-\frac{E_{a1}}{R \cdot T^0}} - A_2 \cdot e^{-\frac{E_{a2}}{R \cdot T^0}}$ - alcohol: $\frac{dP}{dt} = q_{p\max} \cdot \left(\frac{S}{K_{SP}+S}\right) \cdot e^{-K_{pp} \cdot P} \cdot X$ - substrate: $\frac{dS}{dt} = -\left(\frac{1}{Y_{XS}} \cdot \frac{dX}{dt}\right) - \left(\frac{1}{Y_{PS}} \cdot \frac{dP}{dt}\right)$
Decay phase [23–26]	- biomass: $\frac{dX}{dt} = f \cdot X \cdot k;\ k = A \cdot e^{-\frac{Ea}{R \cdot T^0}}$ - alcohol: $P = P_0 + \eta \cdot (S_0 - S)$ - substrate: $\frac{dS}{dt} = -k \cdot S^{\alpha} \cdot P^{\beta}$
All phases	- carbon dioxide concentration: $\frac{dC_{CO_2}}{dt} = g \cdot C_{CO_2} \cdot k \cdot \frac{S}{K_{SP}+S} \cdot \ln\left(k \cdot \frac{S}{K_{SP}+S} \cdot t\right)$
Energetic model	
All phases [23, 26]	- for the bioreactor: $\frac{\Delta Hr \cdot \frac{dS}{dt}}{\rho \cdot c_p} - \frac{K_T \cdot A_T}{V \cdot \rho \cdot c_p}\left(T^0 - T^0_{ag}\right) = \frac{dT^0}{dt}$ - for the bioreactor's jacket: $\frac{F_{ag}}{V_{ag}}\left(T^0_{agi} - T^0_{ag}\right) + \frac{K_T A_T}{V_{ag} \cdot \rho_{ag} \cdot c_{pag}}\left(T^0 - T^0_{ag}\right) = \frac{dT^0_{ag}}{dt}$

Table 3. The model of the alcoholic fermentation process.

http://dx.doi.org/10.5772/intechopen.71711

X	Biomass concentration		g/L
S	Substrate concentration		g/L
P	Alcohol concentration		g/L
k	Kinetic constant		1/h
A	Pre-exponential factor in Arrhenius' equation	148 (calculated using experimental data)	1/h
E_a	Activation energy	21,424 (calculated using experimental data)	J/mol
A_1	Pre-exponential factor in Arrhenius' equation	9.5×10^8 [a]	1/h
E_{a1}	Activation energy	55,000[a]	J/mol
A_2	Pre-exponential factor in Arrhenius' equation	2.55×10^{33}	1/h
E_{a2}	Activation energy	220,000[a]	J/mol
R	Universal gas constant	8.31	J/mol·K
T^0	Temperature in bioreactor	291 and 301	K
K_s	Substrate limitation constant	0.2[a]	g/L
d	Pseudo-constant of the biomass	1.67 (calculated using experimental data)	
f	Pseudo-constant of the biomass	0.34	
α	Pseudo-order of the substrate	0.69[b]	
β	Pseudo-order of the alcohol	0.32[b]	
η	Efficiency in alcohol of fermentation reaction	48[b]	%
S_0	Steady-state operation point of substrate	180	g/L
P_0	Steady-state operation point of alcohol	0	g/L
t	Time		h
μ_{max}	Maximum specific growth rate		1/h
K_P	Alcohol limitation constant	0.14[c]	g/L
q_{pmax}	Maximum specific alcohol production rate	1.02[c]	g/g·cells·h
K_{SP}	Constant in the substrate term for ethanol production	1.68[c]	g/L
K_{PP}	Constant of fermentation inhibition by ethanol	0.07[d]	g/L
Y_{XS}	Ratio of cell produced per glucose consumed for growth	0.607[d]	g/g
Y_{PS}	Ratio of ethanol produced per glucose consumed for fermentation	0.435[c]	g/g

[a][6, 27, 28]
[b][26]
[c][29]
[d][16]

Table 4. Variables and parameters of the kinetic model.

K_T	Heat transfer coefficient	3.6×10^{5} [a]	$J/m^2 \cdot K \cdot h$
A_T	Heat transfer area	0.8[b]	m^2
F_{ag}	Flow of cooling agent	0.01[b]	m^3/h
V_{ag}	Volume of the jacket	0.2[b]	m^3
V	Volume of the mass of reaction	1[b]	m^3
T^0_{agi}	Temperature of cooling agent entering to the jacket	278[b]	K
ΔH_r	Reaction heat of fermentation	−98465[c]	J/mol
ρ	Density of the mass of reaction	1100[b]	kg/m^3
ρ_{ag}	Density of cooling agent	999.8[a]	kg/m^3
c_p	Heat capacity of mass of reaction	3391[b]	$J/kg \cdot K$
c_{pag}	Heat capacity of cooling agent	4217[a]	$J/kg \cdot K$
T^0_{ag}	Temperature of cooling agent in the jacket		K

[a] [17]
[b] experimental data
[c] [30]

Table 5. Parameters of the kinetic model.

3.3. Result and discussion regarding the mathematical model simulation

The nonlinear mathematical model of the batch fermentation process (**Table 3**) used in this work contains the following equations:

- an equation for the latent phase of fermentation that describes the dependence of the phase time of the process temperature;
- the model proposed by Aiba [1] for the growing phase with the three equations of biomass, alcohol production, and substrate consumption;
- the model presented by Bovée-Strehaiano [26] for the decay phase with two equations: one for the substrate consumption and the other for alcohol formation;
- an equation that describes the biomass behavior along the phase no. 3 (the model proposed by Sipos in [23–25]);
- an equation that describes the carbon dioxide concentration behavior along all the phases (the model proposed by Sipos);
- an energy balance model in which the rate of change of the medium's temperature (dT^0/dt) is a result of the balance between the rate of the heat generation due to fermentation and the rate of the heat transfer to the cooling medium inside the bioreactor jacket.

The model proposed by Aiba [1] includes the inhibitory effects of the fermentation product (alcohol). In the growing phase, the value of the maximum specific growth rate of the biomass corresponds to the real one. The non-physiological model proposed by Bovée and Strehaiano [26] was chosen because it accurately describes the substrate consumption and the evolution of the alcohol concentration in the growing and decay phases. This model proposes the use of a semi-empirical model in which the velocity of sugar consumption is described by a chemical law that depends on substrate and product contents. The parameters of the model are adjusted by means of nonlinear programming methods, which compare model predictions with experimental data and minimize errors [23–25]. The Bovée and Strehaiano model is capable of predicting the fermented sugar (and thus thermal planning) within an error of 3.3% [25]. Thus, the model offers a good qualitative and quantitative description of the behavior of the alcoholic fermentation process.

Figures 6–8 show the simulation results of the model presented in **Table 3** considering the following initial values: the initial substrate concentration was 210 g/L and the fermentation temperature was 301 K.

The equation of the latent phase is valid for a time interval [0, 100 h] and the model has been tested for a grape juice variety with an initial concentration of the substrate varying between 180 and 210 g/L, a fermentation temperature between 299 and 303 K and without aeration.

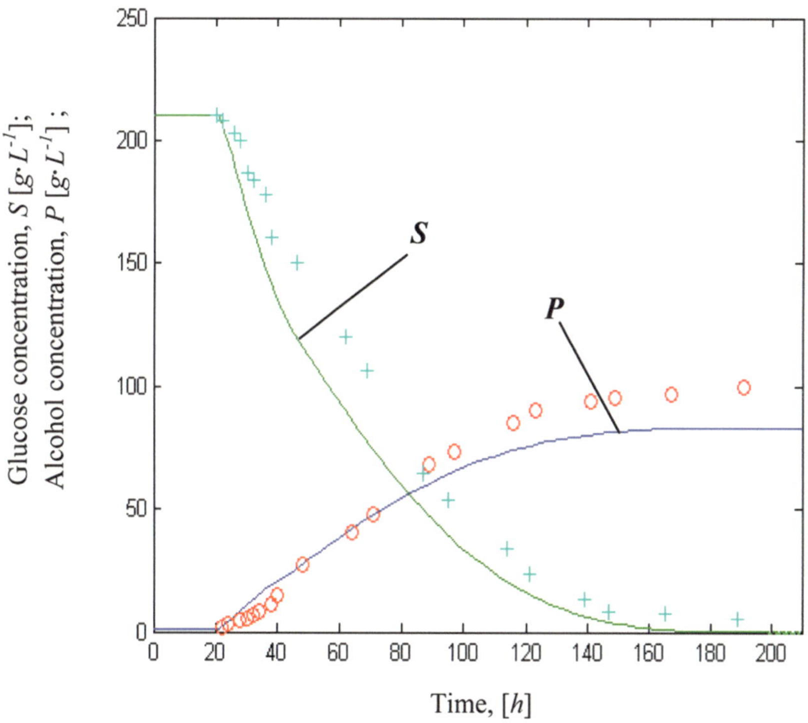

Figure 6. Evolution of glucose and alcohol concentrations; a comparison between experimental values (**o**—glucose and **+**—alcohol) and simulation results (continuous lines).

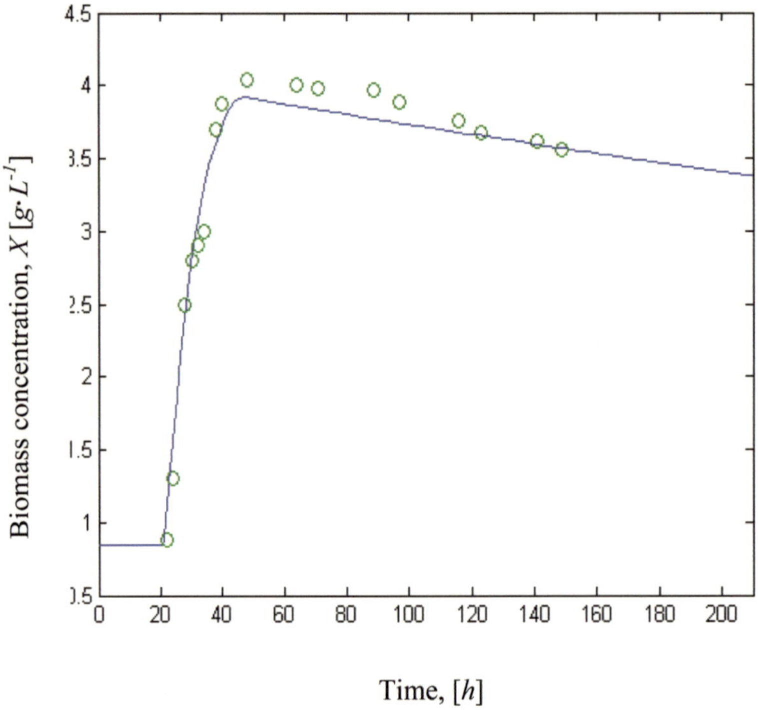

Figure 7. Comparison between the biomass simulation results (continuous line) and experimental data (o).

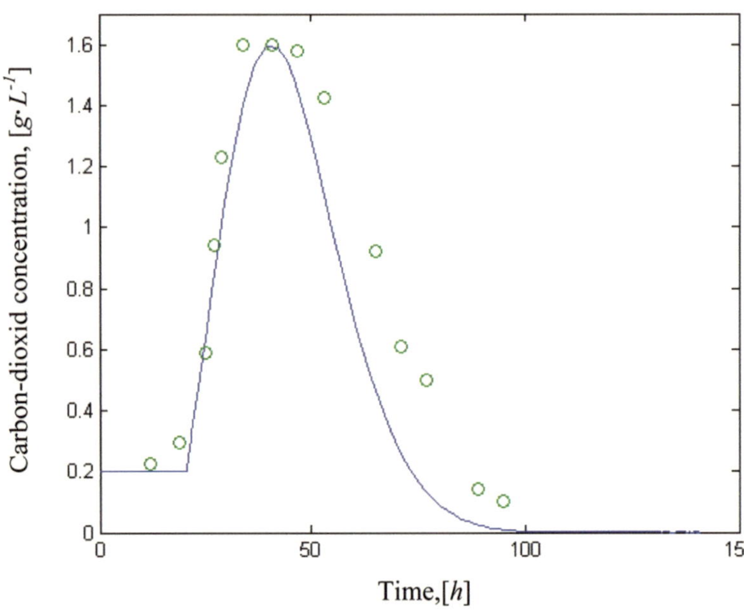

Figure 8. Comparison between the carbon dioxide concentration simulation results (continuous line) and experimental data (o).

Author details

Anca Şipoş

Address all correspondence to: anca.sipos@ulbsibiu.ro

"Lucian Blaga" University of Sibiu, Romania

References

[1] Caraman S, Ceangă E, Frangu L, Mencinicopschi G. Modelarea şi conducerea proceselor biotehnologice (The modelling and control of biotechnological processes). Bucharest: Editura Didactică şi Pedagogică; 2002

[2] Mitrana V. Bioinformatica – unde se întâlnesc: biologia, informatica, matematica? (Bioinformatics - where is met: the biology, informatics and mathematic?). Bucharest: Editura L and S INFOMAT; 1998

[3] Vogel HC, Tadaro CL. Fermentation and biochemical engineering handbook – Principles. 2nd ed. Process Design and Equipment. Noyes: William Andrew Publishing; 1996. Chap.1

[4] Xiong Z, Zhang J. Mint. Modelling and optimal control of fed-batch processes using a novel control affine feedforward neural network. Neurocomputing. 2004;**61**:317-337. DOI: 10.1016/j.neucom.2003.11.006

[5] Tijskens LMM, Hertog MLATM, Nicolaï BM. Food process modeling. Abigton Hall, Cambridge: Woodhead Publishing Series in Food Science, Technology and Nutrition; 2001

[6] Valentinotti S, Srinivasan B, Holmberg U, Bonvin D, Cannizzaro C, Rhiel M, von Stockar U. Optimal operation of fed-batch fermentation via adaptive control of overflow metabolite. Control Engineering Practice. 2003;**11**:665-674. DOI: 10.1016/S0967-0661(02)00172-7

[7] Szederkenyi G, Kristensen NR, Hangos KM, Bay Jorgensen S. Nonlinear analysis and control of a continuous fermentation process. Computers and Chemical Engineering. 2002;**26**:659-670. DOI: 10.1016/S0098-1354(01)00793-1

[8] Zhang Y, Henson MA, Kevrekidis YG. Nonlinear model reduction for dynamic analysis of cell population models. Chemical Engineering Science. 2003;**58**:429-445. DOI: 10.1016/S0009-2509(02)00439-6

[9] Vlassides S, Ferrier JG, Block DE. Using historical data for bioprocess optimization: Modeling wine characteristics using artificial neural networks and archived process information. Biotechnology and Bioengineering. 2001;**73**:55-68. DOI: 10.1002/1097-0290(20010405)73:1%3C55::AID-BIT1036%3E3.0.CO;2-5/full

[10] Banga JR, Balsa-Canto E, Moles CG, Alonso AA. Improving food processing using modern optimization methods; Trends in Food Science and Technology. 2003;**14**:131-144. DOI: 10.1016/S0924-2244(03)00048-7

[11] Harms P, Kostov Y, Rao G. Bioprocess monitoring. Current Opinion in Biotechnology. 2002;**13**:124-127. DOI: 10.1016/S0958-1669(02)00295-1

[12] Karakuzu C, Turker M, Ozturk S: Modelling, on-line state estimation and fuzzy control of production scale fed-batch baker's yeast fermentation. Control Engineering Practice. 2006;**14**:959-974. DOI: 10.1016/j.conengprac.2005.05.007

[13] Kress-Rogers E, Brimelow CJB. Instrumentation and sensors for the food industry. 2nd ed. Abigton Hall, Cambridge: Woodhead Publishing Series in Food Science, Technology and Nutrition; 2001. pp. 714-739

[14] Ferreira LS, De Souza MB Jr, Trierweiler JO, Broxtermann O, Folly ROM, Hitzmann B. Aspects concerning the use of biosensors for process control: experimental and simulation investigations; Computers and Chemical Engineering. 2003;**27**:1165-1173. DOI: 10.1016/S0098-1354(03)00044-9

[15] Ferreira LS, De Souza MB Jr, Folly ROM. Development of an alcohol fermentation control system based on biosensor measurements interpreted by neural networks; Sensors and Actuators B. 2001;**75**:166-171. DOI: 10.1016/S0098-1354(03)00044-9

[16] Torija Ma J, Rozes N, Poblet M, Guillamon JM, Mas A. Effects of fermentation temperature on the strain population of *Saccharomyces cerevisiae*. International Journal of Food Microbiology. 2003;**80**:47-53. DOI: 10.1016/S0168-1605(02)00144-7

[17] El-Mansi EMT, Bryce CFA. Fermentation Microbiology and Biotechnology. London: CRC Press Taylor & Francis Group; 1999

[18] Hall R, Zacune J. Bio-economies: the EU's Real 'Green Economy' Agenda? World Development Movement and the Transnational Institute, Rio de Janeiro; 2012

[19] Bastin G, Dochain D. On-line estimation and adaptive control of bioreactors. In: Process Measurement and Control. Amsterdam: Elsevier Science Publishers B.V.; 1990. Chap. 5

[20] Roels JA. Energetics and Kinetics in Biotechnology. Amsterdam: Elsevier Biomedical Press; 1983. Chap. 4

[21] Ribereau-Gayon P, Dubourdieu D, Doneche B, Lonvaud A. Handbook of Enology. The Microbiology of Wine and Vinifications; vol. I; West Sussex, England: Wiley; 2000

[22] Svendsen C, Skov T, van den Berg FWJ. Monitoring fermentation processes using in-process measurements of different orders. Journal of Chemical Technology and Biotechnology. 2015;90. DOI: 10.1002/JCTB.4483

[23] Sipos A, Meyer XM, Strehaiano P. Development of a non-linear, dynamic mathematical model for the alcoholic fermentation. Acta Alimentaria. 2007;**34**:429-438. DOI: 10.1556/AAlim.2007.0014

[24] Șipoș A, Imre-Lucaci A. Statistical processing and dynamic modeling of an alcoholic fermentation process. Studia Universitatis Babeș-Bolyai Chemia. 2014;**3**:17-28

[25] Șipoș A, Agachi ŞP. Direct sensitivity analysis of a white wine alcoholic fermentation process. Studia Universitatis Babeș-Bolyai Chemia. 2015;**60**:125-141

[26] Bovée JP, Strehaiano P, Goma G, Sevely Y. Alcoholic fermentation: modelling based on sole substrate and product measurement. Biotechnology and Bioengineering. 1984;**26**:328-334. DOI: 10.1002/bit.260260406

[27] Krothapally M, Palanki S. A neural network strategy for end-point optimization of batch processes. ISA Transactions. 1999;**38**:383. DOI: 10.1016/S0019-0578(99)00031-2

[28] Lei F, Rotboll M, Jorgensen SB. A biochemically structured model for *Saccharomyces cerevisiae*. Journal of Biotechnology. 2001;**88**:205-221. DOI: 10.1016/S0168-1656(01)00269-3

[29] Dengfeng L, Ling X, Weili X, Hong-Tao Z, Chi-Chung L, Lihua J, Baoguo X. Hindawi fermentation process modeling with Levenberg-Marquardt algorithm and Runge-Kutta method on ethanol production by *Saccharomyces cerevisiae*. Mathematical Problems in Engineering. 2014;**2014**:1-10. DOI: 10.1155/2014/289492

[30] Costa AC, Atala DIP, Maugeri F, Maciel R. Factorial design and simulation for the optimization and determination of control structures for an extractive alcoholic fermentation. Process Biochemistry. 2001;**37**:125-137. DOI: 10.1016/S0032-9592(01)00188-1

Microbiological, Physical, and Chemical Procedures to Elaborate High-Quality SO_2-Free Wines

Raúl Ferrer-Gallego, Miquel Puxeu, Laura Martín,
Enric Nart, Claudio Hidalgo and Imma Andorrà

Additional information is available at the end of the chapter

http://dx.doi.org/10.5772/intechopen.71627

Abstract

Sulfur dioxide (SO_2) is the most preservative used in the wine industry and has been widely applied, as antioxidant and antibacterial agent. However, the use of sulfur dioxide implicates a range of adverse clinical effects. Therefore, the replacement of the SO_2 content in wines is one of the most important challenges for scientist and winemakers. This book chapter gives an overview regarding different microbiological, physical, and chemical alternatives to elaborate high-quality SO_2-free wines. In the present chapter, original research articles as well as review articles and results obtained by the research group of the Wine Technology Center (VITEC) are shown. This study provides useful information related to this novel and healthy type of wines, highlighting the development of winemaking strategies and procedures.

Keywords: food safety, grape juice, sensory analysis, sulfur dioxide, wine

1. Introduction

In the last decades, the use of the sulfur dioxide (SO_2) has become indispensable in the food industry. This substance is widely applied as antioxidant and antibacterial in many processed foods, being the most preservative used in the wine industry. In wines, SO_2 prevents undesirable sensory properties and the spoilage of wines produced by chemical or microbiological agents. However, in recent times, it has been shown that the intake of SO_2 implicates a wide range of adverse health consequences, such as allergic reactions and cumulative harmful effects [1]. Therefore, negative perceptions toward sulfites have been induced, and a significant increase on the demand of wines with low content of SO_2 has been displayed by consumers in the last years [2]. For this reason, reducing the amount of SO_2 in wines is a decisive strategy for the wine industry and one of the current topics on the oenological science.

In wines, SO_2 is composed by total SO_2, bound SO_2, free SO_2, and molecular SO_2. Proper adjustment of the SO_2 dosage is difficult because it depends on the equilibrium between its free and bound forms. The active form is molecular SO_2, which depends on the concentration of free SO_2 and the pH [3]. This active form has the antimicrobial and antioxidant properties. In terms of antimicrobial, an insufficient addition of SO_2 will not ensure the wine protection, increasing the risk of yeast and bacteria proliferation. In terms of antioxidant, an inadequate dosage will allow an excessive oxidation of aromas and flavors, compromising the quality of wines [4]. Contrary, excessive dosages in wines may cause organoleptic alterations and also health reactions in consumers. Taking this into account, the International Organization of Vine and Wine (OIV) has progressively reduced the maximum limits of the total SO_2 in wines, which is nowadays 150 mg/L for red wines and 200 mg/L for white wines, with some exceptions depending on the sugar content (Regulation (EC) No 607/2009).

Today, there is not a commercial product or recipe able to replace the widespread SO_2 actions. Consequently, diverse technological strategies should be considered by winemakers in each stage of the winemaking process, according to the type of wine to be produced and the winery capabilities. From our point of view, these strategies should be addressed from three joint perspectives; microbiological strategies, physical technologies, and chemical treatments. In this sense, the Wine Technology Centre (VITEC) has been working in this research field since 2012. Our studies have been focused in red and white wines, especially regarding Tempranillo and Albariño grape varieties.

2. Microbiological strategies to elaborate SO_2-free wines

From a microbiological point of view, many factors should be taken into account to reduce the quantity of SO_2 in wines. First, it should be considered that an endogenous content of SO_2 is naturally produced by yeasts during alcoholic fermentation. Second, grape juice composition, yeast nutrition, and fermentation management may strongly influence the ability of yeasts to produce sulfites. Finally, microbiological stability of the SO_2-free wines remains uncertain yet.

As mentioned above, the European Union regulates the levels of total sulfites in wines following the Regulation (EC) 607/2009. Therefore, wines must be labeled with the indication "contains sulfites," when the total content of SO_2 is over 10 mg/L, either exogenous or endogenous. Most organisms produce sulfites as a normal intermediate during digestion or synthesis of the sulfur-containing amino acids, such as methionine and cysteine [5]. Sulfites are minor by-products of yeast fermentation, and therefore, they are natural wine constituents. The ability of yeasts to form SO_2 has been reported in different types of wines and geographical areas, and it was known long time ago and investigated intensively over the years [6, 7].

One of the most important factors to elaborate SO_2-free wines is the choice of the suitable yeast strains used for the development of the alcoholic fermentation. During winemaking process, sulfur (naturally available as sulfate in grape juice) is used by yeasts in the synthesis of amino acids. In particular, *Saccharomyces cerevisiae* produces sulfite as an intermediate product during the assimilatory reduction of sulfate to sulfide, via adenosine-5′-phosphosulfate [6, 8]. The available sulfide (S^{2-}) can be used in the synthesis of amino acids, as well as being excreted as hydrogen sulfide (H_2S). Eventually, the sulfur amino acid biosynthesis (SAAB) pathway plays a crucial role in the active transport of sulfate (SO_4^{2-}) into the cell, as well as in the reduction and production of SO_2 and in the resistance of yeasts against this additive [9]. Yeast strains differ in their capacity to form SO_2, estimating a total average content ranged from 0 to 115 mg/L [10–14]. Most strains of *S. cerevisiae* produce between 10 and 30 mg/L of total SO_2. However, some of them may produce less than 10 mg/L, which were commonly called "low sulfite-forming strains" [6]. On the opposite side, "high sulfite-forming strains" are able to produce more than 100 mg/L. These classifications according to their ability to form SO_2 during the alcoholic fermentation have been reported by several authors over the time [6, 7, 12, 14].

In the last years, the use of yeast strains with a low capacity to produce SO_2 has been one of the most used strategies to reduce the amount of SO_2 in wines [15]. Several studies have compared the amount of SO_2 produced during alcoholic fermentation by different commercial and indigenous yeast strains. In 1985, Suzzi et al. [13] investigated the biological sulfite role in the stabilization of white wines by comparing 1700 strains of *Saccharomyces* isolated from spontaneous fermentations. The majority of them produced less than 10 mg/L of total SO_2, around 350 produced between 10 and 20 mg/L, 52 strains produced between 20 and 40 mg/L, and just two strains produced more than 40 mg/L. More recently, an experiment carried out at industrial scale by Werner et al. [14] showed two distinguishable groups of yeasts, among 22 commercial strains. The first one produced under 10 mg/L of total SO_2, and the second one produced between 10 and 20 mg/L. Significant differences among yeasts strains in production of SO_2 (free and bound-SO_2) were also described by Wells and Osborne [7]. In this case, values ranged from 25 to 60 mg/L of bound-SO_2 were observed. In 2015, Miranda-Castilleja et al. [11] studied the production of total SO_2 of 52 indigenous species of *Saccharomyces* from Querétaro (Mexico), and the obtained results ranged from 37 to 115 mg/L. More recently, VITEC has investigated the natural production of SO_2 of 21 selected yeast strains (commercial and indigenous). Fermentations were conducted using Muscat grape juice at 18 and 25°C. These results showed a total SO_2 production lesser than 10 mg/L in all cases. The results in agreement with other works which also showed diverse yeast strains are able to produce small amounts of total SO_2 (<1.4 mg/L) [16, 17]. Thus, several commercial and indigenous yeast strains have proved to be able to produce SO_2-free wines. However, other considerations should be taking into account, such as the organoleptic properties and microbial stability of this type of wines.

The formation of SO_2 by yeasts is influenced by a complex interaction of genetic, physiochemical, and metabolic factors. H_2S is one of the most undesirable metabolites derived from the alcoholic fermentations due to its unpleasant smell and taste. It should be noted that the biosynthesis and the production of H_2S and SO_2 are linked [18, 19]. As occurs in the case of SO_2, the formation of H_2S varies widely depend on the yeast strains [20, 21]. The release of H_2S

by yeast during the fermentation is a long-standing problem that has been extensively studied in comparison to the SO_2 production. There has been an ever-growing interest in wine yeasts with low production in H_2S. The selection of suitable strains has so far been the principal way of limiting excessive H_2S formation. Other engineering strategies have been used for limiting its production, which generally consisted of overexpression or inactivation of some genes involved in the sulfate reduction pathway [22–24].

Both sulfites and hydrogen sulfides are produced during the biosynthesis of the sulfur containing amino acids, methionine, and cysteine, starting from sulfate assimilation. Given the metabolic link between H_2S and SO_2, such kind of biotechnological and engineering strategies firstly applied to reduce H_2S production could also be applied to decrease SO_2 formation by yeasts. Nonetheless, few works have been aimed to obtain both low SO_2 and low H_2S production. Three strains with low SO_2 production (SO_2 < 10 mg/L) and with reduced H_2S production were selected by De Vero et al [25]. These authors proposed a strategy that combines sexual recombination and specific selective pressure to generate nongenetically-modified *S. cerevisiae* with desired oenological characteristics. More recently, new insight into the regulation of sulfur metabolism in wine yeasts by the identification of variants of MET2 and SKP2 genes within SAAB has been reported to modulate the production of sulfites and sulfides [26]. These results provide novel targets for the improvement of wine yeast strains orientated to produce SO_2-free wines. This knowledge on the sulfate pathway provides a chance to successfully apply engineering strategies to select "low sulfite-forming" yeast strains. However, as we previously highlighted, the production of sulfites by yeast during fermentation not only depend on metabolic factors but also on the environment, including nutrients and fermentation management, among others. Hence, grape juices composition is an imperative factor that should be considered in order to elaborate this type of wines. The insoluble solids contained in the grape juice also appeared to have an effect on the SO_2 content, and wines with the higher insoluble solids obtained lower values of SO_2 [27]. In contrast, results obtained in our experimental cellar showed that grapes with higher content of soluble solids produced higher content of total SO_2 (**Figure 1**). The biplot of the principal component analysis (PCA) shows that the amount of SO_2 produced during the alcoholic fermentation is mainly favored by a high amount of sugars and a low quantity of nitrogen. Furthermore, musts fermented at low temperatures (18°C), and a low titratable acidity may contribute on the production of SO_2.

In addition, the supplementation of musts with amino acids can significantly affect SO_2 and H_2S production depending on the amount added, the time of addition, and the nitrogen concentration [26, 28]. Individual amino acids such as methionine, cysteine, asparagine, and arginine have been shown to influence sulfite formation [18, 28]. Higher the concentration of methionine and cysteine in the grape must, lower the formation of SO_2 [6]. Under ammonia limitations, the addition of nonsulfur amino acids tended to increase the formation of SO_2 (but inhibits the formation of H_2S). The addition of cysteine seems to increase the H_2S content but inhibits the sulfite formation, and the addition of methionine inhibits both SO_2 and H_2S formation [28]. More recently, it was stated that methionine repressed the cysteine-induced increase in the H_2S production but had no effect on the formation of SO_2. Both compounds were produced in greater quantities by yeast when grown in the presence of increasing concentrations of cysteine [18]. It has been reported that yeasts produce higher concentrations

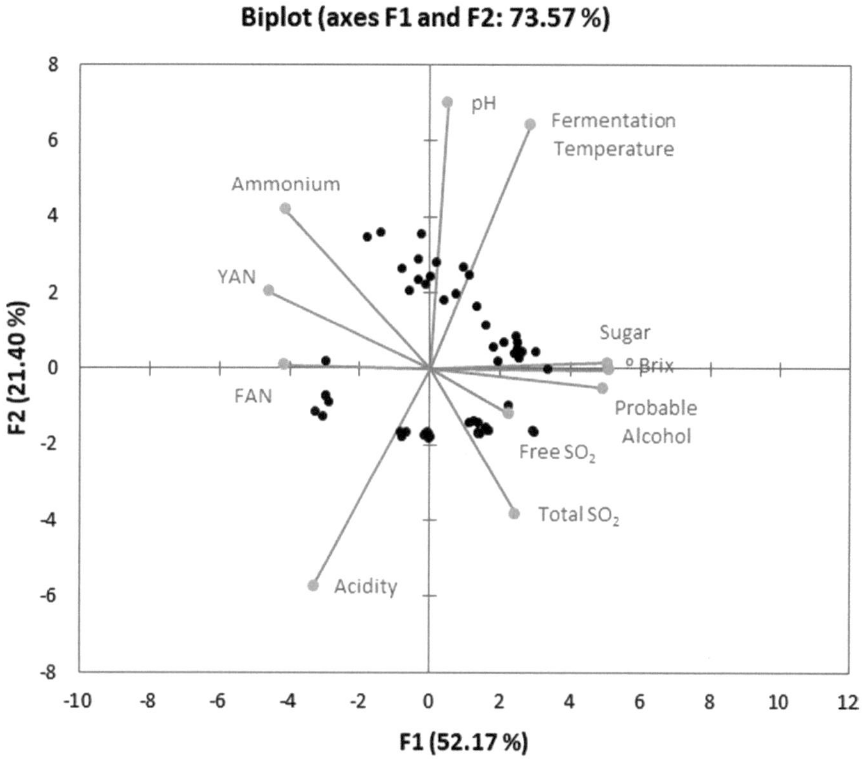

Figure 1. Biplot performed by 74 wines produced from Tempranillo and Albariño musts.

of SO_2 under higher yeast assimilable nitrogen (YAN) quantities [7, 29]. The supplementation on nitrogen using ammonium salts (sulfate or phosphate) allows higher growth rates and biomass yielding and also the stimulation of the fermentative activity [30, 31]. The addition of diammonium phosphate (DAP) significantly decreases H_2S production and improves the kinetics of fermentation and aroma profile of wine [32]. In the last 5 years, VITEC has been studying the effect of ammonium sulfate and DAP addition on the amount of SO_2 produced by yeast along of the alcoholic fermentation. Results obtained showed that the addition of the N-sources slightly increases the total content of SO_2 in wines. The addition of ammonium sulfates and DAP using low sulfite-forming strains to ferment musts showed no significant differences. In the case of musts fermented by "high sulfite-forming" strains, the addition of DAP significantly increased the total content of SO_2 [33].

Other important consideration to elaborate SO_2-free wines is the management of the alcoholic fermentation. In this sense, it has been stated that temperature has several effects on biochemical and physiological properties in yeast cells. Some changes in the sulfur assimilation pathway by *S. cerevisiae* depending on temperature may occur [34]. Our results are in agreement with other authors, who reported that at low temperature, the SO_2 production increases [26]. SO_2 and H_2S production is also affected by pH (acidic pH facilitate SO_2 uptake) and concentration of some minerals (copper and zinc) and vitamins, such as pantothenate or thiamine [9, 26, 35]. Thiamine is a vitamin used as a co-enzyme in the alcoholic fermentation pathway. It stimulates yeast growth, speeds up fermentation, and reduces production of SO_2 binding compounds. Thiamine supplementation allows the transformation of pyruvic acid to acetaldehyde and limits the accumulation of ketonic compounds on wine being considered a factor to reduce the SO_2 amount on wines [36]. A deficiency in thiamine may reduce yeast growth, slow fermentation, and promote the accumulation of pyruvic acid and acetaldehyde, the components responsible of wine oxidation. The effect of major SO_2 binding compounds (acetaldehyde, pyruvic, and α-ketoglutarate) on the production of SO_2 by different yeasts strains is still poorly understood, and more studies should be performed to better understand their role on the SO_2 production [7]. In this way, the results obtained in VITEC are in agreement with the results obtained by Comuzzo and Zironi [33, 36], who showed that the addition of DAP + thiamine reduced the production of α-ketoglutarate.

3. Physical technologies to replace the use of SO_2 in the wine industry

From a physical point of view, different technologies have been used to ensure the wine microbiological stability and to prevent oxidations [37]. The main advantage of using physical methods is the nonaddition of chemical substances that may affect human health. By these technologies, the preservation of the organoleptic properties of wines and the antimicrobial effect should be produced at the same time. Pulsed electric fields (PEF), ultraviolet radiation (UV), high hydrostatic pressure (HHP), and flash-pasteurization lead an antimicrobial result, while the use of ultrasounds (US) or inert gases does not share this property [38–41]. The PEF consists in the application of short electric pulses of high intensity between two electrodes, producing electroporation of the cell membranes increasing their permeability. It has been shown that this technique is effective to inactivate both bacteria and yeasts [42]. Thus, PEF may be applied to eliminate undesirable microorganisms at different winemaking stages, for example, before bottling. It has been stated that the treatments with PEF also reduces the activity of enzymes, such as polyphenol oxidases and peroxidases, increases the extraction of phenolic compounds and affects the aromas of white wines [42, 43]. VITEC has evaluated the antimicrobial effect of PEF, HHP, US, and EMR (electromagnetic radiation). **Figure 2** shows the obtained results after the quantification of viable yeasts and acetic acid bacteria (AAB) in Petri dishes culture. The PEF conditions were electric field 35 kV/cm, voltage 23 kV, pulse rate 0.65 kHz, pulse duration 2.5 μS, initial conductivity 5.04 mS/cm, flow 25 l/h, and initial temperature 20.8°C. The PEF 1 and PEF 2 differed on the final temperature of the treatment which was 23 and 31°C, respectively. Worthy results of PEF as antimicrobial technique were obtained, although high colony-forming units of yeast were observed in the case of PEF 1.

The use of high hydrostatic pressures (HHP) was evaluated in our studies at different pressures (from 400 to 600 MPa) and times (1, 3, 5, and 10 min). HHP results showed that the inhibition of microorganism by this methodology depends not only on the time and pressure

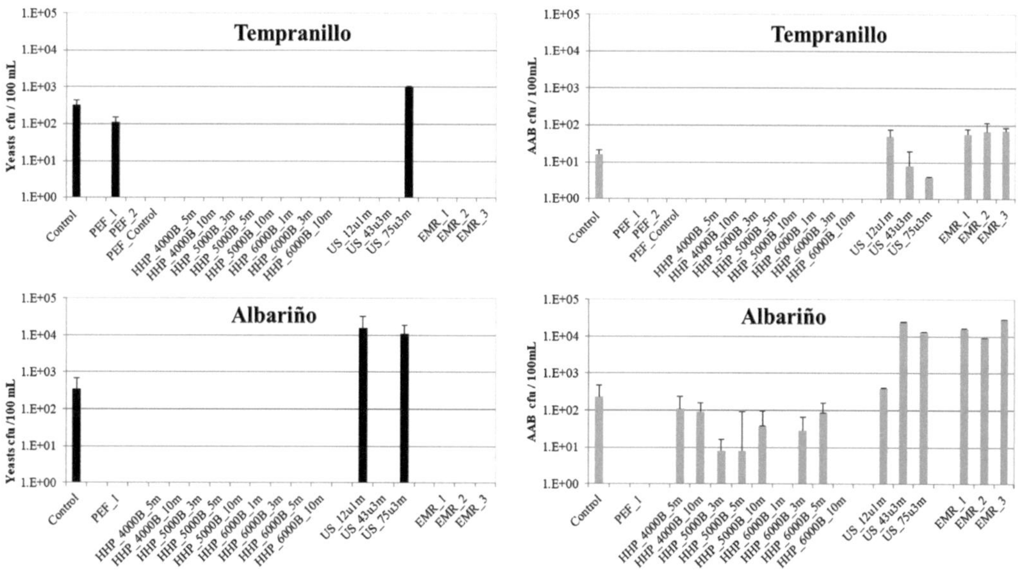

Figure 2. Evaluation of different physical treatments in Tempranillo and Albariño wines (at the end of the alcoholic fermentation) by the quantification of viable yeasts and acetic acid bacteria in Petri dishes culture (cfu, colony forming units).

applied but also on the variety and the type of microorganisms (**Figure 2**). Tempranillo and Albariño yeast growth were inhibited by all pressures and times applied. However, in the case of acetic acid bacteria, the HHP treatment was very efficient for Tempranillo but not for Albariño wines. Even so, low levels of viable AAB (10^2 cfu/100 mL) were found. According to Bartowsky et al. [44], AAB populations from either spoiled or unspoiled wines ranged between 10^2 and 10^3 cfu/mL. According to the literature, pressures above 700 MPa may inhibit the polyphenol oxidase, although lower values of pressure are enough to inactivate yeasts and bacteria [45]. In our experiments, HHP results as a very effective technique against yeast and lactic acid bacteria and a lesser extent against AAB. At the studied conditions, HPP and PEF showed a noteworthy preservation of the organoleptic properties of wines (data not shown), according to other authors [45–47].

Other techniques, such as ultrasounds (US) and EMR, were also evaluated. The EMR is one of the most recent physical technologies evaluated in wines, which has shown a good potential in food processing, such as fruits, vegetables, and juices. This technique allows increasing the wine temperature for a short time period without any external heating source. EMR allows achieving the reduction of microorganisms with low effect on the organoleptic properties of wines, when compared with other heating techniques, such as flash pasteurization. However, recently studies have shown that the application of lower power microwave exposures may increase the growth of *Brettanomyces* cells [48]. In agreement, **Figure 2** shows an increase on AAB after the treatment with EMR in both cases. The application of US at different conditions considering time of application (from 1 to 3 min) and wavelengths (12, 43 and 75 μm) inhibited the yeasts growth but not the bacteria population (**Figure 2**). The effectiveness of US resulted lower than HHP, at least at the experimental conditions studied. As occurred with EMR treatment, an increase on the colony-forming units was observed after the treatment with US. Ultraviolet radiation reduces the population of wine microorganisms, but different resistances to the radiation have been stated depending on species. It appears to be an effective method against *Brettanomyces*, *Saccharomyces*, *Acetobacter*, *Lactobacillus*, and *Pediococcus* [46]. Furthermore, it has been described that phenolic compounds can absorb UV radiation and is therefore less effective in red wines. This technique seems to be more effective in white wines at the end of fermentation, when wines present low turbidity. In order to increase the total polyphenol, it could be also applied at maceration stage [38, 49].

In general, all the physical treatments assessed clearly affect the viability of lactic acid bacteria in Tempranillo and Albariño varieties. In both cases, only viable lactic acid bacteria were detected in the control (data not shown). The employed treatments reduced the viability of yeasts and lactic and acetic acid bacteria. However, in this study, both US and EMR were not effective enough to reduce the population of viable acetic acid bacteria. According to the results, AAB were more resistant to the treatments than lactic acid bacteria (LAB). Regarding techniques, a higher antimicrobial effect of HHP and EMR was observed in comparison to the other methodologies employed. Besides, some wines produced by US and EMR showed oxidation characteristics. As occurred in the antimicrobial assays, the optimization of methods and experimental conditions is an imperative action to avoid adverse effects on the sensory quality of wines. It should be noted that some of these physical techniques are commonly used in food industry, but their implementation on the wine sector is so far to be available for a daily work routine, mainly due to economic and technique questions.

The oxidation is one of the main processes that affect SO_2-free wines. Apart from the mentioned technologies and despite of its antimicrobial effect is limited, the use of inert gases is more and more applied throughout the winemaking process. The oxygen control by the management of the inert gases during the winemaking process must be considered because they have an important impact on the organoleptic properties. Caps are the ultimate physical barrier to preserve wines during storage, and so their oxygen permeability should be considered. The long-term protection is one of the most concerns for wineries in bottled wines with reduced SO_2 content [50]. The assays carried out in VITEC using argon and carbon dioxide showed valuable sensory results (**Figure 3**). The SO_2-free red wines produced by the use of Ar and CO_2 showed higher significant color intensity, tannic intensity, and dryness. Greater aroma intensity and mouthfeel were also found, although values did not show significant differences. In general, Tempranillo-bottled SO_2-free wines obtained higher global punctuations than wines with SO_2 addition.

The oxygen control during all the production process of this type of wines is an imperative engagement. It is important to take into account that wines without sulfite addition are exposed to physicochemical and microbiological alterations. Considering the techniques available in any winery, to avoid microbiological alterations, sterilizing filtration may be an alternative. However, this technique could reduce the sensorial quality of the wine

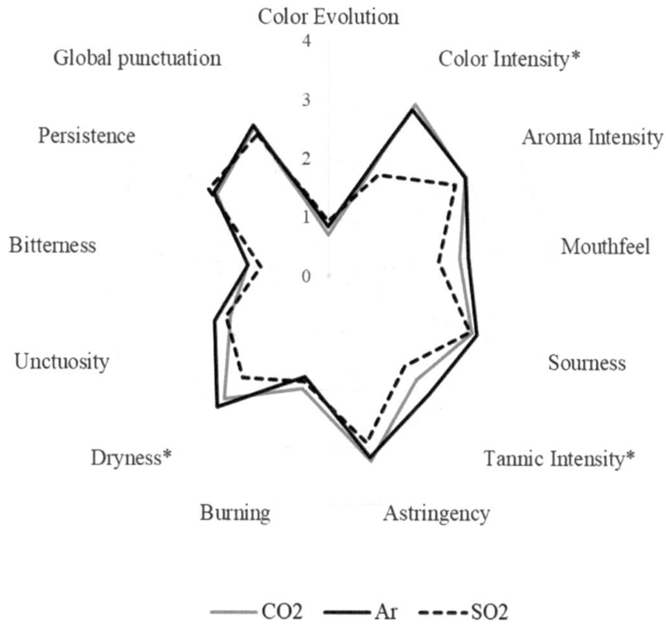

Figure 3. Comparison of the sensory evaluation of Tempranillo wines elaborated using argon (Ar), carbon dioxide (CO_2) and sulfur dioxide (SO_2). * Significant differences by HSD Tukey test ($p < 0.05$).

because it is a very oxidative process. To ensure a correct conservation of the SO_2-free wines, the amount of oxygen incorporated into wine should be controlled, especially at bottling, where concentrations from 0.2 to 4 mg/L may be incorporated, depending on conditions [51]. The amount of oxygen incorporated at bottling is the sum of the dissolved oxygen and the headspace oxygen, which is called TPO (total packaged oxygen). By our experience, between 0.5 and 1.5 mg/L of dissolved O_2 is usually incorporated at this process. Moreover, the oxygen in the headspace changes depending on the type of closure. In submerged caps, the headspace height is commonly 1–2 cm, and the normal values of dissolved oxygen ranged from 0.5 mg/L (with the use of inert gases) to 2 mg/L (without inertization). In the case of screw caps, the headspace height is higher, about 4 to 6 cm, and the oxygen values ranged from 2 to 6 mg/L. In summary, in submerged caps, values of TPO around 1 or 2 mg/L could be optimum, but values over 3 mg/L are not suitable. In screw caps, TPO values around 2.5 mg/L are optimum, but values over 7 mg/L are not suitable. The type of caps employed not only changes the amount of oxygen incorporated at bottling but also is the ultimate barrier physic to protect wines during the storage period. Thus, a correct cap should be selected depending on the type of wine, and also its permeability to oxygen should be measured to estimate the optimum storage period. The measure of the oxygen transmission rate (OTR) helps to carried out these purposes. **Figure 4** shows "high" and "low" oxygen permeability of different types of caps measured in VITEC by the MOCON® equipment. The OTR measurement corresponds to two natural corks stoppers. As can be seen in the figure, the cork stopper represented in green reached the stability of the oxygen permeability at 24 h, while the stopper represented in red did not reach this stability until the third day. Moreover, once reached the stability, the values of OTR

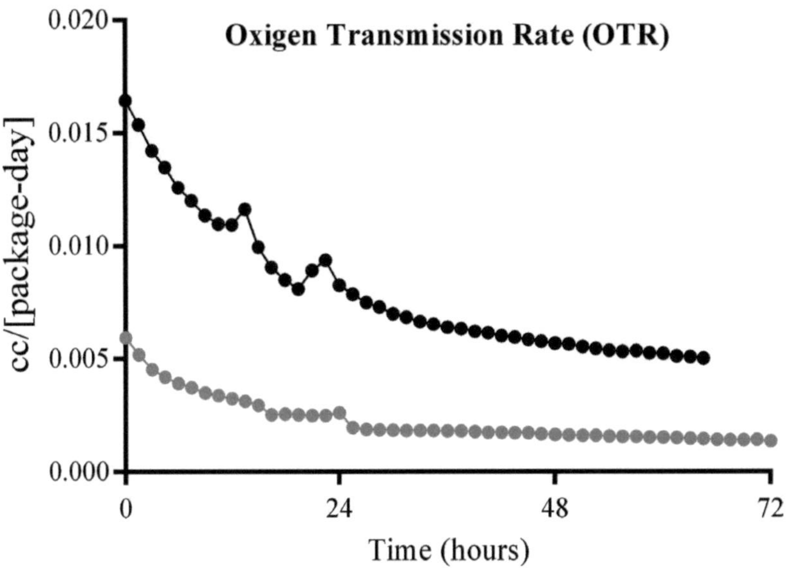

Figure 4. Representative oxygen transmission rate (OTR) of caps with different oxygen permeability.

were 4 times higher for "red" stopper than for "green". It can be also observed a great decrease in the case of the "red" stopper, likely due to higher content of oxygen inside of the cork and therefore higher porosity.

4. Chemical treatments to elaborate SO_2-free wines

The addition of chemical substances to wines is the most used alternative to reduce the SO_2 addition in wines. Over the years, the addition of several chemical substances has been allowed by the OIV with different purposes. Accordingly, new antioxidant and antimicrobial additives have been evaluated as possible alternatives to the use of the SO_2 [37, 52]. Particularly, the addition of dry yeasts enriched in glutathione, chitosan, and dimethyl dicarbonate, and different hydrolyzed and condensed tannins were evaluated by our research group. The most relevant results and some considerations related to these practices are summarized below.

In the last years, the potential application of glutathione (GSH) has increased the attention of many winemakers and researchers. The addition of reduced glutathione to grape juices or wines is allowed by OIV up to 20 mg/L (OIV OENO 445/2015). The use of GSH in the wine production was reviewed in 2013 by several authors [36, 53]. Following studies also demonstrated that the combination of SO_2 and GSH involves a notable protective effect in wines [54]. Recent studies have shown that the addition of glutathione-rich dry inactivated yeast to grape juices modifies the white wine aroma influencing the concentrations of some volatile compounds and precursors with some benefits on its preservation [55–57]. The GSH amount of wine changes depending on the winemaking period. Hence, this compound decreases after wine aging and storage; at pressing could increase its content up to 20 times [58].

Chitosan is a natural polymer formed by deacetylation of chitin, which has a wide range of applications in different field research, such as agriculture, food, and pharmaceutical industry, among others [59]. The use of this polysaccharide in oenology was approved in 2009 by the OIV to fining musts (OIV-OENO 336A-2009). Moreover, it also used as antimicrobial and antioxidant. Chitosan allows the growth of *Saccharomyces* strains but is an antimicrobial against *Brettanomyces*, acetic, and lactic acid bacteria [60–63]. Commonly, it is used to preserve wine from oxidation and also as fining agent for white wine protein stabilization [64, 65]. **Figure 5** shows the potential of chitosan as antimicrobial. In this case, a significant decrease on yeasts, LAB, and AAB after the addition of 10 g/hL of chitosan to Tempranillo wines (after alcoholic fermentation) was observed. This effectiveness was greater for yeasts, decreasing up to 1×10^4 cfu/100 mL.

Dimethyl dicarbonate (DMDC) was also accepted by European Union to be used in wine with a maximum limit amount of 200 mg/L (Regulation (EC) No 643/2006). DMDC is an organic chemical compound, which acts inhibiting the growth of microorganisms [9, 66]. When it is added to wines, it is quickly transformed to methanol and produces certain content on methyl and alkyl carbonates as products reaction by polyphenols or organic acids. These products are usually found at a low concentration, and so the quality of wine, flavors and aromas, should not be affected [67]. DMDC seems to be more effective against yeasts than against

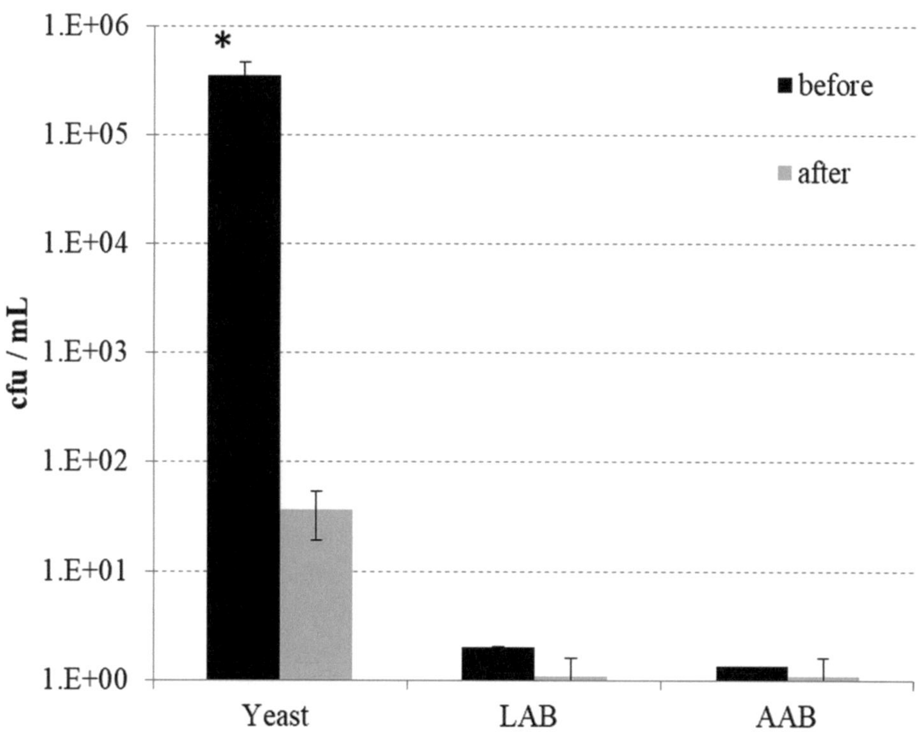

Figure 5. Viable yeasts, lactic acid bacteria (LAB), and acetic acid bacteria (AAB) quantified in Petri dishes culture (cfu; colony-forming units) from Tempranillo wines before and after a treatment with chitosan (10 g/hL). *Significant differences by HSD Tukey test ($p < 0.05$).

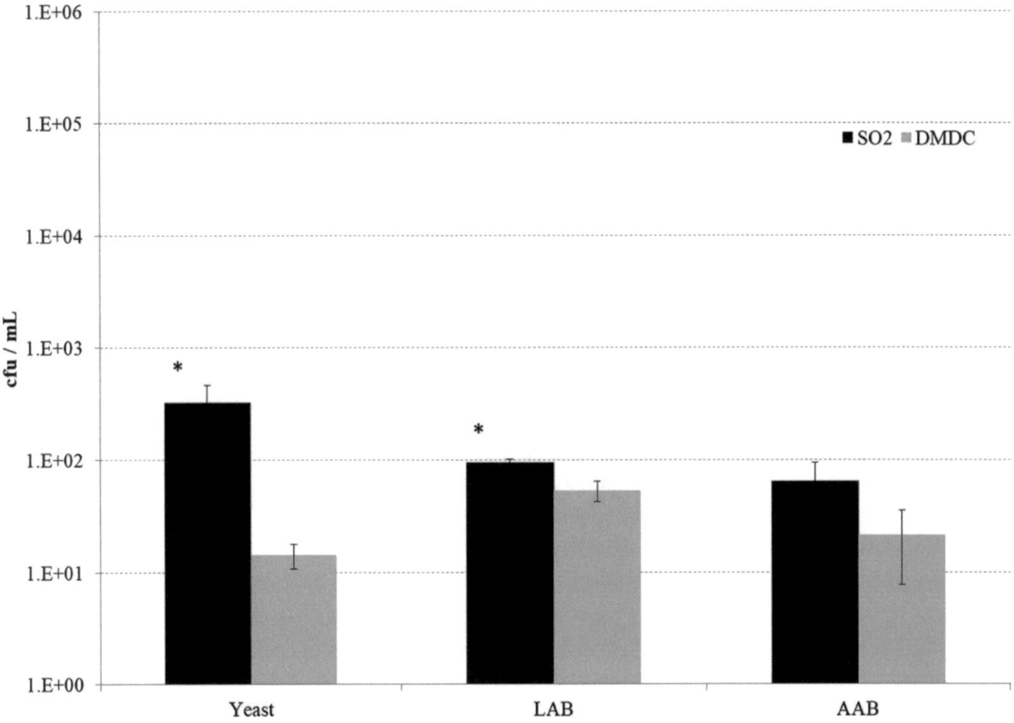

Figure 6. Viable yeasts, lactic acid bacteria (LAB), and acetic acid bacteria (AAB) quantified in Petri dishes culture (cfu, colony-forming units) from Albariño musts treated with dimethyl dicarbonate (DMDC = 20 g/hL). *Significant differences by HSD Tukey test ($p < 0.05$).

bacteria, although its activity depends on several factors, such as the pH [66–68]. In this sense, **Figure 6** shows the results obtained by the addition of DMDC to Albariño musts. The above-mentioned antimicrobial effect can be observed in yeast, LAB, and AAB. However and as occurred with chitosan, DMDC treatment was clearly more effective in yeasts than in bacteria.

The addition of oenological tannins to wine is an accepted practice by the OIV (OENO 12/2002 and revisions OENO 5/2008, OENO 6/2008, OENO 352/2009, and OENO 554/2015), which mainly aims the color stabilization and the improvement of the wine mouthfeel and flavor. Quite a few studies have evaluated the influence of the tannin addition on the chemical and sensory properties of wines. However, the results obtained are not as promising as expected. In 2005, Bautista-Ortiz et al. [69] did not observe any improvement on the chromatic and sensory properties of wines treated with different oenological tannins. Harbertson and co-workers [70] observed that some additions may be unjustified and have limited or negative impacts on the wine quality. A wide range of commercial tannins exists on the market; nonetheless, a lack of information about the composition and origin of the product is a common pattern. This fact could lead to technological problems according to the expected final wine [71]. The antioxidant properties of tannins, with related health beneficial effects, and their benefits when added to wines are also well known [72]. Both characteristics make tannins a very attractive alternative to the use of SO_2 in wine. Some studies showed hopeful results when mixed with antimicrobials, such as lysozyme [17, 73]. The

http://dx.doi.org/10.5772/intechopen.71627

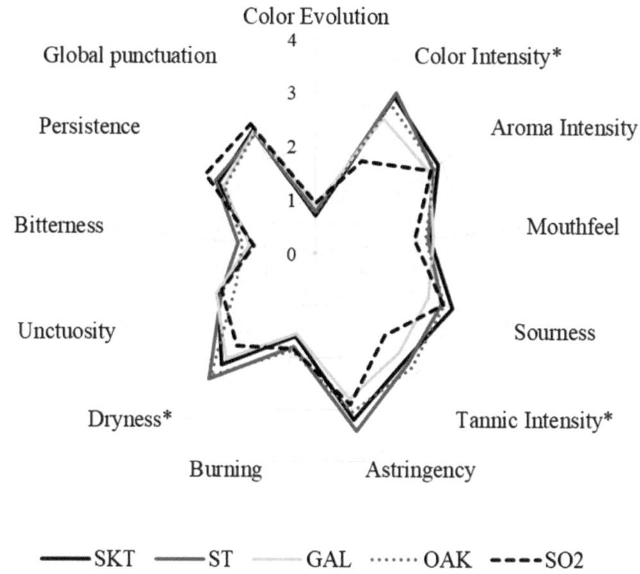

Figure 7. Sensory profile of Tempranillo wines elaborated by different enological tannin additions to grape juices. SO_2: Wine control. ST: Grape seed tannins (40 g/hL), SKT: Grape skin tannins (30 g/hL), GAL: Tara tannin (20 g/hL), OAK: oak tannins (30 g/hL). *Significant differences by HSD Tukey test ($p < 0.05$).

studies carried out in VITEC have recently shown that the addition of tannins mixed with glutathione may be an effective alternative to the use of SO_2 [74]. **Figure 7** shows the sensory analysis of Tempranillo wines with addition of grape seed tannins (ST), grape skin tannins (SKT), oak tannins (OAK), and tara tannins (GAL). In general, the sensory profiles of wines produced with the addition of different tannins were similar (and even better) than wines elaborated by addition of $SO_{2.}$ Significant higher color intensity was observed between control and treated wines. Treated wines also obtained significant dryness and tannic intensity. Astringency and mouthfeel reached higher values but not significant. Lower persistence and higher aroma intensity can also be observed. Low differences between treatments were found, which may be due not only to the different quantity of tannins added but also to their qualitative profile. Recent studies performed by other authors have confirmed the importance of the anthocyanin/tannin ratio on the wine oxidation process and especially on the acetaldehyde formation. Wines with higher tannin addition showed lower production of acetaldehyde [75].

Other chemical substances, such as ascorbic acid and lysozyme, may also be able alternatives to SO_2. Ascorbic acid has the ability to scavenge molecular oxygen before the oxidation of phenolic compounds occurs. It is a highly efficient antioxidant in combination with sulfur dioxide; nonetheless, a pro-oxidation effect may occur when the content of SO_2 and ascorbic acid is low [76]. The reaction between ascorbic acid and oxygen results in dehydroascorbic acid and hydrogen peroxide, which would be removed by sulfites. Under certain conditions, ascorbic acid both accelerates oxygen removal and reduces the O_2:SO_2 molar reaction ratio [4]. In wines, it is generally employed in winemaking stages with high oxygen dissolution, such as grape crushing, after racking or just before bottling. The addition of ascorbic acid in white wines improves color and flavor retention during bottling aging [77]. Certain carbonyl compounds, such as furfural, acetaldehyde, glyoxal, and diacetyl, formed from the oxidation of ascorbic acid may involve the formation of brown pigments by reacting with phenolic compounds. Higher browning was observed in catechin model solutions containing ascorbic acid than in model solutions containing sulfite [78]. These oxidation products of ascorbic acid bind to SO_2 reducing in some extent the ratio between free and total SO_2 content [76]. The mixture of ascorbic acid together with SO_2 seems to be a better antioxidant combination than the use of SO_2 alone, avoiding the oxidation of wine and preserving the aroma profile. In white wines, ascorbic acid provides considerable protection against oxidation under conditions of low oxygen [79]. However, it should be highlighted that the impact of the addition of ascorbic acid to wine composition and sensory characters is far to be clarified [36, 77].

Lysozyme belongs to glycoside hydrolases, which is a type of enzyme that catalyzes the hydrolysis of bonds between N-acetyl muramic acid and N-acetyl-D-glucosamine residues in peptidoglycans, and it is found in the cell walls of bacteria, especially in Gram-positive bacteria. These enzymes are therefore destructive to many bacteria like lactic acid bacteria (LAB). In winemaking, indigenous LAB, such as *Lactobacillus brevis, Oenococcus oeni, Lactobacillus kunkeei, Pediococcus parvulus* and *Pediococcus damnosus,* can be completely inhibited by lysozyme, being this efficacy strongly affected by winemaking and dosage [80, 81]. The addition of lysozyme did not have any negative effect on yeast growth and sugar reduction and may prevent the increase of volatile acidity during the stuck/sluggish of the alcoholic fermentation [17, 81]. This substance had little or no effect on the content of alcohol, titratable acidity, and pH value and did not cause important changes on the sensory characteristics of wines. Nonetheless, it may produce esters in certain wines, contributing to their complexity [73, 82]. Lysozyme may involve changes on yeast nitrogen consumption and the amino nitrogen metabolism, although it does not appear to have an effect on the formation of biogenic amines [16]. The addition of lysozyme may produce a color loss associate with the formation of precipitates in red wines and may induce protein haze in white wines [82]. Lysozyme does not possess an antioxidant activity and therefore does not prevent the wine oxidation. Hence, it becomes necessary the addition of antioxidants, such as proanthocyanidins, in combination with lysozyme to replace the SO_2 actions [16, 73]. A critical point of lysozyme is the safety of wines treated with this additive, since it is an egg allergen (allergen Gal d 4 according to the International Allergen Code) that remains in bottled wine. The OIV issued limitation of 500 mg/L [83], and this quantity is removed by an efficient fining treatment using, for example, bentonite or metatartaric acid [84].

5. Conclusions

The use of yeast strains with a low capacity to produce SO_2, during the alcoholic fermentation is essential to reduce the final amount of SO_2 in wines. Both commercial and indigenous yeasts strains can be used with this purpose. However, factors as grape juice composition, the management of the fermentation, and musts supplementation will be decisive. Different physical technologies and methodologies can be used to elaborate this type of wines. The replacement of the antioxidant and antimicrobial action of the SO_2 is a complex mission. However, the combination of different physical techniques together with a good management of inert gases to control oxygen appears to be a suitable practice to achieve this purpose. In addition, some chemical treatments will help to complete the effects caused by these practices. In general, chemical treatments should be combined at different wine production stages to complete their respective actions. The combination of chemical additions even with SO_2 may help to reduce its use during the winemaking. It should be noted that still today, there is a lack on the knowledge of the microbiological stability of SO_2-free wines during the aging period. Therefore, more research is needed to better understand the effect of the low concentration of SO_2 in wines as well as the use of new additives, especially regarding the wine stability after storage and the effects on the human health.

In summary, multidisciplinary approaches should be considered to elaborate high-quality SO_2-free wines. The combination of microbiological strategies, physical methods, and chemical treatments becomes indispensable to achieve this ambitious purpose. Several yeast strains are able to generate low quantities of SO_2 during alcoholic fermentations (<10 mg/L), and several physical and chemical treatments have shown their antioxidant and antimicrobial effect. Therefore, reducing the SO_2 amount in wine production may be achieved. Nonetheless, more research should be done to adapt winemaking procedures according to the particular working conditions and the desired product of each winery.

Acknowledgements

Thanks are due to the Spanish MICINN for their financial support of VINNO_SO_2 Project (Ref. IPT-2012-0967-060000). The authors also thank AGROVIN S.A. for supplying the yeast strains, Bodegas RODA S.A. (Haro, La Rioja, Spain) and Adegas Valmiñor S.L. (O Rosal, Pontevedra) for supplying the grape samples. We also thank Programa de Desenvolupament Rural de Catalunya 2014–2020 (N° expdte. 56 30032 2017 2A).

Author details

Raúl Ferrer-Gallego*, Miquel Puxeu, Laura Martín, Enric Nart, Claudio Hidalgo and Imma Andorrà

*Address all correspondence to: raul.ferrer@vitec.cat

VITEC, Wine Technology Center, Tarragona, Spain

References

[1] Vally H, Misso NLA, Madan V. Clinical effects of sulphite additives. Clinical & Experimental Allergy. 2009;**39**(11):1643-1651. DOI: 10.1111/j.1365-2222.2009.03362.x

[2] Costanigro M, Appleby C, Menke SD. The wine headache: Consumer perceptions of sulfites and willingness to pay for non-sulfited wines. Food Quality and Preference. 2014;**31**:81-89. DOI: doi.org/10.1016/j.foodqual.2013.08.002

[3] Ribérau-Gayon P, Dubordieu D, Doneche B, Lonvaud E. Handbook of Enology, Vol. 1; The Microbiology of Wine and Vinifications, 2nd Edition. John Wiley & Sons Inc (NYSE:JW.A). 2006

[4] Danilewicz JC. Reaction of oxygen and sulfite in wine. American Journal of Enology and Viticulture. 2016;**67**(1):13-17. DOI: 10.5344/ajev.2015.15069

[5] Thomas D, Surdin-Kerjan Y. Metabolism of Sulfur amino acids in *Saccharomyces cerevisiae*. Microbiology and Molecular Biology Reviews. 1997;**61**(4):503-532

[6] Eschenbruch R. Sulfite and sulfide formation during winemaking – A review. American Journal of Enology and Viticulture. 1974;**25**(3):23-27

[7] Wells A, Osborne JP. Production of SO_2 binding compounds and SO_2 by saccharomyces during alcoholic fermentation and the impact on malolactic fermentation. South African Journal of Enology and Viticulture. 2011;**32**(2):267-279. DOI: doi.org/10.21548/32-2-1387

[8] Donalies UEB, Stahl U. Increasing sulphite formation in *Saccharomyces cerevisiae* by overexpression of MET14 and SSUI. Yeast. 2002;**19**(6):475-484. DOI: 10.1002/yea.849

[9] Divol B, du Toit M, Duckitt E. Surviving in the presence of sulphur dioxide: Strategies developed by wine yeasts. Applied Microbiology and Biotechnology. 2012;**95**(3):601-613. DOI: 10.1007/s00253-012-4186-x

[10] Larsen JT, Nielsen J-C, Kramp B, Richelieu M, Bjerring P, Riisager MJ, et al. Impact of different strains of *Saccharomyces cerevisiae* on malolactic fermentation by *Oenococcus oeni*. American Journal of Enology and Viticulture. 2003;**54**(4):246-251

[11] Miranda-Castilleja DE, Ortiz-Barrera E, Arvizu-Medrano SM, Ramiro-Pacheco J, Aldrete-Tápia JA, Martínez-Peniche RA. Aislamiento, selección e identificación de levaduras Saccharomyces spp. nativas de viñedos en Querétaro, México. Agrociencia. 2015;**49**:759-773

[12] Rankine BC, Pocock KF. Influence of yeast strain on binding of sulphur dioxide in wines, and on its formation during fermentation. Journal of the Science of Food and Agriculture. 1969;**20**(2):104-109. DOI: 10.1002/jsfa.2740200210

[13] Suzzi P, Romano P, Zambonelli C. Saccharomyces strain selection in minimizing SO_2 requirement during Vinification. American Journal of Enology and Viticulture. 1985;**36**(3): 199-202

[14] Werner M et al. Yeasts and natural production of sulphites. Journal of Enology and Viticulture. 2009;(1):2-6

[15] Eglinton JM, Henschke PA. *Saccharomyces cerevisiae* strains AWRI 838, Lalvin EC1118 and Maurivin PDM do not produce excessive sulfur dioxide in white wine fermentations. Australian Journal of Grape and Wine Research. 1996;**2**(2):77-83. DOI: 10.1111/j.1755-0238.1996.tb00098.x

[16] Cejudo-bastante MJ, Sonni F, Chinnici F, Versari A, Perez-coello MS, Riponi C. Fermentation of sulphite-free white musts with added lysozyme and oenological tannins: Nitrogen consumption and biogenic amines composition of final wines. LWT- Food Science and Technology. 2010;**43**(10):1501-1507. DOI: 10.1016/j.lwt.2010.02.011

[17] Sonni F, Bastante MJC, Chinnici F, Natali N, Riponi C. Replacement of sulfur dioxide by lysozyme and oenological tannins during fermentation: Influence on volatile composition of white wines. Journal of the Science of Food and Agriculture. 2009;**89**(4):688-696. DOI: 10.1002/jsfa.3503

[18] Duan W, Roddick FA, Higgins VJ, Rogers PJ. A parallel analysis of H_2S and SO_2 formation by brewing yeast in response to sulfur-containing amino acids and ammonium ions. Journal of the American Society of Brewing Chemists. 2004;**62**(1):35-41

[19] Yoshida S, Imoto J, Minato T, Oouchi R, Kamada Y, Tomita M, et al. A novel mechanism regulates H_2S and SO_2 production in *Saccharomyces cerevisiae*. Yeast. 2011;**28**(2):109-121. DOI: 10.1002/yea.1823

[20] Kumar GR, Ramakrishnan V, Bisson LF. Survey of hydrogen sulfide production in wine strains of *Saccharomyces cerevisiae*. American Journal of Enology and Viticulture. 2010;**61**(3):365-371

[21] Nowak A, Kusewicz D, Kalinowska H, Turkiewicz M, Patelski P. Production of H2S and properties of sulfite reductase from selected strains of wine-producing yeasts. European Food Research and Technology. 2004;**219**:84-89. DOI: 10.1007/s00217-004-0885-6

[22] Cordente AG, Heinrich A, Pretorius IS, Swiegers JH. Isolation of sulfite reductase variants of a commercial wine yeast with significantly reduced hydrogen sulfide production. FEMS Yeast Research. 2009;**9**(3):446-459. DOI: 10.1111/j.1567-1364.2009.00489.x

[23] Huang C, Roncoroni M, Gardner RC. MET2 affects production of hydrogen sulfide during wine fermentation. Applied Microbiology and Biotechnology. 2014;**98**(16):7125-7135. DOI: 10.1007/s00253-014-5789-1

[24] Linderholm A, Dietzel K, Hirst M, Bisson LF. Identification of MET10-932 and characterization as an allele reducing hydrogen sulfide formation in wine strains of *Saccharomyces cerevisiae*. Applied and Environmental Microbiology. 2010;**76**(23):7699-7707. DOI: 10.1128/AEM.01666-10

[25] De Vero L, Solieri L, Giudici P. Evolution-based strategy to generate non-genetically modified organisms *Saccharomyces cerevisiae* strains impaired in sulfate assimilation pathway. Letters in Applied Microbiology. 2011;**53**(5):572-575. DOI: 10.1111/j.1472-765X.2011.03140.x

[26] Noble J, Sanchez I, Blondin B. Identification of new *Saccharomyces cerevisiae* variants of the MET2 and SKP2 genes controlling the sulfur assimilation pathway and the production of undesirable sulfur compounds during alcoholic fermentation. Microbial Cell Factories. 2015;**14**(68):1-16. DOI: 10.1186/s12934-015-0245-1

[27] Liu J, Gallander J. Effect of insoluble solids on the sulfur dioxide content and rate of malolactic fermentation in white table wines. American Journal of Enology and Viticulture. 1982;**33**(4):194-197

[28] Giudici P, Kunkee RE. The effect of nitrogen deficiency and sulfur-containing amino acids on the reduction of sulfate to hydrogen sulfide by wine yeasts. American Journal of Enology and Viticulture. 1994;**45**(1):107-112

[29] Osborne JP, Edwards CG. Inhibition of malolactic fermentation by Saccharomyces during alcoholic fermentation under low- and high-nitrogen conditions: A study in synthetic media. Australian Journal of Grape and Wine Research. 2006;**12**:69-78. DOI: 10.1111/j.1755-0238.2006.tb00045.x

[30] Jiranek V, Langridge P, Henschke PA. Amino acid and ammonium utilization by *Saccharomyces cerevisiae* wine yeasts from a chemically defined medium. American Journal of Enology and Viticulture. 1995;**46**(1):75-83

[31] Kemsawasd V, Viana T, Ardö Y, Arneborg N. Influence of nitrogen sources on growth and fermentation performance of different wine yeast species during alcoholic fermentation. Applied Microbiology and Biotechnology. 2015;**99**(23):10191-10207. DOI: 10.1007/s00253-015-6835-3

[32] Gobbi M, Comitini F, D'Ignazi G, Ciani M. Effects of nutrient supplementation on fermentation kinetics, H2S evolution, and aroma profile in Verdicchio DOC wine production. European Food Research and Technology. 2013;**236**(1):145-154. DOI: 10.1007/s00217-012-1870-0

[33] Nart E, Andorrà I, Puxeu M, Martín L, Hidalgo C, Ferrer-Gallego R. Evaluación de alternativas microbiológicas, físicas y químicas para la elaboración de vinos de alta calidad libres de sulfuroso. La Semana Vitivinícola. 2017;Septiembre(3502):1526-30

[34] García-Ríos E, Ramos-Alonso L, Guillamón JM. Correlation between low temperature adaptation and oxidative stress in *Saccharomyces cerevisiae*. Frontiers in Microbiology. 2016;**7**(AUG):1-11. DOI: 10.3389/fmicb.2016.01199

[35] Eschenbruch R, Bonish P. The influence of pH on sulphite formation by yeasts. Archives of Microbiology. 1976;**107**(2):229-231

[36] Comuzzo P, Zironi R. Biotechnological strategies for controlling wine oxidation. Food Engineering Reviews. 2013;**5**(4):217-229. DOI: 10.1007/s12393-013-9071-6

[37] Santos MC, Nunes C, Saraiva JA, Coimbra MA. Chemical and physical methodologies for the replacement/reduction of sulfur dioxide use during winemaking: Review of their potentialities and limitations. European Food Research and Technology. 2012;**234**(1):1-12. DOI: 10.1007/s00217-011-1614-6

[38] Guerrero RF, Puertas B, Fernández MI, Palma M, Cantos-Villar E. Induction of stilbenes in grapes by UV-C: Comparison of different subspecies of Vitis. Innovative Food Science & Emerging Technologies. 2010;**11**(1):231-238. DOI: 10.1016/j.ifset.2009.10.005

[39] Puértolas E, Saldaña G, Álvarez I, Raso J. Experimental design approach for the evaluation of anthocyanin content of rosé wines obtained by pulsed electric fields. Influence of temperature and time of maceration. Food Chemistry. 2011;**126**(3):1482-1487. DOI: 10.1016/j.foodchem.2010.11.164

[40] Puértolas E, Saldaña G, Condón S, Álvarez I, Raso J. Evolution of polyphenolic compounds in red wine from Cabernet Sauvignon grapes processed by pulsed electric fields during aging in bottle. Food Chemistry. 2010;**119**(3):1063-1070. DOI: 10.1016/j.foodchem.2009.08.018

[41] Puig A, Olmos P, Quevedo JM, Guamis B, Mínguez S. Microbiological and sensory effects of musts treated by high-pressure homogenization. Food Science and Technology International. 2008;**14**(5 suppl):5-11. DOI: 10.1177/1082013208094579

[42] Puértolas E, López N, Condón S, Álvarez I, Raso J. Potential applications of PEF to improve red wine quality. Trends in Food Science & Technology. 2010;**21**(5):247-255. DOI: https://doi.org/10.1016/j.tifs.2010.02.002

[43] Garde-Cerdán T, Marsellés-Fontanet AR, Arias-Gil M, Ancín-Azpilicueta C, Martín-Belloso O. Effect of storage conditions on the volatile composition of wines obtained from must stabilized by PEF during ageing without SO_2. Innovative Food Science & Emerging Technologies. 2008;**9**(4):469-476. DOI: https://doi.org/10.1016/j.ifset.2008.05.002

[44] Bartowsky EJ, Xia D, Gibson RL, Fleet GH, Henschke PA. Spoilage of bottled red wine by acetic acid bacteria. Letters in Applied Microbiology. 2003;**36**(5):307-314. DOI: 10.1046/j.1472-765X.2003.01314.x

[45] Buzrul S. High hydrostatic pressure treatment of beer and wine: A review. Innovative Food Science & Emerging Technologies. 2012;**13**(January):1-12. DOI: 10.1016/j.ifset.2011.10.001

[46] Fredericks IN, du Toit M, Krügel M. Efficacy of ultraviolet radiation as an alternative technology to inactivate microorganisms in grape juices and wines. Food Microbiology. 2011;**28**(3):510-517. DOI: https://doi.org/10.1016/j.fm.2010.10.018

[47] Luo H, Schmid F, Grbin PR, Jiranek V. Viability of common wine spoilage organisms after exposure to high power ultrasonics. Ultrasonics Sonochemistry. 2012;**19**(3):415-420. DOI: https://doi.org/10.1016/j.ultsonch.2011.06.009

[48] Vojisavljevic V, Alsuhaim HS, Pirogova E, editors. Low Power Microwave Exposures at 968MHz Increase the Growth Rate of *Breanomyces bruxellensis* Yeast Cells. 9th International Conference on Microwave and Millimeter Wave Technology, ICMMT 2016 - Proceedings; 2016

[49] Guerrero RF, Puertas B, Fernández MI, Piñeiro Z, Cantos-Villar E. UVC-treated skin-contact effect on both white wine quality and resveratrol content. Food Research International. 2010;**43**(8):2179-2185. DOI: https://doi.org/10.1016/j.foodres.2010.07.023

[50] Pezley M. Production of free sulfur dioxide by wine yeasts. Interdisciplinary Undergraduate Research Journal. Spring; 2015

[51] Biondi Bartolini A, Cavini F, Basquiat M. Oxygen et Vin Du rôle de l'oxygène à la technique de micro-oxygène: Parsec, Florence, Italy; 2008

[52] Guerrero RF, Cantos-Villar E. Demonstrating the efficiency of sulphur dioxide replacements in wine: A parameter review. Trends in Food Science & Technology. 2015;**42**(1): 27-43.DOI: 10.1016/j.tifs.2014.11.004

[53] Kritzinger EC, Bauer FF, du Toit WJ. Role of glutathione in winemaking: A review. Journal of Agricultural and Food Chemistry. 2013;**61**(2):269-277. DOI:10.1021/jf303665z

[54] Tomašević M, Gracin L, Ćurko N, Kovačević Ganić K. Impact of pre-fermentative maceration and yeast strain along with glutathione and SO_2 additions on the aroma of *Vitis vinifera* L. Pošip wine and its evaluation during bottle aging. LWT- Food Science and Technology. 2017;**81**:67-76. DOI: 10.1016/j.lwt.2017.03.035

[55] Gabrielli M, Aleixandre-Tudo JL, Kilmartin PA, Sieczkowski N, du Toit WJ. Additions of glutathione or specific glutathione-rich dry inactivated yeast preparation (DYP) to sauvignon blanc must: Effect on wine chemical and sensory composition. South African Journal of Enology and Viticulture 2017;**38**(1):18-28. DOI:10.21548/38-1-794

[56] Thibon C, Böcker C, Shinkaruk S, Moine V, Darriet P, Dubourdieu D. Identification of S-3-(hexanal)-glutathione and its bisulfite adduct in grape juice from Vitis vinifera L. cv. Sauvignon blanc as new potential precursors of 3SH. Food Chemistry. 2016;**199**:711-719. DOI: 10.1016/j.foodchem.2015.12.069

[57] Šuklje K, Antalick G, Buica A, Coetzee ZA, Brand J, Schmidtke LM, et al. Inactive dry yeast application on grapes modify Sauvignon Blanc wine aroma. Food Chemistry. 2016;**197**:1073-1084. DOI: 10.1016/j.foodchem.2015.11.105

[58] Ferreira-Lima NE, Burin VM, Caliari V, Bordignon-Luiz MT. Impact of pressing conditions on the phenolic composition, radical scavenging activity and glutathione content

of Brazilian *Vitis vinifera* white wines and evolution during bottle ageing. Food and Bioprocess Technology. 2016;**9**(6):944-957. DOI: 10.1007/s11947-016-1680-7

[59] Sigroha S, Khatkar A. Chitosan – A naturally derived antioxidant polymer with diverse applications. Current Organic Chemistry. 2017;**21**(4):333-341. DOI: 10.2174/13852728206 66161018130542

[60] Ferreira D, Moreira D, Costa EM, Silva S, Pintado MM, Couto JA. The antimicrobial action of chitosan against the wine spoilage yeast Brettanomyces/Dekkera. Journal of Chitin and Chitosan Science. 2013;**1**(3):240-245. DOI: 10.1166/jcc.2013.1037

[61] Petrova B, Cartwright ZM, Edwards CG. Effectiveness of chitosan preparations against *Brettanomyces bruxellensis* grown in culture media and red wines. OENO One. 2016;**50**(1):8. DOI: 10.20870/oeno-one.2016.50.1.54

[62] Valera MJ, Sainz F, Mas A, Torija MJ. Effect of chitosan and SO_2 on viability of Acetobacter strains in wine. International Journal of Food Microbiology. 2017;**246**:1-4. DOI: 10.1016/j.ijfoodmicro.2017.01.022

[63] Bağder Elmacı S, Gülgör G, Tokatlı M, Erten H, İşci A, Özçelik F. Effectiveness of chitosan against wine-related microorganisms. Antonie Van Leeuwenhoek. 2015;**107**(3):675-686. DOI: 10.1007/s10482-014-0362-6

[64] Chagas R, Monteiro S, Ferreira RB. Assessment of potential effects of common fining agents used for white wine protein stabilization. American Journal of Enology and Viticulture. 2012;**63**:574-578. DOI: 10.5344/ajev.2012.12016

[65] Nunes C, Maricato E, Cunha A, Rocha MAM, Santos S, Ferreira P, et al. Chitosan-genipin film, a sustainable methodology for wine preservation. Green Chemistry. 2016;**18**(19):5331-5341. DOI: 10.1039/C6GC01621A

[66] Costa A, Barata A, Malfeito-Ferreira M, Loureiro V. Evaluation of the inhibitory effect of dimethyl dicarbonate (DMDC) against wine microorganisms. Food Microbiology. 2008;**25**:422-427. DOI: 10.1016/j.fm.2007.10.003

[67] Delfini C, Gaia P, Schellino R, Strano M, Pagliara A, Ambrò S. Fermentability of grape must after inhibition with dimethyl dicarbonate (DMDC). Journal of Agricultural and Food Chemistry. 2002;**50**(20):5605-5611. DOI: 10.1021/jf0256337

[68] Basaran-Akgul N, Churey JJ, Basaran P, Worobo RW. Inactivation of different strains of *Escherichia coli* O157:H7 in various apple ciders treated with dimethyl dicarbonate (DMDC) and sulfur dioxide (SO_2) as an alternative method. Food Microbiology. 2009;**26**(1):8-15. DOI: 10.1016/j.fm.2008.07.011

[69] Bautista-Ortín AB, Martínez-Cutillas A, Ros-García JM, López-Roca JM, Gómez-Plaza E. Improving colour extraction and stability in red wines: The use of maceration enzymes and enological tannins. International Journal of Food Science and Technology. 2005;**40**(8):867-878. DOI: 10.1111/j.1365-2621.2005.01014.x

[70] Harbertson JF, Parpinello GP, Heymann H, Downey MO. Impact of exogenous tannin additions on wine chemistry and wine sensory character. Food Chemistry. 2012;**131**(3):999-1008. DOI: https://doi.org/10.1016/j.foodchem.2011.09.101

[71] Obreque-Slíer E, Peña-Neira A, López-Solís R, Ramírez-Escudero C, Zamora-Marín F. Phenolic characterization of commercial enological tannins. European Food Research and Technology. 2009;**229**(6):859-866. DOI: 10.1007/s00217-009-1121-1

[72] Versari A, du Toit W, Parpinello GP. Oenological tannins: A review. Australian Journal of Grape and Wine Research. 2013;**19**(1):1-10. DOI:10.1111/ajgw.12002

[73] Chen K, Han SY, Li M, Sheng WJ. Use of lysozyme and Oligomeric Proanthocyanidin to reduce sulfur dioxide and the evolution of volatile compounds in Italian Riesling ice wine during aging process. Journal of Food Processing and Preservation. 2017;**41**(1). DOI: 10.1111/jfpp.12755

[74] Ferrer-Gallego R, Puxeu M, Nart E, Martín L, Andorrà I. Evaluation of tempranillo and albariño SO_2-free wines produced by different chemical alternatives and winemaking procedures. Food Research International. DOI: https://doi.org/10.1016/j.foodres.2017.09.046

[75] Picariello L, Gambuti A, Picariello B, Moio L. Evolution of pigments, tannins and acetaldehyde during forced oxidation of red wine: Effect of tannins addition. LWT-Food Science and Technology. 2017;**77**:370-375. DOI: https://doi.org/10.1016/j.lwt.2016.11.064

[76] Bradshaw MP, Barril C, Clark AC, Prenzler PD, Scollary GR. Ascorbic acid: A review of its chemistry and reactivity in relation to a wine environment. Critical Reviews in Food Science and Nutrition. 2011;**51**(6):479-498. DOI: 10.1080/10408391003690559

[77] Skouroumounis GK, Kwiatkowski MJ, Francis IL, Oakey H, Capone DL, Peng Z, et al. The influence of ascorbic acid on the composition, colour and flavour properties of a Riesling and a wooded chardonnay wine during five years' storage. Australian Journal of Grape and Wine Research. 2005;**11**(3):355-368. DOI: 10.1111/j.1755-0238.2005.tb00035.x

[78] Bradshaw MP, Prenzler PD, Scollary GR. Ascorbic acid-induced browning of (+)-catechin in a model wine system. Journal of Agricultural and Food Chemistry. 2001;**49**(2):934-939. DOI: 10.1021/jf000782f

[79] Barril C, Rutledge DN, Scollary GR, Clark AC. Ascorbic acid and white wine production: A review of beneficial versus detrimental impacts. Australian Journal of Grape and Wine Research. 2016;**22**(2):169-181. DOI: 10.1111/ajgw.12207

[80] Azzolini M, Tosi E, Veneri G, Zapparoli G. Evaluating the efficacy of lysozyme against lactic acid bacteria under different winemaking scenarios. South African Journal of Enology and Viticulture. 2010;**31**(2):99-105

[81] Gao YC, Zhang G, Krentz S, Darius S, Power J, Lagarde G. Inhibition of spoilage lactic acid bacteria by lysozyme during wine alcoholic fermentation. Australian Journal of Grape and Wine Research. 2002;**8**(1):76-83

[82] Bartowsky EJ, Costello PJ, Villa A, Henschke PA. The chemical and sensorial effects of lysozyme addition to red and white wines over six months' cellar storage. Australian Journal of Grape and Wine Research. 2004;**10**(2):143-150

[83] Compendium of International Methods of Wine and Must Analysis. [Internet]. 2016

[84] Peñas E, di Lorenzo C, Uberti F, Restani P. Allergenic proteins in Enology: A review on technological applications and safety aspects. Molecules. 2015;**20**(7):13144

Chapter 10

Influence of Wine Chemical Compounds on the Foaming Properties of Sparkling Wines

Leticia Martínez-Lapuente, Belén Ayestarán and Zenaida Guadalupe

Additional information is available at the end of the chapter

http://dx.doi.org/10.5772/intechopen.70859

Abstract

The foam of a sparkling wine is a key parameter of its quality, and the main characteristic differentiating sparkling wines from the so-called still wines. Both foam formation and duration are directly related to the chemical composition of sparkling wines. This chapter reviews the most recent studies made to determine the influence of chemical compounds on the foamability and foam stability of sparkling wines. Foam properties of sparkling wines are ruled by a large number of molecules, but some compounds seem to be more relevant than others to explain their behavior. The content of total amino acids, polysaccharides, anthocyanins, coumaric acid, and isorhamnetin showed high correlation values with foam quality parameters. The alcohol content and the concentration of acid polysaccharides, proanthocyanidins and free SO_2 are the factors which most negatively affect foam quality. A recent study, by means of prediction models, has concluded that the different forms of malvidin show the highest influence on the foamability parameters in rosé sparkling wines, followed by amino acid compounds, while foam stability model was only predicted by polysaccharides rich in arabinose and galactose. These research findings provide industry with a better understanding of the compositional factors influencing the foam quality of sparkling wines.

Keywords: sparkling wine, foaming properties, quality, chemical composition, predictive models

1. Introduction

Nowadays the economic impact of sparkling wine shows a fast growth in the world wine trade because of its high added value. According to the report published in the year 2014 by the

International Organization of Vine and Wine [1], sparkling wine production increased by 40% in the last decade and by 7% compared to global wine production.

Sparkling wines are obtained after a second fermentation of a base wine that can be carried out in closed bottles or in hermetically sealed tanks. High quality sparkling wines, such as Champagne wines in France, Cava wines in Spain or Talento in Italy, are fermented in closed bottles following the traditional method, and they remain in contact with the yeast lees in a bottle. The first fermentation transforms grape must into base wine. The essence of the traditional method is the second fermentation, which takes place in the bottle and increases the alcohol content and internal bottle pressure (up to 5–7 atmospheres). After this second alcoholic fermentation, the wine is aged on yeast lees for at least 9 months (EC Regulation N° 606/2009) [2]. Autolysis of the yeast occurs during this prolonged contact and involves hydrolytic enzymes that act to release cytoplasmic (peptides, fatty acids, nucleotides, amino acids) and cell wall (mannoproteins) compounds into the wine. This aging on yeast lees leads to significant changes in wine composition and the organoleptic and foam properties of the wine are modified [3].

In sparkling wines the level of dissolved carbon dioxide (CO_2) found in the liquid phase is indeed a parameter of paramount importance. CO_2 is responsible for the visually appealing, and very much sought-after repetitive bubbling process (the so-called effervescence). In fact, foam is the characteristic that differentiates sparkling wines from still wines, being the first sensory attribute that tasters and consumers perceive and that determines the final quality of sparkling wines [4]. Moreover, dissolved CO_2 is also responsible for the very characteristic tingling sensation in aroma and mouthfeel sensations [5]. During carbonated beverage tasting, dissolved CO_2 acts on both trigeminal receptors [6], and gustatory receptors, via the conversion of dissolved CO_2 to carbonic acid [7], in addition to the tactile stimulation of mechanoreceptors in the oral cavity (through bursting bubbles). Moreover, a link has been evidenced between carbonation and the release of some aroma compounds in carbonated waters [8].

The formation of foam, its stability and the size of the bubbles in sparkling wines are directly related to the surface tension. This can be defined as the force per unit area that maintains the bond between the molecules at the surface of the liquid. The presence of surfactants reduces the surface tension of the liquid and allows to the formation and persistence of bubbles. Secondary fermentation in sparkling wines leads to the formation of carbon dioxide, which increases the pressure inside the bottle and causes the gas to dissolve in the liquid. When the bottle is opened, the difference between the pressure in the bottle and that of the atmosphere causes the dissolved gas to spontaneously leave the liquid. Once the pressure on the surface of the liquid has been equalized with the atmospheric pressure, bubbles continue to form inside the liquid [9].

Currently, the quality of sparkling wine is linked to the formation and persistence of foam. Quality foam can be defined as one that causes a slow release of CO_2, in ring shapes from the depths of the liquid, with small bubbles that contribute to the formation of a crown over the surface of the wine, covering it completely, and with bubbles two or three rows deep [10]. Foam duration is directly related to bubble stability, and stability is itself dependent on the

http://dx.doi.org/10.5772/intechopen.70859

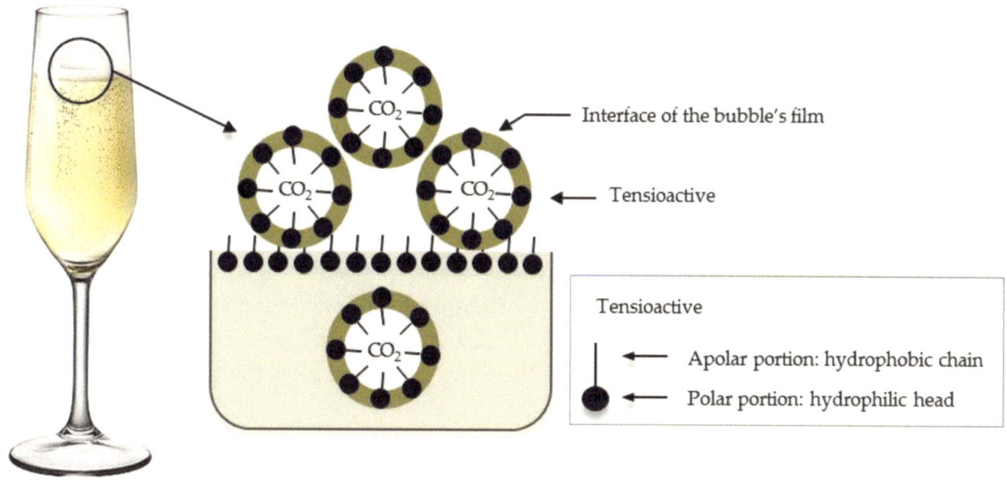

Figure 1. Carbon dioxide/liquid interphase of the bubble's film. Adapted from Ref. [11].

composition of the film that supports it [11]. In sparkling wines, bubbles consist of gas surrounded by a film of wine constituents. These tensioactive components and other substances afford viscosity to the film, giving texture to the bubble (**Figure 1**) [4]. In fact, it was established that foaming properties depend on compounds that decrease surface tension and increase the viscosity of the film between the bubbles. This factor contributes to foam stabilization and renders the bubbles more resistant to coalescence [12, 13].

In brief, foam formation and persistence is directly dependent upon its chemical composition, and the synergistic interaction among numerous foam active compounds which, due to aggregation or complex formation, may modify their surface-active properties. For this reason, and in order to ascertain which compounds affect foam quality, it is necessary to evaluate as many compounds as possible. In this sense, several scientific studies have been carried out in an attempt to determine the wine compounds that could play a role in the foam properties of sparkling wines. Many of these studies carried out in model solutions and in base and sparkling wines, are summarized in the reviews made by several authors [11, 14]. The present chapter increases the knowledge on this topic and reviews the latest studies made to determine the influence of proteins, peptides, amino acids, polysaccharides, phenolic compounds, lipids, organic acids and others on the foamability and foam stability of sparkling wines.

2. Measuring of foaming properties of sparkling wines

Most of the studies published in the literature on sparkling wine foam quality are aimed at establishing the effect of the chemical composition of grape juices, base wines and sparkling wines on their foaming properties. In order to correlate the foaming properties with the physical and chemical characteristics of sparkling wines, much effort has been made to find instrumental techniques that can be used to obtain a quantifiable value for foam quality, and consequently to be able to compare sparkling wines. Among them, methods based on measuring the kinetics of CO_2 discharging, gas sparging methods, and image analysis methods are some of the most often employed [15].

An automated equipment, called "Mosalux" was designed to measure the foaming properties of wines [16]. This apparatus was adapted from that described by Rudin [17] and allows measurement of the increase with time of the height of a wine foam column submitted to a definite effervescence [16]. In fact, this is an objective and normalized method based on the interruption of a beam of ultra red light by the foam produced after the injection of CO_2 into the wine. Three parameters can be measured.

- **HM**: maximum height reached by the foam after carbon dioxide injection through the glass frit, expressed in mm; this could represent the foam-ability, the wine's ability to foam.
- **HS**: foam stability height during carbon dioxide injection, expressed in mm; this could represent the foam stability, the wine's ability to produce stable foam or persistence of foam collar.
- **TS**: foam stability time, until all bubbles collapse, when CO_2 injection is interrupted, expressed in s; this could represent the foam stability time, once effervescence has decreased.

Figure 2 shows the description of the "Mosalux" equipment and an example of the plot generated during a foam measurement.

The "Mosalux" equipment has been the most widely used since 1990 and in addition to research laboratories. It is probably the most used instrumental system in sparkling wine cellars for foam characterization. Moreover, a good relationship has been established between the foaming properties obtained by using "Mosalux" and foam sensory analysis [18]. The "Mosalux" apparatus has also been used to determine other parameters such as the expansion of foam **E**, the Bickerman coefficient **Σ** [19] (lifetime of a bubble in dynamic conditions, when formation and destruction of bubbles are balanced), and the lifetime of foam **LF** [20]. When comparing the different foam parameters (HM, HS, TS, E, LF, and $\sum$) obtained by the gas sparging method, it was concluded that the best parameters to characterize the foam capacities of wines were HM, $\sum$, and TS [21]. Other variation of this system uses an ultrasound emitter-detector and a waveguide to detect foam fluctuations [22, 23] to obtain **Hpeak** (maximum height reached by the foam after air injection through a glass frit). Hpeak has been related to the wine's ability for foaming and **Hplato** (foam height stability during air injection) has been related to the average bubble lifetime. Correlation between the results obtained with this technique and sensory analysis has also been established [24].

http://dx.doi.org/10.5772/intechopen.70859

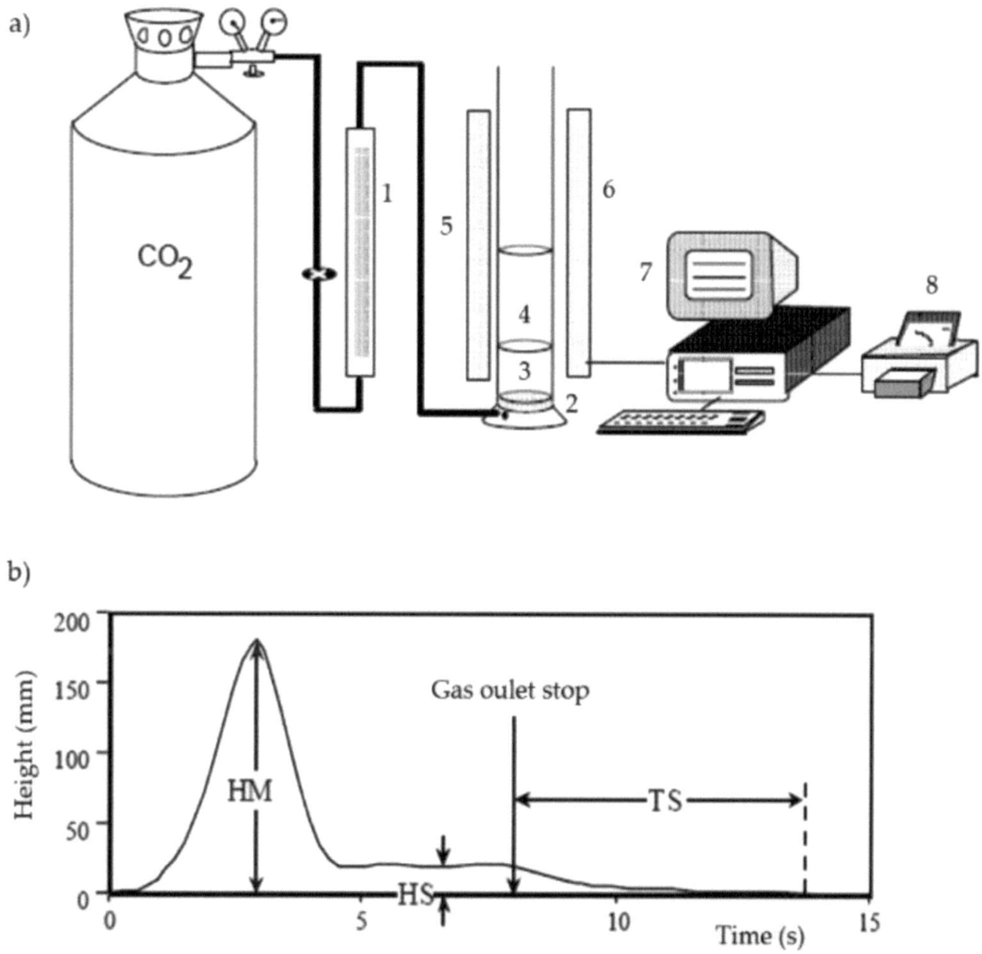

Figure 2. (a) Diagram of the "Mosalux" equipment. (1) Flowmeter, (2) test tube, (3) wine, (4) foam, (5) infrared emitter, (6) infrared receiver, (7) personal computer, (8) printer; (b) example of a foam profile of a sparkling wine.

3. Impact of wine macromolecules on the sparkling wine foam quality

Table 1 includes a summary of the compounds that have been related to foam properties in the different scientific studies published, making reference to the type of sample used: model wine, grape juice, base wine, sparkling wine or isolated foam.

Table 2 shows the correlations (r) at significance level ($p < 0.05$) between parameters that determine foam properties (HM, HS, TS, Peak H and Plateau H) and the chemical composition of grape juices, base wines and sparkling wines.

Compounds	Type of sample	Results	References
Proteins	Model wine	Increase foam	[67]
	Model wine	Increase foam height	[38]
	BW[a]	Increase foam height	[16]
	Separated foam	Increase foam	[34, 35]
	Model wine and BW[a]	Increase foam stability	[20]
	BW[a]	Increase foam height and foam stability	[39]
	BW[a] and SW[a]	Reduce foam height and increase foam stability	[29]
	BW[a]	Increase foam height and reduce foam stability	[25]
	SW[a]	Increase foam height and foam stability	[40]
	Grape juice	Increase foam height	[12]
	BW[a]	Increase foam height and reduce foam stability	[26]
	SW[a]	Increase foam height, foam stability height and decrease foam stability time	[41]
	Grape juice and BW[a]	Increase foam height	[42]
	BW[a] and SW[a]	Increase foam height and foam height stability	[22]
	BW[a] and SW[a]	Increase foam height	[43]
	SW[a]	Increase foam height	[44]
	BW[a]	Increase foam height	[45]
	SW[a]	Increase foam height stability	[24]
	BW[a]	Increase foam height and foam stability	[46]
	SW[a]	Increase maximum height, foam height stability and effervescence	[23]
	BW[a]	Increase foam stability	[70]
	BW[a] and SW[a]	Increase foam height and foam stability height	[47]
	SW[a]	Increase foam height and foam stability height	[48]
	BW[a] and SW[a]	Increase foam height and foam stability height	[27]
	Model wine	Cooperative effects between mannoproteins and the proteins of grape origin to improve foamability	[33]
	BW[a] and SW[a]	Increase foam height	[49]
	BW[a]	Increase foam height	[50]
Peptides	BW[a] and SW[a]	No influence on foam height and foam height stability	[22]
	SWa	Improve foam height stability	[24]
Amino acids	BW[a]	Decrease foam stability time	[25]
	SW[a]	Proline and glutamine increase foam height and foam stability height Decrease foam stability time	[41]
	BW[a] and SW[a]	Increase foam height and foam height stability	[22]
	SW[a]	Increase maximum height, foam height stability and effervescence	[23]
	SW[a]	Increase foam height and foam stability height	[28]

Compounds	Type of sample	Results	References
Polysaccharides	Model wine	Increase foam stability	[38]
	Separated foam	Increase foam	[34]
	BW[a] and SW[a]	Xylose in polysaccharides increase foam stability	[29]
	BW[a]	Increase foam height	[25]
	SW[a]	Increase foam height and stability time	[40]
	BW[a]	Total polysaccharides increase foam height and reduce foam stability time Acid and neutral polysaccharides increase foam height	[26]
	Grape juice and BW[a]	Total and neutral polysaccharides increase foam height	[42]
	BW[a] and SW[a]	Increase foam height and foam height stability	[22]
	BW[a] and SW[a]	Polysaccharides (Mr of 62,000–48,000, 13,000–11,000, and 3000 to 2000) increase foam height, and the Mr. 3000–2000 polysaccharide reduce foam stability	[43]
	SW[a]	Total and acid polysaccharides decrease foam stability time	[66]
	BW[a]	Reduce foam height	[45]
	SW[a]	Mannoproteins increase maximum height, foam height stability and effervescence	[23]
	Model wine and SW[a]	Increase foam height and foam height stability	[30]
	Model wine	Increase foam stability	[56]
	Model wine	Mannoproteins with low content of protein (5%) increase foam stability. Arabinogalactans and hydrophobic low molecular weight fraction (<1 kDa) increase foamability.	[32]
	SW[a]	Mannoproteins, arabinogalactans and pectic polysaccharides (HMW) increase foam height, foam stability height and foam stability time	[31]
	Model wine	Mannoproteins increase foamability	[33]
	SW[a]	Mannoproteins and PRAG increase foam stability time	[28]
	BW[a] and SW[a]	High molecular weight polysaccharides decrease foam height	[49]
	BW[a]	Increase foam stability time	[50]
Polyphenols	Model wine	(+)-catechin increase foamability and foam stability	[61]
	SW[a]	Increase foam height and reduce foam stability	[40]
	Grape juice	Total polyphenol increase foam height Nonflavonoid phenol increase foam height Flavonoid phenol increase foam height	[12]
	BW[a]	Non flavonoids phenols decrease foam stability time	[26]
	Grape juice and BW[a]	Total polyphenols, ortodiphenols, flavonoids and nonflavonoids reduce foam stability time	[42]
	BW[a]	Reduce foam height	[45]
	SW[a]	Anthocyanins increase foam height and foam stability height Proanthocyanidins decrease foam height and foam stability height	[28]
	BW[a]	Increase foam stability time	[50]

Compounds	Type of sample	Results	References
Lipids	BW[a]	C8 and C10 increase collar height and reduce stability foam	[16]
	Model wine and BW[a]	Lipids are only foam active compounds at low alcohol concentration	[64]
	BW[a] and SW[a]	Linoleic acid increase foam stability Palmitic acid increase foam height	[29]
	BW[a] and separated foam	C8, C10, and C12 reduce foam height. Ethyl esters of hexanoic, octanoic, and decanoic acids increase foam height.	[65]
	SW[a]	Monoacylglycerols of palmitic and stearic acids and glycerylethylene glycol fatty acid derivatives increase the promotion and stabilization of foam	[31]
Organic acids	Model wine and BW[a]	Tartaric acid increase foam	[20]
	BW[a] and SW[a]	Tartaric acid increase foam height	[29]
	BW[a]	Malic acid increase foam height Titratable acidity increase foam height Lactic acid decrease foam height Citric and galacturonic acid reduce foam stability time pH reduce foam stability time	[25]
	SW[a]	Malic acid increase foam height and reduce stability foam	
	Grape juice	pH increase foam height Total acidity decrease foam height	[40]
	SW[a]	Galacturonic acid decrease foam stability time	[12]
	BW[a]	Titratable acidity, malic acid increase foam height and reduce foam stability time Lactic acid reduce foam height and increase foam stability time Citric acid and galacturonic acid reduce foam stability time	[41]
	BW[a]	Malic acid increase foam height	[26]
	SW[a]	Tartaric acid increase foam height pH decrease foam height Lactic acid decrease foam stability time	[45]
	BW[a] and SW[a]	Gluconic acid reduce foam height	[49]
Others	Separated foam	Iron increase foam	[34]
	Model wine and BW[a]	Glycerol increase foam	[20]
	BW[a] and SW[a]	Glucose increase foam height Total content of SO_2 reduce foam stability ϒ-butyrolactone increase foam stability	[29]
	BW[a]	Acetaldehyde, ethyl acetate, diacetaldehyde and isoamylic alcohols reduce foam stability time Alcohol content increase foam height and foam stability height Glucose increase foam height and fructose reduce foam height	[25]
	Grape juice	Fructose, glucose and methanol increase foam height Free sulfur dioxide decrease foam height Soluble solid concentration and maturity index increase foam height	[12]
	BW[a]	Alcohol content increase foam height Turbidity increase foam height and reduce foam stability time	[26]

Compounds	Type of sample	Results	References
		Free sulfur dioxide increase foam height and reduce foam stability time Conductivity increase foam height and reduce foam stability time	
	SW[a]	Residual sugars and ethanolamine increase foam height and foam stability height Ethyl acetate decrease foam stability time	[41]
	SW[a]	*Botrytis cinerea* infection decrease foamability	[44]
	BW[a]	Alcohol concentration and total SO_2 reduce foam height	[45]
	SW[a]	Ethanol, volatile acidity and total SO_2 reduce foam height Volatile acidity and total SO_2 reduce foam stability time	[66]
	BW[a]	Lysozyme have a protective effect on foaming properties	[71]
	Model wine	*Botrytis cinerea* protease activity decrease wine foaming properties	[69]
	BW[a]	*Botrytis cinerea* infection decrease foamability and foam stability	[70]
	Model wine	Glycerol and glycerol plus ethyloctanoate increase foam height and foam stability time	[32]
	BW[a] and SW[a]	Ethanol content reduce foam height	[49]

[a]BW: base wines; SW: sparkling wines.

Table 1. Compounds related to foam properties in sparkling wines.

In the majority of the works shown in **Tables 1** and **2**, the chemical compounds have been related to the foam physical parameters obtained by the "Mosalux" device [12, 16, 25–28] or other variations of this method [20, 22–24, 29–33]. All studies have shown that the foam properties of sparkling wines mainly depend on the qualitative composition and quantitative content of surface active substances. The relation found between the foaming properties and the different wine macromolecules is detailed below.

3.1. Proteins

Despite of the low concentration of proteins in sparkling wines (ranging from 4 to 16 mg/L) [14], previous works have shown that these compounds are largely involved in the foaming properties of sparkling wines due to their surfactant properties. Surfactant agents are inferred to stabilize foams by settling at the bubble's edge, with the hydrophobic side interacting with the gas phase and the hydrophilic side interacting with the aqueous liquid phase [34]. The behavior of proteins in the foam depends on their hydrophobicity, solubility (dependent on the isoelectric point and the pH of the wine), and molecular weight [35, 36]. The net charge of macromolecules depends on the pH [37]. The isoelectric point of the wine proteins is between 3.5 and 4.5 [35] and between 4.6 and 5.0 [29]. At the wine pH, 2.9, its proteins would be positively charged and could migrate to the wine/air interphase and to stabilize foam [20]. However, characterization of foaming proteins have showed that foam formation is dependent on protein hydrophobicity but not on their molecular weight or isoelectric point [34].

Compounds	Type of sample	HM	HS	TS	Peak H	Plateau H	References
Proteins	BW[a]	0.32		−0.51			[25]
	Grape juice	0.91					[12]
	SW[a]				0.62	0.49	[22]
	BW[a]	0.31					[45]
	BW[a] and SW[a]	0.58					[49]
	BW[a] and SW[a]	0.44					[43]
	Grape juice	0.75					[42]
Amino acids							
Total amino acids	SW[a]	0.85	0.63				[28]
Acid amino acids	SW[a]	0.82	0.58				[28]
Neutral amino acids	SW[a]	0.85	0.68				[28]
Basic amino acids	SW[a]	0.75	0.62				[28]
Total biogenic amines	SW[a]	0.66	0.64	0.48			[28]
Aspartic acid	SW[a]				0.52	0.67	[22]
	SW[a]	0.86	0.63				[28]
Hydroxyproline	BW[a]			−0.39			[25]
	SW[a]	0.46					[28]
Glutamic acid	BW[a]			−0.50			[25]
	SW[a]				0.66	0.71	[22]
	SW[a]	0.77	0.54	0.46			[28]
Serine	BW[a]			−0.425			[25]
	SW[a]				0.56	0.68	[22]
	SW[a]	0.62	0.59	0.58			[28]
Asparagine	BW[a]			−0.38			[25]
	SW[a]				0.41	0.57	[22]
	SW[a]	0.79	0.68	0.45			[28]
Glycine	BW[a]			−0.39			[25]
	SW[a]				0.41	0.57	[22]
	SW[a]	0.88	0.66	0.35			[28]
Glutamine	BW[a]	0.37		−0.36			[25]
	SW[a]					0.53	[22]
	SW[a]			0.42			[28]
Histidine	SW[a]				0.50	0.48	[22]
Threonine	SW[a]	0.56	0.42				[28]
Proline	BW[a]		0.34		0.58	0.69	[25]
	SW[a]						[22]

Compounds	Type of sample	HM	HS	TS	Peak H	Plateau H	References
	SW[a]	0.82	0.60	0.34			[28]
Histamine	SW[a]	0.39	0.42	0.43			[28]
GABA	BW[a]			−0.38			[25]
	SW[a]				0.52	0.60	[22]
	SW[a]	0.77	0.52				[28]
Arginine	BW[a]			−0.36			[25]
	SW[a]				0.50	0.62	[22]
	SW[a]	0.83	0.65				[28]
α alanine	SW[a]				0.53	0.63	[22]
	BW[a]			−0.37			[25]
	SW[a]	0.83	0.65	0.39			[28]
B alanine	SW[a]	0.92	0.55				[28]
Tyrosine	BW[a]			−0.53			[25]
	SW[a]				0.49	0.63	[22]
	SW[a]	0.81	0.60				[28]
Valine	BW[a]			−0.50			[25]
	SW[a]				0.52	0.67	[22]
Methionine	BW[a]			−0.34			[25]
	SW[a]				0.51	0.63	[22]
	SW[a]	0.89	0.58				[28]
Cysteine	SW[a]	0.79	0.49				[28]
Isoleucine	BW[a]						[25]
	SW[a]	0.67	0.64	0.47			[28]
Leucine	BW[a]			−0.34			[25]
	SW[a]				0.51	0.64	[22]
	SW[a]	0.42	0.55	0.55			[28]
Phenylalanine	BW[a]			−0.29			[25]
	SW[a]				0.42	0.62	[22]
	SW[a]	0.84	0.62	0.36			[28]
Ornithine	BW[a]			−0.31			[25]
	SW[a]	0.79	0.64				[28]
Tryptophan	BW[a]			−0.37			[25]
	SW[a]	0.85	0.59				[28]
Lysine	BW[a]			−0.36			[25]
	SW[a]					0.52	[22]
	SW[a]	0.66	0.61				[28]

Compounds	Type of sample	HM	HS	TS	Peak H	Plateau H	References
Spermidine	SW[a]	0.72	0.41				[28]
Tyramine	SW[a]			0.35			[28]
Putrescine	SW[a]	0.51	0.59	0.43			[28]
Cadaverine	SW[a]	−0.35					[28]
Phenylethylamine	SW[a]		0.60				[28]
Isoamylamine	SW[a]	−0.55					[28]
Polysaccharides							
Total polysaccharides	Grape juice	0.55					[42]
	BW[a]	0.40					[42]
	SW[a]				0.80	0.68	[22]
	SW[a]			0.64			[28]
Polysaccharides from yeasts	SW[a]			0.53			[28]
Polysaccharides from grapes	SW[a]			0.68			[28]
Neutral polysaccharides	Grape juice	0.65					[42]
	BW[a]	0.46					[42]
	SW[a]				0.82	0.71	[22]
Acid polysaccharides	BW[a]	−0.76					[45]
High molecular weight polysaccharides	BW[a] and SW[a]	−0.65					[49]
Polysaccharides Molecular Mass 62,000–48,000	BW[a] and SW[a]	0.51					[43]
Polysaccharides Molecular Mass 13,000–11,000	BW[a] and SW[a]	0.46					[43]
Polysaccharides Molecular Mass 3000–2000	BW[a] and SW[a]	0.32					[43]
Mannoproteins	SW[a]			0.47			[28]
Polysaccharides rich in arabinose and galactose	SW[a]			0.72			[28]
Homogalacturonans	SW[a]			0.58			[28]
Glucans	SW[a]			0.40			[28]
Polyphenols							
Absorbance 520 (nm)	BW[a]			−0.35			[25]
Absorbance 280 (nm)	Grape juice	0.92					[12]
	BW[a]			−0.63			[42]
Total polyphenol	Grape juice	0.76					[12]
	BW[a]			−0.60			[42]
	BW[a]	−0.45					[45]
Total proanthocyanidins	SW[a]	−0.73					[28]

Compounds	Type of sample	HM	HS	TS	Peak H	Plateau H	References
Nonflavonoid phenol	Grape juice	0.59					[12]
	BW[a]			−0.33			[42]
Total flavan-3-ols	SW[a]	0.50		0.42			[28]
Flavonoid phenol	Grape juice	0.52					[12]
	BW[a]			−0.64			[42]
Ortodiphenols	BW[a]			−0.49			[42]
Total monomeric anthocyanins	SW[a]	0.96	0.80				[28]
Non-acylated anthocyanins	SW[a]	0.97	0.81				[28]
Acetyl-glucoside anthocyanins	SW[a]	0.94	0.75				[28]
Coumaryl-glucoside anthocyanins	SW[a]	0.88	0.67				[28]
delphinidin-3-glucoside	SW[a]	0.94	0.71				[28]
cyanidin-3-glucoside	SW[a]	0.84	0.60				[28]
petunidin-3-glucoside	SW[a]	0.95	0.73				[28]
peonidin-3-glucoside	SW[a]	0.87	0.65				[28]
malvidin-3-glucoside	SW[a]	0.98	0.85				[28]
delphinidin-3-(6-acetyl)-glucoside	SW[a]	0.91	0.67				[28]
cyanidin-3-(6-acetyl)-glucoside	SW[a]	0.89	0.62				[28]
petunidin-3-(6-acetyl)-glucoside	SW[a]	0.92	0.69				[28]
peonidin-3-(6-acetyl)-glucoside	SW[a]	0.89	0.65				[28]
malvidin-3-(6-acetyl)-glucoside	SW[a]	0.89	0.92				[28]
delphinidin-3-(6-*p*-coumaryl)-glucoside	SW[a]	0.76	0.52				[28]
cyanidin-3-(6-*p*-coumaryl)-glucoside	SW[a]	0.92	0.68				[28]
petunidin-3-(6-*p*-coumaryl)-glucoside	SW[a]	0.78	0.55				[28]
peonidin-3-(6-*p*-coumaryl)-glucoside	SW[a]	0.91	0.67				[28]
malvidin-3-(6-*p*-coumaryl)-glucoside	SW[a]	0.94	0.76				[28]
cis-caftaric	SW[a]	−0.65					[28]
trans-fertaric	SW[a]	0.35					[28]
coumaric acid	SW[a]	0.77	0.37				[28]
ferulic acid	SW[a]	−0.39		−0.41			[28]
gallic acid	SW[a]	0.62					[28]
(+)-catechin	SW[a]	0.50		0.42			[28]
quercetin-3-rutinoside	SW[a]	−0.43					[28]
myricetin	SW[a]		0.36				[28]
quercetin	SW[a]	0.58					[28]
kaempferol	SW[a]			0.53			[28]
isorhamnetin	SW[a]	0.84					[28]

Compounds	Type of sample	HM	HS	TS	Peak H	Plateau H	References
Lipids							
C8 (n = 28)	BW[a] and separated foam	−0.43					[65]
C10 (n = 28)	BW[a] and separated foam	−0.66					[65]
C12 (n = 28)	BW[a] and separated foam	−0.57					[65]
Ethyl hexanoate (n) 28	BW[a] and separated foam	0.65					[65]
Ethyl octanoate (n) 28	BW[a] and separated foam	0.86					[65]
Ethyl decanoate (n) 28	BW[a] and separated foam	0.90					[65]
Organic acids							
Titratable acidity	BW[a]	0.46					[25]
	Grape juice	−0.59					[12]
pH	BW[a]			−0.32			[25]
	Grape juice	0.71					[12]
Citric acid	BW[a]			−0.38			[25]
Galacturonic acid	BW[a]			−0.42			[25]
Malic acid	BW[a]	0.46					[25]
	BW[a]	0.40					[45]
Lactic acid	BW[a]	−0.43					[25]
Gluconic acid	BW[a] and SW[a]	−0.36					[49]
Others							
Alcohol content	BW[a]	0.47	0.46				[25]
	BW[a]	−0.47					[45]
	BW[a] and SW[a]	−0.92					[49]
Glucose	BW[a]	−0.31					[25]
	Grape juice	0.58					[12]
Fructose	BW[a]	0.56	0.32				[25]
	Grape juice	0.73					[12]
Ethanolamine	BW[a]		0.31				[25]
Acetaldehide	BW[a]			−0.35			[25]
Ethyl acetate	BW[a]			−0.51			[25]
Diacetaldehyde	BW[a]			−0.36			[25]
Isoamylic alcohols	BW[a]			−0.43			[25]

Compounds	Type of sample	HM	HS	TS	Peak H	Plateau H	References
Maturity index	Grape juice	0.78					[12]
Soluble solid concentration	Grape juice	0.75					[12]
Methanol	Grape juice	0.80					[12]
Free sulfur dioxide	Grape juice	−0.65					[12]
Total sulfur dioxide	BW[a]	−0.68					[45]

[a]BW: base wines; SW: sparkling wines.

Table 2. Correlation coefficients (r) at significance levels ($p < 0.05$) between parameters that determine foam properties (HM, HS, TS, Peak H and Plateau H) and the chemical composition of grape juices, base wines and sparkling wines.

Proteins have been the most studied compounds in relation to wine foamability. Most studies indicate a positive influence of protein content on foam height in grape juices, base wines and sparkling wines [16, 20, 22–27, 38–50] (**Tables 1** and **2**). Some of this studies showed positive correlations between proteins and parameter HM [12, 25, 42, 43, 45, 49], Peak H, and Plateau H [22]. The highest correlations between proteins and foamability parameters were observed in juices of white grapes (r > 0.75) [12, 42]. Correlation between proteins and foamability parameters was lower in base wines and sparkling wines [22, 25, 43, 45, 49] (**Table 2**). The work conducted in Spanish sparkling wines was an exception to this because authors observed a negative relation between proteins and foam height [29].

The correlations between proteins and foam stability have shown contradictory results. Therefore, some proteins have been described as good foam formers but poor stabilizers, while others are poor foam formers but good stabilizers [13, 20, 25, 26, 29, 39, 41]. Inverse relationship between HM and TS [16, 25, 26] could justify that proteins may be active agents on foamability but may not sustain a foam collar for a long time.

The influence of specific proteins on foam quality has also been studied in several research papers. Grape invertase is one of the most abundant protein present in wine (from 9 to 14% of the total protein content of a Chardonnay wine) [51]. This grape protein possesses a pI close to the pH of wine and a high hydrophobicity, potentially conferring good surface properties on this protein [51]. Significant decreases in the invertase content in base wines have been shown to correlate with decreases in foam height and foam stability [46]. Other grape proteins, such as thaumatin-like proteins and chinases, did not contribute alone to the formation and stabilization of foam; however, when synergistically acting with mannoproteins, foam height was found to be maximized [33]. On the other hand, the release of proteins from the yeast cells prior to autolysis has also been shown to contribute to foam height and foam stability height in sparkling wines [23, 24, 47].

3.2. Peptides

The hydrophobicity of peptides may be related to the quality of sparkling wine foam [52, 53]. Proteins and peptides with molecular weight within 24–60 kDa, even in low concentrations, provide for the foam formation in base wine [13, 31, 47] since they form adsorption layers with high mechanical strength and, as a result, increase the stability of the sparkling wine foam. In fact, a positive correlation has been reported between polypeptide molecular mass, hydrophobicity and foam stabilizing activity in beer [54, 55]. Although no correlations have been found between peptides and foam properties of sparkling wines [22], bentonite added to the tirage solution produced a reduction in both protein and peptide contents and thereby negatively affected foaming properties [24] (**Table 1**).

3.3. Amino acids

In addition to proteins and peptides, some authors agree in considering amino acids as foaming agents. At wine pH, amino acids carry a net positive charge, so they are surfactant with hydrophilic and hydrophobic groups. This property causes amino acids to be retained in the air/liquid interface, and thus reduces the wine surface tension, improving the sparkling wine ability to foam [28].

Moreno-Arribas et al. [22] showed positive correlations between free amino acids and white sparkling wine foamability (**Tables 1** and **2**). The authors observed that maximum height (Peak H) was significantly correlated with most of the amino-acids, although coefficients of over 0.60 were only found for glutamic acid (r = 0.66). Moreover, Plateau H (associated to bubble lifetime) was highly correlated with glutamic acid (r = 0.71), serine (r = 0.68), valine (r = 0.67) and proline (r = 0.69). Lower positive correlations were found by Andrés-Lacueva et al. [25] in white base wines between glutamine and proline and foamability parameters.

Other study conducted by our research group in white and rosé sparkling wine showed the highest positive correlations between total amino acids and foam height (r = 0.85) and total amino acids and foam stability height (r = 0.63) [28] (**Table 2**). Biogenic amines showed the same behavior as amino acids, although lower correlation values were observed [28] (**Table 2**). When comparing the different amino acids analyzed, glycine, β-alanine, and methionine had the highest correlation with foam height (r > 0.88) [28] (**Table 2**). In general, amino acids with non-polar side chains showed higher values of correlation than amino acids with polar side chains. At wine pH, amino acids are protonated and they act as cationic surfactants according to the hydrophobicity of their side chains. Their amphiphilic character could cause amino acids to become concentrated at the liquid–gas interfaces, improving the sparkling wine foamability [28].

Negative correlations have been observed between amino acids and foam stability time in base wines [25] (**Tables 1** and **2**). However, conflicting results have been published on the influence of amino acid on foam stability of sparkling wines. It has shown that lower levels of amino acids favors a greater stability time of foam [41]; while other authors did not found any influence of these compounds on the foam stability time [28].

It was confirmed that the autolytic capacity of yeast was important for the quality of sparkling wines [23]. The use of a mutant having accelerated autolysis showed that the second fermentation of wines with this mutant improved the foaming properties versus a control strain due to higher increase in both nitrogen compounds (proteins, peptides and amino acids) and polysaccharides [23].

3.4. Polysaccharides

Contradictory results have been published on the effect of polysaccharides on foam quality. Girbau Sola et al. [45] showed a negative influence of acid polysaccharides on foam height in base wines ($r = -076$). The same authors showed that polysaccharides were negatively correlated with foam stability but positively with the average bubble lifetime or Bikerman's coefficient [45]. Similarly, polysaccharides with a molecular weight higher than 180 kDa have also shown a negative influence on foam height ($r = -0.65$), although these authors associated the negative contribution with the presence of β-glucans secreted by *Botrytis cinerea* and stated that other polysaccharides probably would not have a negative effect [49] (**Tables 1** and **2**).

In contrast with the results described above, most studies point to a positive influence of total polysaccharides on both foamability [22, 25, 26, 40, 42] and foam stability [28, 29, 50, 56] (**Table 1**). The relation of the molecular weight of polysaccharides and the foaming properties of wines has also been studied. Polysaccharides of molecular mass of 62 to 48 kDa; 13 to 11 kDa; and 3 to 2 kDa have been demonstrated to be active agents on foamability, and polysaccharides with molecular mass of 3 to 2 kDa might be a foam stability agent, since they were correlated with the Bikerman coefficient [43].

Among polysaccharides, glycoproteins like mannoproteins released by yeast during fermentation and autolysis, have been described as the major compounds affecting foaming properties [13, 23, 30, 33]. The hydrophobic nature of glycoproteins explains why they are better foam stabilizers and foam producers than non-glycosylated proteins. Glycoproteins present a protein moiety with hydrophobic and hydrophilic domains and sugar moieties, which are hydrophilic and they could interact with surface-active materials and be absorbed at the gas/liquid interface. The hydrophilic glycans are located at the liquid layer, among the bubbles, corresponding to the oxidic zone of the protein. Hence, when the layer surrounding the bubbles becomes thinner, the viscosity increases and drainage of the liquid is delayed. The hydrophobic polypeptides increase the surface tension of the bubbles, resulting in more stable foam [13]. In this sense, the literature has tried to explain the influence of mannoproteins on foaming properties. Mannoproteins also influence the viscosity of the bubble wall and reduces the drainage of the liquid [34]. Foaming may be due to their interactions with proteins [36] and their surface properties and capacity to reorientation quickly at the liquid/gas interface in the bubble when the foam is formed [20]. In fact, the proteinaceous fraction of mannoproteins is able to bind to the liquid/air interface and interact with other compounds by means of electrostatic or hydrophobic forces, hydrogen bonds, or covalent linkages [13]. These interactions could lead to the formation of a strong viscoelastic film that could be highly resistant to tension and able to withstand the film's thickness [13], preventing coalescence of bubbles and leading to more stable foams. As a matter of fact the presence of both glycocompounds and protein material deriving from macromolecular fractions of different molecular weights in the adsorption layer of the foam of sparkling wines has been reported [56], and the presence of aggregated materials involving yeast glycoproteins and other unidentified wine components has also been indicated as contributing to the foam stability of sparkling wines [57, 58]. In this sense, reconstitution experiments performed by adding in a model solution different molecular fractions isolated from wine indicated that a synergistic effect in foamability and foam stability exists between high and low molecular weight wine compounds [31]. The fraction most responsible for foam stability was mainly influenced by mannoproteins with low content of protein (5%) and the foamability by arabinogalactans and a hydrophobic low molecular weight fraction (< 1 kDa) [32].

The specific contribution of the different families of wine polysaccharides to the wine foaming properties has been recently studied by our research group [28]. Mannoproteins, glucans, polysaccharides rich in arabinose and galactose, rhamnogalacturonans type II, and homogalacturonans did not show any influence on the foamability of sparkling wines. On the contrary, positive influence was found between foam stability time and all wine polysaccharides, with the exception of rhamnogalacturonans type II. Surprisingly, polysaccharides rich in arabinose and galactose showed higher positive correlations on foam stability (r = 0.72) than mannoproteins (r = 0.47) [28] (**Table 2**).

3.5. Polyphenols

It is widely known that polyphenols are highly reactive compounds. Some authors have tried to establish a correlation between them and the quality of foam in grape juices, base wines and sparkling wines. Polyphenols can interact with proteins and polysaccharides [36, 37], mainly the low molecular weight polyphenols [59], which participate in the hydration layer of the proteins [60]. Moreover, the formation of hydrogen bonds between the hydroxyl groups of the phenolic compounds and the polar head groups of proteins can be particularly relevant for the interaction with the air/liquid interface of the bubble film [61, 62]. These formed compounds could adsorb at the interface and form a stabilizing film around bubbles, which could promote foam formation [28].

Most of the studies carried out to correlate the influence of phenols on foam quality of sparkling wines have shown contradictory results [12, 26, 28, 40, 42, 45, 50, 61] (**Tables 1** and **2**). In fact, total phenolics did not shown correlation with any foam instrumental property in sparkling wines [28], but they showed a negative correlation with foam height in base wines (r = −0.45) [45], and a high positive correlation with foam height in grape juices (r = 0.76) [12]. Moreover, most of studies refer to global measurements of phenolic compounds, which could lead to inaccurate results difficult to understand. A recent study of our group has analyzed the relation of individual phenolics with foam parameters in white and rosé sparkling wines, which could be critical for their production [28].

The study concluded that each phenolic compound exhibits different behavior patterns on foam instrumental properties (**Table 2**). Non acylated, acetyl glucoside and coumaryl glucoside anthocyanins showed the highest positive correlations with foamability, with values

http://dx.doi.org/10.5772/intechopen.70859

ranging from 0.67 to 0.97, but these compounds did not show any effect on the foam stability time. Authors attributed this effect to the interaction of anthocyanins with wine proteins through hydrophobic interactions and hydrogen bonds. Attachment of a long aliphatic chain could confer interesting surfactant behavior on flavylium cations. Therefore, the product formed could be retained in the liquid/air interface, resulting in a reduction of the interfacial tension and an increase in the foamability. On the other hand, total proanthocyanidins showed high negative correlation with sparkling wine ability to foam (r = −0.73). Since proteins play an important role on the foamability of sparkling wines, the negative correlation of proanthocyanidins with foam height could be due to the precipitation of wine proteins by tannins. *Cis*-caftaric was the hidroxicinamic acid most negatively correlated with foam height (r = −0.65), while coumaric acid showed the most positive effect (r = 0.77) and isorhamnetin was the flavonol with a major influence on foam height (r = 0.84).

3.6. Lipids

Some authors describe that lipids can accumulate in the foam, reducing surface tension and stabilizing it [63]. However, the researches made in wines to establish the possible relationships among lipid content, fatty acids, and foam behavior have produced contradictory findings (**Table 1**). The addition of octanoic and decanoic fatty acids to wines had a negative effect on the foam stability time, but it positively influenced foam collar height [16]. However, the addition of a lipid mixture to wine did not affect their foam, but when the ethanol concentration was reduced, authors observed an adverse effect on bubble lifetime [64]. They concluded that linolenic acid and palmitic acid were, respectively, the best indicators of foam stability and foam height in base wines and sparkling wines respectively, both having a positive influence [29].

Moreover, it was studied the influence of fatty acids (free and bound as ethyl esters) on wine foaming in different white wines and separated foam (**Tables 1** and **2**). The free fatty acids C8, C10, and C12 were negatively correlated with foam height with values ranging from 0.43 to 0.66, whereas the ethyl esters of hexanoic, octanoic, and decanoic acids were positively related with values ranging from 0.65 to 0.90. These authors found that the value of foam height was directly proportional to the ratio of esterified to non-esterified fatty acids. So, the higher the coefficient, the greater the foamability; thus, it appeared that it was the esterified forms of fatty acids that increased foam height [65]. It was also shown that monoacylglycerols of palmitic and stearic acids and glycerylethylene glycol fatty acid derivatives were surface active compounds preferentially partitioned by the sparkling wine foam rather than the liquid phase, allowing the inference of their role as key components in the promotion and stabilization of sparkling wine foam [31].

3.7. Organic acids

With regards to organic acids (**Tables 1** and **2**), López-Barajas et al. [12] observed low negative correlations between titratable acidity and foamability in grape juices of white varieties (r = −0.58). However, other studies showed that tartaric acid, titratable acidity and pH increased foam height in grape juices, base and sparkling wines [12, 20, 25, 26, 29, 66]. In fact, pH and foamability in grape juices were highly correlated (r = 0.71) [12], while titratable acidity exhibited lower influence on foam height (r = 0.46) and foam stability time (r = −0.32) in white base wines [25]. In the same way, it was observed that acidity had a marked effect on foam since it modified protein solubility; if the juice acidity was low, protein hydrophobicity would be high, the surface activity could be increased, and then juice would have a higher foamability [35].

Different authors agree in pointing to malic acid as an enhancer of the foam height in base wines and sparkling wines [25, 26, 40, 45], but also stated that malic acid reduces foam stability time [26, 40]. On the contrary, lactic acid exerted the opposite effect on foam height [25, 26]. Malic acid and lactic acid showed low negative correlations with foam height [25, 45], which could indicate that malolactic fermentation is not recommended as a way to maximize foamability in sparkling wines. Moreover, conflicting results have been published on the influence of lactic acid on foam stability time. Some authors have observed a positive influence of lactic acid on foam stability in base wines [26], while others showed the opposite effect in sparkling wines [66]. Other acids such as citric and galacturonic acid reduced foam stability time in base and sparkling wines [25, 26, 41]. Moreover, the presence of gluconic acid due to *Botrytis cinerea* was shown to negatively affect wine foamability (r = −0.36) [49].

3.8. Others

Several authors agree that sulfur dioxide negatively affect the foaming qualities of wines [12, 26, 29, 45, 66] due to SO_2 is a denaturing agent of proteins [16]. In fact, negative correlations have been obtained between free and total sulfur dioxide and foam height in grape juices and base wines [12, 45].

There is some controversy about the effect of ethanol content on foaming quality. Some authors consider that ethanol has beneficial effects on foam [25, 26] while others assign it negative contribution [45, 49, 66]. The negative effect of ethanol on foam seems to be dependent on its content [67]. This could be explained by the ethanol modification of the solvent properties, the interactions between the protein and the solvent, and the structure of the adsorption layer [68]. When the alcohol content is low, other surfactants can be more active and thus more easily adsorbed at the interface, stabilizing the foam formed [20, 64]. In this sense, higher alcohol content was reported to decrease foamability [16]; however, this effect could be counteracted by other compounds produced in the second fermentation. In this regard, juices with a maturation index [ratio between soluble solids (°Brix) and titratable acidity (grams of tartaric acid per litter of juice) ranging from 4 to 5.5 had high foamability [12]. In fact, it was observed a high positive correlation between foam height and maturation index (r = 0.78) [12]. Subsequently, these results were confirmed, showing that maturation indexes for foamability and stability above 5.5 provided the wine with a less optimal alcoholic content for the formation of foam than wine produced from grapes with a maturation index within the stated range [45].

Glycerol is known to contribute to the viscosity of the wines. Due to its tensoactive properties, glycerol has shown a positive influence on foamability in sparkling wines [20, 32]. On the other hand, iron [34] and residual sugars [25, 29, 41] have been related with an improvement of foamability in sparkling wines.

The effect of *Botrytis cinerea* on the foaming characteristics of sparkling wines has also been studied [44, 69, 70]. In these works, it was concluded that this infection can cause a drastic reduction in foamability, since it uses up the proteins in the medium.

Diverse studies have been published about the influence of stabilization treatments, either using clarifiers or filtrations, on the foam quality of wines [20, 24, 41, 46, 47, 71–74]. In all cases, the foams were negatively affected by these treatments, and this was directly correlated with a decrease in the protein concentration. On the contrary, lysozyme additions made to the musts and wines before and after bentonite or charcoal treatments seem to have a protective effect on the wine proteins, and thereby an increase in foamability [71].

Research conducted suggest that many compounds influence foam capacity of sparkling wines (**Tables 1** and **2**); however, the most influencing compounds on the foaming properties have proved to be total amino acids, polysaccharides, anthocyanins, coumaric acid and isorhamnetin, all of them showing correlation coefficients higher than 0.75 (**Table 2**). On the contrary, the alcohol content and the concentration of acid polysaccharides, proanthocyanidins and free SO_2 are the factors which most negatively affect foam quality (**Table 2**).

4. Prediction of foaming properties

In view of the results shown in **Tables 1** and **2**, it can be concluded that foamability and foam stability is a complicated issue. In fact, the foaming capacity of wines depends on a complex equilibrium among all the compounds that favor its formation and stability and those that do not. There is not one compound or group of compounds that is responsible for making and keeping good quality foam. Instead, foam quality depends on a synergistic relationship between many different compounds that when acting together result in the foaming properties.

Foam behavior results from the synergistic interaction between the different foam active compounds which, due to aggregation or complex formation, may modify their surface-active properties. Thus, foaming properties not only are due to the presence or absence of a specific group of compounds but also are influenced by the net balance of the number and type of compounds ranging among different chemical structures. For this reason, and in order to ascertain which compounds have a major influence on the foam quality of sparkling wines, it is necessary to evaluate as many compounds together as possible, and to study the combined effect of all them. In this sense, statistical tools of multiple linear regression [12, 22, 26, 28, 75] and partial least squares regression analysis [29] has been used by several authors in an attempt to predict the foam properties of sparkling wines, and find out the chemical compound that provided the best predictive model of the foam properties.

Most of the studies include in the models all the variables that are usually analyzed in the wineries, and try to predict values for foamability, foam stability and Bickerman coefficient Σ [12, 26, 75]. Results of these researches have shown a great influence of proteins, SO_2, absorbance at 280 nm, glycerol and maturation index. Moreover, stepwise analysis showed that the foam height and Bickerman coefficient of sulphited grape juices could be predicted with a probability higher than 89.97% by the following polynomial equations (Eq. (1) and (2)) [12]:

$$HM = -126.80 + 1.04\text{Combined } SO_2 + 16.85\text{OD}280 + 1.07\text{Proteins} - 44.40\text{Glycerol} \quad (1)$$

$$\sum = 4.76 + 1.68\text{Maturation index} - 5.48\text{Tartaric acid} + 0.34\text{Glucose} \quad (2)$$

Other study conducted by Pueyo et al. [29] applied PLS regression to predict foam height and foam stability in base wines and sparkling wines using 73 chemical variables analyzed. Tartaric acid, glucose, total palmitoleic acid and protein content were the most influent variables in the prediction of foam height in base wines. However, total contents of oleic, palmitic, and stearic acids, and the content of 1-hexanol were the most important variables for predicting foam height in sparkling wines. With regards to foam stability, the variables with high predictive relevance in base wines were the total content of linolenic and undecanoic acids and the free content of undecanoic acid, while the total content of SO_2, the isobutanol, the total acidity, and proteins were the variables with high predictive relevance.

Moreno-Arribas et al. [22] observed that neutral polysaccharides, protein nitrogen and phenylalanine displayed high positive contribution to the prediction of maximum foam height (Peak H), and height at which the foam stabilizes in sparkling wines (Plateau H). The fitted final models, which presented the following adjusted equations (Eq. (3) and (4)), explained 76% of Peak H variation and 70% of the variation of Plateau H.

$$\text{Peak H} = 194.31 + 0.37\text{Neutral polysaccharides} + 59.68\text{Protein nitrogen} \quad (3)$$

$$\text{Plateau H} = 180.45 + 0.17\text{Neutral polysaccharides} + 3.80\text{Phenylanine} \quad (4)$$

A recent work carried out in our group in 2015 used multiple linear regression analysis in white and rosé sparkling wines differentiating models which anthocyanins were included. It was concluded that the different forms of malvidin had the highest influence on the foam height and foam stability height parameters, followed by amino acid compounds ((Eq. (5) to (8)), while foam stability model was only predicted by polysaccharides from grapes, concretely by polysaccharides rich in arabinose and galactose ((Eq. (9) and (10)) [28].

$$\begin{aligned} HM\ (\text{rosé sparkling wines}) &= 84.882 + 0.065\text{Total amino acids} \\ &+ 5.242\text{Non-acylated anthocyanins} - 0.477\text{Total proanthocyanidins}\left(R^2 = 90.2\%\right) \end{aligned} \quad (5)$$

$$HM\ (\text{white sparkling wines}) = 66.997 + 0.206\text{Total amino acids}\left(R^2 = 73.3\%\right) \quad (6)$$

$$\begin{aligned} HS\ (\text{rosé sparkling wines}) &= 9.730 + 0.331\text{Basic amino acids} \\ &+ 2.492\text{Acetyl} - \text{glucoside anthocyanins} + 0.995\text{Total biogenic amines} \\ &+ 0.013\text{Neutral amino acids}\left(R^2 = 97.4\%\right) \end{aligned} \quad (7)$$

$$HS\ (\text{white sparkling wines}) = 13.258 + 2.906\text{Total biogenic amines}\left(R^2 = 19.2\%\right) \quad (8)$$

$$TS\left(\text{rosé sparkling wines}\right) = -22.277 + 0.489\text{Polysaccharides from grapes}\left(R^2 = 46.7\%\right) \quad (9)$$

$$TS\left(\text{white sparkling wines}\right) = -7.348 + 0.359\text{Polysaccharides from grapes}\left(R^2 = 33.9\%\right) \quad (10)$$

5. Conclusions

In conclusion, this work shows that the foam properties of sparkling wines are ruled by a large number of molecules that act in a synergistic way. Nevertheless, some compounds seem to be more relevant than others to explain their foam properties.

Although contradictory results have sometimes been obtained, a high correlation (≥ 0.75) has been found in the literature between the foam properties of sparkling wines and the content of total amino acids, polysaccharides, anthocyanins, coumaric acid and isorhamnetin. On the contrary, the alcohol content and the concentration of acid polysaccharides, proanthocyanidins and free SO_2 are the factors which most negatively affect foam quality.

A recent study, by means of prediction models, has also concluded that the different forms of malvidin shows the highest influence on the foam height and foam stability height parameters, followed by amino acid compounds, while foam stability model was only predicted by polysaccharides from grapes, concretely by polysaccharides rich in arabinose and galactose.

These research findings provide industry with a better understanding of the compositional factors influencing the foam quality of sparkling wines.

Author details

Leticia Martínez-Lapuente, Belén Ayestarán and Zenaida Guadalupe*

*Address all correspondence to: zenaida.guadalupe@unirioja.es

Institute of Vine and Wine Sciences (University of La Rioja, Government of La Rioja and CSIC) Finca La Grajera, Logroño, Spain

References

[1] OIV's Focus. The sparkling wine market. [Internet]. 2017. Available from: http://www.oiv.int/public/medias/3098/les-vins-effervescents-en-complet.pdf [Accessed: 2017-05-15]

[2] Commission Regulation (EC) No 606/2009 of 10 July 2009 laying down certain detailed rules for implementing Council Regulation (EC) No 479/2008 as regards the categories of grapevine products, oenological practices and the applicable restrictions. OJ L 193, 24.7.2009. pp. 1-59

[3] Alexandre H, Guilloux-Benatier M. Yeast autolysis in sparkling wine. A review. Australian Journal of Grape and Wine Research. 2006;**12**:119-127. DOI: 10.1111/j.1755-0238.2006.tb00051.x

[4] Buxaderas S, López-Tamames E. Sparkling wines: Features and trends from tradition. Advances in Food and Nutrition Research. 2012;**66**:1-45. DOI: 10.1016/B978-0-12-394597-6.00001-X

[5] Liger-Belair G. Effervescence in champagne and sparkling wines: From grape harvest to bubble rise. The European Physical Journal Special Topics. 2017;**226**:3-116. DOI: 10.1140/epjst/e2017-02678-7

[6] Dessirier JM, Simons CT, Carstens MI, O'Mahony M, Carstens E. Psychophysical and neurobiological evidence that the oral sensation elicited by carbonated water is of chemogenic origin. Chemical Senses. 2000;**25**:277-284

[7] Dunkel A, Hofmann T. Carbonic anhydrase IV mediates the fizz of carbonated beverages. Angewandte Chemie. 2010;**49**:2975-2977. DOI: 10.1002/anie.200906978

[8] Pozo-Bayon MA, Santos M, Martin-Alvarez PJ, Reineccius G. Influence of carbonation on aroma release from liquid systems using an artificial throat and a proton transfer reaction-mass spectrometric technique (PTR-MS). Flavour and Fragrance Journal. 2009;**24**:226-233. DOI: 10.1002/ffj.1934

[9] Moreno J, Peinado R, editors. Enological Chemistry. 1st ed. San Diego: Elsevier; 2012. p. 320. DOI: 10.1016/B978-0-12-388438-1.00018-2

[10] Martínez-Rodríguez AJ, Pueyo E. Sparkling wines and yeast autolysis. In: Moreno-Arribas MV, Polo C, editors. Wine Chemistry and Biochemistry. New York, NY, USA: Springer; 2009. pp. 61-80. DOI: 10.1007/978-0-387-74118-5

[11] Buxaderas S, López-Tamames E. Managing the quality of sparkling wines. In: Reynolds AG, editor. Managing Wine Quality: Oenology and Wine Quality. 1st ed. Cambridge: Woodhead Publishing Limited; 2010. pp. 164-187. DOI: 10.1533/9781845699987.2.553

[12] López-Barajas M, Viu-Marco A, López-Tamames E, Buxaderas S, de la Torre-Boronat MC. Foaming in grape juices of white varieties. Journal of Agricultural and Food Chemistry. 1997;**45**:2526-2529. DOI: 10.1021/jf9607369

[13] Blasco L, Viñas M, Villa TG. Proteins influencing foam formation in wine and beer: The role of yeast. International Microbiology. 2011;**14**:61-71. DOI: 10.2436/20.1501.01.136

[14] Marchal R, Jeandet P, Robillard B. Macromolecules and champagne wine foaming properties: A review. In: Jeandet P, Clément C, Conreux A, editors. Macromolecules and Secondary Metabolites of Grapevine and Wines. Paris: Intercept Lavoisier; 2007. pp. 349-370

[15] Pozo-Bayón MA, Martínez-Rodríguez A, Pueyo E, Moreno-Arribas MV. Chemical and biochemical features involved in sparkling wine production: From a traditional to an improved winemaking technology. Trends in Food Science and Technology. 2009;**20**:289-299. DOI: 10.1016/j.tifs.2009.03.011

[16] Maujean A, Poinsaut H, Dantan H, Brissonnet F, Cossiez E. Study of the performance and quality of the foam in sparkling wines, II perfecting of measuring technique for foaming ability, performance and stability of the foam in sparkling wines. Bulletin Office Internationale de la Vigne et du Vin. 1990;**711–712**:405-426

[17] Rudin AD. Measurement of the foam stability of beers. Journal of the Institute of Brewing. 1957;**63**:506-509. DOI: 10.1002/j.2050-0416.1957.tb06291.x

[18] Gallart M, Tomas X, Suberbiola G, Lopez-Tamames E, Buxaderas S. Relationship between foam parameters obtained by the gas-sparging method and sensory evaluation of sparkling wines. Journal of the Science of Food and Agriculture. 2004;**84**:127-133. DOI: 10.1002/jsfa.1614

[19] Bikerman JJ. The unit of foaminess. Transactions of the Faraday Society. 1938;**34**:634-638. DOI: 10.1039/TF9383400634

[20] Robillard B, Delpuech E, Viaux L, Malvy J, Vignes-Adler M, Duteurtre B. Improvements of methods for sparkling base wine foam measurements and effect of wine filtration on foam behavior. American Journal of Enology and Viticulture. 1993;**44**:387-392

[21] Gallart M, López-Tamames E, Buxaderas S. Foam measurements in wines: Comparison of parameters obtained by gas sparging method. Journal of Agricultural and Food Chemistry. 1997;**45**:4687-4690. DOI: 10.1021/jf9701244

[22] Moreno-Arribas V, Pueyo E, Nieto FJ, Martín-Álvarez PJ, Polo MC. Influence of the polysaccharides and the nitrogen compounds on foaming properties of sparkling wines. Food Chemistry. 2000;**70**:309-317. DOI: 10.1016/S0308-8146(00)00088-1

[23] Nunez YP, Carrascosa AV, González R, Polo MC, Martínez-Rodríguez AJ. Effect of accelerated autolysis of yeast on the composition and foaming properties of sparkling wines elaborated by a champenoise method. Journal of Agricultural and Food Chemistry. 2005;**53**:7232-7237. DOI: 10.1021/jf050191v

[24] Martínez-Rodriguez AJ, Polo MC. Effect of the addition of bentonite to the tirage solution on the nitrogen composition and sensory quality of sparkling wines. Food Chemistry. 2003;**81**:383-388. DOI: 10.1016/S0308-8146(02)00455-7

[25] Andrés-Lacueva C, López-Tamames E, Lamuela-Raventós RM, Buxaderas S, de la Torre-Boronat MC. Characteristics of sparkling base wines affecting foam behavior. Journal of Agricultural and Food Chemistry. 1996;**44**:989-995. DOI: 10.1021/jf9504879

[26] López-Barajas M, López-Tamames E, Buxaderas S, de la Torre-Boronat MC. Effect of vinification and variety on foam capacity of wine. American Journal of Enology and Viticulture. 1998;**49**:397-402

[27] Cilindre C, Liger-Belair G, Villaume S, Jeandet P, Marchal R. Foaming properties of various champagne wines depending on several parameters: Grape variety, aging, protein and CO_2 content. Analytica Chimica Acta. 2010;**660**:164-170. DOI: 10.1016/j.aca.2009.10.021

[28] Martínez-Lapuente L, Guadalupe Z, Ayestarán B, Pérez-Magariño S. Role of major wine constituents in the foam properties of white and rosé sparkling wines. Food Chemistry. 2015;**174**:330-338. DOI: 10.1016/j.foodchem.2014.10.080

[29] Pueyo E, Martín-Alvarez PJ, Polo MC. Relationship between foam characteristics and chemical composition in wines and Cavas (sparkling wines). American Journal of Enology and Viticulture. 1995;**46**:518-524

[30] Núñez YP, Carrascosa AV, González R, Polo MC, Martínez-Rodríguez A. Isolation and characterization of a thermally extracted yeast cell wall fraction potentially useful for improving the foaming properties of sparkling wines. Journal of Agricultural and Food Chemistry. 2006;**54**:7898-7903. DOI: 10.1021/jf0615496

[31] Coelho E, Reis A, Domingues MRM, Rocha SM, Coimbra MA. Synergistic effect of high and low molecular weight molecules in the foamability and foam stability of sparkling wines. Journal of Agricultural and Food Chemistry. 2011;**59**:3168-3179. DOI: 10.1021/jf104033c

[32] Coelho E, Rocha SM, Coimbra MA. Foamability and foam stability of molecular reconstituted model sparkling wines. Journal of Agricultural and Food Chemistry. 2011; **59**:8770-8778. DOI: 10.1021/jf2010657

[33] Vincenzi S, Crapisi A, Curioni A. Foamability of Prosecco wine: Cooperative effects of high molecular weight glycocompounds and wine PR-proteins. Food Hydrocolloids. 2014;**34**:202-207. DOI: 10.1016/j.foodhyd.2012.09.016

[34] Brissonnet F, Maujean A. Identification of some foam-active compounds in champagne base wines. American Journal of Enology and Viticulture. 1991;**42**:97-102

[35] Brissonnet F, Maujean A. Characterization of foaming proteins in a champagne base wine. American Journal of Enology and Viticulture. 1993;**44**:297-301

[36] Marchal R, Bouquelet S, Maujean A. Purification and partial biochemical characterization of glycoproteins in a champenois chardonnay wine. Journal of Agricultural and Food Chemistry. 1996;**44**:1716-1722. DOI: 10.1021/jf9506592

[37] Vernhet A, Pellerin P, Prieur C, Osmianski J, Moutounet M. Charge properties of some grape and wine polysaccharide and polyphenolic fractions. American Journal of Enology and Viticulture. 1996;**47**:25-30

[38] Gomerieux T. Degassing and maintenance of effervescence in champagne wines. Revue Francaise d'Oenologie. 1989;**116**:15-20

[39] Malvy J, Robillard B, Duteurtre B, Influence d. protéines sur le comportement de la mousse des vins de Champagne. Sciences des Aliments. 1994;**14**:87-98

[40] Andrés-Lacueva C, Lamuela-Raventós RM, Buxaderas S, de la Torre-Boronat MC. Influence of variety and aging on foaming properties of cava (sparkling wine). 2. Journal of Agricultural and Food Chemistry. 1997;**45**:2520-2525. DOI: 10.1021/jf960905p

[41] Puig-Deu M, López-Tamames E, Buxaderas S, Torre-Boronat MC. Quality of base and sparkling wines as influenced by the type of fining agent added pre-fermentation. Food Chemistry. 1999;**66**:35-42. DOI: 10.1016/S0308-8146(98)00093-4

[42] Lao C, Santamaria A, López-Tamames E, Bujan J, Buxaderas S, de la Torre-Boronat MC. Effect of grape pectic enzyme treatment on foaming properties of white musts and wines. Food Chemistry. 1999;**65**:169-173. DOI: 10.1016/S0308-8146(98)00181-2

[43] López-Barajas M, López-Tamames E, Buxaderas S, Suberbiola G, de la Torre-Boronat MC. Influence of wine polysaccharides of different molecular mass on wine foaming. American Journal of Enology and Viticulture. 2001;**52**:146-150

[44] Marchal R, Tabary I, Valade M, Moncomble D, Viaux L, Robillard B, Jeandet P. Effects of *Botrytis cinerea* infection on champagne wine foaming properties. Journal of the Science of Food and Agriculture. 2001;**81**:1371-1378. DOI: 10.1002/jsfa.941

[45] Girbau-Sola T, López-Tamames E, Buján J, Buxaderas S. Foam aptitude of Trepat and Monastrell red varieties in cava elaboration. 1. Base wine characteristics. Journal of Agricultural and Food Chemistry. 2002;**50**:5596-5599. DOI: 10.1021/jf020375z

[46] Dambrouck T, Marchal R, Cilindre C, Parmentier M, Jeandet P. Determination of the grape invertase content (using PTA-ELISA) following various fining treatments versus changes in the total protein content of wine. Relationships with wine foamability. Journal of Agricultural and Food Chemistry. 2005;**53**:8782-8789. DOI: 10.1021/jf051276z

[47] Vanrell G, Canals R, Esteruelas M, Fort F, Canals JM, Zamora F. Influence of the use of bentonite as a riddling agent on foam quality and protein fraction of sparkling wines (cava). Food Chemistry. 2007;**104**:148-155. DOI: 10.1016/j.foodchem.2006.11.014

[48] García MJ, Aleixandre JL, Álvarez I, Lizama V. Foam aptitude of Bobal variety in white sparkling wine elaboration and study of volatile compounds. European Food Research and Technology. 2009;**229**:133-139. DOI: 10.1007/s00217-009-1034-z

[49] Esteruelas M, González-Royo E, Kontoudakis N, Orte A, Cantos A, Canals JM, Zamora F. Influence of grape maturity on the foaming properties of base wines and sparkling wines (cava). Journal of the Science of Food and Agriculture. 2015;**95**:2071-2080. DOI: 10.1002/jsfa.6922

[50] Tkachenko O, Drevova S, Gural L. Research into technologies of obtaining grape must in the production of sparkling base wines. Eastern-European Journal of Enterprise Technologies. 2017;**10**:52-62. DOI: 10.15587/1729-4061.2017.90836

[51] Puff N, Marchal R, Aguié-Béghin V, Douillard R. Is grape invertase a major component of the adsorption layer formed at the air/champagne interface? Langmuir. 2001;**17**:2206-2212. DOI: 10.1021/la0015056

[52] Bartolomé B, Moreno-Arribas V, Pueyo E, Polo MC. On-line HPLC photodiode Array detection and OPA derivatization for partial identification of small peptides from white wine. Journal of Agricultural and Food Chemistry. 1997;**45**:3374-3381. DOI: 10.1021/jf9700844

[53] Moreno-Arribas V, Pueyo E, Polo MC. Peptides in musts and wines. Changes during the manufacture of cavas (sparkling wines). Journal of Agricultural and Food Chemistry. 1996;**44**:3783-3788. DOI: 10.1021/jf960307a

[54] St John Coghlan D, Woodrow CW, Bamforth CW, Hinchliffe E. Polypeptides with enhanced foam potential. Journal of the Institute of Brewing. 1992;**98**:207-213. DOI: 10.1002/j.2050-0416.1992.tb01106.x

[55] Onishi A, Proudlove MO. Isolation of beer foam polypeptides by hydrophobic interaction chromatography and their partial characterization. Journal of the Science of Food and Agriculture. 1994;**65**:233-240. DOI: 10.1002/jsfa.2740650218

[56] Abdallah Z, Aguié-Béghin V, Abou-Saleh K, Douillard R, Bliard C. Isolation and analysis of macromolecular fractions responsible for the surface properties in native champagne wines. Food Research International. 2010;**43**:982-987. DOI: 10.1016/j.foodres.2010.01.003

[57] Senée J, Robillard B, Vignes-Adler M. Films and foams of champagne wines. Food Hydrocolloids. 1999;**13**:15-26. DOI: 10.1016/S0268-005X(98)00061-7

[58] Senée J, Robillard B, Vignes-Adler M. The -potential of the endogenous particles of a wine of champagne in relation to the foaming behavior. Colloids and Surfaces B: Biointerfaces. 2001;**21**:59-67. DOI: 10.1016/S0927-7765(01)00184-9

[59] Correa-Gorospe I, Polo MC, Hernández T. Characterization of the proteic and phenolic fraction in tartaric sediments from wines. Food Chemistry. 1991;**41**:135-136. DOI: 10.1016/0308-8146(91)90041-L

[60] Canals JM, Arola L, Zamora F. Análisis de la fracción proteica de vinos base para cava mediante FPLC. XIV Congreso Internacional del Cava; 1996

[61] Sarker DK, Wilde PJ, Clark DC. Control of surfactant-induced destabilization of foams through polyphenol mediated protein-protein interactions. Journal of Agricultural and Food Chemistry. 1995;**43**:295-300. DOI: 10.1021/jf00050a006

[62] Sausse P, Aguié-Béghin V, Douillard R. Effects of epigallocatechin gallate on b-casein adsorption at the air/water interface. Langmuir. 2003;**19**:737-743. DOI: 10.1021/la026304b

[63] Bosch MP, Comelles F, Castro J. Influència de les propietats superficials dels tensioactius d'orígen natural en l'elaboraciód'un vi de Cava. ACE Revista de Enología. 1989;**16**:10-14

[64] Dussaud A, Robillard B, Carles B, Duteurtre B, Vignes-Adler M. Exogenous lipids and ethanol influences on the foam behavior of sparkling base wines. Journal of Food Science. 1994;**59**:148-151. DOI: 10.1111/j.1365-2621.1994.tb06920.x

[65] Gallart M, López-Tamames E, Suberbiola G, Buxaderas S. Influence of fatty acids on wine foaming. Journal of Agricultural and Food Chemistry. 2002;**50**:7042-7045. DOI: 10.1021/jf0204452

[66] Girbau-Sola T, López-Barajas M, López-Tamames E, Buxaderas S. Foam aptitude of Trepat and Monastrell red varieties in cava elaboration. 2. Second fermentation and aging. Journal of Agricultural and Food Chemistry. 2002;**50**:5600-5604. DOI: 10.1021/jf011209e

[67] Molan PC, Edwards M, Eschenbruch R. Foaming in winemaking. II. Separation and partial characterisation of foam inducing proteins excreted by a pure culture wine yeast. European Journal of Applied Microbiology and Biotechnology. 1982;**16**:110-113. DOI: 10.1007/BF00500736

[68] Puff N, Cagna A, Aguié-Béghin V, Douillard R. Effect of ethanol on the structure and properties of R-casein adsorption layers at the air/buffer interface. Journal of Colloid and Interface Science. 1998;**208**:405-414. DOI: 10.1006/jcis.1998.5846

[69] Marchal R, Warchol M, Cilindre C, Jeandet P. Evidence for protein degradation by Botrytis Cinerea and relationships with alteration of synthetic wine foaming properties. Journal of Agricultural and Food Chemistry. 2006;**54**:5157-5165. DOI: 10.1021/jf060509z

[70] Cilindre C, Castro AJ, Clément C, Jeandet P, Marchal R. Influence of Botrytis Cinerea infection on champagne wine proteins (characterized by two-dimensional electrophoresis/immunodetection) and wine foaming properties. Food Chemistry. 2007;**103**:139-149. DOI: 10.1016/j.foodchem.2006.07.043

[71] Marchal R, Chaboche D, Douillard R, Jeandet P. Influence of lysozyme treatments on champagne base wine foaming properties. Journal of Agricultural and Food Chemistry. 2002;**50**:1420-1428. DOI: 10.1021/jf010780

[72] Viaux L, Morard C, Robillard B, Duteurtre B. The impact of base wine filtration on champagne foam behavior. American Journal of Enology and Viticulture. 1994;**45**: 407-409

[73] Vanrell G, Cabanillas P, Albet S, Canals JM, Arola L, Zamora F. Étude des composés influencant la mousse des Cavas. Revue Francaise d'Oenologie. 2002;**196**:30-36

[74] Vanrell G, Esteruelas M, Canals JM, Zamora F. Influence du type de calrification du vin de base et des adjuvants de tirage sur la qualité de la mousse des vins effervescents. Revue des Oenologues. 2005;**114**:28-30

[75] López-Barajas M, López-Tamames E, Buxaderas S, Tomás X, de la Torre MC. Prediction of wine foaming. Journal of Agricultural and Food Chemistry. 1999;**47**:3743-3748. DOI: 10.1021/jf981201d

Chapter 11

Occurrence and Analysis of Sulfur Compounds in Wine

Daniela Fracassetti and Ileana Vigentini

Additional information is available at the end of the chapter

http://dx.doi.org/10.5772/intechopen.72823

Abstract

Sulfur compounds play an important role in the sensory characteristics of wine. These molecules can derive from the grape, in which the non-volatile forms are usually present as glycosylated molecules, the metabolic activities of yeast and bacteria, the chemical reactions taking place during the wine aging and storage, and the environment. The sulfur compounds include molecules positively correlated to the aromatic profile of wine, namely the volatile thiols, and are responsible for certain defects, imparting notes described as cabbage, onion, rotten egg, garlic, sulfur and rubber. Due to the low concentration of these molecules in wine, their high reactivity and the matrix complexity, the analytical methods which enable their detection and quantification represent a challenge. The solid phase microextraction (SPME) technique has been developed for sulfur compounds associated with off-flavors. The analysis of volatile thiols usually requires a derivatization followed by gas chromatography (GC)-MS or UPLC-MS methods. Besides the sulfur-containing aromas, another sulfur compound that deserves mention is the reduced glutathione (GSH) which has been widely studied due to its antioxidant properties. The analysis of GSH has been proposed using a liquid chromatography technique (HPLC or UPLC) coupled with fluorescence, MS and UV detectors.

Keywords: sulfur off-flavors, volatile thiols, reduced glutathione, sample preparation, analysis, wine

1. Introduction

Sulfur-containing compounds strongly affect the sensory properties of wine and must. They include aromatic molecules, off-flavors and a well-known non-volatile compound with antioxidant properties such as glutathione. The sulfur-containing compounds can be derived from the grape in which the non-volatile forms are usually present, the metabolic activities of yeast and bacteria, the chemical reactions taking place during the wine aging and storage, and the environment [1]. Their presence in wine can be the result of both enzymatic and non-enzymatic mechanisms. In the first case, the sulfur compounds represent the products of metabolic and fermentative pathways whose substrates are both amino acids and some sulfur-containing pesticides. When yeast and bacteria metabolize these thiols, the released sulfur compounds are generally considered off-flavors [2]. Non-enzymatic processes include photochemical, thermal and other chemical reactions of sulfur compounds during winemaking and storage [3]. The reactions involving the sulfur-containing amino acids, in particular, can bring about the light-struck taste in case the bottled white wine is exposed to light greatly affecting the sensory properties [4, 5]. The sulfur-containing off-flavors impart negative notes such as cabbage, onion, rotten egg, garlic, sulfur and rubber [6]. Among them, hydrogen sulfide is probably the best-known sulfur compound in wine. Hydrogen sulfide is a very reactive species and can trigger reactions generating compounds such as mercaptans, dimethyl sulfide and polysulfide, which also have a negative impact on the wine aroma. The long-chain polyfunctional sulfur compounds, also known as volatile thiols, are one of the most important groups of aroma compounds in wine, which confer pleasant aromatic notes at trace levels; at high concentrations, these compounds can be objectionable yet [3, 7–9]. Volatile thiols such as 4-mercapto-4-methylpentan-2-one (4MMP), 3-mercaptohexan-1-ol (3MH) and 3-mercaptohexyl acetate (3MHA) are particularly important, for example, for characterizing the typical aroma of Sauvignon Blanc wine [10]. Moreover, in some wines, 4-mercapto-4-methylpentan-2-ol (4MMPOH) is also detected but its concentration is lower than the thiol-related aromas mentioned above [11].

Besides the sulfur compounds contributing to the sensory characteristics of wine, the glutathione is of particular interest due to its antioxidant properties. This tripeptide can limit the browning of white must as it reduces back the *o*-quinones to the correspondent phenols and, consequently, the formation of the brown polymers are avoided [12].

2. Aromatic compounds containing sulfur

2.1. Sulfur off-flavors

Sulfur-containing compounds are mainly associated with off-odors due to the presence of hydrogen sulfide and mercaptans whose concentrations are typically low as well as their perception threshold [2]. The most abundant sulfur off-flavors found in wine are represented by hydrogen sulfide, methanethiol, ethanethiol, dimethylmercaptans (dimethylsulfide, dimethyldisulfide, dimethyltrisulfide) [13] and other sulfur-containing compounds responsible for off-flavors in wine (**Table 1**). Hydrogen sulfide is the most recognized sulfur compound associated with the rotten egg aroma, it is highly volatile and has a low perception threshold (10–80 μg/L). Even if hydrogen sulfide could be easily removed by copper treatment [14], it is very reactive specie participating in the generation of other mercaptans, such as dimethylsulfide and polysulfide, and ethanthiol, the latter due to its combination with ethanol or acetaldehyde [7]. *Saccharomyces cerevisiae* can release this compound through the degradation of sulfur-containing amino acids and the reduction of elemental sulfur, sulfite or sulfate [15]. However, its synthesis is limited when nitrogen derived from either amino acids (except for cysteine) or ammonium is added [13]. Variable amounts of hydrogen sulfite can be released during the alcoholic fermentation. Smith and co-authors [16] suggest that the sulfur-containing

Compound	Olfactory description	Perception threshold (μg/L)	Range in wine (μg/L)
Hydrogen sulfide	Rotten eggs, reduced taste	0.001–150	0–370
Methanethiol	Cooked cabbage, reduced taste	0.3	0–16
Ethanthiol	Onion, rubber, putrefaction	1.1	0–50
Dimethyl sulfide	Cabbage, asparagus, corn, molasses	10–160	0–910
Carbon disulfide	Cabbage, rubber	>38	0–18
Dimethyl trisulfide	Cabbage, onions, cooked vegetable	0.1	0–111
Diethyl sulfide	Garlic	0.93–18	0–10
Dimethyl disulfide	Cooked cabbage, asparagus, onions	20–45	0–160
Diethyl disulfide	Garlic, onion, burnt rubber	4.3–40	0–160
2-Mercaptoethanol	Barnyard-like, poultry, farmyard	130	0–400
Methylthioacetate	Sulfurous, rotten vegetables	300	0–115
S-Ethylthioacetate	Sulfurous	40	0–180
2-(Methylthio)-1-ethanol	Cauliflower, French bean	250	0–139
2-(Methylthio)-1-propanol	Cauliflower, cooked cabbage	1200	0–5655
2-(Methylthio)-1-butanol	Onion, garlic, earthy	100	0–180
Benzothiazole	Rubber	50–350	0–30
5-(2-Hydroxyethyl)-4-methylthiazole	Green	100–1000	5–50

Table 1. Common fermentative sulfur compounds and off-flavors found in wines [3, 22, 120].

precursors could produce hydrogen sulfite, methanthiol and dimethylsulfide during fermentation and not all these compounds are released in gaseous form. The chemistry behind the reactions and regulation mechanisms involving hydrogen sulfite and other mercaptans produced by the yeast is not well understood [17]. Methionine can be metabolized by yeast to produce fusel alcohol, methionol and 3-methylthio-1-propanol imparting cabbage and cauliflower aromas. Cysteine is a precursor of S-containing heterocycle compounds; these compounds can be metabolized by *Oenococcus oeni* and aroma descriptors such as sulfury, floral, fruity, toasted and roasted can be imparted to the wine [2, 18]. The microbial degradation of these amino acids can lead to the presence of dimethyl sulfide which has shown the property of enhancing the wine aroma due to the interactions with other volatiles, such as esters and norisoprenoids [19, 20]. Nevertheless, dimethyl sulfide imparts notes such as asparagus, corn and molasses which are undesired characters in white wine to a certain extent, although the aroma complexity is increased [21]. The levels of dimethyl sulfide, methionol, diethyl sulfide and diethyl disulfide increase as the aging time and storage temperature increase and they can influence the aroma complexity of aged wine [13, 22]. Mercaptans are responsible for reduced aroma in wine. Different strategies can be used for the removal of sulfur off-flavors. The addition of diammonium phosphate or other yeast-based nutrients leads to an increase of the readily assimilable nitrogen enabling the yeast to convert sulfur-containing amino acids into hydrogen sulfide [23]. The aeration of wine during the alcoholic fermentation as well as wine racking in aerated conditions showed a protective effect against the formation of sulfur off-flavors. The oxygen positively contributes to the fermentative capacity of the yeast [24, 25]. Thereafter, the oxygen presence maintains lower levels of sulfur off-flavors during storage [26] probably due to either the formation molecular mass sulfur compounds or the quinone ability for trapping the sulfur compounds [16]. The aging of the lees can have a positive effect since the mercaptans can link to the cysteinyl residues of the yeast cell wall to give disulfides [27]. Wine treatment with copper or silver showed to be affective for the removal of hydrogen sulfide as copper or silver sulfide precipitates were formed [28]. However, recent studies have indicated that copper addition post-bottling increases the level of sulfur off-flavors during aging [24, 29, 30]. The effect of copper still needs to be clarified as well as its dosages.

The sulfur-containing compounds are associated with the occurrence of light-struck taste, a defect affecting white wine bottled in clear bottles and exposed to light. The photochemical oxidations may affect different wine components including phenols, acids and alcohols [31, 32]. The formation of the light-struck taste is related to the presence of riboflavin, a vitamin highly sensitive to light and methionine. The pathway proposed by Maujean and Seguin [4] involves the photoxidative degradation of methionine to give methional and reduced riboflavin. Methional is unstable when exposed to light and decomposes to acrolein and methanethiol (**Figure 1**). Two molecules of the latter eventually yield dimethyl disulfide (DMDS). The light-struck taste is described as cooked cabbage mainly due to the formation of methanethiol and DMDS though hydrogen sulfide may have a role [33–35]. Light exposure can also lead to the formation of other undesirable compounds, such as furfural [36] which has been positively correlated to the "cooked vegetable" aroma of white wines stored under oxidative conditions [37, 38]. Not all wines are susceptible to developing this defect. The concentration of riboflavin plays the major role [5, 39] and when it is lower than 80–100 μg/L, the risk of the light-struck taste decreases. The riboflavin content in the grape hardly exceeds a few tens of micrograms per liter and the metabolic activities of *S. cerevisiae* affect its levels in wine [40, 41]. The levels of riboflavin can be reduced by means of wine treatment with bentonite or charcoal as well as by using a low producer yeast strain of this vitamin. Finally, nutrients commonly employed for favoring yeast growth could further increase the amount of riboflavin in wine [41]. Moreover, the amounts of methionine could influence the occurrence of light-struck taste as well as the presence of oxygen [42]. The light-struck taste can be limited by the addition of antioxidant compounds. The use of gallic tannins seems to be particularly promising in protecting wine possibly due to the binding of sulfur off-flavors and the quinones present in the polyphenol-based mixture [42].

2.2. Varietal thiols

The varietal thiols including 3-mercaptohexan-1-ol (3MH), 3-mercaptohexylacetate (3MHA) and 4-mercapto-4-methylpentan-2-one (4MMP) are sulfur-containing aromas associated with the typical flavor of Sauvignon Blanc wines [11, 43, 44]. Moreover, in some wine, 4-mercapto-4-methylpentan-2-ol (4MMPOH) was also detected, but its level was lower than the volatile thiols mentioned above [44] (**Table 2**). They are characterized by some fruity aromas, like cassis [45], grapefruit [46], passion fruit [47] and guava [48]. The impact of sulfur compounds on wine aroma has been updated in literature and the varietal character is affected by several of these

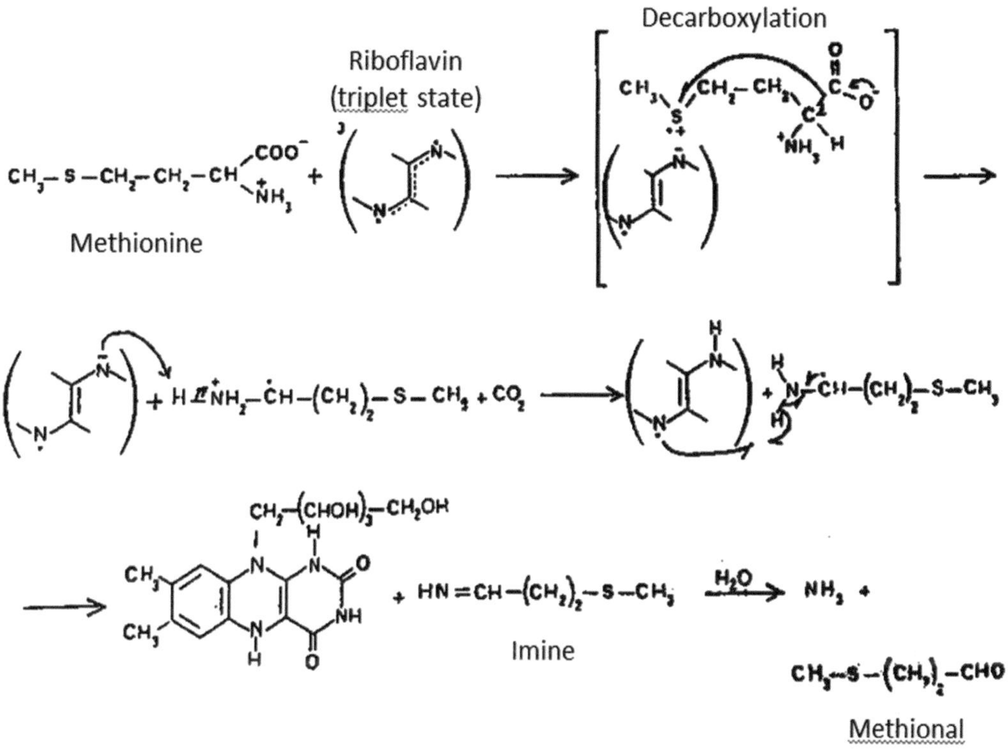

Figure 1. Reaction scheme of methional formation due to light exposure [4].

molecules if their concentration is close to the perception threshold [22, 49, 50]. 3MH and 3MHA have olfactory perception thresholds of 60 and 4 ng/L, respectively and they are responsible for passion fruit-like and grapefruit-like olfactory notes [11]. These compounds were also found in red wines [51]. The 4MMP has an olfactory perception threshold of 0.8 ng/L, and its aroma is described as box tree-like, black currant-like, or even cat urine-like when occurring at high concentration [10]. The 4MMPOH has a perception threshold of 55 ng/L in aqueous alcoholic solution and it is reminiscent of citrus zest and grapefruit [44]. 3MH is the most abundant in wine, in concentrations generally higher than its perception threshold, while 4MMPOH in wine is generally lower than its perception threshold. The level of 4MMP is dependent on the grape cultivar and it changes between different samples in the same cultivar [52]. The concentrations of 3MH and 3MHA in Sauvignon Blanc wines from different countries have been reported to be in the range 688–18,681 and 10–2507 ng/L, respectively [53–55]. Besides Sauvignon Blanc wines, the volatile thiols were detected in white wine made from different *Vitis vinifera* grape varieties, such as Gewürztraminer, Muscat, Riesling, Sylvaner, Pinot Gris, Pinot Blanc, Colombard, Petit Manseng, botrytised Semillon and Grenache [44, 56]. Moreover, these compounds were also found in certain Italian autochthonous varieties including Verdicchio Bianco (also known as Trebbiano di Lugana) [57], Arneis [58], Grillo and Catarratto Bianco Comune [59].

The content of volatile thiols decreases during wine aging, according to the oxidative conditions. Glutathione, sulfur dioxide and anthocyanin content exert a protective effect. In contrast,

Compound	Structure	Olfactory description	Perception threshold (ng/L)[a]	Range in wine (ng/L)[b]
3-mercaptohexan-1-ol (3MH)	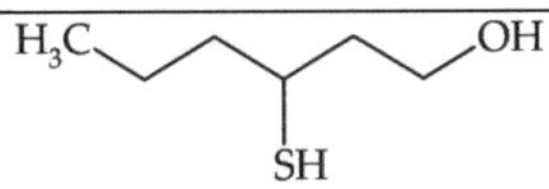	Passion fruit, grape fruit, guava[c]	60	26–18,000
3-mercaptohexylacetate (3MHA)	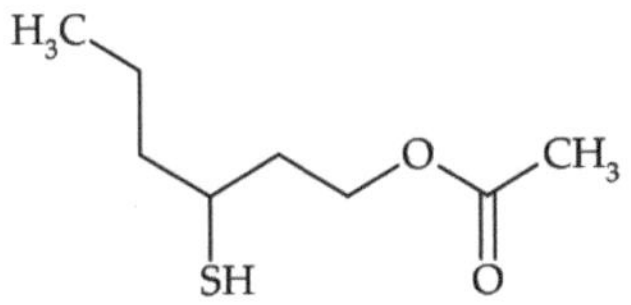	Passion fruit, grape fruit, box tree, guava[d]	4	0–2500
4-mercapto-4-methylpentan-2-one (4MMP)	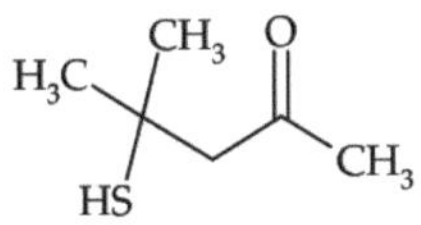	Box tree, black currant, passion fruit[e]	0.8	0–40
4-mercapto-4-methylpentan-2-ol (4MMPOH)	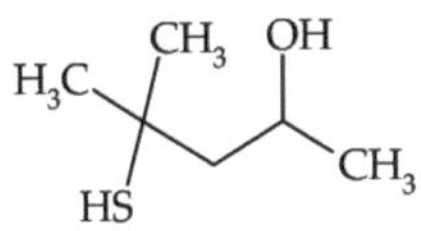	Citrus zest[f]		0–90

[a]In model wine solution [8, 10, 11, 147].

[b]Range *Vitis vinifera* wines [11, 28, 53–55, 92, 148].

[c][43].

[d][8, 11].

[e][8, 147].

Table 2. Contribution of volatile thiols in *Vitis vinifera* wines.

increased contact with oxygen, particularly in the presence of catechin derivatives, promotes their degradation [52, 60, 61]. Three different mechanisms proposed lead to their decay with a different degradation kinetics. The volatile thiols can be easily oxidized when oxygen and iron are present, forming disulfides [62, 63]. Since the volatile thiols act as electrophiles, they can react with phenolic compounds [64]. Moreover, the presence of oxygen and catalyst metals, namely iron and copper, lead the oxidation of phenols into the corresponding *o*-quinones which can bind the volatile thiols [65, 66]. A rapid decline of 3MHA is observed after 3 months of bottle storage, while a much slower decline in 3MH can be noticed. After 1 year of storage, the ester has completely disappeared, while 3MH is still present but its content is halved. The chemical structure of the volatile thiols affects their degradation: 4MMP, a tertiary thiol, results less able to react with *o*-quinones and decreases slower in comparison to 3MH, a secondary thiol, maybe due to its steric hindrance [63]. The wine composition and oxygen exposure are the major reasons for loss of volatile thiols both during winemaking and post-bottling [29, 67]. The presence of antioxidants, such as sulfur dioxide, ascorbic acid and reduced glutathione, limit the polyphenol oxidation either by removing oxygen from wine or by reversing and altering the oxidation process. In particular, the addition of glutathione, up to 20 mg/L before bottling, led to higher 3MH level after 6 months of storage in bottle [29].

2.2.1. *Varietal thiol precursors*

The varietal thiols occur in grape berry as non-volatile cysteinyl- and glutathionyl-conjugated precursors [29, 68]. 3MH bound as cysteinyl-conjugate (Cys-3MH), glutathionyl-conjugate (GSH-3MH) and also cysteinylglycin-conjugate (CysGly-3MH) have been reported [70, 71]. 4MMP occurs in the grape and must as cysteine conjugate (Cys-4MMP) and glutathione conjugate (GSH-4MMP) [72, 73]. Moreover, 3MH can be derived from either (E)-2-hexen-1-ol or (E)-2-hexenal [74, 75]. Through a lipoxygenase/lyase sequence active in the presence of oxygen, (E)-2-hexenal can be obtained from linolenic acid and converted to GSH-3MH by coupling with glutathione [75–77]. (E)-2-hexenal may act as a precursor when hydrogen sulfide is released in the early part of the fermentation [73, 74, 78]. Under aerobic conditions, *S. cerevisiae* is able to oxidize (E)-2-hexen-1-ol into (E)-2-hexenal, and the reverse process can occur under anaerobic conditions [77]. Recently, S-3-(hexan-1-al)-glutathione (GSH-3MHAl) was identified in Sauvignon Blanc juice [79] and it can be considered a precursor of thiol aromas. These compounds are split by the *S. cerevisiae* through the enzymatic activity of carbon-sulfur lyase [10, 80], the volatile thiols are liberated. 3MHA is obtained by yeast activity from 3MH esterified with acetic acid as precursor [7]. This finding correlated the ester and volatile thiol metabolism in yeast for the first time [2]. The capability of *S. cerevisiae* to liberate volatile thiols from their precursors is genetically determined; the yeast selection can represent a useful tool to favor the 3MH and 4MMP release [8, 50]. The fermentation temperature affects the varietal thiol concentrations in the resulting wines: in the range of alcoholic fermentation for white wines, at higher fermentation temperature and higher concentrations of varietal thiols is released. Among the *S. cerevisiae* strains commonly used, VL3 and EG8 release more volatile thiols, in comparison to VL1 and 522 [80]. *S. bayanus* var. *uvarum* strains and hybrids of the latter yeast and *S. cerevisiae* showed a higher ability to release both 3MH and 4MMP when compared to *S. cerevisiae* strain [81]. Furthermore, non-*Saccharomyces* yeasts, like *Torulaspora delbrueckii* and *Pichia kluyveri*, have been proposed as innovative tools to improve wine quality, being able to release varietal thiols [82, 83]. The 3MHA formation from 3MH is also dependent on the strain genetic characteristics [84]. *S. cerevisiae* strains showed the enhanced ability to hydrolyze the S-cysteinyl link rather than the ester synthetic activity. The combined use of different yeast strains, one of them having hydrolyzing ability, with stronger esterification ability, can represent a useful tool to affect the volatile thiol composition [50]. However, several researches have indicated that no clear correlation between precursor concentrations in must and free thiol concentration in wine exist [8, 16, 59, 73, 80, 85–89].

The harvest has a significant effect on the content of thiol precursors in juice. Capone and Jeffery [90] showed that machine harvest and transportation for 12 h of the grapes led to higher levels of thiol precursors in comparison to hand harvest grapes immediately processed. The authors suggested the minimal berry damage results in less formation of the precursors. More than half of the total cysteine conjugates is located in the grape skin; as a consequence, an increased skin contact time augments Cys-3MH in grape musts, while it has little effect on Cys-4MMP and Cys-4MMPOH concentrations [91–93]. Stronger pressing conditions allow a major extraction of thiol precursors in comparison to free run juice. Roland and co-authors [94] demonstrated an easier extraction of Cys-3MH in comparison to GSH-3MH due to their distribution in different parts of the skin. The produced wines contained 3MH at higher concentrations when the juice was collected at the end of the pressing cycle. On the contrary, Patel et al. [95] found lower amounts of 3MH and 3MHA in wine produced from juice obtained from stronger pressing conditions in which the thiol precursors were present in higher amounts. This can be due to the higher extraction of oxidizable phenols inducing the rapid decrease of glutathione and the oxidation of varietal thiols. The pressing performed on an industrial scale caused a strong decrease of thiol precursors. The factors affecting the GSH-portion degradation of the precursors are not completely clear and mechanisms, other than oxidation or proteolysis, could induce the loss of the thiol precursors [59].

3. The reduced glutathione

The reduced glutathione (GSH) is a tripeptide constituted by L-cysteine, γ-glutamic acid and glycine exerting antioxidant and detoxifying activities in the cell [96, 97]. This compound exerts several activities in must and wine. The antioxidant property of GSH is well known: it can reduce the *o*-quinone deriving from the enzymatic oxidation carried out by the polyphenoloxidase enzymes (PPO) on the tartaric esters of hydroxycinnamic acids. This hinders the formation polymers causing the browning of must [12]. During aging, the *o*-quinones are produced as a result of the non-enzymatic oxidation (also known as chemical oxidation) of *o*-diphenols [98]. The level of *o*-diphenols in wine is correlated to the browning of white wines [28, 98, 99]. Caffeoyl-tartaric acid (caftaric acid) and coumaric-tartaric acid (coutaric acid) are some of the most abundant hydroxycinnamic acids in must representing the substrate mainly oxidized by the enzymatic action. The GSH can reduce the oxidized caftaric acid, generating 2-*S*-glutathionyl caftaric acid, also known as Grape Reaction Product (GRP) [100]. The GRP is not a substrate of the PPO and it can trap the *o*-quinone, limiting the formation of brown polymers which are responsible for color changes of white must and wine. GRP can oxide enzymatically by the *Botrytis cinerea* laccase and chemically by the caffeoyl-tartaric acid quinone. In this way, the GRP quinone is originated; it can be a substrate of condensation reaction with the phenols. This molecule is responsible for brown compound formation. The high GSH concentration allows a second nucleophile attack, in position 5 of the benzyl ring. The 2,5-diglutathionyl caftaric acid (GRP2) is formed and is a substrate of the laccase action [12]. With low GSH concentration, the GRP can be oxidized by the excess caftaric acid quinones which can cause an intense browning (**Figure 2**) [101, 102].

GSH is able to limit the loss of the flavoring volatile thiols acting as a competitor for the reduction of the quinones [103]. In fact, GSH concentration is about a thousand times higher than that of volatile thiols which are protected against oxidation. Lavigne and Dubourdieu [103] reported that when the GSH concentration ranged from 6 to 10 mg/L it slowed down the decrease of volatile thiols. Additionally, other aromatic compounds, such as isoamyl acetate (3-methyl-1-butyl acetate), ethyl hexanoate and linalool (3,7-dimethylocta-1,6-dien-3-ol), are better protected during bottle storage [104]. GSH can limit the formation of sotolon (3-hydroxy-4,5-dimethyl-2(5H)-furanone), a compound responsible for the atypical aging character of white wine [103]. It confers aroma descriptors such as dried fig and rancid and its perception threshold is 7 μg/L [105]. Besides sotolon, 2-aminoacetophenone (1-(2-aminophenyl)-ethanone) is also responsible for the atypical aging and it has a lower perception threshold than sotolon, corresponding to 1 μg/L. Both sotolon and 2-aminoacetophenone concentrations increase due to the exposition of wine to oxygen during bottling [106]. GSH can have a protective action on the wine aroma during the oxidative aging. GSH reduces the formation of both these off-flavors during storage. Moreover, GSH can have a positive effect on white wine color which appears to be more stable during aging [103, 107]. The first source of GSH is the grape in which it can exceed 200 mg/L of grape juice according to grape cultivar, environmental conditions and viticultural practices [108]

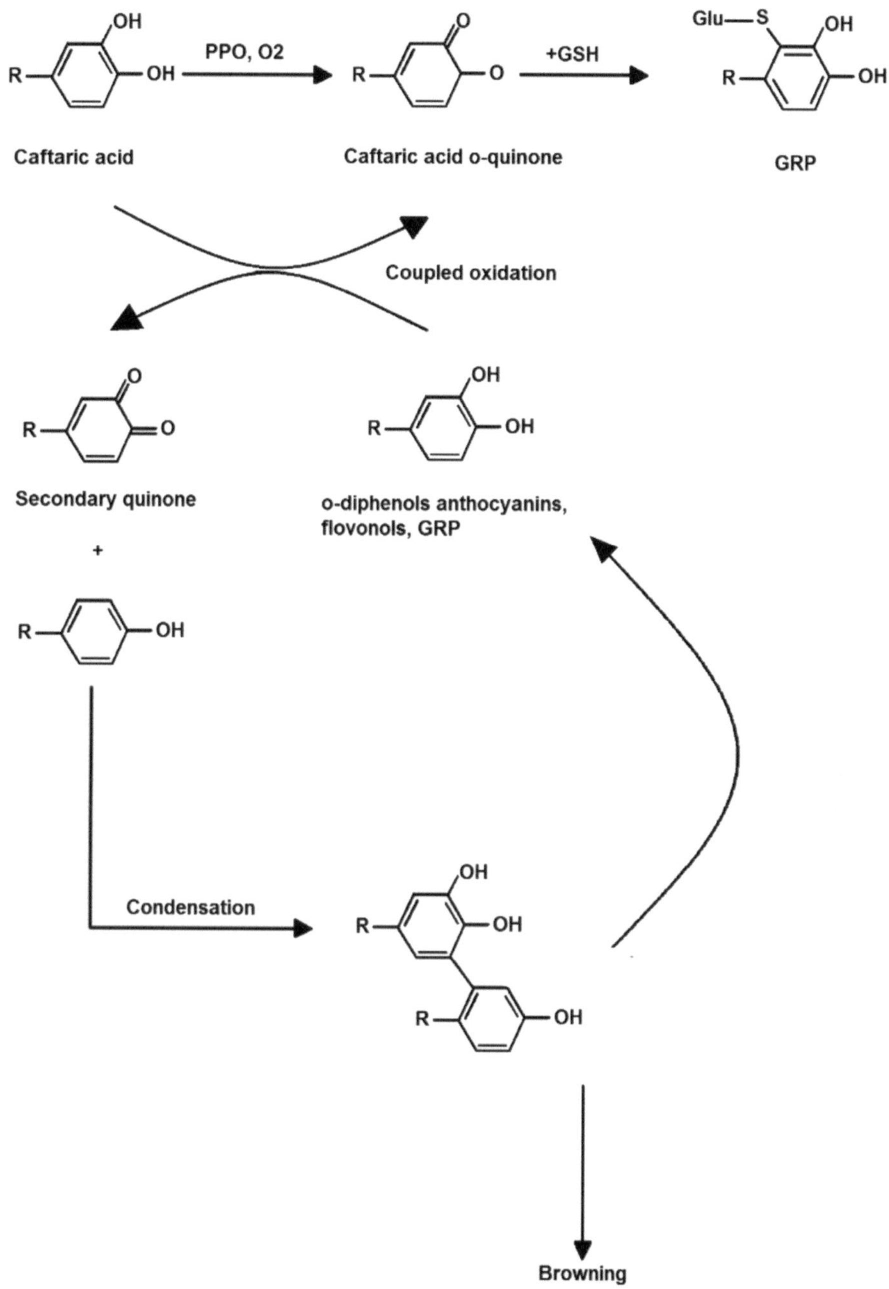

Figure 2. Scheme of GRP and browning compound formation [100–102].

and the amounts of readily assimilable nitrogen in the soil [103]. Lower levels were found in must up to 100 mg/L [108] and it can be affected by exposure to oxygen, tyrosinase activity and pre-fermentative grape skin maceration [93, 109]. The loss of GSH in must production can negatively affect the formation of precursors of the varietal thiol compounds [73] as well as the residual content of GSH during wine aging. Glutathione was reported to be consumed by *S. cerevisiae* at the beginning of the alcoholic fermentation and then to be released by the yeast cell lysis [103, 110]. At the beginning of the alcoholic fermentation GSH almost disappears and then its concentration increases as an effect of the yeast cell synthesis and lysis. However, winemaking on an industrial scale did not preserve GSH in must and the level of this antioxidant increased during the alcoholic fermentation due to the yeast metabolism [111]. Lavigne and Dubourdieu [103] reported the GSH level becomes stable 1 month after the alcoholic fermentation is started. On the contrary, in recent researches, the highest levels of GSH were observed after the racking [111, 112]. The concentration of GSH in wine is lower than in the juice and grapes and it ranges from 3 to 35.5 mg/L [109, 111–113] and can be increased through the choice of an adequate yeast strain [15]. No GSH is released from yeast under nitrogen starvation during the alcoholic fermentation [103, 111]. The GSH concentration decreased after the racking as well as during wine aging on lees [57, 95, 111] maybe due to the adsorption of GSH on the yeast lees as occurs for other low-molecular weight thiols during wine aging [114].

4. Analytical methods

The analytical methods developed for the analysis of sulfur compounds in wine need to overcome the low concentrations and the high reactivity of these molecules as well as the complexity of the matrix. Sulfur compounds in wine are frequently divided into "light" (boiling point < 90°C) and "heavy" (boiling point > 90°C) compounds [3] indicating difficulty in using a relevant common sampling/enrichment technique.

4.1. Sulfur off-flavors

The most common technique employed for the analysis of sulfur compounds associated with off-flavors in wine is static headspace analysis by means of solid phase micro extraction (SPME) combined with gas chromatography (GC) coupled with different detectors (**Table 3**). Methods in dynamic headspace by using Purge and Trap equipment was also described as an alternative to the static headspace. In static headspace technique, the analytes reach the equilibrium state between the liquid and gas phases and then are adsorbed in a fiber. In dynamic space, the analytes in the gas phase into the headspace and the atmosphere around the sample is constantly swept away by a flow of carrier gas, taking volatile analytes with it. Through this technique, the equilibrium state is not reached and, thus, more of the volatile dispersed in the matrix will pass into the headspace whose size results in increased sampling phase. As a consequence, the trapping stage of the analysis offers good sensitivity [43, 115]. The proper combination of sorbent and temperature may permit collection and concentration of specific analytes while venting others. Despite this, the instrumentation requires more complexity and it is more expensive than other sampling techniques, such as the static headspace. Moreover,

Analytical technique		Advantages	Disadvantages
Dynamic headspace	Purge and trap—gas chromatography	Increased sampling phase; good sensitivity	Many sources of errors
Static headspace	Solid phase micro extraction (SPME)—gas chromatography	Cheaper and simpler instrumentation	Compromise between sensitivity and extraction/temperature conditions needs to be properly set
SPME fiber			
	Carboxen-polydimethylsiloxane (CAR-PDMS)[a]	Suggested for sulfur compounds with low boiling point	Less versatile than CAR-PDMS-DVB fiber
	Carboxen-polydimethylsiloxane-divinylbenzene (CAR-PDMS-DVB)[b]	Best extraction yield for the analytical conditions applied	Suitable analytical method setting for increasing the extraction yield (sample temperature, ionic strength, extraction time)

[a][117, 118].
[b][22, 119, 120].

Table 3. Analytical methods for the determination of sulfur off-flavors.

many sources of error in Purge and Trap instruments have been reviewed [116]. As for the dynamic headspace, the choices of the proper fiber as well as sample temperature and the presence of salt increasing the ionic strength can improve extraction yield in static headspace. In particular, the use of carboxen-polydimethylsiloxane fiber (CAR-PDMS) [117, 118] or carboxen-polydimethylsiloxane-divinylbenzene (CAR-PDMS-DVB) [22, 119] has been proposed for the analysis of fermentative sulfur compounds. The use of the latter fiber has been shown to produce good results in terms of repeatability and reproducibility [22]. Moreover, the best modified ionic strength has been obtained by using magnesium sulfate. Due to the different boiling temperature of the fermentative sulfur compounds, a good compromise needs to be set to allow for the adsorption of these compounds through the SPME technique. As reported by Nguyen and co-authors [120], the incubation of the sample was carried out at 45°C for 5 min and the extraction at 45°C for 30 min under agitation. The GC equipment was coupled with MS detector and the compounds of interest were detected in single ion monitoring (SIM) mode and quantified by means of different internal standards properly chosen. The analytical methods described above are suitable for the analysis of the volatiles characterized by low boiling point (lower than 90°C) that is not the case for 3MH, 3MHA and 4MMP.

4.2. Varietal thiols

Among headspace sampling techniques, only purge and trap has been used to analyze 3MH and 3MHA in wines, reaching detection limits closer to their perception threshold [121]. Despite this, the other heavy volatile sulfur aromas were not identified at perception threshold, thus leaving derivatization procedures as the most promising technique for the extraction and analysis of volatile thiols. Due to the low concentration of these compounds, in the order of magnitude of ng/L, the sample preparation has provided a liquid/liquid extraction followed by the evaporation of the organic solvent. As the varietal thiols are present in wine in low concentration as well as being highly reactive, the deuterated analogues are commonly used as internal standards compensating for the eventual loss taking place during the sample preparation [122].

Among the derivatizing compounds, the use of *p*-hydroxymercuribenzoate (pHMB) was first proposed by Tominaga and co-authors [10]. The volatile thiols were derivatized with pHMB, isolated with the use of strong basic anion exchange column followed by the liquid/liquid extraction with dichloromethane. The sample was concentrated and analyzed by GC-MS. As a result of the analytical method improvement, varietal thiols were quantified in several wines produced with *Vitis vinifera* grape varieties [44], suggesting these sulfur compounds play a key role in the aroma of different white wines. Besides 3MH, 3MHA and 4MMP, the method allows the identification and quantification of two other sulfur-containing aromas, such as 2-furanmethanethiol [11] and benzenemethanethiol [123] in wines. The identification of these two latter molecules, together with the identification of ethyl 3-mercaptopropionate, established the role of certain volatile thiols in the bouquet of aged champagne wines [124]. Although very efficient, this procedure is time consuming. In an effort to reduce the time required for the sample preparation, a covalent chromatography was employed for the enrichment of the thiols from the wine extract. Specifically, a cross-linked agarose gel containing phenylmercuric chloride was used where the volatile thiols were trapped after their liquid/liquid extraction with dichlomethane [122]. The ability of some common solid phase extraction sorbents to retain organomercuric salts for selective concentration of thiols in wines was also proposed using styrene-divinylbenzene copolymer sorbent. Nevertheless, the organomercury salt that is formed for the detection of varietal thiols by the analytical methods described (considered as hazardous poison) still remains the key point of this method.

Other analytical approaches employ pentafluorobenzyl bromide as a derivatizing agent, which transforms thiols into their corresponding pentafluorobenzyl derivatives [125–128]. The derivatizing reaction is usually carried out in a purified extract (i.e. water) [128], organic solvent [126], in-cartridge [127], or in-fiber [125]. Another very promising derivatizing agent in the gas chromatography analysis of thiols is ethyl propiolate, which is able to derivatize thiols directly in wine matrix and is a suitable derivatizing reagent for the electron impact mass spectrometry detection system [129].

Moreover, Piano and co-authors [55] proposed an analytical method in which the varietal thiols are detected by liquid chromatography (UPLC) coupled with MS/MS. The sample preparation required several steps in order to protect the thiol aromas from oxidation and the liquid/liquid extraction in order to achieve the concentration of the analytes. The *o*-phthaldialdehyde (OPA) is used as derivatizing agent. The method allows the quantification of 3MH and 3MHA, but the derivatization of 4MMP does not occur. The formation of the OPA derivative of 4MMP was probably prevented by the hydrogen bonding between the thiol group and the carbonyl moiety within the compound itself, and its steric hindrance. Derivatization of 4MMP was an issue when other derivatizing reagents were used [127].

4.2.1. Varietal thiol precursors

The analytical methods proposed for the determination of volatile thiol precursors include both indirect and direct methods [71]. In the first case, the precursors are transformed in volatile compounds and determined by both GC and HPLC coupled with MS detector. In the second case, the precursors are quantified after their proper purification.

In the first method developed, the Cys-4MMP was detected by GC coupled with flame photometric detectors (FDP) by synthetizing it and 4MMP was then detected [69]. Tominaga and co-authors [130] quantified the cysteinylated precursors by GC-MS after their derivatization. The addition of labeled thiol precursors allowed the quantification of those contained in the must by GC-FDP and GC-MS [131]. The GC coupled either with atomic emission detection (AED) [132] or detection-capture mass spectrometry (DCMS) detector [86] were also proposed employing propyl thioacetate as internal standard and derivatizing the released volatile thiols with ethyl chloroformate, respectively. The precursors of 3MH were quantified using the labeled isotopic standard d10-3MH by GC-MS.

Both GC-MS and liquid chromatography techniques were used for the direct determination of thiol precursors. In the case of GC-MS, their derivatization is necessary and different derivatizing agents were proposed [133–135]. In the case of liquid chromatography, both HPLC-MS and HPLC-MS/MS were applied. The use of these types of equipment did not require a derivatization of thiol precursors, but a solid phase extraction (SPE) was applied achieving the sample purification. The quantification was carried out by means of the patterns of labeled compounds [73, 87, 136–139] as well as without them [59, 72]. The quantification by liquid secondary ionization mass spectrometry (LSIMS) was also reported [140].

4.3. The reduced glutathione

Several analytical methods have been proposed for GSH quantification in the grape, must and wine using different analytical techniques [141]. GSH was quantified after a treatment with glutathione reductase enzyme in white wine [113]. In grapevine tissues, the GSH was derivatized with 5,5′-dithiobis-2-nitrobenzoic acid (DTNB) after the enzymatic treatment and the detection was performed spectrophotometrically at 412 nm [142]. The determination of total GSH could be performed after the enzymatic treatment which allowed the breakdown of disulfide bridges. The GSH could be derivatized pre-column with *o*-phthalaldehyde (OPA) and quantified by HPLC; the detection was than conducted by fluorescence [143, 144]. The use of the fluorescence detector allowed the GSH quantification after derivatization with 2,3-naphthalenedialdehyde (NDA), as described by Marchand and de Revel [145]. The technique involved the pre-column derivatization and the separation by HPLC. The determination of oxidized glutathione was also allowed by enzymatic treatment with glutathione reductase. The fluorescence detector was employed for the GSH determination after its derivatization with monobromobimane (mBB); the reaction adduct was identified by capillary electrophoresis [110]. As described above, the mBB is photodegradable and a GSH content underestimation could occur [146]. The method proposed by Du Toit et al. [109] was based on liquid chromatography coupled with a mass spectrometry detector (LC-MS/MS). The determination of both reduced and oxidized glutathione was allowed in the same run. The ethanol present in wine had to be removed prior to the analysis and this step could cause an underestimation of the GSH content. Another analytical method for the GSH quantification was developed using the atomic adsorption spectrometry [147]. The method is based on the reactivity of the mercury toward the thiols. The liquid chromatography (both HPLC and UPLC) coupled with

UV detector allows the quantification of GSH after its derivatization with *p*-benzoquinone [148]. The sample preparation is not required for must and wine, while the juice needs to be obtained from grape under reductive condition prior to the derivatization [111]. Suitable preparative conditions are mandatory for monitoring GSH concentration during must extraction and clarification since such wine making steps are responsible for the chemical and enzymatic oxidations.

5. Conclusions

Several sulfur compounds occur in wine with a strong influence on its aromatic profile. On one hand, some of these molecules impart negative notes and their presence in wine should be counteracted and limited. The formation mechanisms are not completely clear and further researches are needed also to better understand the treatments and/or the winemaking practices potentially decreasing the appearance of these undesired compounds. On the other hand, the varietal thiols are responsible for positive characteristics of wine aroma up to a certain extent. Their protection is essential for the maintenance of the aromatic profile throughout both the winemaking and storage of wine by means of antioxidants of which reduce glutathione can represent a good natural candidate. Further studies will be necessary to investigate the fate of their precursors during winemaking on an industrial scale as well as to increase the aromatic potential of the produced wine. In the last decades, improvements of the analytical methods have been carried out in terms of sensitivity and identification of new odorants has been achieved. The goal for the researches has being the set up more sensitive and less-time consuming methods with a reduced impact on the environment in order to minimize defects and optimize the aromatic profile during winemaking.

Author details

Daniela Fracassetti* and Ileana Vigentini

*Address all correspondence to: daniela.fracassetti@unimi.it

Department of Food, Environmental and Nutritional Sciences (DeFENS), Università degli Studi di Milano, Milan, Italy

References

[1] Landaud S, Helinck S, Bonnarme P. Formation of volatile sulfur compounds and metabolism of methionine and other sulfur compounds in fermented food. Applied Microbiology and Biotechnology. 2008;77:1191-1205. DOI: 10.1007/s00253-007-1288-y

[2] Bartowsky EJ, Pretorius IS. Microbial formation and modification of flavour and off-flavour compounds in wine. In: König H, Unden G, Fröhlich J, editors. Biology of

Microorganisms on Grapes, in Must and Wine. Heidelberg: Springer; 2008. pp. 211-233. DOI: 10.1007/978-3-540-85463-0

[3] Mestres M, Busto O, Guasch J. Analysis of organic sulfur compounds in wine aroma. Journal of Chromatography A. 2000;**881**:569-581. DOI: 10.1016/S0021-9673(00)00220-X

[4] Maujean A, Seguin N. Contribution à l'étude des goûts de lumière dans les vins de Champagne. 3. Les réactions photochimiques responsables des goûts de lumière dans le vin de Champagne. Sciences des Aliments. 1983;**3**:589-601 ISSN: 0240-8813

[5] Mattivi F, Monetti A, Vrhovšek U, Tonon D, Andrés-Lacueva C. High-performance liquid chromatographic determination of the riboflavin concentration in white wines for predicting their resistance to light. Journal of Chromatography A. 2000;**888**:121-127. DOI: 10.1016/S0021-9673(00)00561-6

[6] Vermeulen C, Gijs L, Collin S. Sensorial contribution and formation pathways of thiols in foods: A review. Food Review International. 2005;**21**:69-137. DOI: 10.1081/FRI-200040601

[7] Swiegers JH, Bartowsky EJ, Henschke PA, Pretorius IS. Yeast and bacterial modulation of wine aroma and flavour. In: Blair RJ, Francis ME, Pretorius IS, editors. Advances in Wine Science. Glen Osmond: The Australian Wine Research Institute; 2005. pp. 159-187. DOI: 10.1111/j.1755-0238.2005.tb00285.x

[8] Dubourdieu D, Tominaga T, Masneuf I, des Gachons CP, Murat ML. The role of yeast in grape flavour development during fermentation: The example of Sauvignon blanc. American Journal of Enology and Viticulture. 2006;**57**:81-88

[9] Swiegers JH, Pretorius IS. Modulation of volatile sulfur compounds by wine yeast. Applied Microbiology and Biotechnology. 2007;**74**:954-960. DOI: 10.1007/s00253-006-0828-1

[10] Tominaga T, Murat ML, Dubourdieu D. Development of a method analyzing the volatile thiols involved in the characteristic aroma of wines made from *Vitis vinifera* L. cv. Sauvignon blanc. Journal of Agricultural and Food Chemistry. 1998;**46**:1044-1048. DOI: 10.1021/jf970782o

[11] Tominaga T, Furrer A, Henry R, Dubourdieu D. Identification of new volatile thiols in the aroma of *Vitis vinifera* L. var. Sauvignon blanc wines. Flavour and Fragrance Journal. 1998;**13**:159-162. DOI: 10.1002/(SICI)1099-1026(199805/06)13:3<159::AID-FFJ709>3.0.CO;2-7

[12] Salgues M, Cheynier V, Gunata Z, Wylde R. Oxidation of grape juice 2-s-glutathionyl caffeoyl tartaric acid by *Botrytis cinerea* laccase and characterization of a new substance: 2,5-di-s-glutathionyl caffeoyl tartaric acid. Journal of Food Science. 1986;**51**:1191-1194. DOI: 10.1111/j.1365-2621.1986.tb13081.x

[13] Robinson AL, Boss PK, Solomon PS, Trengove RD, Heymann H, Ebeler SE. Origins of grape and wine aroma. Part 1. Chemical components and viticultural impacts. American Journal of Enology and Viticulture. 2014;**56**:1-24. DOI: 10.5344/ajev.2013.12070

[14] Monk PR. Formation, utilization and excretion of hydrogen sulphide by wine yeast. The Australian & New Zealand Wine Industry Journal. 1986;**1**:10-16

[15] Rauhut D. Usage and formation of sulphur compounds. In: König H, Unden G, Fröhlich J, editors. Biology of Microorganisms on Grapes, in Must and Wine. Heidelberg: Springer; 2008. pp. 181-207. DOI: 10.1007/978-3-540-85463-0

[16] Smith ME, Bekker MZ, Smith PA, Wilkes EN. Source of volatile sulfur compounds in wine. Australian Journal of Grape and Wine Research. 2015;**21**:705-712. DOI: 10.1111/ajgw.12193

[17] Winter G, Henschke PA, Higgins VJ, Ugliano M, Curtin CD. Effects of rehydration nutrients on H_2S metabolism and formation of volatile sulfur compounds by the wine yeast VL3. AMB Express. 2011;**1**:36-47. DOI: 10.1186/2191-0855-1-36

[18] Pripis-Nicolau L, de Revel G, Bertrand A, Lonvaud-Funel A. Methionine catabolism and production of volatile sulphur compounds by *Oenococcus oeni*. Journal of Applied Microbiology. 2004;**96**:1176-1184. DOI: 10.1111/j.1365-2672.2004.02257.x

[19] Segurel MA, Razungles AJ, Riou C, Salles M, Baumes RL. Contribution of dimethyl sulfide to the aroma of Syrah and Grenache Noir wines and estimation of its potential in grapes of these varieties. Journal of Agricultural and Food Chemistry. 2004;**52**:7084-7093. DOI: 10.1021/jf049160a

[20] Escudero A, Campo E, Fariña L, Cacho J, Ferreira V. Analytical characterization of the aroma of five premium red wines. Insights into the role of odor families and the concept of fruitiness of wines. Journal of Agricultural and Food Chemistry. 2007;**55**:4501-4510. DOI: 10.1021/jf0636418

[21] Goniak OJ, Noble AC. Sensory study of selected volatile sulfur compounds in white wine. American Journal of Enology and Viticulture. 1987;**38**:223-227

[22] Fedrizzi B, Magno F, Badocco D, Nicolini G, Versini G. Aging effects and grape variety dependence on the content of sulfur volatiles in wine. Journal of Agricultural and Food Chemistry. 2007;**55**:10880-10887. DOI: 10.1021/jf072145w

[23] Ugliano M, Fedrizzi B, Siebert T, Travis B, Magno F, Versini G, Henschke PA. Effect of nitrogen supplementation and *Saccharomyces* species on hydrogen sulfide and other volatile sulfur compounds in Shiraz fermentation and wine. Journal of Agricultural and Food Chemistry. 2009;**57**:4948-4955. DOI: 10.1021/jf8037693

[24] Viviers MZ, Smith ME, Wilkes E, Smith P. Effects of five metals on the evolution of hydrogen sulfide, methanethiol, and dimethyl sulfide during anaerobic storage of Chardonnay and Shiraz wines. Journal of Agricultural and Food Chemistry. 2013;**61**:12385-12396. DOI: 10.1021/jf403422x

[25] Bekker MZ, Day MP, Holt H, Wilkes E, Smith PA. Effect of oxygen exposure during fermentation on volatile sulfur compounds in Shiraz wine and a comparison of strategies for remediation of reductive character. Australian Journal of Grape and Wine Research. 2016;**22**:24-35. DOI: 10.1111/ajgw.12172

[26] Ferreira V, Bueno M, Franco-Luesma E, Culleré L, Fernández-Zurbano P. Key changes in wine aroma active compounds during bottle storage of Spanish red wines under

different oxygen levels. Journal of Agricultural and Food Chemistry. 2014;**62**:10015-10027. DOI: 10.1021/jf503089u

[27] Lavigne V, Dubourdieu D. Demonstration and interpretation of the yeast lees ability to absorb certain volatile thiols contained in wine. Journal International des Sciences de la Vigne et du Vin. 1996;**30**:201-206 ISSN: 1151-0285

[28] Riberau-Gayon P, Glories Y, Maujean A, Dubourdieu D. Handbook of Enology, the Chemistry of Wine Stabilization and Treatments. 2nd ed. Chichester: John Wiley & Sons Ltd; 2006 ISBN: 13: 978-0-470-01037-2

[29] Ugliano M, Kwiatkowski M, Vidal S, Capone D, Siebert T, Dieval JB, Aagaard O, Waters EJ. Evolution of 3-mercaptohexanol, hydrogen sulfide, and methyl mercaptan during bottle storage of Sauvignon blanc wines. Effect of glutathione, copper, oxygen exposure, and closure-derived oxygen. Journal of Agricultural and Food Chemistry. 2011;**59**:2564-2572. DOI: 10.1021/jf1043585

[30] Vela E, Hernández-Orte P, Franco-Luesma E, Ferreira V. The effects of copper fining on the wine content in sulfur off-odors and on their evolution during accelerated anoxic storage. Food Chemistry. 2017;**231**:212-221. DOI: 10.1016/j.foodchem.2017.03.125

[31] Clark AC, Prenzler PD, Scollary GR. Impact of the condition of storage of tartaric acid solutions on the production and stability of glyoxylic acid. Food Chemistry. 2007;**102**:905-916. DOI: 10.1016/j.foodchem.2006.06.029

[32] Clark AC, Dias DA, Smith TA, Ghiggino KP, Scollary GR. Iron(III) tartrate as a potential precursor of light induced oxidative degradation of white wine: Studies in a model wine system. Journal of Agricultural and Food Chemistry. 2011;**59**:3575-3581. DOI: 10.1021/jf104897z

[33] Haye B, Maujean A, Jacquemin C, Feuillat M. Contribution a l'étude des "goûts de lumière" dans le vin de Champagne. I. Aspects analytiques. Dosage des mercaptans et des thiols dans les vins. Connaissance de la Vigne et du Vin. 1977;**11**:243-254. DOI: 10.20870/oeno-one.1977.11.3.1443

[34] Maujean A, Haye M, Feuillat M. Contribution a l'étude des "goûts de lumière" dans le vin de Champagne. II. Influence de la lumière sur le potentiel d'oxydoreduction. Correlation avec la teneur en thiols du vin. Connaissance de la Vigne et du Vin. 1978;**12**:277-290. DOI: 10.20870/oeno-one.1978.12.4.1427

[35] Maujean A, Seguin N. Contribution à l'étude des goûts de lumière dans les vins de Champagne. 4. Approaches a une solution œnologique des moyens de prévention des goûts de lumière. Science des Aliments. 1983;**3**:603-613

[36] Jung R, Hey M, Hoffmann D, Leiner T, Patz CD, Rauhut D, Schüssler C, Wirsching M. Lichteinfluss bei der Lagerrung von Wein. Mitteilungen Klosterneuburg. 2007;**57**:224-231

[37] Escudero A, Hernández-Orte P, Cacho J, Ferreira V. Clues about the role of methional as character impact odorant of some oxidized wines. Journal of Agricultural and Food Chemistry. 2000;**48**:4268-4272. DOI: 10.1021/jf991177j

[38] Escudero A, Asensio E, Cacho J, Ferreira V. Sensory and chemical changes of young white wines stored under oxygen. An assessment of the role played by aldehydes and some other important odorants. Food Chemistry. 2002;**77**:325-331. DOI: 10.1016/S0308-8146(01)00355-7

[39] Pichler U. Analisi della riboflavina nei vini bianchi e influenza della sua concentrazione. L'Enotecnico. 1996;**32**:57-62

[40] Ournac A. Riboflavine pendant la fermentation du jus de raisin et la conservation du vin sur lies. Annales de Technologie Agricole. 1968;**17**:67-75

[41] Fracassetti D, Gabrielli M, Encinas J, Manara M, Pellegrino L, Tirelli A. Approaches to prevent the light-struck taste in white wine. Australian Journal of Grape and Wine Research. 2017;**23**:329-333. DOI 10.1111/ajgw.12295

[42] Piredda E. Photo-oxidation of methionine in model solution: Effect of hydrolysable tannins [Thesis]. Milan: Università degli Studi di Torino, Milano, Palermo, Sassari, Foggia; 2017.

[43] Darriet P, Tominaga T, Lavigne V, Boidron J, Dubourdieu D. Identification of a powerful aromatic compound of *Vitis vinifera* L. var. Sauvignon wines: 4-Mercapto-4-methylpentan-2-one. Flavour and Fragrance Journal. 1995;**10**:385-392. DOI: 10.1002/ffj.2730100610

[44] Tominaga T, Blanchard L, Darriet P, Dubourdieu D. A powerful aromatic volatile thiol, 2-furanmethanethiol, exhibiting roast coffee aroma in wines made from several *Vitis vinifera* grape varieties. Journal of Agricultural and Food Chemistry. 2000;**48**:1799-1802. DOI: 10.1021/jf990660r

[45] Rigaud J, Etievant P, Henry R, Latrasse A. 4-methoxy-2-methyl-2-mercaptobutane, a major constituent of the aroma of the blackcurrant bud (*Ribes nigrum*). Sciences des Aliments. 1986;**6**:213-220

[46] Demole E, Enggist P, Ohloff G. 1-p-menthene-8-thiol: A powerful flavor impact constituent of grapefruit juice (*Citrus parodisi* Macfayeden). Helvetica Chimica Acta. 1982;**65**: 1785-1794. DOI: 10.1002/hlca.19820650614

[47] Engel KH, Tressl R. Identification of new sulphur-containing volatiles in yellow passion fruit (*Passiflora edulis f. flavicarpa*). Journal of Agricultural and Food Chemistry. 1991;**39**:2249-2252. DOI: 10.1021/jf00012a030

[48] Idstein H, Schreier P. Volatile constituents from guava (*Psidium guajava,* L.) fruit. Journal of Agricultural and Food Chemistry. 1985;**33**:138-143. DOI: 10.1021/jf00061a039

[49] Rauhut D, Kürbel H, MacNamara K, Grossmann M. Headspace GC-SCD monitoring of low volatile sulfur compounds during fermentation and in wine. Analysis. 1998;**26**:142-144. DOI: 10.1051/analusis:1998124

[50] Ugliano M, Henschke PA. Yeast and wine flavor. In: Moreno-Arribas MV, Polo MC, editors. Wine Chemistry and Biochemistry. New York: Springer; 2009. pp. 342-352. DOI: 10.1007/978-0-387-74118-5

[51] Bouchilloux P, Darriet P, Henry R, Lavigne-Cruège V, Dubourdieu D. Identification of volatile and powerful odorous thiols in Bordeaux red wine varieties. Journal of Agricultural and Food Chemistry. 1998;**46**:3095-3099. DOI: 10.1021/jf971027d

[52] Baumes R. Wine aroma precursors. In: Moreno-Arribas MV, Polo MC, editors. Wine Chemistry and Biochemistry. New York: Springer; 2009. pp. 260-263. DOI: 10.1007/978-0-387-74118-5

[53] Lund CM, Thompson MK, Benkwitz F, Wohler MW, Triggs CM, Gardner R, Heymann H, Nicolau L. New Zealand Sauvignon blanc distinct flavor characteristics, sensory, chemical, and consumer aspects. American Journal of Enology and Viticulture. 2009;**60**:1-12

[54] Benkwitz F, Tominaga T, Kilmartin PA, Lund CM, Wohlers M, Nicolau L. Identifying the chemical composition related to the distinct aroma characteristics of New Zealand Sauvignon blanc wines. American Journal of Enology and Viticulture. 2012;**63**:62-72. DOI: 10.5344/ajev.2011.10074

[55] Piano F, Fracassetti D, Buica A, Stander M, du Toit WJ, Borsa D, Tirelli A. Development of a novel liquid/liquid extraction and UPLC-MS/MS method for the assessment of thiols in South African Sauvignon blanc wines. Australian Journal of Grape and Wine Research. 2015;**21**:40-48. DOI: 10.1111/ajgw.12117

[56] Ferreira V, Ortín N, Escudero A, López R, Cacho J. Chemical characterization of the aroma of Grenache rose wines: Aroma extract dilution analysis, quantitative determination, and sensory reconstitution studies. Journal of Agricultural and Food Chemistry. 2002;**50**:4048-4054. DOI: 10.1021/jf0115645

[57] Mattivi F, Fedrizzi B, Zenato A, Tiefenthaler P, Tempesta S, Perenzoni D, Cantarella P, Simeoni F, Vrhovsek U. Development of reliable analytical tools for evaluating the influence of reductive wine making on the quality of Lugana wines. Analytica Chimica Acta. 2012;**732**:194-202. DOI: 10.1016/j.aca.2011.11.051

[58] Piano F, Petrozziello M, Vaudano E, Bonello F, Ferreira V, Zapata J, Hernández-Orte P. Aroma compounds and sensory characteristics of Arneis Terre Alfieri DOC wines: The concentration of polyfunctional thiols and their evolution in relation to different ageing conditions. European Food Research and Technology. 2014;**239**:267-277. DOI: 10.1007/s00217-014-2218-8

[59] Fracassetti D, Stuknytė M, La Rosa C, Gabrielli M, De Noni I, Tirelli A. Thiol precursors in Catarratto Bianco Comune and Grillo grapes and release of varietal thiols in wine under different must clarification conditions of the must. Australian Journal of Grape and Wine Research. 2018;**24**:125-133. DOI: 10.1111/ajgw.12311

[60] Murat M, Tominaga T, Saucier C, Glories Y, Dubourdieu D. Effect of anthocyanins on stability of a key odorous compounds, 3-mercaptohexan-1-ol, in Bordeaux rose wines. American Journal of Enology and Viticulture. 2003;**54**:135-138 ISSN: 0002-9254

[61] Blanchart L, Darriet P, Dubourdieu D. Reactivity of 3-mercaptohexanol in wine: Impact of oxygen, phenolic fractions, and sulfur dioxide. American Journal of Enology and Viticulture. 2004;**55**:115-120 ISSN: 0002-9254

[62] Jocelyn PC. Biochemistry of the SH Group. London, New York: Academic Press; 1972. pp. 63-93

[63] Kotseridis Y, Ray J, Augier C, Baumes R. Quantitative determination of sulfur containing wine odorants at sub-ppb levels. 1. Synthesis of the deuterated analogues. Journal of Agricultural and Food Chemistry. 2000;**48**:5819-5823. DOI: 10.1021/jf0004715

[64] Ribéreau-Gayon P, Glories Y, Maujean A, Dubourdieu D. Handbook of enology, the chemistry of wine stabilization and treatments (2nd ed., Vol. 2). Chichester: John Wiley & Sons Ltd; 2006. pp. 152-172

[65] Danilewicz JC, Seccombe JT, Whelan J. Mechanism of interaction of polyphenols, oxygen, and sulfur dioxide in model wine and wine. American Journal of Enology and Viticulture. 2008;**59**:128-136 ISSN: 0002-9254

[66] Kilmartin PA, Zou H, Waterhouse AL. A cyclic voltammetry method suitable for characterizing antioxidant properties of wine and wine phenolics. Journal of Agricultural and Food Chemistry. 2001;**49**:1957-1965. DOI: 10.1021/jf001044u

[67] Brajkovich M, Tibbits N, Peron G, Lund CM, Dykes SI, Kilmartin PA, Nicolau L. Effect of screw cap and cork closures on SO_2 levels and aromas in Sauvignon blanc wine. Journal of Agricultural and Food Chemistry. 2005;**53**:10006-10011. DOI: 10.1021/jf0512813

[68] Darriet PH, Tominaga T, Timole e, Dubourdieu D. Evidence of the presence of a 4-mercapto-4-methylpentan-2-one precursor in *Vitis vinifera* Sauvignon blanc grape variety. Comptes rendus hebdomadaires des séances de l'Académie des sciences serie III. 1993;**316**:1332-1335

[69] Tominaga T, Masneuf I, Dubourdieu D. Mise en évidence d'un S-conjugué de la cystéine, précursor d'arôme du Sauvignon. Journal International des Sciences de la Vigne et du Vin. 1995;**29**:227-232

[70] Coetzee C, Du Toit WJ. A comprehensive review on sauvignon blanc aroma with a focus on certain positive volatile thiols. Food Research International. 2012;**45**:287-298. DOI: 10.1016/j.foodres.2011.09.017

[71] Peña-Gallego A, Hernández-Orte P, Cacho J, Ferreira V. S-cysteinylated and S-glutathionylated thiol precursors in grapes. A review. Food Chemistry. 2012;**131**:1-13. DOI: 10.1016/j.foodchem.2011.07.079

[72] Fedrizzi B, Pardon KH, Sefton MA, Elsey GM, Jeffery DW. First identification of 4-S-glutathionyl-4-methylpentan-2-one, a potential precursor of 4-mercapto-4-methylpentan-2-one, in Sauvignon blanc juice. Journal of Agricultural and Food Chemistry. 2009;**57**:991-995. DOI: 10.1021/jf802799w

[73] Roland A, Vialaret J, Razungles A, Rigou P, Schneider R. Evolution of S-cysteinylated and S-glutathionylated thiol precursors during oxidation of Melon B. and Sauvignon blanc musts. Journal of Agricultural and Food Chemistry. 2010;**58**:4406-4413. DOI: 10.1021/jf904164t

[74] Schneider R, Charrier F, Razungles A, Baumes R. Evidence for an alternative biogenetic pathway leading to 3-mercaptohexanol and 4-mercapto-4-methylpentan-2-one in wines. Analytica Chimica Acta. 2006;**563**:58-64. DOI: 10.1016/j.aca.2006.01.057

[75] Hatanaka A, Kajiwara T, Matsui K. The biogeneration of green odour by green leaves and it's physiological functions—Past, present and future. Zeitschrift für Naturforschung. Section C, Biosciences. 1995;**50c**:467-472

[76] Allen T, Herbst-Johnstone M, Girault M, Butler P, Logan G, Jouanneau S, Nicolau L, Kilmartin PA. Influence of grape-harvesting steps on varietal thiol aromas in Sauvignon blanc wines. Journal of Agricultural and Food Chemistry. 2011;**59**:10641-10650. DOI: 10.1021/jf2018676

[77] Harsch MJ, Benkwitz F, Frost A, Colonna-Ceccaldi B, Gardner RC, Salmon JM. New precursor of 3-mercaptohexan-1-ol in grape juice: Thiol-forming potential and kinetics during early stages of must fermentation. Journal of Agricultural and Food Chemistry. 2013;**61**:3703-3713. DOI: 10.1021/jf3048753

[78] Pinu FR, Jouanneau S, Nicolau L, Gardner RC, Villas-Boas SG. Concentrations of the volatile thiol 3-mercaptohexanol in Sauvignon blanc wines: No correlation with juice precursors. American Journal of Enology and Viticulture. 2012;**63**:407-412. DOI: 10.5344/ajev.2012.11126

[79] Thibon C, Böcker C, Shinkaruk S, Moine V, Darriet P, Dubourdieu D. Identification of S-3-(hexanal)-glutathione and its bisulfite adduct in grape juice from *Vitis vinifera* L. cv. Sauvignon blanc as new potential precursors of 3SH. Food Chemistry. 2016;**199**:711-719. DOI: 10.1016/j.foodchem.2015.12.069

[80] Murat ML, Masneuf I, Darriet P, Lavigne V, Tominaga T, Dubourdieu D. Effect of *Saccharomyces cerevisiae* strain on the liberation of volatile thiols in Sauvignon blanc. American Journal of Enology and Viticulture. 2001;**52**:136-139

[81] Masneuf-Pomarede I, Murat ML, Naumov GI, Tominaga T, Dubourdieu D. Hybrids *Saccharomyces cerevisiae* x *Saccharomyces bayanus* var. *uvarum* having a high liberating ability of some sulfur varietal aromas of *Vitis vinifera* sauvignon blanc wines. Journal International des Sciences de la Vigne et du Vin. 2002;**36**:205-212

[82] Anfang N, Brajkovich M, Goddard MR. Co-fermentation with *Pichia kluyveri* increases varietal thiol concentrations in Sauvignon Blanc. Australian Journal of Grape and Wine Research. 2009;**15**:1-8. DOI: 10.1111/j.1755-0238.2008.00031.x

[83] Belda I, Ruiz J, Beisert B, Navascués E, Marquina D, Calderón F, Rauhut D, Benito S, Santos A. Influence of *Torulaspora delbrueckii* in varietal thiol (3-SH and 4-MSP) release in wine sequential fermentations. International Journal of Food Microbiology. 2017;**257**:183-191. DOI: 10.1016/j.ijfoodmicro.2017.06.028

[84] Swiegers JH, Kievit RL, Siebert T, Lattey KA, Bramley LR, Francis IL, Kinga ES, Pretorius IS. The influence of yeast strain on the aromatic character of Sauvignon blanc wine. Food Microbiology. 2008;**26**:204-211. DOI: 10.1016/j.ijfoodmicro.2006.01.001

[85] Des Gachons CP, Tominaga T, Dubourdieu D. Sulfur aroma precursor present in S-glutathione conjugate form: Identification of S-3-(hexan-1-ol)-glutathione in must from *Vitis vinifera* L. cv. Sauvignon blanc. Journal of Agricultural and Food Chemistry. 2001;**50**:4076-4079. DOI: 10.1021/jf020002y

[86] Subileau M, Schneider R, Salmon J, Degryse E. New insights on 3-mercaptohexanol (3MH) biogenesis in Sauvignon blanc wines: Cys-3MH and (E)-hexen-2-al are not the major precursors. Journal of Agricultural and Food Chemistry. 2008;**56**:9230-9235. DOI: 10.1021/jf801626f

[87] Grant-Preece PA, Pardon KH, Capone DL, Cordente AG, Sefton MA, Jeffery DW, Sefton MA, Elsey GM. Synthesis of wine thiol conjugates and labeled analogues: Fermentation of the glutathione conjugate of 3-mercaptohexan-1-ol yields the corresponding cysteine conjugate and free thiol. Journal of Agricultural and Food Chemistry. 2010;**58**:1383-1389. DOI: 10.1021/jf9037198

[88] Kobayashi H, Takase H, Kaneko K, Tanzawa F, Takata R, Suzuki S, Konno T. Analysis of S-3-(hexan-1-ol)-glutathione and S-3-(hexan-1-ol)-l-cysteine in *Vitis vinifera* L. cv. Koshu for aromatic wines. American Journal of Enology and Viticulture. 2010;**61**:176-185

[89] Concejero B, Hernandez-Orte P, Astrain J, Lacau B, Baron C, Ferreira V. Evolution of polyfunctional mercaptans and their precursors during Merlot alcoholic fermentation. Food Science and Technology. 2016;**65**:770-776. DOI: 10.1016/j.lwt.2015.09.018

[90] Capone DL, Jeffery DW. Effects of transporting and processing Sauvignon blanc grapes on 3-mercaptohexan-1-ol precursor concentrations. Journal of Agricultural and Food Chemistry. 2011;**59**:4659-4667. DOI: 10.1021/jf200119z

[91] Murat ML, Tominaga T, Dubourdieu D. Assessing the aromatic potential of Cabernet Sauvignon and Merlot musts used to produce rosé wine by assaying the cysteinylated precursor of 3-mercaptohexan-1-ol. Journal of Agricultural and Food Chemistry. 2001;**49**:5412-5417. DOI: 10.1021/jf0103119

[92] Des Gachons CP, Tominaga T, Dubourdieu D. Localization of S-cysteine conjugates in the berry: Effect of skin contact on aromatic potential of *Vitis vinifera* L. cv. Sauvignon blnac must. American Journal of Enology and Viticulture. 2002;**53**:144-146

[93] Maggu M, Winz R, Kilmartin PA, Trought MCT, Nicolau L. Effect of skin contact and pressure on the composition of Sauvignon blanc must. Journal of Agricultural and Food Chemistry. 2007;**55**:10281-10288. DOI: 10.1021/jf072192o

[94] Roland A, Schneider R, Charrier F, Cavelier F, Rossignol M, Razungles A. Distribution of varietal thiol precursors in the skin and the pulp of Melon B. and Sauvignon blanc grapes. Food Chemistry. 2011;**125**:139-144. DOI: 10.1016/j.foodchem.2010.08.050

[95] Patel P, Herbst-Johnstone M, Lee SA, Gardner RC, Weaver R, Nicolau L, Kilmartin PA. Influence of juice pressing conditions on polyphenols, antioxidants, and varietal aroma of Sauvignon blanc microferments. Journal of Agricultural and Food Chemistry. 2010;**58**:7280-7288. DOI: 10.1021/jf100200e

[96] Friedman M. Improvement in the safety of foods by SH containing amino acids and peptides. A review. Journal of Agricultural and Food Chemistry. 1994;**42**:3-20. DOI: 10.1021/jf00037a002

[97] Noctor G, Arisi ACM, Jouanin L, Foyer CH. Manipulation of glutathione and amino acids biosynthesis in the chloroplast. Plant Physiology. 1998;**118**:471-482. DOI: 10.1104/pp.118.2.471

[98] Li H, Guo A, Wang H. Mechanisms of oxidative browning of wine. Food Chemistry. 2008;**108**:1-13. DOI: 10.1016/j.foodchem.2007.10.065

[99] Margalit Y. Concepts in Wine Chemistry. 2nd ed. San Francisco: Wine Appreciation Guild; 2004

[100] Singleton V, Zaya J, Trousdale E, Salgues M. Caftaric acid in grapes and conversion to a reaction product during processing. Vitis. 1984;**23**:113-120 ISSN: 2367-4156

[101] Cheynier V, Van Hulst MWJ. Oxidation of trans-caftaric acid and 2-S-glutathionyl caftaric acid in model solutions. Journal of Agricultural and Food Chemistry. 1988;**36**:10-15. DOI: 10.1021/jf00079a003

[102] Cheynier V, Ricardo da Silva JM. Oxidation of grape procyanidins in model solution containing trans-caffeoyltartaric acid and polyphenol oxidase. Journal of Agricultural and Food Chemistry. 1991;**39**:1047-1049. DOI: 10.1021/jf00006a008

[103] Lavigne V, Dubourdieu D. Affinamento sulle fecce e freschezza dei vini bianchi. VigneVini. 2004;**31**:58-66

[104] Papadopoulou D, Roussis IG. Inhibition of the decrease of volatile esters and terpenes during storage of a white wine and a model wine medium by glutathione and N-acetylcysteine. International Journal of Food Science and Technology. 2008;**43**:1053-1057. DOI: 10.1111/j.1365-2621.2007.01562.x

[105] Cutzach I, Chatonnet P, Dubourdieu D. Study of the formation mechanisms of some volatile compounds during the aging of sweet fortified wines. Journal of Agricultural and Food Chemistry. 1999;**45**:2837-2846. DOI: 10.1021/jf981224s

[106] Lavigne V, Pons A, Darriet P, Dubourdieu D. Changes in the sotolon content of dry white wines during barrel and bottle aging. Journal of Agricultural and Food Chemistry. 2008;**56**:2688-2693. DOI: 10.1021/jf072336z

[107] Hosry LE, Auezova L, Sakr A, Hajj-Moussa E. Browning susceptibility of white wine and antioxidant effect of glutathione. International Journal of Food Science and Technology. 2009;**44**:2459-2463. DOI: 10.1111/j.1365-2621.2009.02036.x

[108] Cheynier V, Souquet JM, Moutounet M. Glutathione content and glutathione to hydroxycinnamic acid ratio in *Vitis vinifera* grapes and musts. American Journal of Enology and Viticulture. 1989;**40**:320-324

[109] Du Toit WJ, Lisjak K, Stander M, Prevoo D. Using LC-MSMS to asses glutathione levels in South African white grape juices and wines made with different levels of

oxygen. Journal of Agricultural and Food Chemistry. 2007;**55**:2765-2769. DOI: 10.1021/jf062804p

[110] Lavigne V, Pons A, Dudourdieu D. Assay of glutathione in must and wines using capillary electrophoresis and laser-induced fluorescence detection: Changes in concentration in dry white wines during alcoholic fermentation and aging. Journal of Chromatography A. 2007;**1139**:130-135. DOI: 10.1016/j.chroma.2006.10.083

[111] Fracassetti D, Tirelli A. Monitoring of glutathione concentration during winemaking by a reliable high-performance liquid chromatography analytical method. Australian Journal of Grape and Wine Research. 2015;**21**:389-395. DOI: 10.1111/ajgw.12139

[112] Andujar-Ortiz I, Pozo-Bayón MÁ, Moreno-Arribas MV, Martín-Álvarez PJ, Rodríguez-Bencomo JJ. Reversed-phase high-performance liquid chromatography-fluorescence detection for the analysis of glutathione and its precursor γ-glutamyl cysteine in wines and model wines supplemented with oenological inactive dry yeast preparations. Food Analytical Methods. 2012;**5**:154-161. DOI: 10.1007/s12161-011-9230-4

[113] Cassol T, Adams DO. Detection of glutathione in white wines using an enzymatic analytical method. American Journal of Enology and Viticulture. 1995;**46**:410

[114] Salmon JM. Interactions between wine lees and polyphenols: Influence of oxygen consumption capacity during simulation of wine aging. Journal of Food Science. 2002;**67**:1604-1609. DOI: 10.1111/j.1365-2621.2002.tb08691.x

[115] Lepine L, Archambault J. Parts-per-trillion determination of trihalomethanes in water by purge and trap gas chromatography with electron capture detection. Analytical Chemistry. 1997;**64**:810-815

[116] Washall JW, Wampler TP. Sources of error in purge and trap analysis of volatile organic compounds. American Laboratory. 1990;**22**:38-43

[117] Mestres M, Sala C, Martí MP, Busto O, Guasch J. Simultaneous analysis of thiols, sulphides and disulphides in wine aroma by headspace solid-phase microextraction-gas chromography. Journal of Chromagraphy A. 1999;**849**:293-297. DOI: 10.1016/S0021-9673(99)00538-5

[118] Segurel MA, Razungles AJ, Riou C, Trigueiro MGL, Baumes RL. Ability of possible DMS precursors to release DMS during wine aging and in the conditions of heat-alkaline treatment. Journal of Agricultural and Food Chemistry. 2005;**53**:2637-2645. DOI: 10.1021/jf048273r

[119] Fedrizzi B, Magno F, Finato F, Versini G. Variation of some fermentative sulfur compounds in Italian "Millesimè" classic sparkling wines during aging and storage on lees. Journal of Agricultural and Food Chemistry. 2010;**58**:9716-9722. DOI: 10.1021/jf101478w

[120] Nguyen DD, Nicolau L, Kilmartin PA. Application of an automated heandspace solid phase micro-extraction for the GC-MS detection and quantification of reductive sulfur compounds in wines. In: Salih B, Çelikbıçak O. editors. Gas Chromatography in Plant

Science. Wine Technology, Toxicology and Some Specific Applications. Rijeka: InTech; 2012. pp. 171-194. DOI: 10.5772/2517

[121] Fedrizzi B, Versini G, Lavagnini I, Badocca D, Nicolini G, Magno F. Hyphenated gas chromatography-mass spectrometry analysis of 3-mercaptohexan-1-ol and 3-mercaptohexyl acetate in wine: Comparison with results of other sampling procedures via a robust regression. Analytica Chimica Acta. 2008;**621**:38-43. DOI: 10.1016/j.aca.2007.10.040

[122] Schneider R, Kotseridis Y, Ray JL, Augier C, Baumes R. Quantitative determination of sulfur-containing wine odorants at sub parts per billion levels. 2. Development and application of a stable isotope dilution. Journal of Agricultural and Food Chemistry. 2003;**51**:3243-3248. DOI: 10.1021/jf0211128

[123] Tominaga T, Guimbertau G, Dubourdieu D. Contribution of benzenmethanethiol to smoky aroma of certain *Vitis vinifera* L. wines. Journal of Agricultural and Food Chemistry. 2003;**51**:1373-1376. DOI: 10.1021/jf020756c

[124] Tominaga T, Guimbertau G, Dubourdieu D. Role of certain volatile thiols in the bouquet of aged champagne wines. Journal of Agricultural and Food Chemistry. 2003;**51**:1016-1020. DOI: 10.1021/jf020755k

[125] Mateo-Vivaracho L, Ferreira V, Cacho J. Automated analysis of 2-methyl-2-furanthiol and 3-mercaptohexyl acetate at ng l-1 level by headspace solid-phase microextraction with on-fibre derivatisation and gas chromatography-negative chemical ionization mass spectrometric determination. Journal Chromatography A. 2006;**1121**:1-9. DOI: 10.1016/j.chroma.2006.04.029

[126] Mateo-Vivaracho L, Cacho J, Ferreira V. Quantitative determination of wine polyfunctional mercaptans at nanogram per liter level by gas chromatography-negative ion mass spectrometric analysis of their pentafluorobenzyl derivatives. Journal Chromatography A. 2007;**1146**:242-250. DOI: 10.1016/j.chroma.2007.02.003

[127] Mateo-Vivaracho L, Cacho J, Ferreira V. Improved solid-phase extraction procedure for the isolation and in-sorbent pentafluorobenzyl alkylation of polyfunctional mercaptans optimized procedure and analytical applications. Journal Chromatography A. 2008;**1185**:9-18. DOI: 10.1016/j.chroma.2008.01.037

[128] Capone DL, Sefton MA, Jeffery W. Application of a modified method for 3-mercaptohexan-1-ol determination to investigate the relationship between free thiol and related conjugates in grape juice and wine. Journal of Agricultural and Food Chemistry. 2011;**59**:4649-4658. DOI: 10.1021/jf200116q

[129] Herbst-Johnstone M, Piano F, Duhamel N, Barker D, Fedrizzi B. Ethyl propiolate derivatisation for the analysis of varietalthiols in wine. Journal Chromatography A. 2013;**1312**:104-110. DOI: 10.1016/j.chroma.2013.08.066

[130] Tominaga T, des Gachons CP, Dubourdieu D. A new type of flavor precursors in *Vitis vinifera* L. cv. Sauvignon blanc: S-cysteine conjugates. Journal of Agricultural and Food Chemistry. 1998;**46**:5215-5219. DOI: 10.1021/jf980481u

[131] Des Gachons CP, Tominaga T, Dubourdieu D. Measuring the aromatic potential of *Vitis vinifera* L. cv. Sauvignon blanc grapes by assaying S-cysteine conjugates, precursors of the volatile thiols responsible for their varietal aroma. Journal of Agricultural and Food Chemistry. 2000;**48**:3387-3391. DOI: 10.1021/jf990979b

[132] Howell KS, Swiegers JH, Elsey GM, Siebert TE, Bartowsky EJ, Fleet GH, Pretorius IS, de Barros Lopes MA. Variation in 4-mercapto-4-methyl-pentan-2-one release by *Saccharomyces cerevisiae* commercial wine strains. FEMS Microbiology Letters. 2004;**240**:125-129. DOI: 10.1016/j.femsle.2004.09.022

[133] Shinkaruk S, Thibon C, Schmitter JM, Babin P, Tominaga T, Degueil M, Desbat B, Jussier C, Bennetau B, Dubourdieu D, Bennetau-Pelissero C. Surprising structural lability of a cysteine-S-conjugate precursor of 4-methyl-4-sulfanylpentan-2-one, a varietal aroma in wine of *Vitis vinifera* L. cv. Sauvignon blanc. Chemistry & Biodiversity. 2008;**5**:793-810. DOI: 10.1002/cbdv.200890076

[134] Thibon C, Shinkaruk S, Tominaga T, Bennetau B, Dubourdieu D. Analysis of the diastereoisomers of the cysteinylated aroma precursor of 3-sulfanylhexanol in *Vitis vinifera* grape must by gas chromatography coupled with ion trap tandem mass spectrometry. Journal of Chromatography A. 2008;**1183**:150-157. DOI: 10.1016/j.chroma.2007.12.082

[135] Thibon C, Shinkaruk S, Jourdes M, Bennetau B, Dubourdieu D, Tominaga T. Aromatic potential of botrytized white wine grapes: Identification and quantification of new cysteine-S-conjugate flavor precursors. Analytica Chimica Acta. 2010;**660**:190-196. DOI: 10.1016/j.aca.2009.10.018

[136] Capone DL, Sefton MA, Hayasaka Y, Jeffery DW. Analysis of precursors to wine odourant 3-mercaptohexan-1-ol using HPLC–MS/MS: Resolution and quantitation of diastereoisomers of 3-S-cysteinylhexan-1-ol and 3-S-glutathionylhexan-1-ol. Journal of Agricultural and Food Chemistry. 2010;**58**:1390-1395. DOI: 10.1021/jf903720w

[137] Luisier JL, Buettner H, Volker S, Rausis T, Frey U. Quantification of cysteine S-conjugate of 3-sulfanylhexan-1-ol in must and wine of petite arvine vine by stable isotope dilution analysis. Journal of Agricultural and Food Chemistry. 2008;**56**:2883-2887. DOI: 10.1021/jf072963o

[138] Roland A, Schneider R, Guernevé CL, Razungles A, Cavelier F. Identification and quantification by LC–MS/MS of a new precursor of 3-mercaptohexan-1-ol (3MH) using stable isotope dilution assay: Elements for understanding the 3MH production in wine. Food Chemistry. 2010;**121**:847-855. DOI: 10.1016/j.foodchem.2009.12.095

[139] Roland A, Vialaret J, Moniatte M, Rigou P, Razungles A, Schneider R. Validation of a nanoliquid chromatography-tandem mass spectrometry method for the identification and the accurate quantification by isotopic dilution of glutathionylated and cysteinylated precursors of 3-mercaptohexan-1-ol and 4-mercapto-4-methylpentan-2-one in white grape juices. Journal of Chromatography A. 2010;**1217**:1626-1635. DOI: 10.1016/j.chroma.2010.01.031

[140] Des Gachons CP, Tominaga T, Dubourdieu D. Sulfur aroma precursorpresent in S-glutathione conjugate form: Identification of S-3-(hexan-1-ol)-glutathione in must

from *Vitis vinifera* L. cv. Sauvignon blanc. Journal of Agricultural and Food Chemistry. 2002;**50**:4076-4079. DOI: 10.1021/jf020002y

[141] Kritzinger EC, Bauer FF, du Toit WJ. Role of glutathione in winemaking: A review. Journal of Agricultural and Food Chemistry. 2013;**61**:269-277. DOI: 10.1021/jf303665z

[142] Adams DO, Liyanage C. Glutathione increases in grape berries at the onset of ripening. American Journal of Enology and Viticulture. 1993;**44**:333-338

[143] Park SK, Boulton RB, Noble AC. Automated HPLC analysis of glutathione and thiol-containing compounds in grape juice and wine using pre-column derivatization with fluorescence detection. Food Chemistry. 2000;**68**:475-480. DOI: 10.1016/S0308-8146(99)00227-7

[144] Janěs L, Lisjak K, Vanzo A. Determination of glutathione content in grape juice and wine by high-performance liquid chromatography with fluorescence detection. Analytica Chimica Acta. 2010;**674**:239-242. DOI: 10.1016/j.aca.2010.06.040

[145] Marchand S, de Revel G. A HPLC fluorescence-based method for glutathione derivatives quantification in must and wine. Analytica Chimica Acta. 2010;**660**:158-163. DOI: 10.1016/j.aca.2009.09.042

[146] Radkowsky AE, Kosower EM, Eisenberg D, Goldberg I. Bimane. 18. (Haloalky1)-1,5-diazabicyclo[3.3.0] octadienediones (Halo-9,lO-dioxabimanes): Sulfur "extraction" by syn-(1–bromoethy l,methyl)bimane. Glutathione "sulfide" from the tripeptide thiol, glutathione. Journal of the American Chememical Society. 1986;**108**:4532-4541. DOI: 10.1021/ja00275a046

[147] Bramanti E, Cavallaro R, Onor M, Zamboni R, D'Ulivo A. Determination of thiolic compounds as mercury complexes by cold vapour atomic absorption spectrometry and its application to wine. Talanta. 2008;**74**:936-943. DOI: 10.1016/j.talanta.2007.07.041

[148] Fracassetti D, Lawrence N, Tredoux AGJ, Tirelli A, Nieuwoudt HH, du Toit WJ. Quantification of glutathione, catechin and caffeic acid in grape juice and wine by a novel ultra-performance liquid chromatography method. Food Chemistry. 2011;**128**:1136-1142. DOI: 10.1016/j.foodchem.2011.04.001

Chapter 12

Characterization of Red Wines from Macaronesia

Jesús Heras-Roger, Carlos Díaz-Romero,
Jacinto Darias-Martín and Domingo Rios-Mesa

Additional information is available at the end of the chapter

http://dx.doi.org/10.5772/intechopen.71795

Abstract

Wines from the Atlantic Islands of Macaronesia come from unusual *terroirs* due to their volcanic soils and the tropical and subtropical climatic conditions from this region. Some of these Islands produce highly appreciated fortified wines traded around the world since the sixteenth century, such as Madeira or Canary. Nowadays their distinct winemaking techniques and sweet wine traditions combine with the production of table wines. Previous studies described peculiar properties in wines from these regions, mostly related with their phenolic content and color, which are particularly important in the less produced red wines. The main purpose of this chapter is to characterize red wines produced in the Atlantic Islands of Macaronesia in terms of oenological and physico-chemical properties. Wines from these islands are extremely atypical, as their climates conditions are exceptional and red grape cultivars are exclusive. Furthermore, specific viticulture techniques are applied in these latitudes to proportionate the unique characteristics outlined in this work. Original experimental data from 300 red wines produced in 8 Atlantic islands from Canary and Cape Verde archipelago and an important reviewing study for Azores and Madeira red wines are considered herein. Results are presented according to archipelago.

Keywords: Macaronesia, red wine, characterization, grape varieties, Canary Islands, Cape Verde Islands

1. Introduction

Macaronesia is a collection of four archipelagos located in the North Atlantic Ocean, in front of the coast of Europe and Africa (see **Figure 1**). The Macaronesian archipelagos from North to South are Azores and Madeira (Portugal), Canary Islands (Spain) and Cape Verde. All the Islands of Macaronesia are volcanic and none of them were part of a continent, so their native plants reached the islands via geographical dispersal but many are endemic [1].

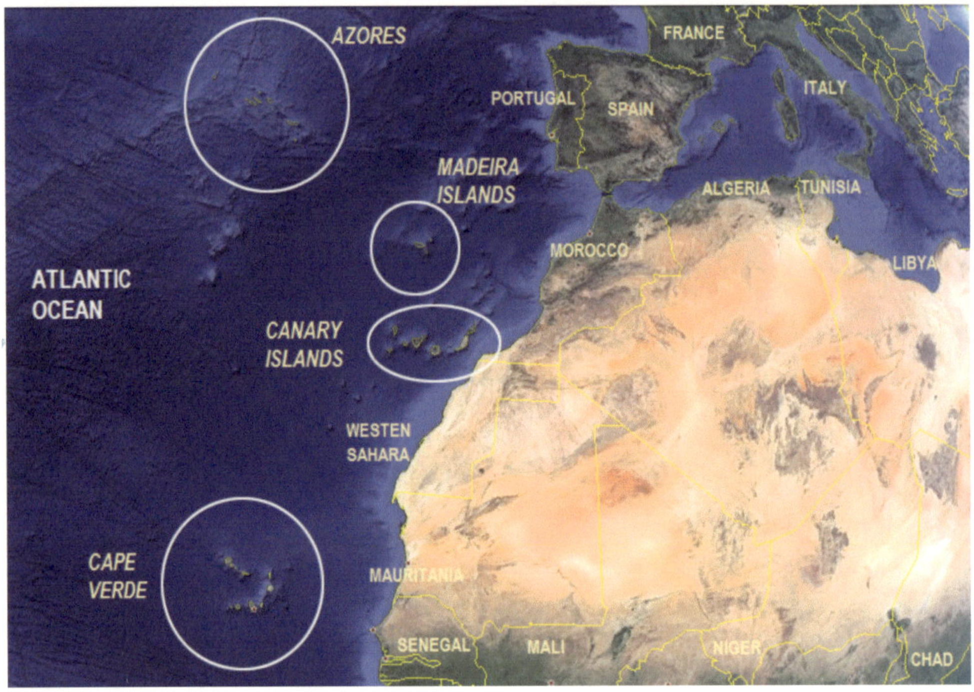

Figure 1. Geographical situation of Macaronesian archipelagos.

Volcanic soils are usually related with highly appreciated wines, as many traditional winemaking regions highlight their volcanic origins like Tokaj in Hungary, Struma valley in Bulgaria, Santorini in Greece or Sicilia in Italy. Nevertheless peculiarities from Macaronesia wines come not exclusively from their volcanic origin, but mostly from their specific *terroirs*. These islands present unusual geographical conditions in terms of soil and location but also atypical traditions and extremely particular microclimates. The climate from these archipelagos ranges from Mediterranean in the Azores and Madeira to arid in some islands of the Canaries (Lanzarote and Fuerteventura) and Cape Verde (Sal and Maio), subtropical in some Canary Islands (Tenerife, La Palma and El Hierro) or even tropical in the case of Fogo in Cape Verde. In general, Portuguese archipelagos enjoy a cooler climate and higher rainfall than Canary Islands and Cape Verde with warmer winters and irregular rainfalls.

Winemaking grapes were introduced in the Macaronesian archipelagos hundreds of years ago by the first Portuguese and Spanish inhabitants. Their volcanic soils are mineral rich and relatively fertile, so it is possible to naturally obtain wines with high acidity. This peculiarity makes most of these islands well suited to elaborate fortified and white wines. Moreover wines from these areas benefit from an acid/sugar balance not easily achieved in other regions. Traditional grape cultivars used in Macaronesia are characterized by a high potential alcoholic content and a valuable acidity ideal to produce fortified wines [2].

Macaronesia archipelagos are similar because of their volcanic origin and oceanic influences but very different between themselves. There is also high contrast inside the archipelagos. Azores, Madeira Island or west Canary Islands (La Palma and El Hierro) are extremely mountainous, with deep valleys and steep slopes where the vines grow on little terraces rich in organic matter. Nevertheless there are also plain arid islands like Porto Santo in Madeira or the east Canary Islands (Fuerteventura and Lanzarote) where grape cultivation is highly adapted. An exceptional climatic contrast is present in Tenerife (Canary Islands), Pico (Azores) and Fogo (Cape Verde), which are the most important wine producers in their, respectively, archipelagos and also islands characterized by steep volcanoes. The fertile terraces from each of these islands volcanoes permit to cultivate quality winemaking grapes from near to the see, as in Pico or Fuerteventura, to very high altitudes with an important day/night temperature contrast as in Tenerife or Fogo.

Macaronesia climatic conditions combined with rugged terrain, the potential for volcanic activity, extreme winds, difficulties for mechanical harvesting and inconveniences derived from fungal attacks do not exactly add up to the most ideal winemaking environment. Moreover vines need warm and dry condition combined with regular winters to grow well, but in these areas humidity from the ocean is constant and winters are extremely mild, therefore grape cultivars have adapted and only pursued the less sensible to fungal diseases.

Red wine is produced nowadays in all Macaronesian archipelagos with several varieties. Most of the production is quality wine bottled by firms with certified origin protection figures, as *DOP* or *Vinho Regional*. Nevertheless, there are also artisanal productions for self-consumption through all the productive islands. The peculiarities from these regions in terms of volcanic soil, hours of sun, humidity and climatic conditions proportionate specific attributes to the red wines produced. This research is particularly original as the vast majority of quality winemaking regions lay between 30 and 50° latitude and there are only a few subtropical and tropical winemaking regions producing red wines. Macaronesian wines have been previously researched, but results are rarely compared.

2. Red wine production in Macaronesia

Macaronesian archipelagos have a long winemaking tradition associated with fortified and white wines, thus these wines are the most studied [3–10]. Furthermore the production of red wine in Macaronesia is uneven. While some islands have a long winemaking history related to fortified wines and started red winemaking just some years ago, other islands produce red wines protected by European quality figures such as DOP for decades now. Moreover some regions export red wine to international markets whereas in other areas it is only produced artisanally for self-consumption. This section briefly presents the most important conditions associated with wine production and the current red winemaking state-of-the-art in each one of the Macaronesian archipelagos from North to South. In general, these regions present relatively mild winters with irregular rainfalls, therefore viticulture has been adapted to adequately handle the sprout of the vine.

2.1. Azores

The archipelago of Azores is an autonomous region of Portugal located on a line between Lisbon and New Jersey, about 1400 km west of Portugal and about 1925 km southeast of Canada. All islands have high mountains, being mount Pico, on the island of Pico, the highest point in Portugal at 2351 m. All populated islands show conditions to elaborate quality wines, but most of the production is found in the islands of Pico, Graciosa and Terceira in the center group and São Miguel in the east group.

The climate of the Azores is mild and moist all year, as it is influenced by the distance to continents and the passing Gulf Stream. Due to the marine influence, temperatures remain mild year-round being climate generally wet and cloudy. The archipelago's mid-Atlantic location means that the greatest threats to the grapes are wind and sea water, not harsh temperatures. Wind effects have been traditionally solved by building peculiar stonewalls of volcanic rocks around the vines; these walls give protection from ocean winds and radiate heat at night. Humidity and cloudy consequences are controlled continuously evaluating the risk of fungal attack [11].

Azores winemaking history starts soon after the settlement of Portuguese conquerors in the mid-fifteenth century. Winemaking traditions from this area are related to fortified wine styles, as it might be expected from a region with its oceanic mild climate. Azores fortified wine resembles Porto and Madeira wines, but its production is characterized by the vines present in the island and the distance from the continent. Azores was affected by the *phylloxera* [12].

There are no *DOP* regions in this archipelago but there are three *IPR* regions: Pico, Biscoitos and Graciosa. Pico and Biscoitos traditionally produced fortified wines but today the three regions are mostly focused on unfortified white wines. Red wine has no tradition, as the cool oceanic climate lend to produce white wines with an adequate natural acidity. Nevertheless red wines are produced as table wines classified as *Açores Vinho Regional* using European well-known cultivars as Syrah, Merlot and Cabernet Sauvignon but also Touriga Nacional, Castelão and Complexa.

2.2. Madeira

The archipelago of Madeira is an autonomous region of Portugal located about 1000 km southwest of Portugal, and about 400 km north of the Canary Islands. In Madeira and Porto Santo Islands, grapes are cultivated to elaborate quality wines. The terrain of the mountainous volcanic island of Madeira is difficult to cultivate and vineyards are planted on terraces of basaltic bedrock. Their climate is oceanic with tropical influences, such as high rainfall. The archipelago enjoys mild temperatures strongly influenced by the ocean throughout the year.

The islands of Madeira have a long winemaking history. Madeira is internationally famous for the eponymous fortified wine produced there and extensively researched [6, 13]. As in most European countries, the *phylloxera* reached the island by the nineteenth century and Madeira vines were hardly affected. Their climatic conditions also mean constant viticulture hazards related to fungal grape diseases.

Not fortified wines production in Madeira have been relatively rare, but in the last decades these volcanic and fertile lands produced red wines that break the traditional fortified wine image of the industry from this region. These wines are labeled as *Terras Madeirenses*, a Portuguese wine region for both islands classified as *Vinho Regional* or table wines for non-fortified wines. The most used red cultivar is Tinta Negra Mole (also known as Negramoll or Mulata), but Bastardo (also known as Trousseau or Maturana Tinta), Complexa and Touriga Nacional are also present in red wine blendings. Some winemakers introduce international varieties such as Cabernet Sauvignon or Merlot.

2.3. Canary Islands

The archipelago of Canary Islands is an autonomous community of Spain located about 100 km west of Africa. Red wine is produced in all the islands with European figures of protection, as the archipelago counts with 10 *DOP*. The largest of the archipelago's islands, Tenerife, is the only island with more than one *DOP* and also home to the majority of the region's production, particularly for export, but most Canary table wines are present in their local markets. These islands are the largest and most populated archipelago of Macaronesia.

These islands have a subtropical climate with long hot summers and moderately warm winters. The absence of low temperatures during winter involves problems for the sprout of the vine, as cold hours are usually not enough to obtain a regular vine cycle. Therefore grape ripening can be extremely irregular even for the same vine plant depending on the microclimate. Moreover precipitation levels and maritime influences are highly dependent on location and elevation. Thus completely fertile green areas with high humidity as well as deserts can be found on the same archipelago and sometimes even in the same island.

Wine-growing in this fertile islands dates back to the Spanish conquest in the fifteenth century. Tenerife was the most important centre of "*Malmsey*" (*Malvasia*) production in the past with a strong reputation for fortified white wines as well as other Macaronesian regions. White sweet wines from these islands were greatly appreciated in England and the States during the sixteenth century, being known as "*Canary*". Rivalry between kingdoms restrained their commerce in the eighteenth century and red winemaking grapes were then introduced to elaborate similar products to Madeira fortified wines, being this fact the first step toward the red winemaking tradition from Tenerife [14].

Canary Islands grape producers' adapted viticulture to their own specific conditions. For instance, in the arid and windy Lanzarote *DOP*, they found an indigenous way of cultivating vines on the arid ground known as *"gerias"*. This grape production system consists on funnel-shaped hollows dug into thick layers of coarse volcanic ashes with only one vine per hollow. Hollows are filled with soil and poured thick layers of volcanic ashes over it, as the porous volcanic granules known as *"picón"* retain the night humidity to feed the plants during the day and concentrates heat at night. Furthermore producers built low semicircular walls around vines to protect them from the constant winds. Another example can be found in *"La Orotava" DOP* with a unique system of vine training called *"cordón trenzado"*, literally meaning braided or plaited cord. This method of vine training evolved as a result of constraints on the areas used for cultivation as the island present limited space. Traditionally, this vine training method allowed the braids to be easily moved and thus allow the soil below them to be used for growing additional crops.

In general vines from Canary Islands are old when compared to other winemaking regions as the *phylloxera* never reached the archipelago. Most vines are not planted on rootstocks but directly on the ground and draw water solely from the terrain as usually they are not irrigated. Mechanical harvesting is impossible given the specific methods of vine training and the small pieces of land. Vineyards are fragmented and can be found near the coast or at elevated altitudes. Most vineyards are planted on the gentle lower slopes of the mountains present where humid mists created by the winds moderate the hot maritime climate.

Red winemaking is important in the Canary Islands but some islands traditionally produce more than others. Most of these red wines are fresh and young because of the natural acidity balance of their soils but some regions are also specialized in oak-aged red wines. These red wines are usually elaborated with indigenous grapes present for centuries in the islands, but international grape varieties (Syrah, Merlot and Cabernet) can also be used. The most widespread red grape cultivar is Listán Negro (also known as Almuñeco), a variety well suited to Beaujolais-like wine style made using carbonic maceration. The workhorse grape of Madeira Tinta Negra is also present in Canary Islands known as Negramoll. The varietal richness of these islands is amazingly high, as many other red grapes can be found exclusively vinified by some winemakers, such as Tintilla, Bastardo (also known as Trousseau), Baboso (also known as Alfrocheiro or Tinta Preta in Cape Verde), Listán Prieto (also known as Mission Grape in the United States of America), Vijariego Negro or Castellana among others. Not just the grapes from ungrafted and often extremely senior vines make Canary red wines atypical. The combination of little-known grapes, high-elevation vineyards and volcanic soils made these red wines distinctive.

Red wine is produced in the 10 *DOP*. Tenerife houses half of the region's *DOP* with Abona, Tacoronte-Acentejo, Valle de Guimar, Valle de la Orotava and Ycoden-Daute-Isora. The remaining designations cover other islands in their entirety such as El Hierro, Gran Canaria, La Gomera, La Palma and Lanzarote *DOP*s. There is also a general *DOP* including all islands (also Fuerteventura). Each area has a unique microclimate and different soil composition, lending to distinctive wines with mineral notes. Some areas even present clear subzones inside the same *DOP*. For instance, only in the North area of La Palma Island a peculiar red wine called *"Vino de Tea"* is produced. These wines are sold straight from atypical barrels not made from oak but from *pinus canariensis.* These traditional red wines share organoleptical characters with Greek retsina wines, as their introduce pine in their winemaking techniques giving them a peculiar resin aroma.

2.4. Cape Verde

The archipelago of Cape Verde is a sovereign nation located about 650 km west of the coast of Africa and about 1300 km south of Canary Islands. Fogo is the only island producing grapes in sufficient quantity to elaborate quality wines because it registers the greatest precipitations. Fogo is characterized by its particular steep volcano covered with ash where vegetation and cultivation of the vine is extremely complicated.

As in the rest of Macaronesia, the culture of vines was started by the first settlers during the sixteenth century. Fogo wine was even exported to Guinea and Brazil during the eighteenth century. Wines from this area are mostly white as grapes cultivated there produce refreshing wines with pleasant mineral notes. There is also a strong tradition of fortified sweet wines locally known as *"Pasito"*, as the homonym wine from Italy. Red wines are not the most produced in the island but they can be easily found in restaurants and local markets even though they have no origin protection figure. In fact wine tourism is taking an important role in this island [15, 16].

The importance of latitude and altitude is one of the most relevant aspects when it comes to understanding the *terroir* from this region. Fogo is located much closer to the equator than most of the world's key wine-growing regions and its vineyards are generally more than 1500 m above sea level. These conditions temper the high temperatures and the heat associated with its latitude. Cooler nights delay grape ripening and sunlight absorbed by black soils during the day keeps the vines isolated at night. In this sense, climate cannot be strictly described as tropical. Due to its altitude and high irregular rainfall, it is possible to obtain quality wines, even with the possibility of a second harvest in the month of January given the natural vine cycle in this environment.

In Fogo, vine cultivation is not exclusive and is normally associated with other products such as corn or sweet potatoes. At the foot of the volcano is the wine-growing community of Chã das Caldeiras where most of the samples from this study come from. Chã is located at 1700 m altitude and characterized by small vineyards around the active volcano erupted several times recently (1995, 2000 and 2014). This area enjoys a microclimate with thermal amplitude, good rainfall and rich volcanic soils, being all these factors important for quality winemaking production using adequate viticulture practices [17]. Oidium is the only fungal disease detected and the *phyloxera* never reached Cape Verde.

A range of European grape varieties are cultivated in the Island. The most used red cultivar is Tinta Preta (also known as Baboso Negro, Bruñal or Alfrocheiro Preto), but also other red cultivars are present such as Castelão (also known as Periquita), Bastardo (also known as Trousseau or Maturana Tinta) and Moscatel Negro (also known as Black Hamburg).

Most of the samples from this study are red wines from the vineyards located around the caldera and produced by Chã das Caldeiras wine cooperative. These wines are mainly elaborated Tinta Preta and destined to early consumption without periods of guard. In addition to the bottled quality wines from Chã our study also consider sweet red wines produced artisanally from other vineyards situated on the volcanic soils inside the caldera. These wines are less representative, produced in small volumes and mostly destined for self-consumption soon after the fermentation. These traditional homemade wines are known as *"manecom"*. *"Manecoms"* can be both white or red but sweet red wines are the most popular and the only ones considered in our study. Although homemade, some of these wines are intended for both personal consumption and to be marketed in bulk.

In addition to Chã Caldeiras winery whose vineyards are in the caldera, there are other wineries in the island like Monte Barro or Achada Grand producing bottled red wines. Most probably these wines differ in the physico-chemical characteristics due to the differences in the vine cultivation area, related to both altitude (1700 m vs. 700 m) and rainfall.

3. Macaronesian red wine characterization

3.1. Material and methods

Experimental results from 300 red wines produced in Canary Islands and Fogo (Cape Verde) are described herein together with scientific literature from red wines produced in Azores and Madeira. The 250 samples from Canary Islands and 40 red wines from Fogo were bottled as quality wine destined to the regional market. The remaining 10 samples of red wine from Fogo were *"manecom"*; this means red wine elaborated for self-consumption.

Samples come from different harvest (2003–2014). Most analysis was performed using International Organisation of Vine and Wine (OIV) reference methods [18]. A phenolic and metallic profile was also obtained. Color characteristics using CIELab and magnitude of co-pigmentation are important in red wines and thus were analytically quantified. **Table 1** presents the methods applied with the uncertainty associated to each parameter. All data have been grouped and compared according to the archipelago of origin introducing similar data from published scientific works when available. Significant differences between islands from the same archipelago or even between regions from the same island have been previously detailed in the literature due to particular microclimates influences. Nevertheless differences inside archipelagos are not considered in the present chapter as they are not part of the aim of this study.

3.2. Oenological parameters

Results for the most important oenological parameters in red wines are summarized in **Table 1**. Canary wines are characterized by a great variability and heterogeneity, as these wines come from different harvest and are elaborated with different varietals in various

Wine characteristics	Canary (bottled with DOP)		Cape Verde/Fogo	
	(n = 250)	(n = 60) [19, 20]	Chã Caldeiras bottled (n = 40)	"Manecom" (n = 10)
Alcoholic content (%v/v)	13.72 ± 1.20 (11.00–19.54)	12.28 ± 0.60	15.58 ± 0.61 (14.70–17.66)	16.67 ± 0.87 (16.06–17.29)
Density (g/cm³)	0.9941 ± 0.006 (0.9893–1.0513)	0.9922 ± 0.001	0.9896 ± 0.0116 (0.9370–0.9944)	1.0130 ± 0.0096 (1.0062–1.0198)
pH	3.74 ± 0.18 (3.23–4.54)	3.55 ± 0.17	3.99 ± 0.18 (3.46–4.23)	3.94 ± 0.02 (3.93–3.95)
Titrable acidity (g. tart./l)	5.16 ± 0.72 (3.93–7.59)	5.10 ± 0.54	4.95 ± 0.71 (4.15–6.62)	7.89 ± 1.61 (6.75–9.03)
Ethanal (mg/l)	16 ± 39 (n.d.–246)	30 ± 24	30 ± 45 (n.d.–150)	50 ± 65 (15–310)
Glucose + fructose (g/l)	1.90 ± 6.90 (n.d.–59.00)	1.95 ± 0.95	1.70 ± 2.21 (n.d.–5.9)	43.05 ± 31.75 (20.60–65.5)
Free SO_2 (mg/l)	16.0 ± 8.0 (5.0–48.0)	16.0 ± 9.0	24.4 ± 13.8 (7.0–47.0)	1.5 ± 0.71 (1.0–2.0)
Total SO_2 (mg/l)	78.0 ± 34.0 (11.0–184.0)	57.0 ± 32.0	83.4 ± 27.6 (26.0–129.0)	12.5 ± 3.5 (10.0–15.0)
Nitrogen ammonia (g/l)	38.0 ± 29.0 (n.d.–207.0)	–	27.6 ± 19.1 (10.3–59.8)	53.3 ± 20.4 (20.1–60.8)
Glycerol (g/l)	11.0 ± 2.9 (0.4–20.2)	–	7.6±0.3 (7.4–7.8)	8.0 ± 0.1 (7.8–8.2)
IPT DO 280 nm (UA)	52.7 ± 12.9 (14.6–103.1)	35.0 ± 5.6	68.3 ± 8.7 (56.7–89.3)	70.6 ± 16.2 (59.1–82.0)
Folin–Ciocalteu Index	50 ± 11 (27–83)	–	62 ± 15 (32–93)	32 ± 25 (15–76)
Total tannins (g/l)	2.00 ± 0.80 (0.20–4.40)	1.30 ± 0.40	4.78 ± 0.61 (3.97–6.25)	4.94 ± 1.13 (4.14–5.74)
Total anthocyanins (mg/l)	296.1 ± 178.8 (29.2–798.4)	108.2 ± 55.2	350 ± 202 (33.7–815.3)	123.1 ± 198.9 (12.2–489.5)

Results expressed as X ± σ (Min.–Max. value if available). n.d. = not detected.

Table 1. Conventional red wine parameters from Canary and Cape Verde islands.

islands. Most red wines considered were produced to be consumed in the short term, so they have typical young wines values in terms of phenolics. Nevertheless some samples present high potential for aging, as they are distinguished by high total phenol index (IPT).

All samples conformed to the established legal wine standards as far as analytical parameters are concerned. Canary Islands results are within the usual ranges and comparable to those published for the same geographic origin [19, 20], although pH is slightly higher than usual [21]. This can be due to their characteristic soils and warm climate conditions, as it have been previously demonstrated that wines produced in warm areas trend to present higher pHs than those produced in colder regions [22]. Moreover the tannin content is also higher than the values described for French vines [21], although this can be due to an over-quantification, since the method counts with limited specificity.

Most wines from Cape Verde present high ethanol content, probably due to the natural conditions affecting this archipelago and to late grape picking. The lowest ethanol content in Fogo attained 14.7%vol, higher than the average in Canary. *"Manecoms"* show an alcoholic graduation even higher than bottled wines, probably because Fogo inhabitants are used to wines with high alcoholic content.

Fogo wines also present unusual high pHs. These values lead to color losses in red wines due to anthocyanins chemical form and their greater oxidation [23]. High pH also reduces the protective action from SO_2. pH depends on various factors from soil, climate or grape ripening to winemaking techniques and the values observed might be due to regional characteristics, such as soil potassium content, climate and skins macerations. In this sense Fogo winemakers could diminish wine pH by reducing maceration periods, with earlier harvest or introducing specific rootstocks. The relationship between high pH in these wines and potassium content are further discussed in the metallic profile section.

The residual nitrogen ammonia for most *"manecoms"* samples is high in relation with conventional levels [21]. This nitrogen bioavailability leads to a greater ease of wine diseases increased by their high pH and low SO_2 protection. Nevertheless Fogo wines avoid many microbiological problems due to their high alcoholic content. In general, these wines contain low levels of ethanol, as most samples are not subjected to micro oxygenation. Their high pHs discourage winemakers to elaborate oak-aged wines, as color stability would be hardly affected by time [24]. Red wines from both archipelagos were also characterized by high glycerol values in agreement with their high alcohol content.

3.3. Acid profile

Wine acids are classified in two groups. The first one includes those whose concentrations evolve during grape ripening, such as tartaric, malic and citric acid. In the second group are those derived from fermentations such as lactic, acetic and succinic acid. Gluconic acid is out of this classification because of its fungal origin. These acids content is detailed on **Table 2**.

None bottled wine exceeds the legislated 1.2 g/l value for volatile acidity. Those presenting greater values correspond to sweet wines with greater limits. *"Manecoms"* show extremely high values due to their artisan winemaking. The initial malic acid content in wine depends on grape ripening, but its concentration drastically reduces after malolactic fermentation. This fermentation was completed in most red wines, even though some samples present high values above 4 g/l. These wines acid content is within the values considered normal [21], although some Canary wines present lactic acid contents above the average.

The very low gluconic acid in Fogo wines reveals none important fungal incidences, probably due to the extremely dry climatic conditions derived from latitude and altitude. Nevertheless

	Canary (bottled with DOP)		Cape Verde/Fogo		Madeira
	(n = 250)	**(n = 60) [19]**	**Chã Bottled (n = 40)**	***"Manecom"* (n = 10)**	**(n = 12) [25]**
Tartaric acid (g/l)	2.50 ± 1.00 (1.10–4.90)	2.08 ± 0.42	1.60 ± 0.61 (1.20–2.30)	1.95 ± 0.92 (1.30–2.60)	1.97 ± 0.33 (1.50–2.73)
Lactic acid (g/l)	1.78 ± 1.00 (0.07–5.53)	–	2.40 ± 0.26 (2.20–2.70)	1.35 ± 1.39 (0.10–2.70)	–
Acetic acid (g/l)	0.60 ± 0.24 (0.15–1.53)	0.50 ± 0.14	0.81 ± 0.14 (0.53–1.19)	2.09 ± 1.56 (0.99–3.20)	–
L-malic acid (g/l)	0.38 ± 0.71 (n.d.–4.21)	0.24 ± 0.34	0.56 ± 1.11 (0.10–3.72)	0.27 ± 0.31 (0.05–0.50)	–
Gluconic acid (g/l)	0.36 ± 0.42 (n.d.–2.29)	–	0.05 ± 0.05 (n.d.–0.1)	0.02 ± 0.04 (n.d.–0.10)	–
Citric acid (mg/l)	140 ± 102 (n.d.–620)	–	70 ± 42 (n.d.–150)	30 ± 12 (n.d.–50)	445 ± 220 (141–707)

Results expressed as X ± σ (Min. – Max. value if available). n.d. = not detected.

Table 2. Red wine acid profile from canary, Cape Verde and Madeira islands.

grapes fungal attacks are common in Canary Islands according to our results. Madeira's citric acid content is unusually high and extremely low in Fogo. In Canary Islands, the citric acid content highly varies probably due to the different elaborations considered.

3.4. Metallic content

Some scientist previously addressed wine metallic differentiation depending on the Canary Island of origin [26–28] and studied the content from Azores and Madeira wines [29]. Differences between *DOP* and islands are in some cases highly significant. The cited references are useful to study these specific profiles but this is not our aim, as it introduces other considerations. **Table 3** resumes experimental data and average results from literature.

The metallic content conforms to commercial standards though potassium concentration is slightly higher than values previously reported [30]. Its source in wine is diverse; on one hand the addition of potassium during winemaking can increase it, but soil content due to other crops enrichment or fertilizers is also relevant. However, soil characteristics considered alone cannot explain its high content. It can be also increased by winemaking techniques, such as excessive macerations with grape skins for instance. Most probably its concentration is related to the grapevine needs, as in warm, dry and windy climates vine plants need more water as it is constantly evaporating from their leaves. Therefore grapevines breathe intensively in order to absorb soil water in a greater extent than those plants located in cold or humid climates. Greater water absorption contributes to a greater mineral absorption. Thus in warm climates plants absorb more water and by addition more potassium, concentrating

	Canary (bottled with DOP)		Fogo Chã (n = 40)	Madeira (n = 36) [29]	Azores (n = 28) [29]
	(n = 250)	(n = 249) [30]			
K	1428 ± 459 (531–3727)	835 ± 333	1270 ± 247 (988–1849)	936 ± 215	923 ± 217
Na	98 ± 57 (19.0–351.2)	92.0 ± 38.2	23.6 ± 3.6 (17.1–29.0)	51.0 ± 30.1	57.7 ± 30.5
Ca	69.3 ± 22.1 (24.1–83.2)	68.7 ± 15.0	50.5 ± 35.4 (30.5–90.3)	73.1 ± 23.8	80.5 ± 15.7
Mg	129.0 ± 27.0 (65.0–263.1)	84.1 ± 15.4	115.0 ± 40.0 (75.0–125.6)	101.2 ± 13.9	112.5 ± 41.8
Fe	1.7 ± 1.0 (0.3–7.4)	2.6 ± 1.1	2.6 ± 1.9 (0.9–7.3)	2.3 ± 1.4	1.5 ± 1.0
Cu	0.20 ± 0.70 (n.d.–6.70)	0.17 ± 0.20	0.02 ± 0.01 (n.d.–0.1)	0.63 ± 0.66	0.22 ± 0.41
Zn	0.26 ± 0.15 (n.d.–0.53)	0.56 ± 0.31	0.18 ± 0.10 (0.10–0.20)	1.02 ± 0.60	0.63 ± 0.55
Mn	1.3 ± 0.7 (n.d.–5.1)	1.0 ± 0.4	0.4 ± 0.1 (0.3–0.5)	1.9 ± 0.8	0.7 ± 0.2

Results expressed as X ± σ (Min. – Max. value if available). n.d. = not detected.

Table 3. Metallic content (mg/l) from canary, Cape Verde, Madeira and Azores islands.

this metal in grapes and latter in wine. The greater or lesser concentration might depend on the climatic conditions of each harvest, on the grapevine situation as greater winds or dryness enhances it, and even on the vine itself, as roots absorb potassium differently depending on the cultivar. This climate influence might explain the potassium increase in Canary Islands wine, as the literature cited is almost 15 years old and today vines are exposed to warmer conditions due to the temperature increase from the last decade.

Potassium concentrations correlate with high wine pHs, a peculiar characteristic already observed in the previous section for Fogo wines. Since high pHs lead to a greater wine sensitivity to microbiological diseases, it is advisable to control potassium content. Obviously potassium is not the only factor conditioning the final wine pH, although it is one of the most important according to our experimental results.

The sodium content for all Macaronesian regions except Fogo is higher than those from other origins [31], probably due to oceanic winds and the marine aerosol. Fogo samples contain significantly less than most Canary, Azorean and Madeiran red wines because of the high altitude of Chã vineyards. In fact Fogo sodium content is similar to continental wines, where the marine aerosol is non-existent and its origin is essentially due to soil and agronomic practices. Some Canary red wines also presented low sodium content when compared with most samples from the same region, coming these wines from Tenerife vineyards located at high altitudes, which are presumably less influenced by oceanic winds.

The content of iron is noteworthy in all the archipelagos probably because of their volcanic soils, as it is difficult to suggest any other cause affecting all islands. Copper concentration is unusually variable between regions, relatively high in Madeira but extremely low in Fogo. Similarly, zinc and manganese content are low in Fogo wines when compared to other Macaronesia regions. Calcium and magnesium content are similar between archipelagos. The different concentration of all these minerals might be an interesting source for wine characterization according to geographical area using them as potential origin markers.

3.5. Volatile compounds

Wines from Madeira are mostly characterized by 2-ethylhexan-1-ol, 3,5,5-trimethylhexan-1-ol, ethyl 2-methylbutanoate, ethyl DL-2-hydroxycaproate, decanoic acid and 2-ethoxythiazole, whereas wines from Azores and Canary Islands are mainly characterized by 3-ethoxypropan-1-ol, 1-octen-3-ol, (Z)-3-hexenyl butanoate, 2,3-dihydrobenzofuran and 4-(methylthio)-1-butanol [32]. This latter compound is particularly important in Fogo, where red wines present an unusual high concentration of heavy sulfur volatiles in its aroma [33]. These concentrations might be due to the active volcano where grapevines are located. Fogo wines also present the highest concentration in the minor volatiles fatty acids. Results for the most common volatile compounds in red wines are summarized in **Table 4**.

Higher alcohols followed by ethyl esters and fatty acids are the most predominant chemical groups among the volatiles from red wines elaborated in Macaronesia. In fact hexanol-derived compounds with decaonic acids and whiskey lactones discriminate Madeira red wines from the rest of Macaronesian archipelagos [32] and a characteristic profile for alcohols, ethyl esters and fatty acids has been obtained for these wines [13]. Similarly, the concentration of esters in Azores wines revealed to be significantly lower than in red wines from the other Macaronesian archipelagos. The unique exception is the ethyl hexanoate compound, which seems to be present in a greater extent than in other islands [33–36].

Canary wines studies conclude that 3-metyl-1-butanol, 2,2-butanodiona and ethyl butirate were the main odorants [37, 38]. Moreover specific studies have been conducted to analyze the volatiles from red wines aged in pine casks from La Palma Island [39].

3.6. Phenolic content

Phenolic compounds are responsible for the most important sensory attributes of red wines. Phenolic content depends on the winemaking, which are influenced by factors such as variety, ripening, cultivation techniques and climatic conditions among other considerations [40]. Thus the phenolic composition of Macaronesian red wines is potentially different from other regions because of their specific *"terroir"*. Gallic and caftaric acids are the most concentrated hydroxybenzoic and hydroxycinnamic acid respectively in Canary wines [41] but there is high heterogeneity. Previous studies for the same region confirmed that these wines polyphenolic composition is highly heterogenic and depends on various factors such as local climate

Compound group	Volatile compound	Canary (n = 6) [34]}	Fogo (n = 4) [33]	Madeira (n = 48) [35]}	Azores (n = 3) [36]
Alcohols (μg/l) (50–60%)	Hexanol	–	0.214 ± 0.011 (0.167–0.256)	8.359 ± 0.469 (7.604–8.779)	0.076 ± 0.03 (0.042–0.1)
	Benzyl alcohol	–	0.131 ± 0.01 (0.032–0.184)	0.137 ± 0.007 (0.106–0.224)	0.113 ± 0.129 (0.022–0.204)
	Phenylethanol	1.086 ± 0.122 (0.961–1.313)	7.320 ± 0.739 (6.250–8.300)	12.348 ± 1.011 (9.706–18.005)	0.745 ± 0.181 (0.574–0.935)
	2-(Methylthio) ethanol	0.139 ± 0.012 (0.091–0.175)	54.333 ± 10.204 (25.500–74.500)	0.002 ± 0.001 (0.001–0.005)	–
	Linalool	2.522 ± 0.678 (0.151–5.021)	3.664 ± 0.461 (0.475–6.301)	–	0.147 ± 0.031 (0.120–0.180)
Esters (mg/l) (30–40%)	Ethyl lactate	–	4.138 ± 0.710 (2.250–6.370)	5.300 ± 0.175 (3.088–7.206)	0.004 ± 0.001 (0.003–0.005)
	Ethyl hexanoate	3.348 ± 0.288 (2.751–4.560)	2.383 ± 0.155 (2.050–2.760)	1.788 ± 0.404 (1.544–2.271)	4.402 ± 1.615 (2.55–5.516)
	Diethyl succinate	–	1.479 ± 0.088 (0.957–2.180)	0.566 ± 0.041 (0.184–1.049)	0.005 ± 0.002 (0.003–0.007)
	Ethyl octanoate	1.223 ± 1.014 (0.848–2.351)	2.530 ± 0.133 (2.150–2.930)	34.506 ± 6.598 (20.069–54.823)	0.186 ± 0.119 (0.066–0.304)
	Phenylethyl acetate	0.945 ± 0.451 (0.316–1.641)	1.565 ± 0.104 (1.160–1.770)	3.584 ± 0.113 (2.993–5.090)	–
	Ethyl decanoate	1.184 ± 0.751 (0.561–1.905)	3.985 ± 0.152 (3.300–4.730)	0.248 ± 0.014 (0.181–0.380)	0.197 ± 0.025 (0.176–0.224)
	Isoamyl acetate	–	2.048 ± 0.091 (1.350–2.660)	0.542 ± 0.343 (0.491–0.654)	0.038 ± 0.019 (0.022–0.059)
Fatty acids (μg/l) (5%)	Butanoic acid	0.065 ± 0.154 (0.026–0.073)	–	0.026 ± 0.004 (0.019–0.040)	0.100 ± 0.127 (0.010–0.190)
	Decanoic acid	0.194 ± 0.088 (0.098–0.286)	2.750 ± 0.781 (2.500 – 3.000)	0.694 ± 0.080 (0.098–1.607)	–
	Hexanoic acid	9.674 ± 0.546 (8.486–11.056)	n.d.	1.537 ± 0.208 (1.486–1.625)	0.123 ± 0.145 (0.030–0.290)

Results expressed as X ± σ (Min. – Max. value if available). n.d. = not detected.

Table 4. Volatiles compounds from Canary, Cape Verde, Madeira and Azores islands.

	Canary wines		Fogo (n = 4) [33]	Madeira		Azores (n = 3) [36]
	(n = 250) [41]	(n = 55) [43]		(n = 12) [25]	(n = 5) [44]	
Gallic acid	41.8 ± 24.0 (3.6–125.8)	21.1 ± 11.5 (5.6–44.7)	–	–	398.1 ± 34.5 (341.2–429.0)	–
Syringic acid	7.9 ± 2.8 (2.1–20.4)	2.0 ± 0.8 (0.9–4.0)	10.8 ± 3.4 (6–13.7)	–	18.6 ± 8.8 (4.8–28.6)	–
Coumaric acid	9.7 ± 9.8 (<0.9–67.8)	2.1 ± 2.3 (0.1–6.8)	10.9 ± 5.5 (7.4–19.1)	–	8.1 ± 5.4 (4.5–16.1)	–
Catechin	85.9 ± 29.9 (6.6–199.7)	20.2 ± 8.5 (9.4–38.4)	6.9 ± 2.5 (3.8–10.0)	0.7 ± 0.1 (0.5–0.9)	–	–
Resveratrol	5.1 ± 3.0 (<0.7–13.3)	3.3 ± 1.1 (0.2–5.7)	–	0.3 ± 0.1 (0.2–0.5)	24.5 ± 25.4 (4.5–57.7)	4.2 ± 2.1 (2.8–5.7)
Quercetin	2.8 ± 2.5 (n.d.–13.7)	17.5 ± 11.48 (1.9–49.8)	4.1 ± 0.5 (3.4–4.5)	–	55.7 ± 18.0 (35.9–79.1)	5.6 ± 4.1 (3.0–12.0)
Malv.3gluc.	92.7 ± 79.4 (1.5–371.2)	–	68.5 ± 39.6 (19.6–116.0)	–	–	–

Results expressed as X ± σ (Min. – Max. value if available). n.d. = not detected.

Table 5. Phenolic compounds (mg/l) from Canary, Fogo, Madeira and Azores islands.

and vinification conditions [42]. Results for the most common phenolic compounds in red wines are summarized in **Table 5** for all Macaronesian archipelagos.

Quercetin is the most concentrated flavonol and the content of this compound group, that is flavonols, in Azores [36], Madeira [44], Fogo [33] and Canary [41] is unusually higher than in other winemaking areas [45, 46]. Stilbenes content in Macaronesia red wines has been also described as higher than in continental wines [25, 36, 43] but similar to those from Greek red wines [47]. These differences may be due to the sunny climate of Macaronesia. Flavonol and stilbene content depends on sun exposure because these phenolics protect against solar radiation, thus Macaronesian vines might increase their synthesis to combat UV radiation [48]. Anthocyanins content is also in the upper quartile when compared with other winemaking regions probably because of these solar radiation influences, which are more important in the most southern islands. In fact it has been considered that the content of catechins and proanthocyanidins in Canary wines is higher than in Madeira red wines [44].

3.7. Color

Phenolic reactions are responsible for the colorimetric changes observed while wine aging [24]. **Table 6** summarizes the main colorimetric indexes for Canary and Fogo red wines. Wines from both archipelagos present high chromacity, as the average color intensity for Canary is 9 Units of Absorbance (U.A.) and *"manecom"* even achieves 17.7 U.A. Hue color values lead to low perceptions of oxidation.

	Canary (n = 250)	Fogo Chã (n = 40)	*Manecom* (n = 10)
Color intensity $A_{420} + A_{520} + A_{620}$ (U.A.)	9.00 ± 3.60 (1.41–24.15)	10.73 ± 4.00 (2.10–18.00)	17.67 ± 0.80 (17.10–18.23)
Hue color (A_{420}/A_{520})	0.75 ± 0.16 (0.51–1.28)	0.69 ± 0.15 (0.41–0.99)	0.73 ± 0.11 (0.48–0.105)
Lightness, L* (C.U.)	19.90 ± 10.12 (1.42–68.02	6.87 ± 2.58 (3.87–10.14)	1.07 ± 1.10 (0.29–1.85)
Chroma, C* (C.U.)	53.71 ± 11.12 (9.91–70.60)	36.15 ± 6.68 (27.14–43.39)	7.41 ± 7.69 (1.98–12.85)
Hue angle, h* (C.U.)	28.91 ± 6.51 (13.72–44.03)	18.56 ± 3.95 (14.24– 23.62)	14.45 ± 0.15 (14.35–14.56)
Redness-greenness axis a* (C.U.)	41.61 ± 12.83 (5.40–60.43)	34.09 ± 5.58 (26.31–39.75)	7.18 ± 7.45 (1.92–12.45)
yellowness-blueness axis b* (C.U.)	29.81 ± 11.83 (2.52–56.01)	11.82 ± 4.42 (6.68–17.39)	1.84 ± 1.90 (0.50–3.18)
Saturation, S (C.U.)	3.52 ± 1.22 (0.61–7.33)	5.61 ± 1.16 (4.28–7.01)	6.91 ± 0.07 (6.86–6.96)
Co-pigmentation (parts per unit)	0.18 ± 0.11 (0.05–0.41)	0.11±0.09 (0.02–0.35)	0.09 ± 0.08 (0.02–0.25)
Polymeric pigments (parts per unit)	0.41 ± 0.15 (0.06–0.80).	0.40 ± 0.14 (0.10–0.60)	0.49 ± 0.12 (0.19–0.72)

Results expressed as X ± σ (Min.–Max. value if available).

Table 6. Colorimetric red wines parameters from canary and Cape Verde islands.

The minimum values for coordinates a* and b* in CIELab Units (C.U.) are lower than those obtained in wines from the mainland of Spain [49]. Chroma (C*) and saturation (S*) reveal great heterogeneity. Fogo red wine lightness (L*) is significantly lower than Canary red wines giving a darker color, probably due to a higher anthocyanin extraction derived from the wine-making techniques applied to maximize color and alcoholic content.

One of the most important pigment interactions is co-pigmentation, which occurs when anthocyanic pigments associates with itself or with another substance known as co-pigment, these compounds can be very variable, from other colorless anthocyanin to phenolic acids or flavonols depending on the red wine considered [40]. The percentage of co-pigmentation in parts per unit provides an estimation of co-pigments and their influence in color. The average co-pigmentation in Canary wines is 18 ± 11% of color, ranging from 0.5 to 40%. Co-pigmentation in Fogo wines is lower in disagreement with studies where warm regions produced wines with significantly higher co-pigmentation than cold areas [50]. This low co-pigmenation is related with the high amount of polymeric pigment present in Fogo red wines.

The percentage of polymeric pigments in the young Fogo wines is similar to the percentage quantified for Canary wines where also oak-aged red wines are considered. Short-aged red wines do not normally present so many polymeric pigments. Its content in Fogo wines might be related to late harvest, excessive grape ripening and peculiar winemaking techniques which maximize oxidation and thus increase phenolic polymerizations during winemaking and storage.

4. Conclusion

Macaronesian red wines present peculiar characteristics and wide diversity. Their detailed study is still a current research issue with a promising future as these winemaking regions are mostly known because of fortified and white wines. Their atypical *terroir* combined with traditions and exclusive grape cultivars makes them a prospective red winemaking region. Red wines from Macaronesia can be grouped according to archipelago of origin but are also highly heterogeneous due to microclimates. Significant differences have been obtained in terms of oenological properties, acids, minerals, volatiles, phenolics and color influencing their organoleptic peculiarities. These wines are highly affected by the unusual circumstances of their regions making them unique. For instance, their volcanic soil might increase the iron concentration, whereas oceanic winds modify their sodium content. Moreover the warm and dry climates from these islands determine higher water needs from grapevines, which carry out a greater potassium absorption and thus a greater wine content increasing pH. Furthermore solar radiation involves a higher flavonols and stilbenes biosynthesis in grapes. In summary, Macaronesian red wines need further research but most probably their valorization as atypical food products from an exclusive environment would be enhanced during the coming years.

Author details

Jesús Heras-Roger[1*], Carlos Díaz-Romero[2], Jacinto Darias-Martín[2] and Domingo Rios-Mesa[1]

*Address all correspondence to: jherasr@gmail.com

1 Excelentísimo Cabildo Insular de Tenerife, Tenerife, Spain

2 Departamento de Ingeniería Química y Tecnología Farmacéutica, Área de Ciencia y Tecnología de Alimentos, Universidad de La Laguna, La Laguna, Spain

References

[1] Petit J. Chapter 5. Macaronesia. In: Petit J, Prudent G, editors. Climate Change and Biodiversity in the European Union Overseas Entities. Switzerland. IUCN; 2010. pp. 122-135. https://portals.iucn.org/library/sites/library/files/documents/2010-064.pdf

[2] Marques JC. Vinhos da Macaronésia: conhecer e diferenciar para competir. Seminário Vitivinicultura Atlântica. Available from: https://vitiviniculturaatlantic.0a.picowines.net

[3] Câmara JS, Alves MA, Marques JC. Changes in volatile composition of Madeira wines during their oxidative ageing. Analytica Chimica Acta. 2006;**563**(1):188-197. DOI: 10.1016/j.aca.2005.10.031

[4] Câmara JS, Marques JC, Alves MA, Silva Ferreira AC. 3-Hydroxy-4,5-dimethyl-2 (5H)-furanone levels in fortified Madeira wines: Relationship to sugar content. Journal of Agricultural and Food Chemistry. 2004;**52**(22):6765-6769. DOI: 10.1021/jf049547d

[5] Reader H, Dominguez M. Fortified wines: Sherry, port and Madeira. Fermented Beverage Production: Springer. 1995:159-207. DOI: 10.1007/978-1-4757-5214-4_7

[6] Pereira V, Albuquerque F, Cacho J, Marques JC. Polyphenols, antioxidant potential and color of fortified wines during accelerated ageing: The Madeira wine case study. Molecules. 2013;**18**(3):2997-3017. DOI: 10.3390/molecules18032997

[7] Pereira V, Albuquerque F, Ferreira A, Cacho J, Marques J. Evolution of 5-hydroxymethylfurfural (HMF) and furfural (F) in fortified wines submitted to overheating conditions. Food Research International. 2011;**44**(1):71-76. DOI: 10.1016/j.foodres.2010.11.011

[8] López R, Ortín N, Pérez-Trujillo JP, Cacho J, Ferreira V. Impact odorants of different young white wines from the Canary Islands. Journal of Agricultural and Food Chemistry. 2003;**51**(11):3419-3425. DOI: 10.1021/jf026045w

[9] Pérez-Olivero S, Pérez Trujillo J, Conde J. Minor volatile compounds in white wines from Canary Islands, Madeira, and Pico (Azores) by headspace solid-phase microextraction-gas chromatography-mass spectrometry: A qualitative study. ISRN Analytical Chemistry. 2013. Article ID 529306, 9 pages. DOI: 10.1155/2013/529306

[10] Darias-Martín JJ, Andrés-Lacueva C, Díaz-Romero C, Lamuela-Raventós RM. Phenolic profile in varietal white wines made in the Canary Islands. European Food Research and Technology. 2008;**226**(4):871-876. DOI: 10.1007/s00217-007-0609-9

[11] Zorman M, Lopes DJH, Martins JO, Prendes C, Lorenzo CD, Cabrera R. Incidence of downy mildew *Plasmopara viticola* (Berk. Et Curtis ex. de Bary) Berl. Et de Toni in Terceira island, Azores. Revista de Ciências Agrárias. 2008;**31**(2):134-138

[12] de Sequeira MM, Santos-Guerra A, Jarvis CE, Oberli A, Carine MA, Maunder M, et al. The Madeiran plants collected by Sir Hans Sloane in 1687, and his descriptions. Taxon 2010;**59**(2):598-612

[13] Câmara JS, Alves MA, Marques JC. Multivariate analysis for the classification and differentiation of Madeira wines according to main grape varieties. Talanta. 2006;**68**(5):1512-1521. DOI: 10.1016/j.talanta.2005.08.012

[14] Francisco-Ortega J, Santos-Guerra A. Early evidence of plant hunting in the Canary Islands from 1694. Archives of Natural History. 1999;**26**(2):239-267. DOI: 10.3366/anh.1999.26.2.239

[15] López-Guzmán T, Castillo Canalejo AM, Cerezo López JM. Analysing the potential of wine tourism of the island of Fogo (Cape Verde). Tourismos. 2014;**9**(2):241-258

[16] Castillo Canalejo A, López Guzmán T. Wine tourism and economic development. A case study in Green Cape (Africa). Papeles de Geografia. 2011;**53/54**:65-76

[17] Mota Gomes A. A problemática da Geologia e dos Recursos Hídricos na Ilha do Fogo. Relatório inédito, Praia, Cabo Verde 2006

[18] Oiv. Compendium of International Methods of Wine and Must Analysis. Paris, France: International Organisation of Vine and Wine; 2009. pp. 154-196. Available in http://www.oiv.int

[19] Díaz C, Conde JE, Claverie C, Díaz E, Trujillo JPP. Conventional enological parameters of bottled wines from the Canary Islands (Spain). Journal of Food Composition and Analysis. 2003;**16**(1):49-56. DOI: 10.1016/s0889-1575(02)00134-5

[20] Díaz C, Conde JE, Méndez JJ, Trujillo JPP. Volatile compounds of bottled wines with denomination of origin from the Canary islands (Spain). Food Chemistry. 2003;**81**(3):447-452. DOI: 10.1081/AL-120018257

[21] Blouin J, Cruège J. Analyse et composition des vins. Paris: Editorial Dunod; 2003

[22] Kontoudakis N, Esteruelas M, Fort F, Canals JM, Zamora F. Comparison of methods for estimating phenolic maturity in grapes: Correlation between predicted and obtained parameters. Analytica Chimica Acta. 2010;**660**(1):127-133. DOI: 10.1016/j.aca.2009.10.067

[23] Heras-Roger J, Diaz Romero C, Darias-Martin J. What gives a wine its strong red color? Main correlations affecting copigmentation. Journal of Agricultural and Food Chemistry. 2016;**64**(34):6567-6574. DOI: 10.1021/acs.jafc.6b02221

[24] Heras-Roger J, Alonso-Alonso O, Gallo-Montesdeoca A, Díaz-Romero C, Darias-Martín J. Influence of copigmentation and phenolic composition on wine color. Journal of Food Science and Technology. 2016;**53**(6):2540-2547. DOI: 10.1007/s13197-016-2210-3

[25] Rudnitskaya A, Rocha S, Legin A, Pereira V, Marques JC. Evaluation of the feasibility of the electronic tongue as a rapid analytical tool for wine age prediction and quantification of the organic acids and phenolic compounds. The case-study of Madeira wine. Analytica Chimica Acta. 2010;**662**(1):82-89. DOI: 10.1016/j.aca.2009.12.042

[26] Frias S, Trujillo JP, Peña E, Conde JE. Classification and differentiation of bottled sweet wines of Canary Islands (Spain) by their metallic content. European Food Research and Technology. 2001;**213**(2):145-149. DOI: 10.1007/s002170100344

[27] Frías S, Conde JE, Rodríguez-Bencomo JJ, García-Montelongo F, Pérez-Trujillo JP. Classification of commercial wines from the Canary Islands (Spain) by chemometric techniques using metallic contents. Talanta. 2003;**59**(2):335-344. DOI: 10.1016/S0039-9140(02)00524-6

[28] Frías S, Conde JE, Rodríguez MA, Dohnal V, Pérez-Trujillo JP. Metallic content of wines from the Canary Islands (Spain). Application of artificial neural networks to the data analysis. Molecular Nutrition & Food Research. 2002;**46**(5):370-375. DOI: 10.1002/1521-3803(20020901)46:5<370::AID-FOOD370>3.0.CO;2-F

[29] Trujillo JPP, Conde JE, Pont MLP, Câmara J, Marques JC. Content in metallic ions of wines from the Madeira and Azores archipelagos. Food Chemistry. 2011;**124**(2):533-537. DOI: 10.1016/j.foodchem.2010.06.065

[30] Conde J, Estevez D, Rodriguez-Bencomo J, García Montelongo F, Perez-Trujillo J. Characterization of bottled wines from the Tenerife Island (Spain) by their metal ion concentration. Italian Journal of Food Science. 2002;**14**(4):375-388

[31] Vrček IV, Bojić M, Žuntar I, Mendaš G, Medić-Šarić M. Phenol content, antioxidant activity and metal composition of Croatian wines deriving from organically and conventionally grown grapes. Food Chemistry. 2011;**124**(1):354-361. DOI: 10.1007/s00217-011-1535-4

[32] Perestrelo R, Silva C, Câmara JS. A useful approach for the differentiation of wines according to geographical origin based on global volatile patterns. Journal of Separation Science. 2014;**37**(15):1974-1981. DOI: 10.1002/jssc.201400374

[33] Pereira Dilson da Cruz Fernandes. Analysis of Phenolic, Sulfur and Volatiles Aroma Compounds in Wines of Fogo Island Cape Verde [Phd These]. Portugal. FCUP; 2017. https://sigarra.up.pt/fcup/en/teses.tese?P_ALUNO_ID=105209&p_processo=24050

[34] González JC, Delgado MR, Bencomo JR, Valido HC, Trujillo JP. Determinación de volátiles mayoritarios en vinos tintos de las islas canarias. In: Delgado-Díaz S, editor. Aportaciones al conocimiento del vino canario. La Laguna: Editorial Instituto de Estudios Canarios; 2010

[35] Perestrelo R, Fernandes A, Albuquerque F, Marques J, Câmara J. Analytical characterization of the aroma of Tinta Negra mole red wine: Identification of the main odorants compounds. Analytica Chimica Acta. 2006;**563**(1):154-164. DOI: 10.1016/j.aca.2005.10.023

[36] Baptista JA, da P Tavares, Joaquim F, Carvalho RC. Comparison of polyphenols and aroma in red wines from Portuguese mainland versus Azores Islands. Food Research International. 2001;**34**(4):345-355. DOI: 10.1016/S0963-9969(00)00174-5

[37] Culleré L, Escudero A, Pérez-Trujillo JP, Cacho J, Ferreira V. 2-Methyl-3-(methyldithio) furan: A new odorant identified in different monovarietal red wines from the Canary Islands and aromatic profile of these wines. Journal of Food Composition and Analysis. 2008;**21**(8):708-715. DOI: 10.1016/j.jfca.2008.05.004

[38] Perfil aromático de diferentes vinos tintos jóvenes procedentes de las Islas Canarias. Avances en ciencias y técnicas enológicas: Transferencia de tecnología de la red GIENOL al sector vitivinícola; 2007

[39] Díaz Díaz E, García Pérez A, Bhethencourt Piñero F, González Díaz A, Darias Martín J. Pitch Pine Wine in La Palma Island [Spain]. Viticultura Enología Profesional; Spain. 2000. http://www.latindex.org/latindex/ficha?folio=6579

[40] Heras-Roger J, Díaz-Romero C, Darias-Martin J. Phenolic compounds in wine: Types, color effects and research. In: Garde-Cerdán T, editor. Phenolic Compounds: Types, Effects and Research. New York: Nova Science Publishers; 2017. pp. 133-1780

[41] Heras-Roger J, Díaz-Romero C, Darias-Martín J. A comprehensive study of red wine properties according to variety. Food Chemistry. 2016;**196**:1224-1231. DOI: 10.1016/j.foodchem.2015.10.085

[42] Silva CL, Pereira J, Wouter VG, Giró C, Câmara JS. A fast method using a new hydrophilic–lipophilic balanced sorbent in combination with ultra-high performance liquid chromatography for quantification of significant bioactive metabolites in wines. Talanta. 2011;**86**:82-90. DOI: 10.1016/j.talanta.2011.08.007

[43] Rodrıguez-Delgado M, González G, Pérez-Trujillo J, Garcıa-Montelongo F. Trans-resveratrol in wines from the Canary Islands (Spain). Analysis by high performance liquid chromatography. Food Chemistry. 2002;**76**(3):371-375. DOI: 10.1016/S0308-8146(01)00258-8

[44] Paixão N, Pereira V, Marques JC, Câmara JS. Quantification of polyphenols with potential antioxidant properties in wines using reverse phase HPLC. Journal of Separation Science. 2008;**31**(12):2189-2198. DOI: 10.1002/jssc.200800021

[45] Li Z, Pan Q, JinG Z, Mu L, Duan C. Comparison on phenolic compounds in *Vitis vinifera* cv. Cabernet sauvignon wines from five wine-growing regions in China. Food Chemistry. 2011;**125**(1):77-83. DOI: 10.1016/j.foodchem.2010.08.039

[46] Ginjom I, D'Arcy B, Caffin N, Gidley M. Phenolic compound profiles in selected Queensland red wines at all stages of the wine-making process. Food Chemistry. 2011;**125**(3):823-834. DOI: 10.1016/j.foodchem.2010.08.062

[47] Kallithraka S, Tsoutsouras E, Tzourou E, Lanaridis P. Principal phenolic compounds in Greek red wines. Food Chemistry. 2006;**99**(4):784-793. DOI: 10.1016/j.food-chem.2005.07.059

[48] Winkel-Shirley B. Biosynthesis of flavonoids and effects of stress. Current Opinion in Plant Biology. 2002;**5**(3):218-223. DOI: 10.1016/S1369-5266(02)00256-X

[49] Negueruela A, Echávarri J, Pérez M. A study of correlation between enological colorimetric indexes and CIE colorimetric parameters in red wines. American Journal of Enology and Viticulture. 1995;**46**(3):353-356

[50] Heras-Roger J, Pomposo-Medina M, Díaz-Romero C, Darias-Martín J. Copigmentation, color and antioxidant activity of single-cultivar red wines. European Food Research and Technology. 2014;**239**(1):13-19. DOI: 10.1007/s00217-014-2185-0

Chapter 13

Novel Memory-Based Sensory Approach to Assess Large-Scale Typicality: The Case of Mainland Portugal Red Wines

Anibal Jose-Coutinho, Patricia Avila and Jorge M. Ricardo-da-Silva

Additional information is available at the end of the chapter

http://dx.doi.org/10.5772/intechopen.73039

Abstract

Twenty professional wine experts were asked to describe their prototypical construct of a representative young red wine from each of the 12 Protected Geographical Indications (PGI) of mainland Portugal. No samples were assessed; the experiment was based on memory alone by completion of 12 extended self-reported sensory questionnaires. Four large-scale areas were differentiated, the typicality being statistically validated and described from a sensory standpoint. Alcohol, acidity, bitterness, and astringency were cross-linked; the respective variations were correlated with published literature and expressed as key factors for the regional macroscale area differentiation. Bitterness and astringency were found to be sensory different and related on a geographical scale, as bitterness was primarily affected by inland/coastal influence; while astringency confirmed its customary north/south dependence that finding is to be considered a new understanding. Moreover, with the proposed methodology, it was possible to achieve a novel nationwide sensory characterization of PGIs, overcoming present day limitations on macroscale sensory research and sample representativeness. Results by uncalibrated prototypical memory assessment of single PGI Beira Atlântico were compared with the outcome of calibrated wine sampling assessment by local experts, using the same sensory questionnaire, and were found significantly correlated. The need for a calibration stage was found uneven regarding the overall group of scrutinized wine descriptors.

Keywords: red wine, sensory, typicality, astringency, bitterness

1. Introduction

The geographical origin of wine as well as its price has a strong impact on consumers. Wine is a product in which the concept of typicality is prevalent, especially in Europe [1]. Protected geographical indications (PGI) and protected designation of origin (PDO) products are the base of European artisan production, maintaining sensory characteristics, assuring consumer confidence, and preserving market position [2].

A large number of sensory methodologies are based on the use of professional tasters to show the sensory properties of a wine category (wine variety or PDO). Scientists have tested different sensory approaches, but the perceptual sample-tasting methods may be considered the most commonly used. These approaches work correctly, if the samples are grouped in well-separated regions; however, in more complex distributions, their classification power becomes poor [3]. Concerns regarding the scarcity and representation of samples that are assessed have always undermined typicality research. A recent study stated that a sample-tasting panel did not discriminate among the wines for astringency and bitterness, probably because the samples were inexpensive wines with very similar phenolic contents [4].

The role of expertise, as well as the importance of experience and long-term memory (LTM) for extraordinary performance, is widely documented. Wine professionals (wine-makers, wine journalists, etc.) are often nonavailable for testing [5]; hence, our experts were not formally trained, as usually done in quantitative descriptive analysis (DA). Due to their high level of long-term professional expertise, we hypothesized that the abstraction of sensory prototypes is memorized and could be considered as a synthesis of all previous wine tasting experiences from the category. Descriptions stored in LTM can be used to generate images of objects and scenes [6]. Recognition of meaningful stimuli such as words or objects rapidly activates conceptual information and leads to the retrieval of additional relevant information from LTM [7]. Neuroscience research stated that new information can come from different sources, including sensory stimuli (SS) or LTM; however, no (brain) regions were found to be more active during updating from SS than updating from LTM [8]. In the case of sensory analysis, flavor may be processed by tasters as a psychological construct, the data being more consistent with a perceptual/cognitive process rather than a consequence of rating strategies [9].

Our nationwide study, a novel macroscale approach on wine sensory typicality, was built on the case that senior experts have common mental representations of typical wines from each PGI, resulting in a feasible assessment of predetermined attributes and respective lexicon [10] which can be used to describe the sensory characteristics of a typical exemplar of the category or prototype. Clear agreement between experts concerning typicality scores [11] and existence of shared cognitive constructs of typicality [12] has been demonstrated. The literature does not always clearly delineate what constitutes training and what is experience. Experience relates to a familiarity with a product class resulting from long-term exposure to a wide variety of members representing that class [13]. Qualitative methods can be used to enhance quantitative studies, and with time, labor or financial limitations, may be considered a valid substitute for quantitative methods with heavy training/calibration requirements. Sensory methods are constantly evolving, becoming simpler, faster, and highly reviewed [14–20]. If neuroscience and sensory science results suggest that expertise may be more a cognitive skill rather than a perceptual one, and if all of the perceptually based techniques have a persistent low wine sampling hazard due to the logistical and/or time-consuming factors, then why are conceptual techniques, based on the long-term memory from professional wine experts and *keepers of memory* [1], not accepted as a sensory methodology?

Currently, there are 12 Protected Geographical Wine Indications (PGI) in mainland Portugal (**Figure 1**). For such a small country, does the actual range of 12 PGI's offer 12 single and

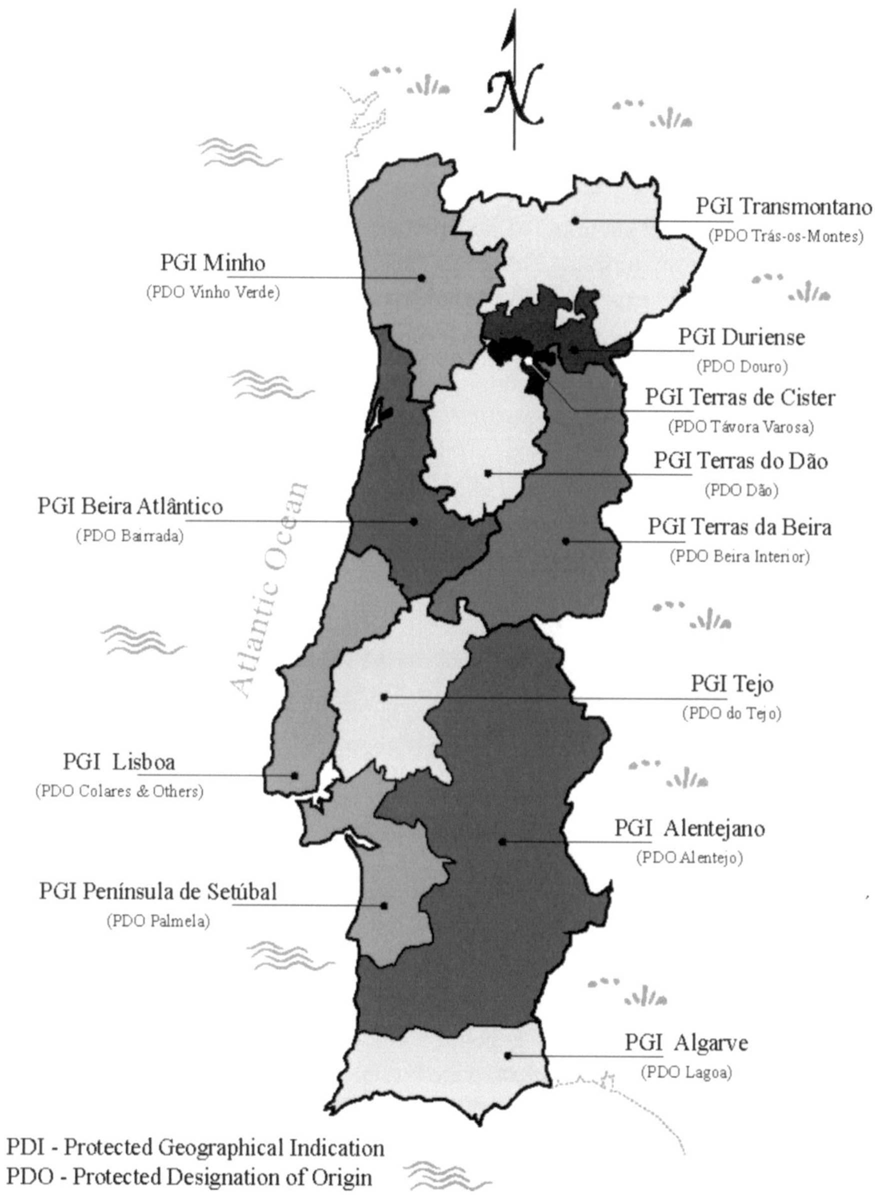

Figure 1. Protected geographical indications (and respective inner protected designations of origin) in mainland Portugal.

typical wine sensory profiles, so different from one another that typicality governs consumers on their purchasing decisions?

This study builds on the suggestion, which is commonly expressed in Portuguese and Anglophone wine press, that typicality assessment of PGI-certified still red wines of mainland Portugal does not match 12 exclusive sensory profiles, 1 for each Geographical Indication [21]. The supporting evidence of typicality similitude among certain PGIs can be found in bioclimatic [22–24] and biogeographical zoning studies [25]. Moreover, the phenolic composition [26, 27] and aroma [28] exhibited zonal differences.

Following our recently published white wine nationwide study [29], encompassing a novel memory-based DA, each member of our panel of professional experts was asked to mentally retrieve and assess his prototypical sensory profile of each of the 12 mainland Portugal PGI-certified young red wines. To our knowledge, this is the first macroscale sensory study covering a whole wine-producing country.

Finally, the outcome of our sample-free and uncalibrated DA methodology was compared with the results of a classic sampling DA single PGI study [30] assessed by a calibrated panel of PGI Beira Atlântico local experts, both studies sharing the same sensory questionnaire.

2. Materials and methods

2.1. Subjects

When compared with previous studies [31], rigorous selection criteria were applied, as the definition of a wine expert considered the following categories:

- Renowned wine-makers with tasting activity of >4 times per week in more than one Portuguese wine region;
- Oenology researchers and scholars who were regularly involved in wine-making and/or wine evaluation;
- Wine professionals (e.g., *Master of Wine,* wine judges, wine writers, wine buyers);
- Wine experts with an extensive (more than 15 years) history of wine involvement.

As this method is mostly dependent on indisputable knowledge brought by well-known senior Portuguese professionals, graduate wine students, wine consumers and enthusiasts, even with a long-term wine involvement, were not eligible for group of wine experts. The sensory raw data of this work was generated by 20 renowned wine-makers, referenced scholars and professional opinion leaders, wine buyers, and distributors from the Portuguese wine industry. Recent literature showed that the number of panelists is well above average to ensure a statistically robust outcome [4].

http://dx.doi.org/10.5772/intechopen.73039

2.2. Questionnaire

The questionnaire was designed according to the classical three-tier sensory method of assessment: visual, aromatic, and gustatory/tactile descriptors (full questionnaire in supporting information). With the exception of hue, the questionnaire used an 11-integer point rating scale. On the left was written: "no trace of the attribute" (score 0) and on the right was written: "extreme intensity of the attribute" (score 10).

The color was assessed in its intensity and hue. Results from this last descriptor were standardized to match the same scale because the previously reported research justified the use of a 4-integer point scale with 1—Violet-Purple, 2—Purple-Ruby, 3—Ruby-Garnet, and 4—Garnet-Brick [32]. The aroma was evaluated with 1 overall intensity measure and 18 aromatic categories that summarize a significant amount of scientific research [33], much of which has already been used in the form of aroma wheels [34]. The taste of red wines was evaluated using 14 measures, including essential tastes (sour, sweet, salty, and bitter), mouthfeel categories resulting from the various types of astringency [35], prickling, weight, persistence, and mouthcoat sensation [10].

2.3. Procedure

Respondents were invited and briefed by the same experimenter. They received oral instructions on how to fill the questionnaire. The experts were told to quantify their prototypical memory activated by the following question: "How would you sensory assess a typical young commercial red wine (released during the first year following harvest) from this particular PGI?" Despite some reasonable doubts on similar cognitive construct within the panel, no training phase or leveling pre-stage was performed. No wine samples were assessed. All experts independently completed individual self-reported questionnaires by scoring several sensory attributes that characterized their own expert construct for what is a typical PGI young red wine of each of the 12 Portuguese wine regions.

2.4. Statistical data analysis

Descriptive statistics (central tendency and dispersion) were computed for each sensory attribute, using IBM SPSS Statistics 20.0. Using the same software, bivariate statistics such as one-way ANOVA were performed in order to analyze the relation between the sensory attributes and 12 PGIs. Moreover, this tool reveals whether or not significant differences are found and also the strength of the association. This was followed by the completion of principal components analysis (PCA) with a Varimax rotation which sought to explore the possibility of reducing the initial amount of sensory variables into fewer dimensions—the principal components (PC). PCA of sensory data permitted the differentiation among geographic areas [28]. The number of retained components was based on the Kaiser criterion—according to which the components with an eigenvalue greater than or equal to 1 are retained—and occasionally, the next principal component was also retained. Latent variables encompassing different initial variables found to be highly correlated on each PC were estimated, named accordingly, and new descriptive statistics (mean and standard deviation) were computed on the basis of such aggregate variables (three initial variables were kept isolated: color intensity, color tonality (hue), and aromatic intensity).

Given that the central hypothesis of this study anchored on similarity of several PGIs typical wine sensory profiles, global research of this proximity was assessed by hierarchical cluster analysis (HCA), which is an exploratory data analysis tool that aims at sorting different objects (the 12 PGIs) into groups in a way that the similarity between two objects is maximal if they belong to the same group and minimal otherwise. The squared Euclidean distance was used as measure of proximity, and the complete linkage method algorithm was applied to group the PGIs. Raw data from the above-mentioned three initial sensory variables excluded from the PCA, as well the means of the latent sensory variables that emerged from the PCA outcome, were lined up as the HCA input.

3. Results and discussion

Twelve Protected Geographical Indications, memory assessed by the typicality of their respective young commercial red wines, were sensory validated, since all PGIs showed differences with significant statistical meaning (at least in 1 attribute) which was verified by the ANOVA method, considering one fixed factor: the Geographical Indication. Relative standard deviation, also known as coefficient of variation [36], showed a lower percentage for the majority of the sensory attributes, which indicates low variability in the data set means (**Table 1**).

3.1. Representation of aroma and taste

Except for the aromatic intensity assessment, the 18 initial aroma attributes were grouped into 5 principal components (**Table 2**) that were named according to the previously reported literature on aromatic series: **PC1** ***Aredfruitwoody*****, PC2** ***Aripefruit*****, PC3** ***Agreenchemical*****, PC4** ***Aoverripeness*** **and PC5** ***Aflorcitrusmineral*****.**

It was possible to find an aggregate sensory measurement that explained 66.7% of the total variance. The simplification procedure that was based on the expert panel assessments may be optimal for macroscale area profiling purposes in contrast to the previously reported use of highly specific, isolated, and less obvious sensory descriptors.

The application of PCA to all tactile/textural and mouthfeel categories simplified the characterization of the taste of the PGI young red wines.

The 14 initial taste attributes were grouped into 4 principal components (**Table 3**) that were named according to the classical tastes and contributing research on mouthfeel perceptions, reported above: **PC1** ***Tdryastringent*****, PC2** ***Tsweetviscous*****, PC3** ***Tbittersalty*** **and PC4** ***Tfullpersistent*****.**

We found an aggregate sensory measurement that explained 75% of the total variance, again supporting the role of the simplified procedure on the tannins tactile range of descriptors [35].

http://dx.doi.org/10.5772/intechopen.73039

		Hue	Color intensity	Aroma intensity	(PC1) Aredfruitwoody	(PC2) Aripefruit	(PC3) Agreenchemical	(PC4) Aoverripeness	(PC5) Aflor citrusmineral	(PC1) Tdryastringent	(PC2) T sweetviscous	(PC3) T bittersalty	(PC4) T fullpersistent
Algarve	Mean	3.00	6.19	6.13	4.50	5.77	2.80	2.84	2.77	3.31	6.48	1.54	5.38
	SD	0.37	0.98	1.02	1.69	1.52	1.18	1.61	1.35	1.58	0.56	0.98	0.85
Alentejano	Mean	2.70	7.30	7.15	4.79	6.95	3.48	3.85	2.87	3.43	6.68	1.57	6.00
	SD	0.47	0.73	0.88	1.45	1.10	1.31	1.90	1.06	1.05	0.72	0.97	1.04
Beira Atlântico	Mean	2.00	5.65	5.30	4.07	3.28	4.93	3.33	3.80	5.73	3.24	3.12	5.60
	SD	0.46	1.39	1.08	1.13	1.22	1.51	1.40	1.25	1.34	0.63	1.43	0.95
Duriense	Mean	1.65	7.55	6.95	4.71	6.04	3.81	3.48	4.77	4.61	5.66	1.58	7.05
	SD	0.59	1.00	0.89	1.34	1.15	1.48	1.35	1.63	1.26	0.81	1.09	0.81
Lisboa	Mean	1.82	6.35	5.35	4.25	4.13	4.31	2.91	3.61	5.18	4.18	2.63	5.35
	SD	0.64	1.46	0.79	1.36	1.17	1.25	1.28	1.58	1.12	0.89	1.05	0.70
Minho	Mean	1.11	8.58	5.84	2.93	3.24	5.26	1.89	3.35	6.43	2.43	3.79	5.03
	SD	0.32	1.64	1.71	1.19	1.39	1.73	1.14	1.81	1.43	0.87	1.45	1.52
Península Setúbal	Mean	2.76	6.24	6.12	4.84	5.46	3.87	3.56	3.47	3.66	5.93	2.31	5.50
	SD	0.44	0.90	0.86	1.38	1.44	1.53	1.69	1.73	1.33	0.69	1.49	0.85
Tejo	Mean	2.84	6.32	6.05	4.73	5.09	3.95	2.95	3.32	4.26	5.37	1.79	5.47
	SD	0.37	0.58	0.91	1.35	0.92	1.48	1.61	1.68	1.64	0.62	1.25	0.75
Terras da Beira	Mean	2.32	6.16	5.74	4.13	4.21	3.96	3.08	4.39	5.07	4.52	1.95	6.08
	SD	0.67	1.17	0.93	1.22	1.13	1.57	1.56	1.42	1.28	0.76	1.22	0.69
Terras de Cister	Mean	2.60	5.10	5.20	3.74	3.45	3.73	2.70	3.77	5.03	3.94	2.20	5.55
	SD	0.52	0.99	1.03	1.19	1.14	1.08	0.98	1.24	1.42	0.67	1.04	0.76
Terras do Dão	Mean	2.00	6.60	6.90	4.57	4.19	3.94	3.63	5.23	5.03	4.52	1.92	6.70
	SD	0.46	1.10	1.02	1.47	1.56	1.30	1.64	1.62	1.11	0.73	1.18	0.83
Transmontano	Mean	2.38	6.25	5.38	3.98	4.91	3.91	3.66	3.83	4.63	4.71	1.88	6.00
	SD	0.50	0.86	0.81	1.45	1.19	1.42	1.71	1.50	1.20	0.77	1.06	0.89
Total	Mean	2.23	6.60	6.07	4.29	4.77	4.02	3.18	3.79	4.71	4.82	2.19	5.84
	SD	0.73	1.38	1.20	1.43	1.68	1.52	1.58	1.64	1.57	1.44	1.36	1.07
	% RSD	33%	21%	20%	33%	35%	38%	50%	43%	33%	30%	62%	18%

Table 1. Mean and standard deviation of scores on all final sensory attributes by PGI, after PCA. Includes relative standard deviation.

AROMA	Principal component (PC)				
	PC1	PC2	PC3	PC4	PC5
Woody	**0.814**	0.054	0.085	0.151	0.111
Spicy	**0.674**	0.02	0.236	0.154	0.206
Red fruit	**0.644**	0.139	0.088	−0.253	0.346
Bread and pastry	**0.595**	0.185	0.012	0.426	0.04
Caramelized	**0.492**	0.442	0.083	0.433	−0.219
Black fruit	0.058	**0.846**	−0.131	−0.14	0.249
Stone fruit	−0.027	**0.809**	0.054	0.167	0.048
Dried fruit	0.251	**0.717**	−0.092	0.351	−0.23
Jam and jelly fruit	0.513	**0.65**	−0.2	0.203	−0.086
Chemical	−0.036	0.064	**0.757**	−0.026	−0.095
Vegetable	0.292	−0.192	**0.742**	−0.157	0.177
Animal	−0.012	−0.024	**0.733**	0.307	0.109
Herbal	0.231	−0.16	**0.709**	0.094	0.243
Dry flowers	0.157	0.058	0.189	**0.763**	0.234
Nut fruit	0.118	0.206	−0.009	**0.691**	0.216
Floral	0.304	0.046	−0.079	0.077	**0.784**
Mineral	0.11	−0.091	0.218	0.208	**0.697**
Citrus fruit	−0.001	0.177	0.381	0.29	**0.574**
Variance explained (cumulative)	**15.2 (15.2)**	**15.1 (30.3)**	**14.1 (44.5)**	**11.2 (55.6)**	**11.1 (66.7)**

Major loadings on each PC are highlighted in bold.

Table 2. Loadings of 18 aromatic attributes in the first 5 principal components named PC1 *Aredfruitwoody*, PC2 *Aripefruit*, PC3 *Agreenchemical*, PC4 *Aoverripeness* and PC5 *Aflorcitrusmineral*.

Wine phenolic compounds may either taste astringent, bitter or both [37]. Furthermore, mutual suppression is a fundamental property among all tastes. However, this does not occur with astringency [38] because this tactile sensation plays a strong role as a key sensory profiler. Despite the close relationship between bitterness and astringent [39], our panel of professional experts placed major factor loadings for each one in different principal components, hence endorsing the greater sensitivity and lingual acuity of our super tasters [40]. Although the significant grouping of the originally segmented terms may lead to doubt about their usefulness for providing quality data regarding the astringency of the wines [35], our results suggest that the variation in the PC1 *Tdryastringent* that was elicited by the red wines can be differentiated and rated.

3.2. Sensory differentiation

Multivariate statistical methods showed that evidence of sensory aggregations between PGIs and extended geographical clusters or macroscale areas with similar sensory profile was

http://dx.doi.org/10.5772/intechopen.73039

Taste	Principal component (PC)			
	PC1	PC2	PC3	PC4
Grain (Mouthfeel)	**0.879**	−0.073	0.071	0.122
Rough (Mouthfeel)	**0.862**	−0.216	0.154	0.075
Astringent (Global)	**0.812**	−0.341	0.186	0.165
Dry (Mouthfeel)	**0.787**	−0.045	0.247	−0.186
Sweet	−0.071	**0.886**	0.071	−0.092
Alcohol	−0.299	**0.799**	−0.113	0.168
Smooth (Mouthfeel)	−0.163	**0.787**	−0.045	0.206
Oily (Mouthcoat)	0.044	**0.622**	−0.166	0.583
Acid (Sour)	0.492	**−0.621**	0.363	0.187
Salty	0.084	0.025	**0.864**	0.009
Bitter	0.393	−0.011	**0.715**	0.053
Bubbly	0.167	−0.395	**0.572**	−0.209
Length	0.188	−0.107	0.077	**0.862**
Full body	−0.078	0.466	−0.1	**0.714**
Variance explained (cumulative)	**24.2 (24.2)**	**24.1 (48.3)**	**13.6 (61.9)**	**13.1 (75.0)**

Major loadings on each PC are highlighted in bold.

Table 3. Loadings of 14 tactile/textural and mouthfeel descriptors in the first 4 principal components named PC1 *Tdryastringent*, PC2 *Tsweetviscous*, PC3 *Tbittersalty* and PC4 *Tfullpersistent*.

proposed. The experts' typicality construct for the PGI red wines from mainland Portugal led to a sensory aggregation around four clusters (**Figures 2** and **3**) that were named according to their respective geography and a combination of geomorphological and bioclimatic characteristics [22]:

- **Cluster #1: SOUTHERN** (PGI Pen. Setubal (PS) + PGI Tejo (T) + PGI Alentejano (AL) + PGI Algarve (AG))
- **Cluster #2: CENTRAL VALLEYS** (PGI Duriense (D) + PGI Terras do Dão (TD))
- **Cluster #3: CENTRAL COAST and PERIPHERAL VALLEYS** (PGI Beira Atlântico (BA) + PGI Lisboa (L)) and (PGI Terras Beira (TB) + PGI Transmontano (TM) + PGI Terras Cister (TC))
- **Cluster#4: (Single) PGI MINHO (M)**

Clustered groups were, finally, sensory described on the basis of mean scores comparison as the ANOVA results confirmed the statistical robustness of our design (**Table 4**). A radar chart of such means is presented, such representation being straightforward and relatively effortless to analyze (**Figure 4**).

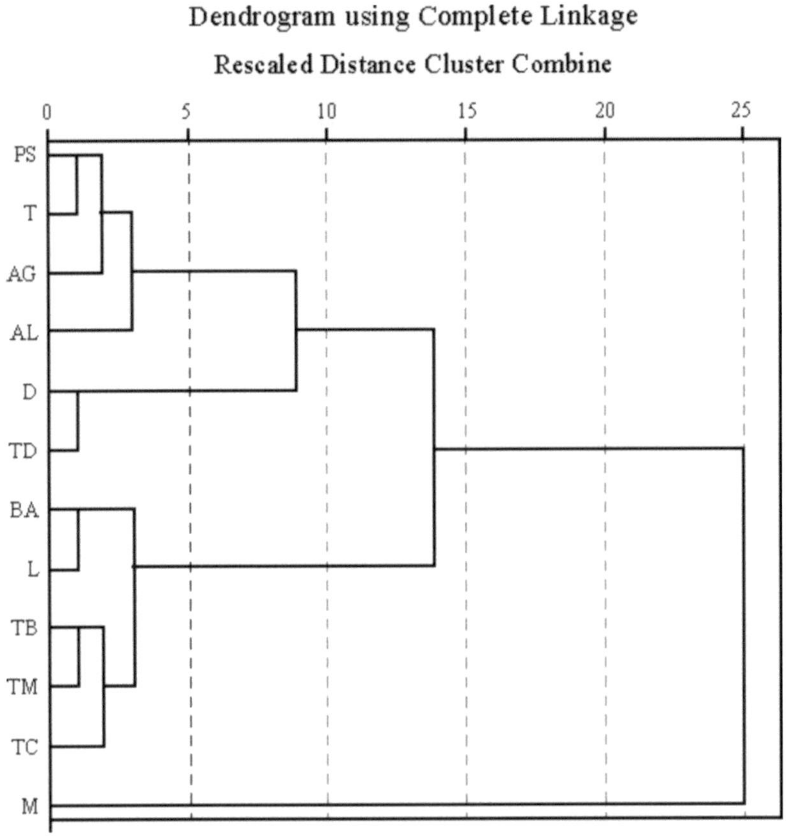

Figure 2. Dendogram showing the distance between PGIs (PS, peninsula de Setubal; T, Tejo; AG, Algarve; AL, Alentejano; D, Duriense; TD, Terras do Dão; BA, Beira Atlântico; L, Lisboa; TB, Terras da Beira; TM, Transmontano; TC, Terras de Cister; M, Minho).

3.3. Typical wine profiles of the new-found macroscale areas

3.3.1. Typical young red wine profile of the SOUTHERN macroscale area

The final sensory attributes reveal some characteristics that sustain the fast clustering dynamics that encompass the four SOUTHERN PGIs (PGI Peninsula de Setubal, PGI Tejo, PGI Alentejano and PGI Algarve), thus enabling the creation of a typical sensory profile for this macroscale area (**Figure 4**). In agreement with our results, similar geoviticultural (temperate hot climate, temperate warm nights and a moderately strong drought) and sensory analysis (high alcohol content and intense aromatic ripe fruity wines) groupings were reported for these same four PGIs [41].

The color hue got the highest assessment (**Table 4** and **Figure 4**), indicating a Ruby-Garnet hue, and color intensity rated second lowest. Temperatures of 30°C and higher might lead to lower anthocyanin synthesis [42]. Lower acidity levels are usually correlated with higher grape pH, although the relation is affected by potassium accumulation, which is itself temperature dependent. At a higher pH, the chalcone form (slightly yellow) increases progressively over

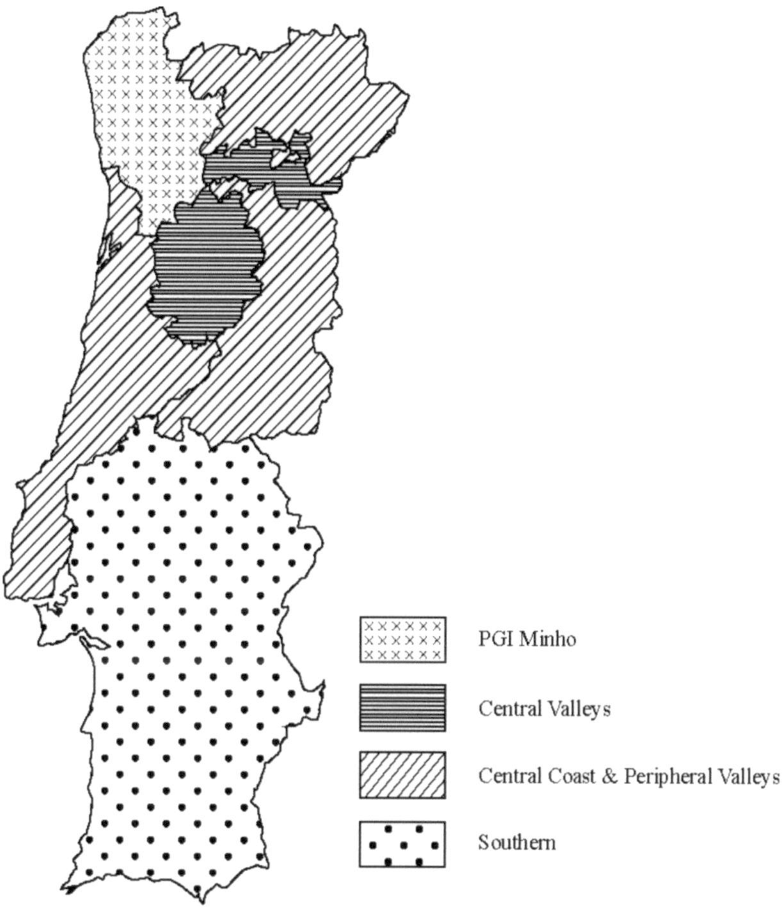

Figure 3. Four large-scale clusters in mainland Portugal originated from experimental design on sensory typicality.

the hemiketal (colorless), the quinoidal (blue) and the flavylium cation (red) anthocyanin structures [43, 44]. One study showed that raisining decreased the lightness and increased the color saturation and hue [45]. These findings may explain the ruby-garnet dominant hue and the low color intensity of typical young red wines from the SOUTHERN macroscale area.

For the aroma, the lowest assessments were given to PC3 *Agreenchemical* and PC5 *Aflorcitrusmineral*. Inversely, PC1 *Aredfruitwoody* and PC2 *Aripefruit* received higher scores (**Table 4** and **Figure 4**). Published results indicate that a masking effect of vegetative aromas by fruit aromas may occur [46].

The aroma profile showed an intense raisiny and jammy character with dominant black and stone fruit; however, the experts noted that the aromas were also woody, spicy, and of young yeasty red fruits. Red wine fruitiness may correlate directly with the ethanol enhancement because studies on red wine aroma have confirmed that when ethanol is enhanced, the intensity of the fruity odor increases [47]. Additionally, the existence of inodorous constituents in red grape skins, which are extractible by ethanol and transformed by yeasts, produced a

Final clustered zonings vs. final sensory attributes		Hue*	Color intensity	Aroma intensity	(PC1) Aredfruitwoody	(PC2) Aripefruit	(PC3) Agreenchemical	(PC4) Aoverripeness	(PC5) Aflor citrusmineral	(PC1) Tdryastringent	(PC2) Tsweetviscous	(PC3) Tbittersalty	(PC4) Tfullpersistent
Southern	Mean	**6.09**	6.51	6.36	**4.71**	**5.82**	*3.52*	3.30	*3.11*	*3.67*	**6.11**	1.80	5.59
CENTRAL VALLEYS	Mean	2.75	7.08	**6.93**	4.64	5.11	3.88	**3.55**	**5.00**	4.82	5.09	*1.75*	**6.88**
Central coast and peripheral valleys	Mean	4.08	*5.90*	*5.39*	4.03	3.99	4.17	3.13	3.88	5.12	4.12	2.35	5.72
PGI MINHO	Mean	*0.35*	**8.58**	5.84	*2.93*	*3.24*	**5.26**	*1.89*	3.35	**6.43**	*2.43*	**3.79**	*5.03*
Univariate ANOVA results (By Attribute)	Sigma	0.001	0.010	0.003	0.000	0.031	0.049	0.047	0.001	0.001	0.002	0.016	0.002
	Eta Squared	0.806	0.641	0.730	0.910	0.518	0.459	0.466	0.810	0.818	0.770	0.599	0.754

*Hue was standardized to encompass the 0–10 points integer scale.
Minimal scores in italic; maximal scores in bold.
Summary of ANOVA results.

Table 4. Mean scores of final sensory attributes regarding the four clustered zonings of mainland Portugal.

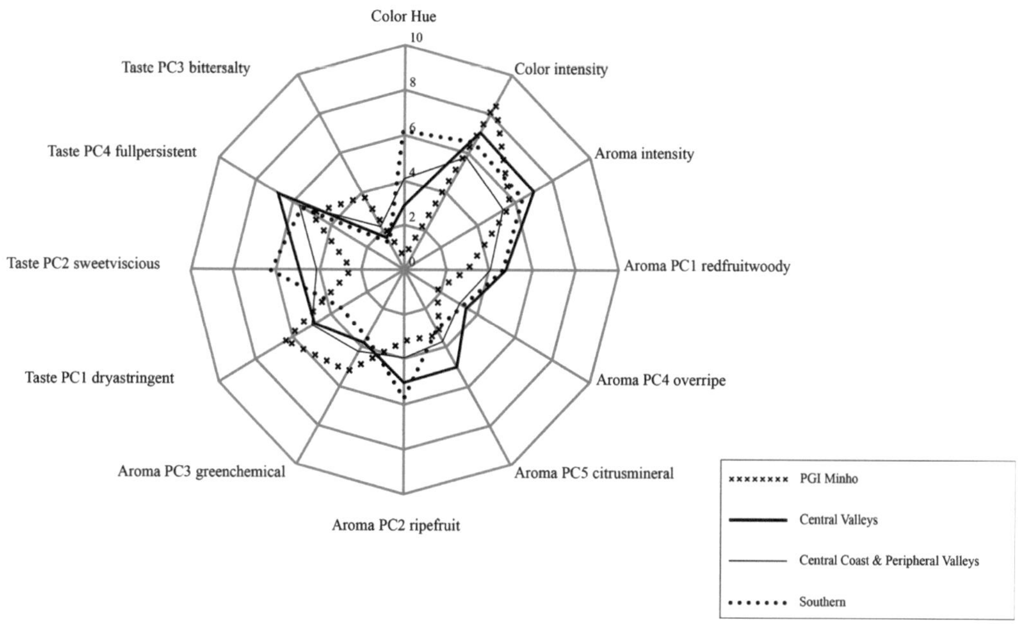

Figure 4. Radar chart of the red wine sensory aggregate attribute means per immediate cluster.

specific aroma of red-berry and black-berry fruit in the finished wines [48]. The SOUTHERN macroscale area ranked second on the aromatic intensity scale and received the lowest scores for mineral, floral, and citrus fruit. Chemical and vegetable notes were almost nonexistent.

For the taste, the lowest assessment was given to PC1 *Tdryastringent*, and the highest assessment was given to PC2 *Tsweetviscous* (**Table 4** and **Figure 4**). The taste characterization demonstrated a clear typicality for the sweet and oily intensive textures, which are higher alcohol-driven tactile sensations, with a weakness in the fresh acidic taste and tannin smoothness. Lower levels of dry, astringent, rough, and grain tannin-related mouthfeels are frequent in these medium-bodied immediate southern red wines. An oily viscosity enhancement results in a smoother palate texture [49] and reduces the astringency [38]. Wines with high levels of polysaccharides tend to decrease the impact of the bitterness and astringency [50]. The phenolic composition may be due to macroscale area differentiation because the amount of epigallocatechin (EGC) was negatively correlated with the perceived astringency [51] and could decrease the coarse perception [52]. PC2 *Tsweetviscous* encompassed alcohol, sweet taste, smooth mouthfeel, and oily mouthcoat sensations as positively correlated; however, other published results indicate that ethanol and glycerol do not significantly contribute to viscous mouthfeel [53].

3.3.2. Typical young red wine profile of the CENTRAL VALLEYS macroscale area

A portion of inner center-northern Portugal was sensory-zoned. PGI Terras do Dão and PGI Terras do Douro were immediately clustered and named CENTRAL VALLEYS because of their centered geography on the mountainous continental plateau (**Figure 4**).

The CENTRAL VALLEYS macroscale area was presented as an immediate clustering output. According to the cluster analysis method and based on the related isolated and aggregate means, the sensory profile has the following characterization:

The color got the second lowest assessment for its hue, indicating a Purple-Ruby hue, and the second highest assessment was for *C Intensity* (**Table 4** and **Figure 4**). These results are characteristic of fresh/cool continental regions with extreme thermal amplitudes, and this explains the high-color intensities [54].

For the aroma, there was a strong typicality with the three highest assessments for *Aintensity, Aoverripeness*, and *Aflorcitrusmineral* (**Table 4** and **Figure 4**). The red wine aromatic intensity (*Aintensity*) was the highest for the CENTRAL VALLEYS macroscale area. Additionally, the floral and citrus fruit notes (PC5 *Aflorcitrusmineral*) may be related to the *Touriga Nacional* cultivar (representative of PGI Terras do Dão and PGI Duriense) because they exhibited maximum scores [55]. The highest mineral assessment [56] and the exhibited essence of dried fruits and flowers from PC4 *Aoverripeness* are attributes that may result from the heterogeneous mountain that is known to have significant gaps in altitude and semi-arid locations, especially near the Spanish border [23]. This may explain the observed proximity with the Southern regions, as shown in the clustering dendogram (**Figure 2**).

For the taste, the lowest assessment was for PC3 *Tbittersalty*, and the highest assessment was for PC4 *Tfullpersistent* (**Table 4** and **Figure 4**). The taste characterization exhibited a clear typicality based on the extremely long persistence, which relates with the highest assessment of fullness. This feature may relate to a higher phenolic content of such high density wines which has shown to enhance flavors and mouthfeel persistence [57, 58]. Both the main yeast polysaccharides (mannoproteins) and the principal grape polysaccharides (arabinogalactan-proteins and rhamnogalacturonans) increase the perception of body [52]. The evaluation of wine finish duration by trained panelists indicated the finish of the high ethanol wines lasted longer than the ones with low ethanol [59]. Similar to the Southern reds, wines from the CENTRAL VALLEYS exhibited the lowest salty, bitter, and bubbly sensation levels, suggesting the partial proximity. The bitterness intensities elicited by ethanol-sugar mixtures are lower than those elicited by unmixed ethanol solutions [60], these findings in perfect alignment with our results.

3.3.3. Typical young red wine profile of the CENTRAL COAST and PERIPHERAL VALLEYS macroscale area

For the remaining continental PGIs, which are listed under the PERIPHERAL VALLEYS cluster, they were sensory assessed as similar to the CENTRAL COAST PGIs, and the data were combined (**Figure 3**). Interestingly, the late PGI Beiras, which encompass the administrative merging of coastal and inner wine regions (PGI Terras da Beira (TB), PGI Terras de Cister (TC), and PGI Beira Atlântico (BA)), was officially extinct after the harvest of 2011.

The hue of the red wines was rated from Purple/Ruby to Ruby/Garnet halfway between typical fresh and warm region hues. Indeed, this peripheral setting includes a vast macroscale area border that may explain the wider values observed in our results. The color intensity received the lowest score (**Table 4** and **Figure 4**).

The aromatic intensity of PERIPHERAL VALLEYS and CENTRAL COAST red wines was the lowest among all of the macroscale areas (**Table 4** and **Figure 4**), which is another difference with the neighboring Central Valleys and is a similarity with the Northern-coastal PGI Minho.

The taste characterization showed long mouthfeel persistence, and it ranked second highest for the assessment of full bodied taste, a feature that is similar to the neighboring Central Valleys. The levels of salty, bitter, and tactile bubbly were slightly increased, suggesting a similarity with the Northern-coastal PGI Minho and some typicality in the acidic-fresh sensory profile (**Table 4** and **Figure 4**).

3.3.4. *Typical young red wine profile of the single-clustered PGI MINHO*

The northern-coastal PGI MINHO is a wine region, where the world-renowned PDO Vinho Verde is located. This region presented the highest primary typicality assessments according to the stand-alone cluster analysis results (**Figure 2**). This result was strongly supported by 10 of the 12 analyzed attributes that had extreme scores (**Table 4** and **Figure 4**). Similar observations were previously reported by other authors [41], which were based on their geoviticultural multicriteria climatic classification system [61]. The sensory attributes revealed characteristics that allowed clustering dynamics and helped establish a typical sensory profile:

For the color, the highest assessment was for *C Intensity*, and the lowest assessment was for *C Hue*, indicating a Violet-Purple hue. The high intensity is related to the inky local red cultivar, *Vinhão*, which has been shown to have anthocyanin levels that are 24 times higher than other cultivars [62]. A correlation between the anthocyanic content and the bitterness that is elicited was recently published [63]. Moreover, the lowest assessment for PC2 *Tsweetviscous* correlates with the highest acidity levels, and it thus correlates with the lowest pH, explaining the highest content of the colored anthocyanins [43]. The hue assessment that indicated a violet-purple hue is aligned with the first-year consumption of extremely young reds [32], which is typical of PGI Minho and its inner PDO Vinho Verde.

The aroma was considered to be typical as the experts scored one highest assessment for PC3 *Agreenchemical* and three lowest assessments for PC1 *Aredfruitwoody*, PC2 *Aripefruit*, and PC4 A*overripeness*.

The aromatic profile is built on several extreme assessments. The high levels of green and chemical notes and the low levels of ripe fruit may relate to the cool climate and the low ripening conditions [64]. An average intensity was found for the PGI MINHO red wines with an extremely low presence of woody, spicy, and young yeasty red fruit notes.

The taste was considered to be typical as the experts scored the highest assessment for PC1 *Tdryastringent* and PC3 *Tbittersalty* and the lowest assessment for PC2 *Tsweetviscous* and PC4 *Tfullpersistent.*

Located on the extreme northwest corner of mainland Portugal, the PGI MINHO sensory profile exhibited typical gustatory characteristics, such as the highest intensity of dryness, astringency, roughness, and grain tannin mouthfeel, which are related to the greener ripeness [26]. Polyphenols are recognized as substances that provide astringency sensation. However, other substances such as organic acids, sugars, and ethanol can also influence this sensation [65]. The fresh profile was amplified to its peak by the highest presence of salty, bitter, and bubbly sensations. The bubbly sensations were due to typical carbonic gas addition. Inversely, these red wines were assessed as light bodied with a fair persistence, low alcohol-driven tactile sensations, and reduced tannin smoothness. The lowest assessments were for sweet and oily textures. The highest assessment of astringency occurred for the low alcoholic wines [66]. The maximal rate of fresh acidic taste was found for the PC2 *Tsweetviscous,* and this is also typical of the PGI MINHO profile (**Figure 4**). Furthermore, the lack of ethanol and proanthocyanidins greatly increased the perceived acidity [67]. The combination of the effect of excess acidity and astringency was termed *Green* by a panel of experienced wine tasters [35]. Acidity reduces the perception of body [68]. The results of this study contradict the reports of enhanced bitterness by alcohol [69], although some authors suggest that alcohol may suppress the bitter taste when held in the mouth [70].

3.4. Astringency vs. bitterness: A novel macroscale orthogonal approach

The fact that four PCs—**PC1 *Tdryastringent*, PC2 *Tsweetviscous*, PC3 *Tbittersalty* and PC4 *Tfullpersistent***—explained 75% of the total variance of our experiment, justified further

Correlations

	(PC1) Tdryastringent	(PC2) Tsweetviscous	(PC3) Tbittersalty	(PC4) Tfullpersistent
(PC1) Tdryastringent	1	−0.957**	0.434	0.154
(PC2) Tsweetviscous	**−0.957****	1	−0.592*	−0.017
(PC3) Tbittersalty	0.434	**−0.592***	1	−0.502
(PC4) Tfullpersistent	0.154	−0.017	−0.502	1
N	12	12	12	12

*Correlation is significant at the 0.05 level.
**Correlation is significant at the 0.01 level.

Table 5. Pearson correlation matrix of four aggregate tactile attributes.

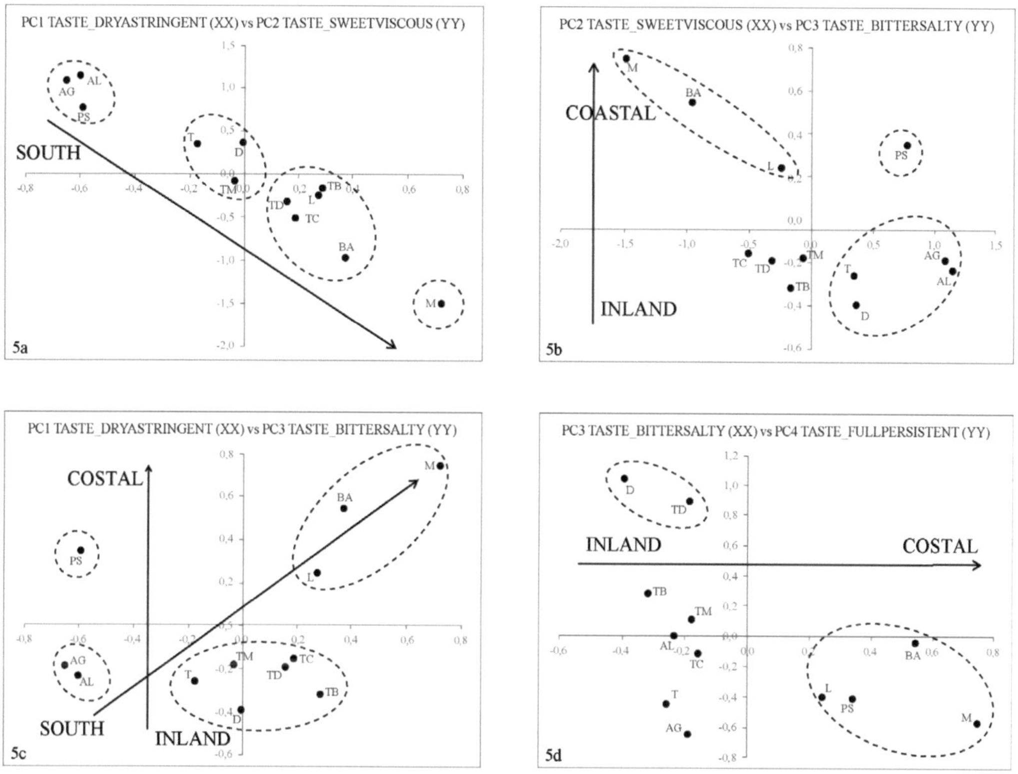

Figure 5. Standardized score bi-plots of the four aggregate tactile descriptors over 12 PGIs (PS, peninsula de Setubal; T, Tejo; AG, Algarve; AL, Alentejano; D, Duriense; TD, Terras do Dão; BA, Beira Atlântico; L, Lisboa; TB, Terras da Beira; TM, Transmontano; TC, Terras de Cister; M, Minho).

research on potential correlation with an orthogonal mapping of the leading sensory attributes of each PC, envisaging red wine PGI typicality. Pearson Correlation was computed between those four PCs (**Table 5**), and results showed expected significant correlation between PC1 and PC2.

The coefficient of −0.957 (close to maximum −1) between PC1 *Tdryastringent* and PC2 *Tsweetviscous* reflected a strong linear mapping of mainland Portugal PGIs, enabling a clear sensory profiling statement, virtually with a North-South reading, as shown in **Figure 5a**.

Astringency was assessed as drier and rougher in the north and smoother in the South. The bi-plot in **Figure 5a** also showed a clear northward decrease in alcohol content. Conversely, acidity was enhanced northward. The sensory measure PC3 *Tbittersalty* included bitterness, saltiness, and fizziness (natural or added CO_2) perceptions as its leading scores. The interpretation must be merely indicative, as the explained variance was weaker than PC1 and PC2 levels; however, those three aggregate perceptions seemed to act as key profilers for sensory differentiation, as showed in **Figure 5b** and **c**. This finding is statistically significant, the negative correlation of PC3 *Tbittersalty* with PC2 *Tsweetviscous* suggesting an inland to coastal increase in bitterness, leading to a well-defined cluster including super humid coastal PGI Minho and temperate coastal PGI Beira Atlântico and PGI Lisboa, as well as a surprising stand-alone southern coastal region (PGI Peninsula de Setubal), which was replaced, in its customary cluster, by a sweet, lowest bitter-salty, north-continental PGI Duriense (**Figure 5c**). The nationwide differentiation of astringency and bitterness introduced an orthogonal reading (N north-S south vs. E inland-W coastal, respectively) rather than a linear one, these findings add novelty to sensory research. The weakest contributing PC4 *Tfullpersistent* presented several inverse scores when compared to PC3 *bittersalty* (**Figure 5d**), and this finding, lacking statistical significance, may justify new and comprehensive studies to confirm the following orthogonal sensory mapping **PC1 (northward) – PC2 (southward) – PC3 (coastal westward) – PC4 (inland eastward)**, the geographical reading showed in **Figure 6**.

3.5. Classic sampling vs. prototypical memory

Among the scientific community, our novel, nationwide, sample-free, noncalibrated, expert memory-based method was accepted [29, 71], as it was challenged on the basis that the data relied on respondents' perceptions rather than direct experience. In this new research, we compare the results mentioned before in this study—signaled as Experiment 1 (LTM)—with the completion of the same questionnaire (Supporting Information) by 19 local experts from the single PGI Beira Atlântico—Experiment 2 (WS)—as a result of a blind tasting of 15 young red wine samples, representing the leading supermarket brands of this wine region.

3.5.1. Experiment 2

Results of the wine tasting assessment by local PGI Beira Atlântico wine experts were organized in order to fully compare with extracted results of experiment 1 regarding the same PGI (**Figure 6**). The panel is widely calibrated, as these professionals belong to the certification panel which must approve typicality and quality of all submitted PGI Beira Atlântico wines. This panel complies with an annual program of inner and inter-laboratorial calibration on aromatic thresholds, triangular tests, and tastant solutions (control standards of sucrose, NaCl, citric acid, quinine, and others).

The assessment of the same sensory questionnaire by the nationwide professional uncalibrated panel and the local professional PGI Atlântico calibrated panel showed a significant Pearson positive correlation of 0.669, this to be considered promising and a green flag for newcoming comparative studies regarding other PGIs.

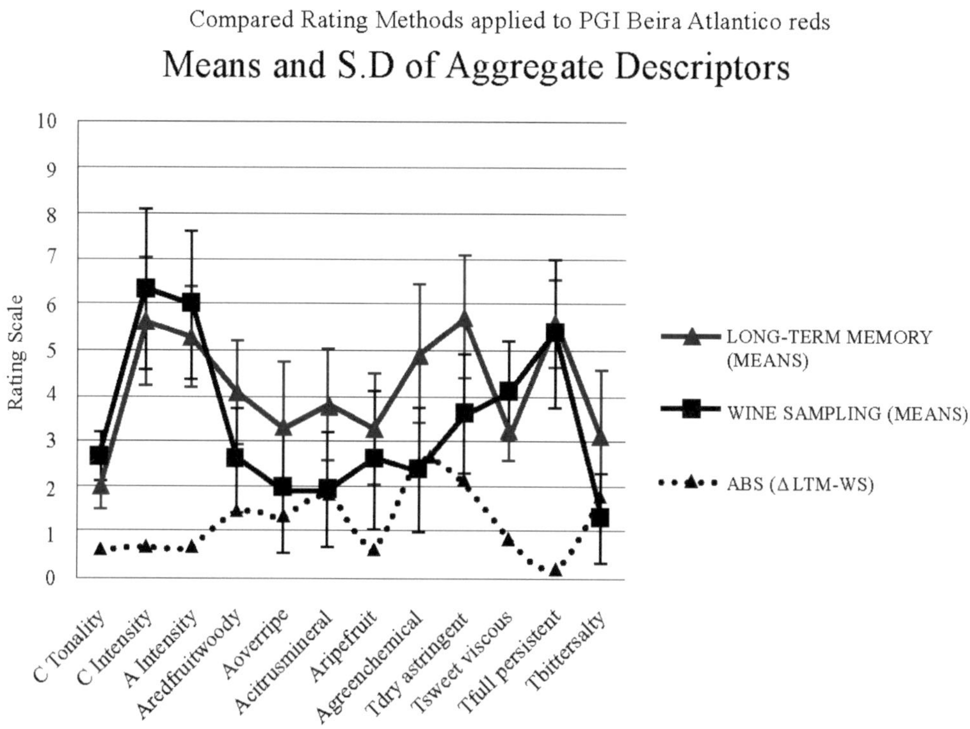

Figure 6. Mean differences between long term memory (LTM) and wine sampling (WS) methods. ABS scores represent absolute differences between both methods.

Primary	**Aggregate**	**(PD Means)**	**(PD Means)**	**ABS ()**
Descriptor (PD)	**Descriptor**	**Long-term memory**	**Wine sampling**	**Δ (LTM-WS)**
		Experiment 1	**Experiment 2**	
Aroma vegetable	Aroma-greenchemical	6.25	2.36	3.89
Taste grain astringency	Taste-dryastringent	5.00	2.19	2.81
Aroma herbal	Aroma-greenchemical	5.95	3.21	2.74
Aroma red fruit	Aroma-redfruitwoody	5.55	2.85	2.70
Taste acid	Taste-sweetviscous (inverted)	7.00	4.46	2.54
Aroma mineral	Aroma-citrusmineral	5.15	2.66	2.49
Taste salty	Taste-bittersalty	3.65	1.27	2.38
Taste rough astringency	Taste-dryastringent	5.60	3.24	2.36
Aroma animal	Aroma-greenchemical	4.25	1.91	2.34
Taste bitter	Taste-bittersalty	4.60	2.30	2.30

Table 6. Top 10 underperfoming primary descriptors.

Focusing on the major mean differences, results show that the local experts tend to evaluate with higher appreciation their local wines, regarding negative sensory descriptors such as *PC3 Agreenchemical* and *PC1 Tdryastringent*.

3.5.2. Uneven calibration needs

There is an ongoing line of research focusing on cost-benefit of calibration stages [72]. Our findings emphasize uneven calibration needs. In **Table 6**, we have highlighted the initial sensory descriptors which could gain accuracy with a calibration stage, such as *mineral aroma* [56], as opposed to other attributes that, according to our results, need no calibration among wine experts, hence, saving time and money [73].

4. Conclusions

In this sensory study on wine typicality, four large-scale areas were identified. PGI MINHO was found to be the most typical of all PGIs, with several extreme rates for the Color, Aroma and Taste. Encompassed by the four Mediterranean PGIs (PGI Peninsula de Setubal, PGI Tejo, PGI Alentejano and PGI Algarve), the SOUTHERN cluster presented several extreme sensory assessments that were essentially opposed to the single-clustered PGI MINHO's profile. Alcohol, acidity, bitterness and astringency were cross-linked; the respective variations were correlated with published literature and expressed as key factors for the regional macroscale differentiation. Bitterness and astringency were found to be sensory different and related on a geographical scale, as bitterness was primarily affected by inland/coastal influence, while astringency confirmed its customary north/south dependence, this finding to be considered new. Moreover, with the proposed methodology, it was possible to achieve a novel nationwide sensory characterization of PGIs, overcoming present day limitations on macroscale sensory research and sample representativeness.

Results by memory of (1 out of 12) extracted PGI Beira Atlântico were compared with the outcome of wine sampling assessment by local experts, using the same sensory questionnaire, and were found significantly correlated. Major differences between results by memory and by sampling were found mostly on unpleasant (negative-prone) sensory descriptors, the local experts showing greater sympathy for local wines. The need for a calibration stage was found uneven regarding the overall group of scrutinized wine descriptors, envisaging potential accuracy on final results only with a selective calibration phase on fewer descriptors.

These findings may lead to condensed information on typicality, which may contribute to a feasible macroscale regulation of small-sized wine regions and allow Portuguese red wines to be readily understood and sorted by a larger group of consumers.

Supporting information

Sensory Questionnaire (**Table 7**).

http://dx.doi.org/10.5772/intechopen.73039

STILL RED WINES SENSORY QUESTIONNAIRE

PGI (REGION)

PLEASE ENTER ONE RESPONSE TO EACH QUESTION BY PLACING A TICK IN THE APPROPRIATE BOX.
PLEASE ENSURE ALL QUESTIONS ARE ANSWERED.

DO YOU RESIDE IN THIS REGION? YES ☐ NO ☐

HOW WELL DO YOU KNOW/UNDERSTAND THE SENSORY PROFILE OF THIS PGI WINES?

NOT WELL AT ALL (PLACE A TICK) — PERFECTLY WELL

0	1	2	3	4	5	6	7	8	9	10

YOU WILL BE ANSWERING ABOUT PGI WINES, EXCLUSIVELY.
BUT HOW DIFFERENT WOULD YOUR ASSESSMENT BE IF THIS QUESTIONNAIRE REFERRED TO ITS INNER PDO SENSORY PROFILE?

TOTALLY EQUAL (PLACE A TICK) — TOTALLY DIFFERENT

0	1	2	3	4	5	6	7	8	9	10

CHARACTERIZATION OF TYPICAL **COLOR** (CHOOSE ONLY ONE OF THIS COLOR CATEGORIES, ACCORDING TO PGI TYPICITY)

VIOLET - PURPLE ☐
PURPLE - RUBY ☐
RUBY - GARNET ☐
GARNET - BRICK RED ☐

REGARDING THE CHOSEN COLOR CATEGORY, PLEASE TICK ONLY ONE CASE ACCORDING TO ITS PGI TYPICAL INTENSITY

EXTREMELY DILUTED (PLACE A TICK) — EXTREMELY INTENSE

0	1	2	3	4	5	6	7	8	9	10

CHARACTERIZATION OF TYPICAL **AROMA**
(PLEASE ENSURE ALL QUESTIONS ARE ANSWERED, EVEN IF YOU DO NOT ENVISAGE ONE OR MORE AROMATIC CATEGORIES WITHIN PGI TYPICITY)

GLOBAL AROMATIC INTENSITY **PGI's typicity on overall aromatic intensity**

NO TRACE — EXTREMELY INTENSE

0	1	2	3	4	5	6	7	8	9	10

FLORAL AROMAS **Intensity of rose, orange blossom, violet, carnation, ...**

NO TRACE — EXTREMELY INTENSE

0	1	2	3	4	5	6	7	8	9	10

DRY FLOWER AROMAS **Intensity of hay, dried rose, dried chamomile, ...**

NO TRACE — EXTREMELY INTENSE

0	1	2	3	4	5	6	7	8	9	10

HERBAL AROMAS **Intensity of minths, thyme, grass, infusions, anise, mediterranean bush, ...**

NO TRACE — EXTREMELY INTENSE

0	1	2	3	4	5	6	7	8	9	10

VEGETABLE AROMAS **Intensity of red peppers, green peppers, olives, ...**

NO TRACE — EXTREMELY INTENSE

0	1	2	3	4	5	6	7	8	9	10

MINERAL AROMAS **Intensity of slate, clay, chalk, graphite, hot earth, silex, petroly, mineral-water-like, ...**

NO TRACE — EXTREMELY INTENSE

0	1	2	3	4	5	6	7	8	9	10

CITRUS FRUIT AROMAS **Intensity of orange, lemon, bergamot, grapefruit, ...**

NO TRACE — EXTREMELY INTENSE

0	1	2	3	4	5	6	7	8	9	10

0	1	2	3	4	5	6	7	8	9	10

NO TRACE — EXTREMELY INTENSE

0	1	2	3	4	5	6	7	8	9	10

BLACK FRUIT AROMAS Intensity of blackcurrant, blueberry, mulberry, ...

NO TRACE											EXTREMELY INTENSE
	0	1	2	3	4	5	6	7	8	9	10

STONE FRUIT AROMAS Intensity of black plum, red plum,dark cherry, red cherry, ...

NO TRACE											EXTREMELY INTENSE
	0	1	2	3	4	5	6	7	8	9	10

DRIED FRUIT AROMAS Intensity of raisin, dried plum, dried fig, dried banana, ...

NO TRACE											EXTREMELY INTENSE
	0	1	2	3	4	5	6	7	8	9	10

NUT FRUIT AROMAS Intensity of almond, walnut, chestnut, pine nut, ...

NO TRACE											EXTREMELY INTENSE
	0	1	2	3	4	5	6	7	8	9	10

JAM & JELLY FRUIT AROMAS Intensity of red fruit or black fruit preserve, with or without alcohol, ...

NO TRACE											EXTREMELY INTENSE
	0	1	2	3	4	5	6	7	8	9	10

BREAD & PASTRY AROMAS Intensity of bread, butter, egg pastry and bakery, vanilla, ...

NO TRACE											EXTREMELY INTENSE
	0	1	2	3	4	5	6	7	8	9	10

SPICY AROMAS Intensity of pepper, clove, nutmeg, cinnamon, cocoa bean, coffee bean, ...

NO TRACE											EXTREMELY INTENSE
	0	1	2	3	4	5	6	7	8	9	10

CARAMELIZED AROMAS Intensity of caramel, honey, molasses, pollen, ...

NO TRACE											EXTREMELY INTENSE
	0	1	2	3	4	5	6	7	8	9	10

WOODY AROMAS Intensity of fresh (eucalyptus, cedar, resin, ...), smoked and burnt wood (ink, toasted bread, grilled nuts, toasted coffee, chocolate, ...)

NO TRACE											EXTREMELY INTENSE
	0	1	2	3	4	5	6	7	8	9	10

CHEMICAL AROMAS (POSITIVE) Intensity of tooth paste, medicinal, glue, canned fruit, metalic, ...)

NO TRACE											EXTREMELY INTENSE
	0	1	2	3	4	5	6	7	8	9	10

ANIMAL AROMAS (POSITIVE) Intensity leather, meat, bacon, musk, wet fur, ...

NO TRACE											EXTREMELY INTENSE
	0	1	2	3	4	5	6	7	8	9	10

CHARACTERIZATION OF TYPICAL **TASTE**

(PLEASE ENSURE ALL QUESTIONS ARE ANSWERED, EVEN IF YOU DO NOT ENVISAGE ONE OR MORE TASTANT CATEGORIES WITHIN PGI TYPICITY)

BUBBLY Intensity of bubbly, carbonated mouthfeel

NO TRACE											EXTREMELY INTENSE
	0	1	2	3	4	5	6	7	8	9	10

SWEET Intensity of sweet, sugary taste

NO TRACE											EXTREMELY INTENSE
	0	1	2	3	4	5	6	7	8	9	10

http://dx.doi.org/10.5772/intechopen.73039

ACID (SOUR) — Intensity of sour, acidic, fresh taste, as in lemon juice

NO TRACE												EXTREMELY INTENSE
	0	1	2	3	4	5	6	7	8	9	10	

SALTY — Intensity of salty, sodium or calcium tartrate water

NO TRACE												EXTREMELY INTENSE
	0	1	2	3	4	5	6	7	8	9	10	

BITTER — Intensity of bitter taste, as in coffee, resin, quinine

NO TRACE												EXTREMELY INTENSE
	0	1	2	3	4	5	6	7	8	9	10	

DRY (MOUTHFEEL) — Intensity of lack of lubrication, desiccation in the mouth

NO TRACE												EXTREMELY INTENSE
	0	1	2	3	4	5	6	7	8	9	10	

SMOOTH (MOUTHFEEL) — Intensity of supple, velvet, silk, smooth coating of mouth surfaces

NO TRACE												EXTREMELY INTENSE
	0	1	2	3	4	5	6	7	8	9	10	

ROUGH (MOUTHFEEL) — Intensity of grippy, chewy, hard, full, fleshy, harsh mouthfeel

NO TRACE												EXTREMELY INTENSE
	0	1	2	3	4	5	6	7	8	9	10	

GRAIN (MOUTHFEEL) — Intensity of clay, talk, chalky, grainy, dusty matter brushing against mouth surfaces

NO TRACE												EXTREMELY INTENSE
	0	1	2	3	4	5	6	7	8	9	10	

ASTRINGENT (GLOBAL) — Intensity of global mouthfeel impact, from light, diluted to puckery, chewy interaction

NO TRACE												EXTREMELY INTENSE
	0	1	2	3	4	5	6	7	8	9	10	

FULL BODY — Intensity of weight, dry extract, fullness

NO TRACE												EXTREMELY INTENSE
	0	1	2	3	4	5	6	7	8	9	10	

ALCOHOL — Intensity of caustic, burning, heat sensations

NO TRACE												EXTREMELY INTENSE
	0	1	2	3	4	5	6	7	8	9	10	

OILY (MOUTHCOAT) — Intensity of oily, viscous, round textures

NO TRACE												EXTREMELY INTENSE
	0	1	2	3	4	5	6	7	8	9	10	

LENGTH (PERSISTENCE) — Intensity of persistence, length

NO TRACE												EXTREMELY INTENSE
	0	1	2	3	4	5	6	7	8	9	10	

PLEASE DOUBLE-CHECK THAT ALL QUESTIONS HAVE BEEN ANSWERED.

THANK YOU FOR YOUR KIND COOPERATION.

Table 7. Sensory Questionnaire.

Author details

Anibal Jose-Coutinho[1]*, Patricia Avila[2] and Jorge M. Ricardo-da-Silva[1]

*Address all correspondence to: ajosecoutinho@gmail.com

1 Laboratory Ferreira Lapa, Instituto Superior de Agronomia, Linking Landscape Environment Agriculture and Food (LEAF), University of Lisbon, Lisbon, Portugal

2 Department of Social Methods Research, University Institute of Lisbon (ISCTE-IUL), Lisbon, Portugal

References

[1] Maitre I, Symoneaux R, Jourjon F, Mehinagic E. Sensory typicality of wines: How scientists have recently dealt with this subject. Food Quality and Preference. 2010;**21**(7):726-731. DOI: 10.1016/j.foodqual.2010.06.003

[2] Endrizzi I, Aprea E, Biasioli F, Corollaro ML, Demattè ML, Penasa M, Bittante G, Gasperi F. Implementing sensory analysis principles in the quality control of PDO products: A critical evaluation of a real-world case study. Journal of Sensory Studies. 2013;**28**(1):14-24. DOI: 10.1111/joss.12018

[3] Piclin N, Pintore M, Lanza C, Scacco A, Guccione S, Giurato L, Chrétien JR. Sensory analysis of red wines: Discrimination by adaptive fuzzy partition. Journal of Sensory Studies. 2007;**23**(2008):558-569. DOI: 10.1111/j.1745-459X.2008.00172.x

[4] Heymann H, Machado B, Torri L, Robinson a L. How many judges should one use for sensory descriptive analysis? Journal of Sensory Studies. 2012;**27**(2):111-122. DOI: 10.1111/j.1745-459X.2012.00373.x

[5] Perrin L, Symoneaux R, Maître I, Asselin C, Jourjon F, Pagès J. Comparison of three sensory methods for use with the napping® procedure: Case of ten wines from Loire valley. Food Quality and Preference. 2008 Jan [cited 2014 Jun 15];**19**(1):1-11. DOI: 10.1016/j.foodqual.2007.06.005

[6] Tempere S, Hamtat ML, Bougeant JC, de Revel G, Sicard G. Learning Odors: The impact of visual and olfactory mental imagery training on odor perception. Journal of Sensory Studies. 2014;**29**(6):435-449. DOI: 10.1111/joss.12124

[7] Potter MC. Very short-term conceptual memory. Memory & Cognition. 1993;**21**(2):156-161. DOI: 10.3758/BF03202727

[8] Roth JK, Courtney SM. Neural system for updating object working memory from different sources: Sensory stimuli or long term memory. NeuroImage. 2007 Nov 15;**38**(3):617-630. DOI: 10.1016/j.neuroimage.2007.06.037

[9] Prescott J. Flavour as a psychological construct: Implications for perceiving and measuring the sensory qualities of foods. Food Quality and Preference. 1999 Jul;**10**(4–5):349-356. DOI: 10.1016/S0950-3293(98)00048-2

[10] Lawless LJR, Civille GV. Developing lexicons: A review. Journal of Sensory Studies. 2013;**28**(4):270-281. DOI: 10.1111/joss.12050

[11] Ballester J, Patris B, Symoneaux R, Valentin D. Conceptual vs. perceptual wine spaces: Does expertise matter? Food Quality and Preference. 2008;**19**(3):267-276. DOI: 10.1016/j.foodqual.2007.08.001

[12] Urdapilleta I, Parr W, Dacremont C, Green J. Semantic and perceptive organisation of sauvignon blanc wine characteristics: Influence of expertise. Food Quality and Preference. 2011 Jan;**22**(1):119-128; [cited 2014 Jun 3]. DOI: 10.1016/j.foodqual.2010.08.005

[13] Gawel R, Campus W, Osmond G. The use of language by trained and untrained experienced wine tasters. Journal of Sensory Studies. 1997;**12**(8):267-284. DOI: 10.1111/j.1745-459X.1997.tb00067.x

[14] Lawless HT, Heymann H. Sensory Evaluation of Food: Principles and Practices. Vol. XXIII. New York: Springer-Verlag; 2010. p. 596. DOI: 10.1007/978-1-4419-6488-5

[15] Stone H. Rapid Sensory Profiling Techniques. Elsevier; 2015. pp. 27-51. DOI: 10.1533/9781782422587.1.27

[16] Stone H, Sidel J, Oliver S, Woolsey A, Singleton RC. Sensory evaluation by quantitative descriptive analysis. In: Descriptive Sensory Analysis in Practice [Internet]. Food & Nutrition Press, Inc; 2008. pp. 23-34. DOI: 10.1002/9780470385036.ch1c

[17] Dehlholm C, Brockhoff PB, Meinert L, Aaslyng MD, Bredie WLP. Rapid descriptive sensory methods - comparison of free multiple sorting, partial napping, napping, flash profiling and conventional profiling. Food Quality and Preference. 2012;**26**(2):267-277. DOI: 10.1016/j.foodqual.2012.02.012

[18] Courcoux P, Qannari EM, Faye P. Rapid Sensory Profiling Techniques. Elsevier; 2015. pp. 153-185. DOI: 10.1533/9781782422587.2.153

[19] Thuillier B, Valentin D, Marchal R, Dacremont C. Pivot profile: A new descriptive method based on free description. Food Quality and Preference. 2015;**42**:66-77. DOI: 10.1016/j.foodqual.2015.01.012

[20] O'Sullivan MG. A Handbook for Sensory and Consumer-Driven New Product Development. Woodhead Publishing; 2017. p. 370. DOI: 10.1016/B978-0-08-100352-7.12001-9

[21] Mayson R. The Wines and Vineyards of Portugal. Mitchell Beazley; 2006. p. 368

[22] Fraga H, Malheiro AC, Moutinho-Pereira J, Jones GV, Alves F, Pinto JG, Santos JA. Very high resolution bioclimatic zoning of Portuguese wine regions: Present and future scenarios. Regional Environmental Change. 2013 Jun 6;**14**(1):295-306. DOI: 10.1007/s10113-013-0490-y

[23] Magalhães N, Oliveira A, Caeiro L. Contributo para a caracterização climática das Regiões Vitícolas de Portugal Continental. In: 3º Simpósio de Vitivinicultura do Alentejo. Évora; 1995. pp. 81-90

[24] Tonietto J, Sotes Ruiz V, Celso Zanus M, Montes C, Uliarte EM, Bruno LA, Climaco P, Pena A, Crivellaro Guerra C, Catania CD, Kohlberg EJ, Pereira GE, Ricardo-Da-Silva JM, Ragout JV, Navarro LV, Laureano O, De Castro R, Del Monte RF, Del Monte SA, Gomez-Miguel V, Carbonneau A. The effect of viticultural climate on red and white wine typicity. Characterization in Ibero-American grape-growing regions. Journal International des Sciences de la Vigne et du Vin. 2014;(Special Laccave):19-23

[25] Costa JC, Aguiar C, Capelo JH, Lousã M, Neto C. Biogeografia de Portugal Continental. Quercetea; 1990

[26] Sun B, Ricardo-Da-Silva JM, Spranger MI. Quantification of catechins and proanthocyanidins in several Portuguese grapevine varieties and red wines. Ciência e Técnica Vitivinícola. 2001;**16**(1):23-34

[27] Sun B, Sa M, Leandro C, Caldeira I, Duarte F, Spranger MI. Reactivity of polymeric proanthocyanidins toward salivary proteins. Journal of Agricultural and Food Chemistry. 2013;**61**:939-946

[28] Vilanova M, Soto B. The impact of geographic origin on sensory properties of Vitis Vinifera cv. Mencía. Journal of Sensory Studies. 2005;**20**:503-511. DOI: 10.1111/j.1745-459X.2005.00046.x

[29] Jose-Coutinho A, Avila P, Ricardo-Da-Silva JM. Sensory profile of Portuguese white wines using long-term memory: A novel Nationwide approach. Journal of Sensory Studies. 2015;**30**:381-394. DOI: 10.1111/joss.12165

[30] Gomes LS, Jose-Coutinho A, da Silva AG, Ricardo-da-Silva JM. Sensory profile characterization and typicality assessment of PDO "Bairrada" and PGI "Beira Atlântico" red wines. Ciência e Técnica Vitivinícola. 2016;**31**(2):73-87. DOI: 10.1051/ctv/20163102073

[31] Parr WV, Heatherbell D, White KG. Demystifying wine expertise: Olfactory threshold, perceptual skill and semantic memory in expert and novice wine judges. Chemical Senses. 2002;**27**(8):747-755

[32] Hernández B, Sáenz C, de la Hoz JF, Alberdi C, Alfonso S, Diñeiro JM. Assessing the color of red wine like a taster's eye. Color Research and Application. 2009 Apr;**34**(2):153-162. DOI: 10.1002/col.20475

[33] Ballester J, Abdi H, Langlois J, Peyron D, Valentin D. The odor of Colors: Can wine experts and novices distinguish the Odors of white, red, and Rosé wines? Chemosensory Perception. 2009 Oct 20;**2**(4):203-213. DOI: 10.1007/s12078-009-9058-0

[34] Noble AC, Arnold RA, Buechsenstein J, Leach EJ, Schmidt JO, Stern PM. Modification of a standardized system of wine aroma terminology. American Journal of Enology and Viticulture. 1987;**38**(2):143-146

[35] Gawel R, Iland PG, Francis IL. Characterizing the astringency of red wine: A case study. Food Quality and Preference. 2001;**12**(1):83-94. DOI: 10.1016/s0950-3293(00)00033-1

[36] Abdi H. Coefficient of variation. In: Salkind NJ, Dougherty DM, Frey B, editors. Encyclopedia of Research Design. Sage; 2010. pp. 169-171

[37] Valentova H, Skrovankova S, Panovska Z, Pokorny J. Time-intensity studies of astringent taste. Food Chemistry. 2002;**78**(1):29-37. DOI: Pii s0308-8146(01)00330-210.1016/s0308-8146(01)00330-2

[38] Gawel R. Red wine astringency: A review. Australian Journal of Grape and Wine Research. 1998;**4**:74-95. DOI: 10.1111/j.1755-0238.1998.tb00137.x

[39] Lesschaeve I, Noble AC. Polyphenols: Factors influencing their sensory properties and their effects on food and beverage preferences. The American Journal of Clinical Nutrition. 2005 Jan;**81**(1 Suppl):330S-335S

[40] Pickering GJ, Robert G. Perception of mouthfeel sensations elicited by red wine are associated with sensitivity to 6-N-propylthiouracil. Journal of Sensory Studies. 2006;**21**(3):249-265. DOI: 10.1111/j.1745-459X.2006.00065.x

[41] Climaco P, Ricardo-da-Silva JM, Laureano O, De Castro R, Tonietto J. O clima viticola das principais regioes produtoras de uvas para vinho de Portugal. In: Clima, zonificación y tipicidad del vino en regiones vitivinícolas iberoamericanas. Madrid: CYTED; 2012. pp. 313-357

[42] Mira de Orduña R. Climate change associated effects on grape and wine quality and production. Food Research International. 2010 Aug;**43**(7):1844-1855; [cited 2014 May 27]. DOI: 10.1016/j.foodres.2010.05.001

[43] Kontoudakis N, Gonzalez E, Gil M, Esteruelas M, Fort F, Miquel Canals J, Zamora F. Influence of wine pH on changes in color and polyphenol composition induced by micro-oxygenation. Journal of Agricultural and Food Chemistry. 2011;**59**(5):1974-1984. DOI: 10.1021/jf103038g

[44] Alcalde-Eon C, Escribano-Bailon MT, Santos-Buelga C, Rivas-Gonzalo JC. Changes in the detailed pigment composition of red wine during maturity and ageing - a comprehensive study. Analytica Chimica Acta. 2006;**563**(1–2):238-254. DOI: 10.1016/j.aca.2005.11.028

[45] Figueiredo-Gonzalez M, Cancho-Grande B, Simal-Gandara J. Evolution of colour and phenolic compounds during Garnacha Tintorera grape raisining. Food Chemistry. 2013;**141**:3230-3240. DOI: 10.1016/j.foodchem.2013.05.142

[46] Hein K, Ebeler SE, Heymann H. Perception of fruity and vegetative aromas in red wine. Journal of Sensory Studies. 2009;**24**(3):441-455. DOI: 10.1111/j.1745-459X.2009.00220.x

[47] Escudero A, Campo E, Fariña L, Cacho J, Ferreira V. Analytical characterization of the aroma of five premium red wines . Insights into the role of odor families and the concept of fruitiness of wines. Journal of Agricultural and Food Chemistry. 2007;**55**:4501-4510. DOI: 10.1021/jf0636418

[48] Pineau B, Barbe JC, Van Leeuwen C, Dubourdieu D. Contribution of grape skin and fermentation microorganisms to the development of red- and black-berry aroma in merlot wines. Journal International des Sciences de la Vigne et du Vin. 2011;**45**:27-37. DOI: 10.20870/oeno-one.2011.45.1.1485

[49] Jones PR, Gawel R, Francis IL, Waters EJ. The influence of interactions between major white wine components on the aroma, flavour and texture of model white wine. Food Quality and Preference. 2008;**19**(6):596-607. DOI: 10.1016/j.foodqual.2008.03.005

[50] Carvalho E, Mateus N, Plet B, Pianet I, Dufourc E, De Freitas V. Influence of wine pectic polysaccharides on the interactions between condensed tannins and salivary proteins. Journal of Agricultural and Food Chemistry. 2006;**54**:8936-8944. DOI: 10.1021/jf061835h

[51] Quijada-Morín N, Regueiro J, Simal-Gándara J, Tomás E, Rivas-Gonzalo JC, Escribano-Bailón MT. Relationship between the sensory-determined astringency and the flavanolic composition of red wines. Journal of Agricultural and Food Chemistry. 2012;**60**:12355-12361. DOI: 10.1021/jf3044346

[52] Vidal S, Francis L, Williams P, Kwiatkowski M, Gawel R, Cheynier V, Waters E. The mouth-feel properties of polysaccharides and anthocyanins in a wine like medium. Food Chemistry. 2004;**85**:519-525. DOI: 10.1016/S0308-8146(03)00084-0

[53] Runnebaum RC, Boulton RB, Powell RL, Heymann H. Key constituents affecting wine body - an exploratory study. Journal of Sensory Studies. 2011;**26**(1):62-70. DOI: 10.1111/j.1745-459X.2010.00322.x

[54] Mateus N, Machado JM, de Freitas V. Development changes of anthocyanins in *Vitis vinifera* grapes grown in the Douro Valley and concentration in respective wines. Journal of the Science of Food and Agriculture. 2002;**82**(14):1689-1695. DOI: 10.1002/jsfa.1237

[55] Guedes-de-Pinho P, Falqué E, Castro M, E Silva HO, Machado B, Ferreira ACS. Further insights into the floral character of Touriga Nacional wines. Journal of Food Science. 2007 Aug 1;**72**(6):S396-S401. DOI: 10.1111/j.1750-3841.2007.00405.x

[56] Heymann H, Hopfer H, Bershaw D. An exploration of the perception of minerality in white wines by projective mapping and descriptive analysis. Journal of Sensory Studies. 2014;**29**(1):1-13. DOI: 10.1111/joss.12076

[57] Kontoudakis N, Esteruelas M, Fort F, Canals JM, De Freitas V, Zamora F. Influence of the heterogeneity of grape phenolic maturity on wine composition and quality. Food Chemistry. 2011;**124**(3):767-774. DOI: 10.1016/j.foodchem.2010.06.093

[58] Fernandes P, Ricardo-da-Silva JR, Castro R. Comportamento Agronómico e Enológico das Castas Syrah e Touriga Nacional em seis "Terroirs" de Portugal. In: 8° Simpósio de Vitivinicultura do Alentejo. Évora. 2010. pp. 143-153

[59] Baker AK, Ross CF. Sensory evaluation of impact of wine matrix on red wine finish: A preliminary study. Journal of Sensory Studies. 2014;**29**:139-148. DOI: 10.1111/joss.12089

[60] Nurgel C, Pickering G. Modeling of sweet, bitter and irritant sensations and their interactions elicited by model ice wines. Journal of Sensory Studies. 2006;**21**(5):505-519. DOI: 10.1111/j.1745-459X.2006.00081.x

[61] Tonietto J, Carbonneau AA. Multicriteria climatic classification system for grape-growing regions worldwide. Agricultural and Forest Meteorology. 2004 Jul;**124**(1–2):81-97. DOI: 10.1016/j.agrformet.2003.06.001

[62] Dopico-García MS, Fique A, Guerra L, Afonso JM, Pereira O, Valentão P, Andrade PB, Seabra RM. Principal components of phenolics to characterize red Vinho Verde grapes: Anthocyanins or non-coloured compounds? Talanta. 2008 Jun 15;**75**(5):1190-1202. DOI: 10.1016/j.talanta.2008.01.012

[63] Gonzalo-Diago A, Dizy M, Fernández-Zurbano P. Contribution of low molecular weight phenols to bitter taste and mouthfeel properties in red wines. Food Chemistry. 2014;**154**:187-198. DOI: 10.1016/j.foodchem.2013.12.096

[64] Gómez-Míguez MJ, Cacho JF, Ferreira V, Vicario IM, Heredia FJ. Volatile components of Zalema white wines. Food Chemistry. 2007;**100**:1464-1473. DOI: 10.1016/j.foodchem.2005.11.045

[65] Goldner MC, Zamora MC. Effect of polyphenol concentrations on astringency perception and its correlation with gelatin index of red wine. Journal of Sensory Studies. 2010;**25**(5):761-777. DOI: 10.1111/j.1745-459X.2010.00304.x

[66] Fontoin H, Saucier C, Teissedre P-L, Glories Y. Effect of pH, ethanol and acidity on astringency and bitterness of grape seed tannin oligomers in model wine solution. Food Quality and Preference. 2008 Apr;**19**(3):286-291. DOI: 10.1016/j.foodqual.2007.08.004

[67] Gonzalo-Diago A, Dizy M, Fernandez-Urbano P. Taste and Mouthfeel properties of red wines Proanthocyanidins and their relation to the chemical composition. Journal of Agricultural and Food Chemistry. 2013;**61**:8861-8870. DOI: 10.1021/jf401041q

[68] Jackson RS. Wine Science, Principles and Applications. Academic Press; 2014. p. 984. DOI: 10.1016/B978-0-12-381468-5.00001-4

[69] Fischer U, Noble AC. The effect of ethanol, Catechin concentration and pH on sourness and bitterness of wine. American Journal of Enology and Viticulture. 1994;**45**(1):6-10

[70] Le Berre E, Atanasova B, Langlois D, Etiévant P, Thomas-Danguin T. Impact of ethanol on the perception of wine odorant mixtures. Food Quality and Preference. 2007 Sep;**18**(6): 901-908. DOI: 10.1016/j.foodqual.2007.02.004

[71] Jose-Coutinho A, Avila P, Ricardo da Silva JM. Finding sensory profilers amongst red wine composition: A novel Nationwide approach. Ciencia e Tecnica VItivinicola. 2015;**30**(2): 69-83. DOI: doi.org/10.1051/ctv/20153002069

[72] Lesschaeve I. Sensory evaluation of wine and commercial realities: Review of current practices and perspectives. American Journal of Enology and Viticulture. 2007;**58**(2):252-258

[73] Findlay CJ, Castura JC, Schlich P, Lesschaeve I. Use of feedback calibration to reduce the training time for wine panels. Food Quality and Preference. 2006;**17**(3):266-276. DOI: 10.1016/j.foodqual.2005.07.005

Section 3

Economic Analysis and Valorization of Wine Products

Chapter 14

Wine Firm's Size and Economic Performance: Evidence from Traditional Portuguese Wine Regions

João Rebelo, Sofia Gouveia,
Lina Lourenço-Gomes and
Ana Alexandra Marta-Costa

Additional information is available at the end of the chapter

http://dx.doi.org/10.5772/intechopen.71320

Abstract

The wine market is an excellent example of monopolistic competition, demonstrating both vertical and horizontal product differentiation. The propensity toward monopolistic competition and the asymmetric preferences of wine consumption can justify the coexistence of different sized wineries. The main goal of this chapter is to analyze the economic performance of wineries, using indicators widely employed in economic and business literature, and to assess the potential influence of size on firm's performance. To achieve this goal, different statistical tests are applied to firms' data collected from two representative Portuguese demarcated wine regions (Douro and Vinho Verde). The results confirm that the wine firm's performance is a multidimensional construct, exhibiting similarities and differences, according to the index used. The size of firms increases their export performance but exhibits a nonstatistical effect on the financial performance. The indices of productivity and productive efficiency show that there is no standard behavior. The results of this study can be useful both for managers and public decision-makers, given the importance of controlling economic performance in order to guarantee the industry's competitiveness in an increasingly globalized wine business.

Keywords: competitiveness, demarcated wine regions, economic performance, size, wine industry, monopolistic competition

1. Introduction

Over recent decades, the wine industry has been under an intense process of globalization, with an impressive increase in competition as a consequence of the new suppliers entering the market: new world wine producing countries (such as Australia, New Zealand, South Africa,

Argentina, Chile, and more recently China), and at the same time, the decreasing consumption of the old world traditional wine-producing countries (namely, Italy, France, Spain, and Portugal). The opening up and globalization of the wine markets raise both challenges and opportunities to the wine regions and consequently to the respective wine firms, which must adopt strategies that address the consumer's demand in an economic and sustainable way.

The wine market reveals itself to be an excellent example of monopolistic competition, wherein the market is determined by significant variety, differentiated by growing regions, and consumer preference/demand for taste, price, and advertising. There is not only one market for wine, but many different ones depending on price, tastes, and brands, with consumers selecting within red or white wines, or even within grape varieties. This is why "wine" is a bundle of monopolistic and competitive markets and not a single competitive or monopoly undifferentiated market.

Following Church and Ware [1], the assumption of symmetry and consumers′ taste for variety supports the essential feature of monopolistic competition, wherein every brand is in competition with each other. Consequently, the focus of the models of monopolistic competition is not typically on strategic decisions regarding to product specification or design, since the products are assumed to be equally differentiated. Instead, the model focuses on the issue of the extent of variety: the number of products available in the market. Two conditions must be satisfied to achieve equilibrium: profit maximization and free-entry condition. In the process of profit maximization, it is expected that the group of larger firms has a behavior referred to as true or perfect monopolistic competition and the group of small firms as oligopolistic, in which the equilibrium is attained whether the firms compete over price or quantity. The free-entry condition implies that the profit must be an incentive to enter or exit. The profit of an additional entrant must be nonpositive and the profit of a firm in the industry non-negative. Ignoring the integer constraint, these two considerations imply that the profit of firms in the industry must be zero, at equilibrium. Moreover, in this market structure, the number of firms depends on product elasticity substitution and on the extent of scale economies. As elasticity substitution increases, products become less differentiated, and as result, the demand becomes more elastic, reducing the market power of the firm and the equilibrium prices tend toward a state of perfect competition. This squeezes price-cost margins, reducing profitability. The opposite occurs if the elasticity substitution decreases. Otherwise, the greater the extent of economies of scale, the smaller the number of firms that will be in the industry and hence varieties, leading to an increase in prices and in the market share at which firm's breakeven. Conversely, reduction in the extent of scale economies reduces the minimum price markup and the market share for a firm to break even, resulting in an increase in the number of firms where equilibrium occurs and in the variety of products.

Theoretically, the model of monopolistic competition justifies the coexistence of wineries of different sizes and production technologies, charging different prices for distinct wines, according to the consumer's willingness to pay. In this scenario, the profitability could be independent of the firm's size.

Portugal seems to be a good example of a traditional wine producing country, with a strong presence in the international market and a market structure of monopolistic competition.

In 2015, there were 39,727 farms of grape vine cultivation that represented a planted area of 201,445 ha [2], the fourth largest in European Union (EU). The total production of 2015 was 6,205,756 hl [2] (fifth in the EU, after France, Italy, Spain, and Germany and 11th in the world wine production). From the total wine production (77.76% protected Designation of Origin), 70.5% is red wine, 29.5% white wine, 13.2% fortified wine, namely Port wine. The domestic consumption reached 4.94 million hectoliters (79.6% of production), and 2.8 million hectoliters were exported (45.11% of production), meaning that Portugal is both an importer and exporter [2]. The average price of exporting was 2.63 Euro/Liter, against the 0.55 Euro/Liter for importing, meaning that Portugal exports wine with a higher added value [2].

Due to the soil, climatic, social, and economic factors, Portugal is a heterogeneous wine country, with very specific conditions as testified by one of the currently most recognized Portuguese winemakers (Dirk Nieeport in [3]): *Portugal has one thing that only few countries nowadays have - traditions, history, old vines, old and diversified varieties, and personalized methods of vinification*. Due to its diversity, the country has long since been divided into various wine regions, there currently being fourteen demarcated wine regions, each one with the own *terroir* characteristics. The Douro and Vinho Verde demarcated regions are representatives of the Portuguese wine *terroir* model, located in the Northeast and Northwest of Portugal, respectively. In 2015, the Douro region had an area of 43,659 ha (21.7% of the Portugal area) and produced 1,407,006 hl (22.7% of the total Portuguese production), shared by Port wine (almost 50%) and still wines. The Vinho Verde region extends along an area of 15,821 ha (7.9% of the Portuguese area) with a production of 693,026 hl (6.38% of the Portuguese production), only still wines [2].

In a microeconomic level, the Portuguese wine supply is characterized by the presence of a diversified typology of economic agents [2]: warehouse, distiller, bottler, exporter/importer, producer of vinegar and wine, merchant without establishment, preparer, producer, winemaker, and winemaker bottler. Considering the classification of economic activities in the EU (NACE) in the code 11021, production of still and liquors wines, in 2015, there were 1246 active firms, according to the database *racius* [4].

Similarly to other traditional wine countries [5], it is known that Portuguese wine firms, including the ones located in the Douro and Vinho Verde regions, embrace different marketing strategies that range from the low cost or mass market types to those based on product differentiation or a market niche. In general, the first strategy is adopted by the largest firms, which prefer to offer standardized wines and to cater to large distribution channels. On the other hand, small firms prefer to satisfy the needs of narrow markets, emphasizing the importance of differentiation based on the *terroir*, *appellation*, and geographic identity in order to enhance perceived wine quality, which endorses the monopolistic competition model. Given this multifaceted market structure, the importance of monitoring and controlling economic performance becomes clear, as such entrepreneurial strategies may guarantee survival in the long term. The assessment of a wine firm's performance is a critical managerial issue [6], despite its measure is a multidimensional construct, whose indicators to be used depend on the stakeholder perspective and on the outcomes to be achieved.

Researchers in economics and management strategy showed a long interest in understanding the main determinants of the firm performance in different market structures. Two theoretical alternatives approaches have been used [7] to analyze the importance of the industry and the business unit factors in the firm performance: the paradigm structure-conduct-performance (SCP) of the industrial organization [8, 9], suggesting that the industry structure is the main determinant of performance differences across firms, and the business strategy [10, 11], assuming that the resources and capabilities are the main determinants of differences of profitability across firms.

Some studies on the agrifood sector conclude that there is a direct relation between firm size and financial performance in the food industry [e.g., 7, 12]. The findings suggest a positive relationship between firm size with return on assets (ROA) [7, 12], return on equity (ROE), and return on investment (ROI) [12] and based in arguments like greater resources and market opportunities of large firms, which enables to be more efficient in the use of inputs, to benefit from economies of scale and to a stronger market power. However, other studies [13] reported a negative relation, showing that smaller firms can obtain a better financial performance. Thus, the empirical evidence is inconclusive on the effect of firm size on the financial performance of agrifood firms, suggesting the need of additional empirical research, that takes into account both the market structure and different indexes of firm performance.

Based on information collected from the Douro and Vinho Verde wine regions, the main goal of this chapter is to analyze the economic performance of wineries, using indicators widely employed in economic and management literature and to assess the potential influence of size on economic performance. To achieve this goal, different nonparametric and parametric tests are applied to indices related to export performance, profitability, productivity, productive efficiency, and solvency.

2. Methodology and sampling

2.1. Methodology

In line with the aims of the chapter in the first step, based on information included in firm's annual financial reports, indices are computed related to export performance, profitability, solvency ratios, labor productivity, and economic efficiency.

The index of export performance (EP), measured by the ratio between firm exports and turnover, reflects the firm's ability to be competitive in international markets. Size is an important factor in shaping the firm's potential for internationalization. In free markets, the size of a firm can reveal its potential market power, capability, and scale economies: all essential elements in gaining an international market. Vivas and Sousa [14] identify internationalization strategies of firms in the wine industry in Portugal, as well as the option to export to a wide range of countries, pointing out a lack of size of the production units, coupled with poor investment in the commercial area. This framework constraints the ability to compete with larger producers, more experienced in international markets and holding a more consistent commercial attitude. Therefore, it is expected that EP will be positively correlated with size.

Profitability indices measure the extent to which a business generates a profit from the resources applied to the business. Traditionally, to compute profitability, indices such as ROA (ratio between total earnings before interests and taxes and total assets), ROE (ratio between net income and total equity), and ROI (ratio between total earnings before interests and taxes and the sum of plants, equipment, and stocks) are used. In a market structure, such as monopolistic competition, where firms of different sizes coexist and survive, it is expected that there should be no strong statistical relationships between profitability and size, namely on the ROE.

Complementary to the profitability indices, the solvency ratios gauge the firms' ability to pay all financial obligations if all assets are sold or to continue operations after financial adversity. Two ratios can be used: the leverage or debt to equity ratio (Lev) and financial autonomy or equity to asset ratio (FA). Since microfirms have difficulties in accessing market capital and/or to bank loans, it is expected that these ratios change according to size.

To estimate productivity, two measures specifically relating to labor are used: turnover per employee (T/E) and added value per Employee (AV/E). In the profit maximization strategy, it is expected that managers try to optimize the use of labor, independently of size. Consequently, the indices of productivity may or may not be related to size.

The indices of economic (productive) efficiency allow us to get information about the relative performance of a firm in comparison with the optimum (frontier) that it is possible to achieve. Based on the firms included in a technologically homogeneous sample, the frontier and, consequently, productive efficiency can be computed through parametric or nonparametric approaches. In our case, productive efficiency is computed using the data envelopment analysis (DEA), a nonparametric approach initially proposed by several authors [15, 16]. This approach has the advantage of not needing to assume a functional form to the production (cost or profit) function, since the frontier surface is constructed through linear programming. The main disadvantage is that any deviation between observed and estimated value is considered as (in)efficiency, or even as a consequence of random factors. A firm is efficient if no other firm is able to produce a higher level of output from the same input (output oriented), or a firm is efficient if it produces the same output from less input (input oriented). Following the methodology described by Sellers and Alampi-Sottini [6], we apply an input-oriented model, since the inputs are under the control of firms more than the output. Three controllable inputs (cost of raw materials, number of employees, and value of debts[1]) have been chosen. Moreover, the output data on annual turnover have been considered. Assuming $j = 1, \ldots, n$ firms, each one using vector of m inputs, $X_j = (x_{1j}, x_{2j}, \ldots, x_{mj})$, to produce vector of outputs, $Y_j = (y_{1j}, y_{2j}, \ldots, y_{sj})$ and linear variable returns to scale (VRS). The programming model is given by Eq. (1):

$$\text{Max } z_0 = \theta + \varepsilon \sum_{r=1}^{s} s_r^+ + \varepsilon \sum_{i=1}^{m} s_i^-$$

subject to:

[1]This variable is included because access to loans and the correspondent cost is a crucial issue in the firm performance.

$$\sum_{j=1}^{n} x_{ij}\lambda_j + s_i^- = x_{r0}$$

$$\sum_{j=1}^{n} y_{rj}\lambda_j - s_r^+ = \theta y_{i0}$$

$$\sum_{j=1}^{n} \lambda_j = 1$$

$$\lambda_j, s_r^+, s_i^- \geq 0$$

$$j = 1, \ldots, n, r = 1, \ldots s; i = 1, \ldots, m. \quad (1)$$

where θ is the measure of efficiency for each unit. A firm is efficient if $\theta^*=1$ and all the slacks are equal to zero, and $\varepsilon > 0$ is a non-Archimedean element that is defined to be smaller than any positive real number. Since the indices of productive efficiency summarize the relative management performance in the inputs allocation, they may or not be related to size, although a positive relation stands out.

In the second step, the data are analyzed using descriptive statistics and both nonparametric and parametric tests, in order to test if there is an association between indices and size. More specifically, since we have different groups of observations for each variable, the Kruskal-Wallis equality-of-population rank test is used. To test if there are differences between means of groups in the same variable, the Hotelling's T2 is used. Furthermore, to check the robustness of the nonparametric tests, ordinary least squares (OLS) regression models are estimated, where the dependent variables are the indices and the explanatory variables are dummies according to size.

2.2. Sampling

In accordance with the aim of the study, as well as to assure the technological homogeneity of the sample, two demarcated wine regions are considered: the Douro region and the Vinho Verde region, both in the North of Portugal. The Douro is one of the oldest world wine regions, fitting well the *terroir* model [17]. It is characterized by the production of two different products: the fortified Port wine and still wines. The Vinho Verde region only produces still wines. In both regions, there are a very large number of wine brands that are sold in different markets at different prices.

Given the economic and technological differences between the two regions, it is recommended that each one of them is analyzed as a separate sample. Therefore, in order to define the respective sample of each region, we start by selecting the firms that are registered as a winery in the regulator of Port and/or still Douro wines (Port and Douro Wines Institute: www.ivdp.pt) and in the regulator of Vinho Verde (Commission of Viticulture from Vinhos Verdes region: www.vinhoverde.pt). Moreover, to overcome any possible effects of seasonality, we decided to collect economic and financial data that are available for the last 2 years, 2014 and 2015.

This means that this study is carried out with a pooled sample data of firms and years, and the databases are obtained from the Financial and Economic Entrepreneurial Data Base provided by Informa Dun & Bradstreet (D&B) and reporting wine firms included in the 11,021 NACE-2009 code (production of still and liquors wines). Considering that not all the financial reports include the information required for this research, we eventually got a sample database that includes: for the Douro region, 434 pooled observations, 218 for 2014 and 216 for 2015[2], with 204 wine firms observed for both years and 26 for only 1 year, and for the Vinho Verde region, the sample includes a pool of 299 observations, 154 of 2014 and 145 of 2015, in a total of 162 wine firms, 137 observed for both years and 25 for only 1 year.

Table 1 shows the main descriptive statistics of the variables used, for both wine regions. The range between the minimum and the maximum, together with the comparison between the standard deviation (SD) and mean, allows us to conclude that each sample reflects a heterogeneous market structure, typical of monopolistic competition. According to the value of the mean of the variables, the firms of the Douro region are bigger than those of the Vinho Verde region (**Table 1**).

In order to build a detailed basis of information about the market structure of each wine region, we have classified firms according their size, adopting the European Commission recommendation [18][3]. This recommendation is used to obtain criteria to define the size of the firm, the number of employees, the turnover, or total assets on the balance sheet (**Table 2**). Consequently, these criteria should be used to analyze any industry, including in the wine business. These criteria also suggest the use of financial and productive indicators to assess the performance of the industry.

3. Results

In this section, performance indices (export performance, profitability, solvency, productivity, and efficiency) are presented, both for the Douro and Vinho Verde regions (**Table 3**). As can been seen, the statistical descriptive measures show a highly heterogeneous industry in and for both wine regions.

For the Douro region, the average of the index of EP is 18%. The profitability ratios are ROA = 0.9%, ROE = 20.5%, and ROI = 68.1%. Regarding solvency, the mean of leverage (Lev) is 2.25 and the equity/asset (AF) ratio is 28%. The productivity indices show that on average, the amount sold per employee (T/E) is 121.5 thousand Euro and the added value per employee (AV/E) is 31.5 thousand Euro. Comparing the standard deviation (SD) with the mean, we conclude that there is a high relative dispersion between observations, which is confirmed through the values of the weighted means of the indices: EP = 58%, ROA = 3.9%, ROE = 4.4%,

[2] Considering that the total national number of firms active in 2015 is 1246, the sample of Douro represents 17.33% and Vinho Verde 11.64%.

[3] The classification of firms by the EU, according their size, aims to improve the consistency and effectiveness of policies targeting small and medium enterprises and would, to this end, homogenize the definition of the size of the firm between member states, to limit the risk of distortion of competition resulting from different levels of public support.

Variables	Minimum	Maximum	Mean	SD
Douro region				
Assets (10^3 euro)	3.000	257,010.000	7,886.000	29,006.000
Equity (10^3 euro)	−2060.000	184,547.000	4249.000	18,806.000
Investments (10^3 euro)	0.100	220,150.000	6231.000	23,708.000
Turnover (10^3 euro)	0.100	132,503.000	2881.000	12,199.000
Added value (10^3 euro)	−1,012.000	46,618.000	735.000	3299.000
Employees (#)	1	593	16.000	53.000
Vinhos verdes region				
Assets (10^3 euro)	6.900	50,727.000	2137.000	5167.000
Equity (10^3 euro)	−1,277.000	31,492.000	966.000	2898.000
Investments (10^3 euro)	0.100	41,583.000	1594.000	3973.000
Turnover (10^3 euro)	0.400	31,999.000	996.000	3130.000
Added value (10^3 euro)	−660.000	13,401.000	265.000	1112.000
Employees (#)	1	148	7.000	14.000

Legend: SD – Standard Deviation.

Table 1. Descriptive statistics of the variables used for Douro (434 observations) and Vinhos Verdes (299 observations) regions (years 2014 and 2015).

Size	Criteria
Micro	Employment: < 10 workers; and turnover or assets: < 2 million Euro
Small	10 ≤ workers < 50; and 2 ≤ turnover or assets < 50 million Euro
Medium	50 ≤ workers < 250; and or 50 ≤ turnover or assets < 250 million Euro
Big	Workers ≥ 250; and or turnover or assets ≥ 250 million Euro

Table 2. Criteria to identify size of the firms using European Commission recommendation [18].

ROI = 4.94%, Lev = 0.85, FA = 0.54, T/E = 181.8 thousand Euro, AV/E = 46.4 thousand Euro, a picture that is different of that one given by the arithmetic mean and calls for a group analysis. The average of economic efficiency (EF) is 0.454 meaning that the Douro wine industry could use 55.6% fewer inputs to obtain the same level of output, if all the firms adopted the observed best productive practice. The averages of technical efficiency (TE) and scale efficiency (SE) are 0.692 and 0.692, respectively, suggesting that the deviation from the production frontier is due to a similar extent to the inefficient use of inputs (technical efficiency) and to firms not operating at an optimum size (scale efficiency).

Regarding the Vinho Verde region, the arithmetic means of the sample are: EP = 11.8%, ROA = −0.5%, ROE = −8.2%, ROI = −0.07%, Lev = 7.52, FA = 0.25, T/E = 101.3 thousand Euro, AV/E = 24.3 thousand Euro. However, as with the Douro region, the simple means compares

Indices	Minimum	Maximum	Mean	SD
Douro region				
EP	0.000	0.930	0.180	0.250
Profitability				
ROA	−8.415	1.060	0.009	0.430
ROE	−34.850	41.170	0.205	3.180
ROI	−10.020	165.840	0.681	8.920
Solvency				
Lev	−258.000	204.720	2.250	20.890
FA	−14.490	0.970	0.280	0.940
Productivity				
T/E (10^3 Euro)	80.000	2799.500	121.500	227,917.000
AV/E (10^3 Euro)	−120.706	305.200	31.500	37,534.000
Efficiency				
EF	0.000	1.000	0.454	0.250
TE	0.083	1.000	0.692	0.282
SE	0.000	1.000	0.693	0.267
Vinhos Verdes region				
EP	0.000	0.993	0.118	0.203
Profitability				
ROA	−1.116	0.899	−0.005	0.163
ROE	−21.665	8.634	−0.082	1.800
ROI	−10.295	4.511	−0.073	0.904
Solvency				
Lev	−18.731	897.680	7.517	57.255
FA	−5.165	0.998	0.252	0.706
Productivity				
T/E (10^3 Euro)	0.398	3053.000	101.337	263.676
AV/E (10^3 Euro)	−83.004	410.333	24.287	38.608
Efficiency				
EF	0.008	1.000	0.417	0.227
TE	0.084	1.000	0.678	0.285
SE	0.008	1.000	0.675	0.283

Legend: SD – Standard Deviation; EP - Export Performance; ROA - Return on Assets; ROE - Return On Equity; ROI - Return on Investment; Lev - Leverage; FA - Financial Autonomy; T/E - Turnover per Employee; AV/E - Added Value per Employee; EF - Economic Efficiency; TE - Technical Efficiency; SE - Scale Efficiency.

Table 3. Indices of performance of Douro (434 observations) and Vinho Verde regions (years 2014 and 2015).

poorly with the weighted means, which have the following values: EP = 30.7%, ROA = 4.4%, ROE = 5.3%, ROI = 5.9%, Lev = 1.21, FA =0.45, T/E = 134.4 thousand Euro, and AV/E = 35.7 thousand Euro. The average of EF is 0.417, meaning that Vinho Verde firms could use 59.3% fewer inputs if all of them were on the production frontier. The averages of TE and SE are 0.678

	EP	ROA	ROE	ROI	Lev	FA	T/E	AV/E	EF	TE	SE
EP	1										
ROA	0.036	1									
ROE	−0.052	0.085***	1								
ROI	−0.051	0.134*	0.022	1							
Lev	0.022	0.029	−0.009	−0.007	1						
FA	0.143*	0.549*	−0.022	−0.044	0.004	1					
T/E	0.057	0.058	0.039	−0.015	−0.003	−0.052	1				
AV/E	0.244*	0.303*	0.111*	0.026	0.019	0.004	0.632*	1			
EF	0.177*	0.199*	0.086***	0.045	0.017	0.235*	0.538*	0.619*	1		
TE	0.038	−0.015	0.057	0.078	−0.065	0.026	0.258*	0.238*	0.499*	1	
SE	0.164*	0.200*	0.016	−0.021	0.092**	0.254*	0.319*	0.412*	0.601*	−0.346*	1

*, **, *** statistically significant at 1%, 5% and 10%, respectively.
Legend: EP - Export Performance; ROA - Return on Assets; ROE - Return On Equity; ROI - Return on Investment; Lev - Leverage; FA - Financial Autonomy; T/E - Turnover per Employee; AV/E - Added Value per Employee; EF - Economic Efficiency; TE - Technical Efficiency; SE - Scale Efficiency.

Table 4. Pearson correlation coefficients of indices of performance of Douro region (434 observations, years 2014 and 2015).

and 0.675, respectively, indicating that the causes of productive inefficiency are due to a similar extent to the inefficient use of inputs and to firms not operating at an optimum size.

In order to analyze the linear relation between the indices of performance, coefficients of Person correlation are computed for both regions (**Tables 4** and **5**) and special attention is given to the signal and statistical significance of these coefficients. For the Douro region (**Table 4**), the EP shows a highly positive correlated with FA, AV/E, EF, and SE. The ROA is positively correlated with ROE, ROI, FA, AV/E, EF, and SE. The ROE is positively correlated only with ROA, AV/E, and EF. The ROI is only positively related to ROA. The Lev is only positively related to SE. The FA is positively related to EP, ROA, EF, and SE. The T/E is positively correlated with ROA, AV/E, EF, TE, and SE. The AV/E is positively correlated with EP, ROA, ROI, T/E, and with the three indices of efficiency. The EF is positively correlated with the other indices of efficiency, and with EP, ROA, ROE, ROE, FA, T/E, and AV/E. The TE is positively correlated with T/E, AV/E, EF, and negatively with SE. The SE is positively correlated with EP, ROA, Lev, FA, T/E, FA, and negatively with TE. This last result means that firms being close to the optimum size do not mean they are efficient in the use of inputs. From the 55 linear correlation coefficients computed, 43.6% are statistically significant and only the correlation between TE and SE is negative. ROI and Lev are the indexes that present more linear independence to the other indices.

Regarding to the Vinho Verde region (**Table 5**), the EP is positively correlated with ROA, FA, AV/E, EF, and SE; the ROA is positively correlated ROI, FA, T/E AV/E, EF; the ROE is negatively correlated with ROI and Lev; the ROI is positively correlated with ROA, AV/E, EF, and SE and negatively correlated with ROE; the Lev is negatively correlated with ROE, EF, and

	EP	ROA	ROE	ROI	Lev	FA	T/E	AV/E	EF	TE	SE
EP	1										
ROA	0.165*	1									
ROE	0.018	0.006	1								
ROI	0.083	0.602*	−0.108***	1							
Lev	−0.045	0.002	−0.779*	0.019	1						
FA	0.127**	0.437*	0.003	0.378*	−0.029	1					
T/E	0.063	0.135**	0.032	0.078	−0.029	0.055	1				
AV/E	0.179*	0.358*	0.055	0.170*	−0.042	0.213	0.474*	1			
EF	0.177*	0.374*	0.079	0.116**	−0.127**	0.312*	0.404*	0.534*	1		
TE	0.006	0.265*	0.090	0.039	0.046*	0.017	0.182*	0.236*	0.401*	1	
SE	0.182*	0.158*	0.017	0.108***	−0.172*	0.314*	0.186*	0.258*	0.544*	−0.502*	1

*, **, *** statistically significant at 1%, 5% and 10%, respectively.
Legend: EP - Export Performance; ROA - Return on Assets; ROE - Return On Equity; ROI - Return on Investment; Lev - Leverage; FA - Financial Autonomy; T/E - Turnover per Employee; AV/E - Added Value per Employee; EF - Economic Efficiency; TE - Technical Eficiency; SE - Scale Efficiency.

Table 5. Pearson correlation coefficients of indices of performance Vinho Verde (299 observations, years 2014 and 2015).

SE; the FA is positively correlated with EP, ROA, ROI, EF, and SE; the T/E is positively correlated with ROA, AV/E, EF, TE, and SE; the AV/E is positively correlated with EP, ROA, ROI, T/E, and with the indices of efficiency; the EF is positively correlated with EP, ROA, ROI, FA, T/E, AV/E, TE, and SE and negatively with leverage; the TE is positively correlated with ROA, Lev, TE/E, AV/EEF, and negatively with SE; and the SE is positively correlated with EP, ROA, ROI, FA, T/E, AV/E, and EF and negatively with Lev and TE. For this region, from the 55 linear correlation coefficients computed, 45.5% are statistically significant, with 5 of them being negative (ROE-ROI, ROE-Lev, Lev-EF, Lev-SE, and SE-TE). For instance, the negative correlation Lev and EF implies that the more indebted the firm is, the less efficient it is, both in terms of technical efficiency and scale efficiency. Moreover, the negative signal of the correlation coefficient between SE and TE shows that the firms′ optimum size does not lead to the better use of inputs, as has also been demonstrated to be the case in Douro region.

Summarizing, the statistical significance of the Pearson correlation coefficients confirms the assumption that the performance of the company should not be assessed on one index, highlighting the importance computing, and analyzing alternative performance indices, in order to reinforce the results. Moreover, both in terms of signals and statistical significance, there are differences between the correlation coefficients of the Douro and Vinho Verde regions, giving yet another clear reason to analyze them as separate samples.

Therefore, in order to ascertain whether there are differences in the performance of the wine firms, according to their size, each sample was divided into different groups according to the referred criteria of **Table 2**, which are the number of employees, total turnover, and total assets. Since the results are very similar, we only report the results ordered according to the number of employees.

	EP	ROA	ROE	ROI	Lev	FA	T/E (10^3 €)	AV/E (10^3 €)	EF	TE	SE
Micro (n = 312)											
Mean	0.107	0.002	0.278	0.930	2.363	0.223	99.247	27.298	0.437	0.709	0.666
SD	0.189	0.506	3.734	10.515	24.462	1.099	144.965	35.737	0.260	0.296	0.299
Small (n = 100)											
Mean	0.306	0.024	0.014	0.038	2.223	0.399	171.342	37.889	0.465	0.610	0.759
SD	0.281	0.063	0.600	0.096	5.454	0.208	386.509	37.780	0.213	0.239	0.144
Medium (n = 17)											
Mean	0.636	0.051	0.055	0.059	0.758	0.608	214.696	68.480	0.631	0.808	0.779
SD	0.163	0.024	0.037	0.031	0.472	0.160	189.556	46.094	0.158	0.150	0.109
Big (n = 5)											
Medium	0.750	0.044	0.044	0.052	0.592	0.592	193.381	39.880	0.714	0.866	0.808
SD	0.134	0.019	0.029	0.022	0.024	0.024	63.845	32.928	0.197	0.155	0.098
Statistical F	48.870	2.680	0.700	0.670	1.500	6.040	239.800	2.750	49.200	10.520	3.050
(Sig.)	0.020	0.283	0.633	0.647	0.423	0.145	0.004	0.278	0.020	0.088	0.257
KW	92.610	4.720	3.280	2.280	9.490	19.210	33.050	31.480	19.080	15.310	1.160
(Sig.)	0.000	0.192	0.350	0.517	0.023	0.000	0.000	0.000	0.000	0.002	0.763
OLS regressions											
Constant	0.750	0.043	0.044	0.052	0.689	0.592	193.381	39.880	0.712	0.866	0.808
(Sig.)	(0.000)	(0.620)	(0.750)	(0.560)	(0.000)	(0.000)	(0.000)	(0.160)	(0.000)	(0.000)	(0.000)
Micro	−0.643	−0.041	0.232	0.878	1.673	−0.369	−94.113	−12.581	−0.275	−0.156	−0.142
(Sig.)	(0.000)	(0.640)	(0.360)	(0.090)	(0.180)	(0.000)	(0.060)	(0.660)	(0.000)	(0.000)	(0.000)
Small	−0.443	−0.019	−0.031	−0.013	1.533	−0.193	−22.039	1.990	−0.247	−0.266	−0.049
(Sig)	(0.000)	(0.830)	(0.840)	(0.890)	(0.120)	(0.000)	(0.690)	(0.940)	(0.000)	(0.000)	(0.160)
Medium	−0.115	0.007	0.010	0.009	0.067	0.016	21.315	28.600	−0.008	−0.058	−0.029
(Sig)	(0.190)	(0.930)	(0.940)	(0.930)	(0.890)	(0.900)	(0.810)	(0.380)	(0.340)	(0.290)	(0.550)
Statistical F	97.030	0.420	0.520	1.040	1.380	12.560	3.750	4.320	10.590	17.850	7.940
(Sig.)	(0.000)	(0.740)	(0.670)	(0.370)	(0.250)	(0.000)	(0.010)	(0.010)	(0.000)	(0.000)	(0.000)

Legend: EP - Export Performance; ROA - Return on Assets; ROE - Return On Equity; ROI - Return on Investment; Lev - Leverage; FA - Financial Autonomy; T/E - Turnover per Employee; AV/E - Added Value per Employee; EF - Economic Efficiency; TE - Technical Efficiency; SE - Scale Efficiency; SD – Standard Deviation; Sig - significance; KW - Kruskal-Wallis test; OLS - Ordinary Least Squares.

Table 6. Indices of performance of Douro region (434 observations, years 2014 and 2015) ordered according to the number of employees.

Table 6 includes the results of the Douro region. Taking into account the number of observations for each size, we observe that almost 72% of the firms are micro, 23% are small, 3.9% are medium, and only 1.1% are big, confirming the hypothesis of a heterogeneous market structure that underlies the monopolistic competition approach. Assuming the value and significance of the Kruskal-Wallis test, since the null hypothesis of equality of population rank test is rejected, the results indicate a positive relationship between size and the indices of EP, Lev, FA, T/E, AV/E, EF, and TE. This is not the case for the profitability indicators (ROA, ROE, and ROI) and SE. However, the Hotelling statistical F is applied to examine that the means of the four sizes are the same. The results only reject this hypothesis for the EP, T/E, and EF (at a 5% level of significance) and TE (at 10% of level of significance).

http://dx.doi.org/10.5772/intechopen.71320

In order to check the robustness of these results, the OLS regressions, whose dependent variable is the respective index of performance, are estimated. The explanatory variables are dummy variables of the size, with "big" being omitted. The results of these models for the Douro region are also included in **Table 6**. The signal of the coefficients and, namely, the global significance (values of F) confirm that there are differences in the performance of firms dependent on their respective size. Specifically, it is noted that export performance, financial autonomy, productivity, and efficiency are lower in micro and small wine enterprises. However, this difference seems not to have consequences on the profitability and leverage indices, since the respective regressions are not statistically significant.

Concerning the Vinho Verde region (**Table 7**), according to the number of employees, there are no big wine firms, with 77.6% being micro, 21.4% small, and 1% medium, confirming the dominance of very small companies. The Kruskal-Wallis test allows us to infer that there is only a positive relationship between size and the indices of EP, T/E, AV/E, and TE and that there are no statistically significant relationships with the other indices. The Hotelling F test

	EP	ROA	ROE	ROI	Lev	FA	T/E (10^3 €)	AV/E (10^3 €)	EF	TE	SE
Micro (n = 232)											
Mean	0.102	−0.011	−0.096	−0.100	8.875	0.226	93.541	22.251	0.414	0.702	0.650
SD	0.202	0.179	2.033	1.022	64.888	0.764	292.740	40.684	0.239	0.283	0.310
Small (n = 64)											
Mean	0.157	0.010	−0.043	0.015	2.973	0.349	126.066	29.720	0.427	0.587	0.765
SD	0.183	0.083	0.424	0.139	6.121	0.439	114.987	28.139	0.178	0.273	0.134
Medium (n = 3)											
Mean	0.533	0.130	0.155	0.160	1.359	0.483	216.533	69.279	0.551	0.880	0.623
SD	0.193	0.113	0.159	0.139	1.139	0.185	26.386	43.879	0.131	0.183	0.029
Statistical F	44.230	0.710	0.620	0.700	0.620	3.880	65.340	10.420	28.260	55.230	639.950
(Sig.)	(0.110)	(0.640)	(0.670)	(0.650)	(0.670)	(0.340)	(0.090)	(0.210)	(0.130)	(0.090)	(0.030)
KW	20.970	3.890	0.920	2.950	2.710	0.990	23.650	15.570	3.240	10.770	2.270
(Sig.)	(0.000)	(0.140)	(0.630)	(0.230)	(0.260)	(0.620)	(0.000)	(0.000)	(0.190)	(0.000)	(0.320)
OLS regressions											
Constant	0.533	0.128	0.155	0.159	1.359	0.483	216.533	69.279	0.551	0.880	0.623
(Sig.)	(0.000)	(0.470)	(0.800)	(0.670)	(0.680)	(0.060)	(0.000)	(0.230)	(0.000)	(0.000)	(0.000)
Micro	−0.432	−0.139	−0.251	−0.260	7.516	−0.256	122.992	47.027	−0.137	−0.178	0.027
(Sig.)	(0.010)	(0.440)	(0.690)	(0.490)	(0.140)	(0.330)	(0.010)	(0.419)	(0.180)	(0.090)	(0.370)
Small	−0.376	−0.119	−0.193	−0.147	1.521	−0.147	−92.933	39.720	−0.129	−0.300	0.144
(Sig)	(0.020)	(0.500)	(0.760)	(0.690)	(0.680)	(0.579)	(0.130)	(0.490)	(0.210)	(0.010)	(0.000)
Statistical F	5.080	0.890	0.120	1.470	1.240	1.260	3.260	1.020	0.920	8.430	25.540
(Sig.)	(0.010)	(0.410)	(0.880)	(0.230)	(0.290)	(0.280)	(0.040)	(0.360)	(0.400)	(0.000)	(0.000)

Legend: EP - Export Performance; ROA - Return on Assets; ROE - Return On Equity; ROI - Return on Investment; Lev - Leverage; FA - Financial Autonomy; T/E - Turnover per Employee; AV/E - Added Value per Employee; EF - Economic Efficiency; TE - Technical Efficiency; SE - Scale Efficiency; SD – Standard Deviation; Sig - significance; KW - Kruskal-Wallis test; OLS - Ordinary Least Squares.

Table 7. Indices of performance of Vinho Verde (299 observations, years 2014 and 2015) ordered according the number of employees.

only permits the rejection of the null hypothesis of equality means of the three groups, for the variables T/E, TE (both at 10%), and SE (at 5%). The results of the OLS regressions show that the indices EP, T/E, and TE increase positively with the size, but negatively with SE. This last result implies that the larger firms tend to deviate from the optimal size, although they are more technical efficient in the use of inputs.

Generally speaking, we can summarize that Douro wine firms are more heterogeneous than their Vinho Verde counterparts, which can be explained by comparing the technologies and market structures of the two markets. In both cases, the relationship between size and performance depends on the index of performance used, and it becomes clear that organizational performance is a multidimensional construct that should take into account the different dimensions of economic performance and even the respective regions[4] (spatial differentiation) in which they are located.

4. Conclusions

The wine industry neatly illustrates the market structure of monopolistic competition, exhibiting as it does varying degrees of product differentiation and free entry into the industry, with competition between groups of firm's size. Some of them search for market niches acting as oligopolies and others for mass markets acting as in perfect competition. In this context of horizontal product differentiation, notwithstanding the globalization of the wine market, a better understanding of the performance of the wine industry requires a microeconomic analysis based on measures at both the level of the firm and that of the wine region.

Based on data for the Douro and Vinho Verde regions, two representative Portuguese wine regions that fit well into the *terroir* model, and where horizontal product differentiation is clearly present, the main goal of this research has been to estimate the economic performance of wineries using different approaches and to assess the influence of the respective sizes of firms on performance indices, including ratios of traditional profitability, solvency, labor productivity and export performance, and productive efficiency. The results show that the wine firm's performance is a multidimensional construct and there are expected differences between wine regions, which are well predicted by the monopolistic competition model specifically regarding horizontal product differentiation.

Nonetheless, in both regions, export performance increases with size, which can be explained by the fact that big firms have at their disposal greater marketing resources and the ability to make substantial investment in the search for new international markets. Relative to the profitability indices, in both regions, there is no statistically significant relationship to size, and there coexist different typologies of firms in the market, according to the monopolistic

[4]In our case (results not reported in the chapter but provided by the authors) in order to test if there are differences between the two wine regions, the estimation of OLS regression, where the only explanatory dummy variable is the region, indicates that EP, ROI, AV/T, and EF are higher for the Douro region and not statistically different for the other variables.

http://dx.doi.org/10.5772/intechopen.71320

competition model. The indicators of productivity show that there is no standard result as these differ by region, demonstrating that the productivity of labor is higher in the bigger firms of the Douro region, but not in the Vinho Verde region where the turnover per employee is higher for microfirms, there being no statistical differences between groups.

The indices of productive efficiency of Douro firms show that the bigger firms are empowered to adopt management strategies that improve the use of economic resources. This is a consequence of a better use of inputs and the adoption of an appropriate size. In opposite, the firms of the Vinho Verde region show a different result, in which the economic efficiency is independent of the firms' size. This could be a consequence of the better use of inputs by bigger wineries to be annulled by the deviation from the optimal size.

Summing up, the results do not show in a clear way that bigger wine firms outperform small ones, depending on the conclusions drawn from the indices that are used. The results also highlight the diversity among wine firms and the need to understand why they are different. Consequently, future research will be directed toward understanding which specific resources constitute a source of competitive advantage, expressed in variables that are not considered in this study and that are only available through the use of firm's enquires.

Regarding policy implications, mainly in the areas of public support and regulation, this research shows that micro and small wineries are the most common types in the Portuguese wine industry and that, in the short run, this situation will not change. Moreover, assuming that globalization and international competition continue to increase in the wine industry, the search for new international markets should be aware of the importance of scale economies. This is an aim that can be achieved through organic growth or strategic alliances, either along the whole production and distribution chain or exclusively in the area of marketing.

Acknowledgements

This work is supported by the project NORTE -01-0145-FEDER-000038 (INNOVINE & WINE: Innovation Platform of Vine & Wine) and by European and Structural and Investment Funds in the FEDER component, through the Operational Competitiveness and Internationalization Program (COMPETE 2020) [Project No 006971 (UIC/SOC/04011)]; and national funds are received through the FCT: Portuguese Foundation for Science and Technology under the UID/SOC/04011/2013.

Author details

João Rebelo*, Sofia Gouveia, Lina Lourenço-Gomes and Ana Alexandra Marta-Costa

*Address all correspondence to: jrebelo@utad.pt

Department of Economics, Sociology and Management (DESG), Centre for Transdisciplinary Development Studies (CETRAD), University of Trás-os-Montes and Alto Douro (UTAD), Vila Real, Portugal

References

[1] Church J, Ware R. Industrial Organization: A Strategic Approach. Irwin, Lisbon: McGraw-Hill; 2000

[2] Instituto da Vinha e do Vinho – IVV. Vinhos e Aguardentes de Portugal 2015 [Internet]. 2016. Available from: www.ivv.min-agricultura.pt [Accessed: 25–05-2017]

[3] Coelho A. Temos que fazer coisas com personalidade e acreditar nelas. Opinião de Dirk Nieeport. Jornal Público. 2017;**8** june:39

[4] Observatório Racius. Estatísticas sobre o mundo empresarial em Portugal [Internet], 2014–2015. Available from: https://www.racius.com/observatorio/filtrar/ [Accessed: 01–06-2017]

[5] Hammervoll T, Mora P, Toften K. The financial crisis and the wine industry: The performance of niche firms versus mass-market firms. Wine Economics and Policy. 2014;**3**:108-114. DOI: 10.1016/j.wep.2014.11.001

[6] Sellers R, Alampi-Sottini V. The influence of size on winery performance: Evidence from Italy. Wine Economics and Policy. 2016;**5**:33-41. DOI: 10.1016/j.wep.2016.03.001

[7] Chaddad FR, Mondelli MP. Sources of firm performance differences in the US food economy. Journal of Agricultural Economics. 2013;**64**(2):382-404. DOI: 10.1111/j.1477-9552.2012.00369.x

[8] Bain JS. Industrial Organization. 2nd ed. New York: Willey; 1968

[9] Schamlensee R. Do markets differ much? American Economic Review. 1985;**75**:341-351

[10] Wernerfelt B. A resource-based view of the firm. Strategic Management Journal. 1984;**5**:171-180. DOI: 10.1002/smj.4250050207

[11] Mahoney J, Pandian J. The resource based view within the conversation of strategic management. Strategic Management Journal. 1992;**13**:363-380. DOI: 10.1002/smj.4250130505

[12] Adenaeuer L, Heckelei T. Foreign Diret investment and the performance of European agribusiness firms. Journal of Agricultural Economics. 2011;**62**(3):639-654. DOI: 10.1111/j.1477-9552.2011.00300.x

[13] Garcia-Fuentes PA, Ferreira GFC, Kennedy PL. Economic performance of US multinational agribusiness: Foreign direct investment and firm strategy. Agribusiness. 2013;**29**: 242-255. DOI: 10.1002/agr.21316

[14] Vivas C, Sousa A. Globalization and international expansion strategies of the wine sector companies in Portugal. In: American Association of Wine Economists-Proceedings of 6th Annual Conference; 7–10–06-2012; Princeton, New Jersey (USA). 2012. p. 1-27

[15] Charnes A, Cooper WW, Rhodes E. Evaluating program and managerial efficiency: An application of data envelopment analysis to program follow through. Management Science. 1981;**27**(6):668-697. DOI: 10.1287/mnsc.27.6.668

[16] Charnes A, Cooper WW, Rhodes E. Measuring the efficiency of decision making units. European Journal of Operational Research. 1978;**2**(6):429-444

[17] Rebelo J, Caldas J. The Douro wine region: A cluster approach. Journal of wine research. 2013;**24**(1):19-37. DOI: 10.1080/09571264.2012.717220

[18] European Commission. Commission recommendation of 06-05-2003 concerning the definition of micro, small and medium size enterprises (2003/361/EC). Official Journal of the European Union. 2003;**L124**:36-41

Chapter 15

Potential for Use of the Residues of the Wine Industry in Human Nutrition and as Agricultural Input

Renato Vasconcelos Botelho,
Gabriela Datsch Bennemann,
Yohandra Reyes Torres and
Alessandro Jefferson Sato

Additional information is available at the end of the chapter

http://dx.doi.org/10.5772/intechopen.73132

Abstract

The use of underutilized resources, with the aim of increasing productivity and creating wealth, will increasingly deserve the attention of the wine sector. The treatment of agricultural by-products will increasingly enter the priority agenda of the agribusiness sector, with a view to its use, the environment's re-cleanliness and, in many cases, whenever possible, for both purposes. Solid waste from the process of grape industrialization, when not adequately disposed, is aggressive to the environment. Such residues release significant amounts of liquid effluents when disposed in the soil, and this liquid contains high content of nutrients, organic matter, and other elements that can change the environment, especially of streams and sources, causing the death of aquatic beings. However, if properly used, they can be used as raw materials for other purposes. The solid residues of industrially processed grapes, which may have some potential economic interest, are pomace, seeds, liquid (lees), and other materials. In this context, this chapter presents the description of these by-products and their potential for use.

Keywords: *Vitis* spp., by-products, pomace, winery, flour, polyphenols, antioxidants, seed oil

1. Introduction

Approximately 75,100 million tons of grapes are produced annually in the world and are destined mostly to wine production or in natura consumption, with Italy, France, Spain, the United States, Australia, China, Chile, South Africa, Argentina, and Germany being the main producers. These countries produce about 275.7 million hectoliters of wine.

This results in approximately 13 million tons of residues, which are normally used as fertilizer or simply discarded [1].

Although not widely used, the residues from the wine industry have high fiber content and antioxidant substances that could be beneficial for use in food, but there are few reports about this potential. One of the best applications for grape residues, especially the pomace, could be to obtain flour, which can be used in the preparation of biscuits, breads, cereal bars, homemade pasta, and juices [2, 3].

The phenols are defined as the largest group of natural antioxidants, with about 8000 different compounds. They are distributed in several plant foods, being secondary metabolites of these and considered fundamental for the proper development of the plant, defense against environmental injuries, and infectious processes. Among the different vegetables, grapes are considered one of the major sources of phenolic compounds; however, it is known that there is considerable diversity among cultivars, and this results in grapes with different characteristics, such as flavor and color, which is certainly associated with the content and profile of polyphenols [4].

Grape seeds usually contain 8–20% of oil, which represent about 5% of the fruit weight, with about 3 million tons of grape seeds discarded annually in the world [5]. The seeds constitute about 20% of the gross weight of the fruit and calculated on dry matter, representing between 40 and 60% [6]. The complete use of grapes, including seeds, is considered an important economic and sustainability factor, since the oil has a pleasant and neutral taste and has a high concentration of linoleic acid and natural vitamin E, which provides considerable oxidative stability to the product. Grape seeds also have a considerable content of phenolic compounds (about 60–70% of their content), with smaller percentages found in other parts of the fruit such as 28–35% in the skin and approximately 10% in the pulp [6].

These compounds present in grapes are recognized for their role in modulating the expression of antioxidant enzymes [7], protection against oxidative damage in rat brain cells [8], and some anti-inflammatory effects. Some studies have shown anticholinergic effects of grape seed oil, with a proven reduction of low-density lipoprotein (LDL) and increases in high-density lipoprotein (HDL) levels, characteristics related to cardioprotective effects [9].

Some phenolic compounds were identified as the monomers, gallic acid, (+)-catechin, (+)-epicatechin, and epicatechin-3-*O*-gallate, and a large variety of procyanidins oligomers in skin and grape seeds [10, 11]. These high contents of bioactive compounds in both flour and grape seed oil characterize them as a functional food, widely disseminated in current nutritional practices.

Due to its abundance and its richness in compounds with bioactive properties, the study of the use of grape residues as an agricultural input, aiming at the management of plant cover, pests, and diseases, represents a great potential, especially considering the issues related to sustainability, impacts on the environment, and production costs.

In viticulture, intensive management practices and the use of agrochemicals are adopted on a large scale to control pests, diseases, and weeds. These practices tend to have significant impacts on agroecosystems, since they alter the characteristics of the habitat and the composition of the trophic chain. Thus, in recent years, there has been a growing demand for alternative management methods that reduce impacts on nontarget organisms and are satisfactorily effective against target organisms.

In this context, the use of winery residues is indicated as an important perspective in the management of grapevine, once the action of its components, such as phenolic acids and flavonoids, have already been reported as efficient against microorganisms [12] and also weeds [13]. The isolated effect of some compounds present in this residue has already been characterized on some insects, such as the flavonoids, which caused inhibited development of *Phyllotreta cruciferae* (Coleoptera: Chrysomelidae) [14].

2. Wine production and waste generation

The main by-products of winemaking are separated during the crushing and pressing stages of the grapes, and only small quantities of these residues are availed. Recovery of compounds from the continuous waste produced by the wine industry could represent a significant step forward in maintaining the balance of the environment, as the large quantities of waste generated in the wineries present serious problems of storage, processing, or disposal in ecological and economic terms. This situation explains the growing interest in exploring the by-products of winemaking.

Solid residues from the grape industrialization process, when disposed outdoor, are aggressive to the environment, releasing significant amounts of liquid effluents, which contain high nutrient content, organic matter, and other elements that, in contact with the soil and sources of water, can cause death of living organisms. However, adopting some technologies, they can be used as raw material for other purposes. Solid residues of industrially processed grapes, which may have some potential economic interest, are stalks, pomace, seeds, liquid (lees), and other materials. Although many polyphenols are transferred from the grapes to the wine during the maceration process, the residual seeds have been the focus of studies that relate them as good sources of phenolic compounds [7].

The pomace consists of the skin, the seeds, and the remains of the pulp of the grape, resulting from the crushing of the berries through a process of separation of the juice or must. Under normal conditions, the pomace is equivalent to 15% of the weight of the berries.

Other solid wastes are lees and tartar. The sludge originates in the bottom of the kites, is dense, and comes from the processes of purification of the wine stored. The tartar is solid and deposits on the walls of the containers (pipes) used to age the wine. Liquid wastes result from washes and spills of raw material. These materials should be submitted to the treatment of effluents in canteens. Regulation (EC) No 1493/99 defines wine sludge as the residue which is deposited in receptacles containing wine after fermentation, or at the time of storage, or after authorized treatment, as well as the residue obtained by the filtration and/or centrifugation of this wine product [12]. Also considered as wine lees is waste that is deposited in containers containing grape must. The amount of lees obtained annually depends on several factors, namely those inherent to the constitution of the grape varieties, maturation stage and phytosanitary state of the berries, climatic factors, and the winemaking techniques adopted. In general, these represent about 5% of the wine volume. A quantity of 140 kg of grapes produces approximately 1 hL of wine, giving 5.5 kg of liquid lees with 4.5% alcohol.

Pomace is the main by-product of winemaking, not only because of its alcoholic and tartaric richness but also because of the economic interest of some of its physical components. Pomace, as is well known, is the product resulting from the pressing of the wine masses, consisting of the solid parts of the grapes and the must or the wort/wine assembly that soaks them. Regulation (EC) No 1493/1999 defines it as the residue of pressing of fresh grapes, fermented or not. The pomace consists mainly of water, wines, and lees—these being dependent on the pressing; alcohols, especially ethanol, and also methanol and glycerol; aldehydes, esters, volatile acids, polyphenols, proteins, cellulose, pectins, mineral salts, and sugar residues [12].

3. Composition of grapes pomace and its potential for use

Botanically, the grape can be classified as a fruit of the vine, belonging to the family Vitaceae and to the genus *Vitis*, with the following main species: *Vitis vinifera, Vitis labrusca, Vitis rupestris, Vitis aestivalis, Vitis riparia, Vitis cinerea*, and *Vitis rotundifolia* [15]. In this sense, the existence of numerous grape varieties causes differences in their chemical composition, which allows to select the most suitable cultivars for both industrialization and in natura consumption.

The composition of the grape berry is generally formed by 6–12% of skin, 2–5% of seed, and 85–92% of pulp. The pulp, which constitutes the main part of the grape, is composed of the following constituents: 65–85% water, 12–25% reducing sugars, 0.6–1.4% organic acids, 0.25–0.5% of mineral substances, and 0.05–0.1% of nitrogenous compounds, besides several water-soluble and fat-soluble vitamins [16].

Like the other nutrients presented, the mineral composition of the grape may vary according to the conditions provided for growth, such as the composition of the soil and the use of fertilizers and herbicides. The nutritional relevance of the juice is mainly due to the high content of potassium, calcium, iron, magnesium and phosphorus, and low levels of sodium. Thus, fresh grape juice presents from 2.5 to 3.5 g L^{-1} of mineral substances [17].

Pomace (a mixture of grapes skin and seed) accounts for about 16% of the total processed grape and is one of the most abundant residues in the wine industry, and this material is known to be rich in many compounds such as phenolic acids, flavonoids, tannins, and saponins and, therefore, has a great utility potential for a variety of purposes, including pests, diseases, and weed control in agricultural crops, and the use of grape extract as a corrosion inhibitor by the metallurgical industry due to its high antioxidant capacity. The pomace contains another coproduct with high added value: the seed, which generates vegetable oil [4, 5].

In a research carried out in Brazil [17], it was analyzed for the mineral content (N, P, K, S, Ca, Fe, Mg, Mn, Fe, and Zn), anthocyanins, and phenolic compounds in flours produced from residues of different grape cultivars from different wineries (**Tables 1** and **2**). Mineral analysis showed a significant difference for all grape cultivar, with the exception in phosphorus content. Residues from cv. Seibel showed higher levels of N, Cu, and Mg. The cultivars Ancelotta, Tannat, and Ives present higher contents of K, Zn, Mn, Fe, and Ca. For the concentration of anthocyanins, cultivars Cabernet Sauvignon (114.7 mg/100 g), Tannat (88.5 mg/100 g), and Ancelotta (33.8 mg/100 g) had the highest concentrations. The cultivars Pinot Noir (7.0 g AGE/100 g), Tannat (4.3 g AGE/100 g),

http://dx.doi.org/10.5772/intechopen.73132

Grape variety	Phosphor (mg/100 g)	Nitrogen (mg/100 g)	Sulfur (mg/100 g)	Potassium (mg/100 g)	Zinc (mg/100 g)
x ± DP					
Cabernet Sauvignon	$39.59 \pm 1.3^{n.s.}$	21.83 ± 0.99^{a}	52.27 ± 1.64^{ab}	13.77 ± 2.02^{a}	NS
Tannat	33.05 ± 0.99	22.09 ± 1.31^{a}	38.53 ± 2.5^{b}	23.52 ± 3.14^{b}	16.35 ± 0.44^{d}
Ancelotta	43.88 ± 1.33	22.09 ± 1.06^{a}	85.4 ± 4.32^{c}	25.82 ± 0.01^{b}	33.59 ± 1.66^{e}
Pinot Noir	49.76 ± 9.81	22.45 ± 0.59^{a}	56.49 ± 2.69^{ab}	13.44 ± 1.49^{a}	6.7 ± 0.50^{b}
Malbec	42.43 ± 2.01	26.17 ± 0.87^{b}	53.94 ± 3.07^{ab}	19 ± 0.94^{c}	9.84 ± 0.81^{c}
Merlot	39.96 ± 3.42	3.92 ± 0.72^{c}	54.92 ± 1.46^{ab}	10.84 ± 0.56^{a}	0.08 ± 0.02^{a}
Seibel	44.1 ± 1.77	30.93 ± 1.45^{e}	67.47 ± 0.58^{ab}	5.96 ± 1.12^{d}	6.29 ± 0.35^{b}
Ives	35.65 ± 0.86	25.05 ± 0.91^{b}	87.88 ± 0.98^{c}	25.15 ± 0.69^{b}	17.47 ± 0.24^{d}
Cabernet Franc	45.14 ± 5.3	17.77 ± 0.74^{d}	63.12 ± 1.25^{ab}	18.66 ± 0.56^{c}	NS
Variety	**Copper (mcg/100 g)**	**Manganese (mg/100 g)**	**Magnesium (mg/100 g)**	**Iron (mg/100 g)**	**Calcium (mg/100 g)**
x ± DP					
Cabernet Sauvignon	86.52 ± 11.8^{ad}	24.87 ± 5.3^{abc}	62.21 ± 1.96^{b}	209.78 ± 7.02^{ab}	284.22 ± 3.59^{b}
Tannat	63.97 ± 19.18^{ac}	36.64 ± 0.26^{d}	104.32 ± 3.03^{a}	286.5 ± 6.05^{c}	429.5 ± 4.74^{d}
Ancelotta	89.94 ± 0.71^{ad}	32.27 ± 1.25^{bd}	86.93 ± 1.75^{ac}	212.89 ± 8.48^{ab}	357.02 ± 4.2^{c}
Pinot Noir	52.95 ± 1.44^{abc}	25.84 ± 1.37^{abc}	105.51 ± 3.62^{a}	164.84 ± 4.71^{a}	362.84 ± 1.67^{c}
Malbec	125.06 ± 2.57^{d}	19.91 ± 3.15^{c}	75.49 ± 1.95^{c}	250.06 ± 18.54^{bc}	312.13 ± 1.21^{b}
Merlot	9.25 ± 0.22^{b}	9.02 ± 1.25^{e}	58.73 ± 2.94^{b}	169.26 ± 11.66^{a}	158.04 ± 4.63^{a}
Seibel	296.42 ± 44.61^{e}	30 ± 6.58^{abd}	154.7 ± 20.42^{d}	166.36 ± 10.37^{a}	356.21 ± 13.68^{c}
Ives	36.38 ± 1.11^{bc}	29.26 ± 2.32^{abd}	102.87 ± 5.70^{a}	241.54 ± 19.81^{bc}	308.8 ± 3.96^{b}
Cabernet Franc	42.64 ± 1.64^{abc}	21.09 ± 0.78^{ac}	101.36 ± 1.73^{a}	155.3 ± 9.05^{a}	296.21 ± 7.84^{b}

Means followed by different letters on the same column are significantly different according to NSK at $P < 0.05$. NS, not significant.

Table 1. Mineral content in grape pomace flour in southern Brazil [18].

Grape variety	Phenolic compounds (g AGE/100 g)	Anthocyanins (mg/100 g)
Cabernet Sauvignon	1.7 ± 0.29[a]	114.7 ± 0.15[f]
Tannat	4.3 ± 0.18[e]	88.5 ± 2.02[e]
Ancelotta	3.8 ± 0.04[d]	33.8 ± 2.4[d]
Pinot Noir	7.0 ± 0.17[f]	3.48 ± 0.35[bc]
Malbec	2.3 ± 0.31[b]	12.89 ± 0.51[ac]
Merlot	1.7 ± 0.02[a]	15.78 ± 0.38[a]
Seibel	5.9 ± 0.39[c]	18.35 ± 2.1[a]
Ives	2.4 ± 0.10[b]	19.86 ± 6.46[a]
Cabernet Franc	1.5 ± 0.12[a]	0.9 ± 0.96[b]

Means followed by different letters on the same column are significantly different according to NSK at $P < 0.05$. NS, not significant.

Table 2. Anthocyanin content and phenolic compounds in grape pomace flour in southern Brazil [18].

and Ancelotta (3.9 g AGE/100 g) had the highest content of phenolic compounds. Considering these results, the potential of using the residue of winemaking to produce flour for human consumption became evident, highlighting the grapes 'Tannat' and 'Ancelotta'.

The main phenolic compounds presented in the pomace of 'Pinot Noir' grapes were (1) flavonoids: flavan-3-ols (catechin, epicatechin, epicatechin gallate, procyanidins A and B), flavonols (quercetin and quercetin methyl glucoside), and anthocyanins (delphinidin-3-glucoside, 3-glucoside cyanidin-3-glucoside, petunidin-3-glucoside, peonidin-3-glucoside, malvidin-3-glucoside, vitisin) and (2) non-flavonoids: hydroxybenzoate (gallic acid) [5].

Grape pomace also showed a great wealth of biologically active substances in the study that were developed to determine the chemical composition of seeds and skin of grape pomace of different cultivars of *V. vinifera* grapes both white and red in Italy and in California [19]. These authors verified that in the seeds and in the skin, the Italian samples presented higher content of organic matter—lignin and copper. In addition, the contents of K, Fe, and Zn were higher in grapes from California. The Italian white grapes had a higher content of saponins in the pell and tended to have a higher phenolic content in both the skin and the seeds.

A study with grape pomace extracts (GPEs) of the grapevines 'Niagara' and 'Isabella' (*V. labrusca*) was carried out to evaluate their effects on oxidative stability and quality of chicken meat. The use of the grape pomace extract was efficient to maintain the lipid stability of the chicken meat, presenting results compatible with those exhibited by the synthetic antioxidant butyl hydroxytoluene (BHT) [14].

Due to the richness of bioactive components, the use of grape pomace flour has been tested in different food product such as biscuits [3], muffins [20], and cookies [21]. These studies showed that, in general, the addition of flour does not negatively affect the preference for food. Moreover, the presence of bioactive compounds was verified, indicating the possibility of adding nutritional value to the food with the supplementation of flour of grape pomace.

A study with grape pomace of the cv. Marselan (*V. vinifera*) was carried out to evaluate the effect of its inclusion in extruded snack with respect to nutritional, technological, and sensory parameters [22]. The microbiological determinations of coliforms at 45°C and *Salmonella* in the flours were also performed. The centesimal composition presented fiber (58.01%), carbohydrates (17.62%), and ashes (12.46%) as the main constituents. Resveratrol (6.14 mg g^{-1}), luteolin (5.16 mg g^{-1}), and kaempferol (3.01 mg g^{-1}) were the phenolic compounds detected in greater quantity in the flour. The fiber formulation containing 9% (5% fiber) of flour presented better acceptance results with regard to color, aroma, and texture attributes compared to the standard snack formulation. According to these authors [22], for the nutritional enrichment (fibers and phytochemicals) and for adding value to the agro-industrial residue discarded by the wineries, the addition of pomace flour in extruded snacks is viable and quite interesting.

Grape pomace is known to be rich in many compounds such as phenolic acids, flavonoids, tannins, and saponins presenting a high potential for pests, diseases, and weed control in agricultural crops, including vineyards. The antimicrobial effect of 'Pinot Noir' grape pomace extracts against *Staphylococcus aureus* and *Candida albicans* was already reported [18].

Considering the composition of the grape pomace, one of the possibilities is the induction of resistance of plants to pathogens. In this context, a study was carried out to evaluate the effect of autoclaved grape pomace extract (GPE) in the induction of phytoalexin deoxyanthocyanidine in sorghum [23]. To obtain the extract, ground dry pomace was macerated in water at 70°C for approximately 12 h in the dark. The GPE was sterilized by autoclaving for 20 min at 120°C under a pressure of 1.1 kgf cm^{-2}, followed by fractionation in the tested doses. These authors reported the effect of GBE on the deoxythyanocyanidine synthesis, with a higher accumulation at the 3% dose.

The downy mildew (*Plasmopara viticola*) is responsible for qualitative and quantitative losses in grapevines production. In this context, a study was carried out aiming to verify the action of grape pomace extract (GPE) on sporangia germination of *P. viticola* and the severity of mildew on grapevines cv. Carmen [24]. These authors verified reduction of germination of the pathogen and incidence of mildew in grapevines. The highest dose of the extract (12%) controlled mildew severity at 36%, demonstrating the potential of the residue extract for the organic system.

4. Composition of grape seeds oil

Pomace represents approximately 16% of the grapes, of which about 20–26% are seeds [25]. Although winemakers have traditionally considered these wastes an economic and environmental problem, they are gradually being considered as a potential to add value to the products. Grape seeds usually contain 8–20% oil, with about 3 million tons of grape seeds being discarded annually worldwide [26].

The complete use of grapes, including seeds, is considered an important economic and sustainability factor, since the oil has a pleasant and neutral taste and has a high concentration of linoleic acid and natural vitamin E, which provides oxidative stability to the product.

These properties are related to the reduction of low-density lipoprotein (LDL) and increases in high-density lipoprotein (HDL) levels—characteristics related to cardioprotective effects [26].

The oils present in grape seeds have numerous health benefits, especially vitamin E and essential fatty acids. These fatty acids are considered as protectors of cardiovascular diseases, while vitamin E has antitumor and neuroprotective properties, is able to lower cholesterol levels, and has antioxidant activity [26]. Additionally, in the extraction of the oil, part of it still remains in the residue and could be used for animal nutrition.

Seeds are rich sources of polyphenol components and have been widely studied by several groups [27–29]. Oils that are not refined may contain tocopherol and other active compounds with antioxidant properties [10].

The mineral content of grape seeds prior to the winemaking process and wine manufacturing residues has been shown to be an important source of nutrients and essential elements. In some grape seeds collected from different locations in Turkey, the mineral contents (Al, B, Ca, Co, Mo, Cr, Fe, K, Mg, Mn, Na, P, S, Se, and Zn) were determined [28]. Ca, K, Mg, Na, and P were the main minerals contained in grape seeds. While the contents of vitamins A and B1 and of the minerals Fe, Mn, and Zn were elevated in all seeds, Cu, Mo, and Cr contents were very low.

A trial was carried out to study the composition of fatty acids, oxidative stability, and antioxidant and antiproliferative activity of cold pressed oils and flours extracted from the pomace of grape cultivars Chardonnay, Concord, and Rubi [29]. In the seeds, the phenolic profiles were also measured. The most abundant fatty acid in the oils was linoleic acid, which ranged from 66.0 g 100 g^{-1} of total fatty acids, in cv. Rubi; to 75.3 g 100 g^{-1} in 'Concord' grape seed oil. The oils also had high levels of oleic acid and low levels of saturated fat. Seed oil of cv. Rubi had the highest oxidative stability index. The total phenolic content was up to 100 times lower in oils than in flours. Lutein, zeaxanthin, cryptoxanthin, β-carotene, and α-tocopherol levels were also measured. The antioxidant activity evaluated by the 2,2-diphenyl-1-picrylhydrazyl (DPPH) radical elimination capacity ranged from 0.07 to 2.22 mmol Trolox (TE) g^{-1} oil equivalents and 11.8 to 15.0 mmol TE g^{-1} flour. In this study, the antiproliferative activity against HT-29 colon cancer cells was also tested. Seed meal and grape seed oil recorded significant ($p < 0.05$) inhibition of cancer cell growth, demonstrating the potential for the development of value-added applications for these seed oils and flours as dietary sources of natural antioxidants and anti-proliferative agents for health.

Studies on vitamin E extractions with oil of seeds of different grape varieties showed variations in concentrations. In the study with grape cultivars Barbera, Malbec, Gamay, Cabernet Sauvignon, Pinot Noir, Merlot, Cabernet Franc, and Syrah extracted with supercritical fluid, CO_2, and petroleum ether, the authors found values between 3.58 mg 100 g^{-1} and 30.9 mg 100 g^{-1} of vitamin E [29]. The result of the analysis of grapes grown in Brazil presented as results the oil of the 'Isabella' grape seed containing values lower than 1 mg of tocopherol in 100 g of oil. The best results were obtained for the cvs. Cabernet Sauvignon and Merlot (5.67 and 7.03 mg 100 g^{-1}) [30].

5. Conclusions

Considering the aspects presented in this chapter, it was evident to the innumerable nutritional and pharmaceutical properties of the residues of the wine industry. Both pomace and grape seed are materials that are considered as industrial waste but are rich in vitamins, polyphenols, unsaturated fatty acids, and other important components for nutrition and with positive effects on human health. Considering the increasing need for sustainability of agricultural and industrial activities, the use of these wine by-products may represent an advance in the production chain from an economic, social, and environmental point of view.

Author details

Renato Vasconcelos Botelho[1]*, Gabriela Datsch Bennemann[2], Yohandra Reyes Torres[3] and Alessandro Jefferson Sato[4]

*Address all correspondence to: rbotelho@unicentro.br

1 Department of Agronomy, State University of Midwestern of Paraná (Unicentro), Guarapuava, Paraná, Brazil

2 Department of Nutrition, State University of Midwestern of Paraná (Unicentro), Guarapuava, Paraná, Brazil

3 Department of Chemistry, State University of Midwestern of Paraná (Unicentro), Guarapuava, Paraná, Brazil

4 Federal University of Paraná (UFPR), Palotina, Paraná, Brazil

References

[1] OIV. Organisation International de la Vigne et du Vin [Internet]. 2016. Available from: http://www.oiv.org/uk/accueil/ndex.php [Accessed: 2017/04/20]

[2] Paz MF, Marques RV, Schumann C, Corrêa LB, Corrêa ÉK. Características tecnológicas de pães elaborados com farelo de arroz desengordurado. Brazilian. Journal of Food Technology. 2015;**18**(2):128-136. DOI: 10.1590/1981-6723.6014

[3] Piovesana A, Bueno MM, Klajn VM. Elaboração e aceitabilidade de biscoitos enriquecidos com aveia e farinha de bagaço de uva. Brazilian Journal of Food Technology. 2013;**16**(1):68-72. DOI: 10.1590/S1981-67232013005000007

[4] Poudel PR, Tamura H, Kataoka I, Mochioka R. Phenolic compounds and antioxidant activities of skins and seeds of five wild grapes an two hybrids native to Japan. Journal of Food Composition and Analysis. 2008;**21**:622-625. DOI: 10.1016/j.jfca.2008.07.003

[5] Yilmaz Y, Göksel Z, Erdog-an SS, Oztürk A, Atak A, Ozer C. Antioxidant activity and phenolic content of seed, skin and pulp parts of 22 grape (*Vitis vinifera* L.) cultivars (4 common and 18 registered or candidate for registration). Journal of Food Processing and Preservation. 2015;**39**(6):1682-1691. DOI: 10.1111/jfpp.12399

[6] Matthäus B. Virgin grape seed oil: Is it really a nutritional highlight? European Journal of Lipid Science and Technology. 2008;**110**:645-650

[7] Puiggròs F, Llópiz N, Ardévol A, Bladé C, Arola L, Salvadó MJ. Grape seed procyanidins prevent oxidative injury by modulating the expression of antioxidant enzyme systems. Journal of Agricultural and Food Chemistry. 2005;**53**:6080-6086

[8] Guo L, Wang LH, Sun B, Yang JY, Zhao YQ, Dong YX. Direct in vivo evidence of protective effects of grape seed procyanidin fractions and other antioxidants against ethanol-induced oxidative DNA damage in mouse brain cells. Journal of Agricultural and Food Chemistry. 2007;**55**:5881-5891

[9] Beveridge THJ, Girard B, Kopp T. Yield and composition of grape seed oils extracted by supercritical carbon dioxide and petroleum ether: Varietal effects. Journal of Agriculture Food Chemistry. 2005;**53**:1799-1804

[10] Maier T, Schieber A, Kammerer DR, Carle R. Residues of grape (*Vitis vinifera* L.) seed oil production as a valuable source of phenolic antioxidants. Food Chemistry. 2009;**112**: 551-559

[11] Selani MM, Contreras-Castillo CJ, Shirahigue LD, Gallo CR, Plata-Oviedo M, Montes-Villanueva ND. Wine industry residues extracts as natural antioxidants in raw and cooked chicken meat during frozen storage. Meat Science. 2011;**88**:397-403. DOI: 10.1016/j.meatsci.2011.01.017

[12] Cheng VJ, Bekhit AEA, Mcconnell M. Effect of extraction solvent, waste fraction and grape variety on the antimicrobial and antioxidant activities of extracts from wine residue from cool climate. Food Chemistry. 2012;**134**:474-482

[13] Cayuela ML, Millner PD, Meyer SLF, Rioig A. Potential of olive mill waste and compost as biobased pesticides against weeds, fungi and nematodes. Science of the Total Environment. 2008;**399**:11-18

[14] Onyilagha JC, Gruber MY, Hallett RH. Constitutive flavonoids deter flea beetle insect feeding in *Camelina sativa* L. Biochemical Systematics and Ecology. 2012;**42**:128-133

[15] Ishimoto EY. Efeito hipolipemiante e antioxidante de subprodutos da uva em hamsters [thesis]. São Paulo: Universidade de São Paulo

[16] Santana MTA, Siqueira HH, Lacerda RJ, Lima LCO. Physical chemistry and enzymatic characterization of grape 'Patricia' cultivated in primavera do Leste – MT. Ciência e Agrotecnologia. 2008;**32**(1):186-190. DOI: 10.1590/S1413-70542008000100027

[17] Malacrida CR, Motta S. Compostos fenólicos totais e antocianinas em suco de uva. Ciência e Tecnologia de Alimentos. 2005;**25**(4):659-664

[18] Bennemann, GD, Assis CF, Moreira GCRC, Lima LH, Karolyne Kruger Carvalho KK, Torres YR, Botelho, RV. 39th World Congress of Vine and Wine. 2016; BIO Web of Conferences 7, 04007. DOI: 10.1051/bioconf/20160704007

[19] Spanghero M, Salem AZM, Robinson PH. Chemical composition, including secondary metabolites, and rumen fermentability of seeds and pulp of Californian (USA) and Italian grape pomaces. Animal Feed Science and Technology. 2009;**152**:243-255. DOI: 10.1016/j.anifeedsci.2009.04.015

[20] Bennemann GD, Nezello MC, Eing KKC, Novello D, Schwarz K, Botelho RV. Desenvolvimento e aceitabilidade de muffins adicionados de farinha de casca de uva das cultivares Ancelotta e Bordô. Revista da Universidade Vale do Rio Verde. 2016;**14**(2):864-874

[21] Karnopp AR, Figueroa AM, Los PR, Teles JC, Simões DRS, Barana AC, Kubiaki FT, Oliveira JGB, Granato D. Effects of whole-wheat flour and bordeaux grape pomace (*Vitis labrusca* L.) on the sensory, physicochemical and functional properties of cookies. Food and Technology. 2015;**35**(4):750-756

[22] Bender ABB, Luvielmo MM, Loureiro BB, Speroni CS, Boligon AA, Silva LP. Obtenção e caracterização de farinha de casca de uva e sua utilização em snack extrusado. Brazilian Journal of Food Technology. 2016;**19**:1-9. DOI: 10.1590/1981-6723.1016

[23] Horst MV, Leite CD, Bortuli D, Santos I, Botelho RV, Faria CMDR. Induction of deoxiantocianidine in sorghum with the use of grape pomace autoclaved extract. In: Anais do VII Reunião sobre Indução de Resistência de plantas a patógenos. 2014; Maringá, Paraná State, Brazil: Universidade Estadual de Maringá

[24] Leite CD, Horst MV, Maia, AJ, Santos I, Botelho RV, Faria CMDR. Extato de bagaço de uva no controle do míldio da videira. In: Anais do 47° Congresso Brasileiro de Fitopatologia, 17-21 August 2014, Londrina, Paraná, Brazil. Londrina: Universidade Estadual de Londrina

[25] Rice AC. Solid-waste generation and by-product recovery pontential from winery residues. American Journal of Enology and Viticulture. 1976;**27**:21-26

[26] Choi Y, Lee J. Antioxidant and antiproliferative properties of a tocotrienol-rich fraction. Food Chemistry. 2009;**114**:1386-1390

[27] Fuleki T, Ricardo-da-Silva JM. Catechin and procyanidin composition of seeds from grape cultivars grown in Ontario. Journal of Agricultural and Food Chemistry. 1997;**45**: 1156-1160

[28] Lutterodt H, Slavin M, Whent M, Turner E, Yu L. Fatty acid composition, oxidative stability, antioxidant and antiproliferative properties of selected cold-pressed grape seed oils and flours. Food Chemistry. 2011;**128**:391-399

[29] Oomah BD, Liang J, Godfrey D, Mazza G. Microwave heating of grape seed: Effect on oil quality. Journal of Agricultural Food Chemistry. 1998;**46**:4017-4021

[30] Freitas LS, Jacques RA, Richter MF, Silva AL, Caramão EB. Pressurized liquid extraction of vitamin E from Brazilian grape seed oil. Journal of Chromatography. 2008;**120**:80-83

Chapter 16

Revalorization of Grape Seed Oil for Innovative Non-Food Applications

Juan Carlos de Haro, Juan Francisco Rodríguez,
Manuel Carmona and Ángel Pérez

Additional information is available at the end of the chapter

http://dx.doi.org/10.5772/intechopen.71710

Abstract

Grape processing produces a substantial amount of residues that are highly polluting and expensive to treat, being grape seed one of the main by-products with high commercial interest. During the extraction process of grape seed oil, most of the nutraceutical compounds remain on the solid cake. This book chapter resumes the potential utilization of grape seed oil for producing biobased materials through environmentally friendly processes that could substitute petroleum-derived products. Special attention is given to transesterification and epoxidation processes. The transesterification of grape seed oil in presence of methanol drives to the production of a biodiesel with excellent low-temperature properties. According to EN 14214, grape seed oil-based biodiesel presents a slightly lower cetane number than the specified limit. In addition, this biodiesel presents a low oxidation stability which can be improved by the incorporation of oxidation stabilizer. Attending to the epoxidation of grape seed oil, short reaction times and high temperatures are advised. Epoxidized grape seed oil can be used for the synthesis of biobased polyols and its further application on the synthesis of polyurethane compounds.

Keywords: biodiesel, biopolyol, epoxidation, grape seed oil, polyurethane foam, transesterification

1. Introduction

Grapes and grape-related products are among one of the most important horticultural products and therefore they are of high commercial interest. According to the Food and Agriculture Organization of the United Nations (FAO), the grape farming occupies 7.12 million hectares and approximately 74.49 million tonnes were produced in 2014. The top five countries producer of grapes are China (12.6 MTon), USA (7.15 MTon), Italy (6.93 MTon), Spain (6.22 MTon), and France (6.17 MTon) [1]. Despite grapes can be commercialized as fresh fruit, dried (raisins) and in the form of jellies and jams, beverages (wine and juice) are the most important derived products. As an estimation performed by the International Organization of Vine and Wine (OIV), 276 million hL were globally produced in 2015.

The socioeconomic activity of grape pressing for producing grape juice and wine imply the generation of a considerable amount of solid wastes, since approximately 30% (w/w) of the material used become a waste. There are several environmental problems associated with wineries, including water pollution, soil degradation, damage to natural vegetation, odors, and air emissions [2].

The main solid by-products and residues produced during wine production are vine shoots, grape marc or pomace (composed by skins, seeds, and stems), and wine lees.

- Vine shoots are a non-wood lignocellulosic residue generated during the vine pruning. Manzone et al. [3] quantified the pruning residue between 1850 and 5360 kg/ha, depending on the weather conditions during the year and the structure of plantation. This residue, which is mainly composed by cellulose (34%), hemicelluloses (19%), and lignin (27%), have been used for the extraction of phenolic acids [4], lactic acid [5], and also for production of biosurfactants [6] and energy [7, 8].
- Grape marc is the main by-product of the wine industry. It is composed by stems, skins, and seeds that remains after pressing the grapes. Taking into account that the marc represents 15–20% of grapes' weight [9], around 13.03 Mt of grape marc are produced annually. The main chemical components of grape marc are phenols, which confer a high antioxidant capacity, peptic substances, cellulose, and lignin [10].
    - Grape skins can represent up to 20% of grape pomace. This winery by-product was mainly used as compost or fuel as unusable waste, overall on the small wineries. However, grape skin represents nowadays a valuable source of biologically active phytochemicals due to the high amount of phenolic compounds that contents [11].
    - Grape stems are the structure of grape bunch and represent around 3% of grape marc mass. These are composed mainly of cellulose (30.3%), hemicelluloses (21.0%), lignin (17.4%), tannins (15.9%), and proteins (6.1%). However, the compounds that can be found in grape stems depend on several factors, such as geographic origin, time of harvest, and climatic conditions [12].
    - Grape seeds are one of the main by-products from grape processing industries. An individual grape berry typically contains two/three seeds and it constitutes around 4% of grape marc. Grape seeds contain 13–19% oil, which is rich in essential fatty acids, about 11% protein, 60–70% of non-digestible carbohydrates, non-phenolic antioxidants, such as tocopherols and beta-carotene [10], and also phenolic compounds with antioxidant capacity [13].
- Wine lees, which are generated during the fermentation and aging processes of the wine, are defined as the residue formed at the bottom of recipients containing wine, after

fermentation, during storage or after authorized treatments, as well as the residue obtained by the filtration or centrifugation of this product. The lees consists of a solid phase, mainly composed by microorganisms (yeast and bacteria), insoluble carbohydrates from the cellulosic and hemicellulosic fractions, phenolic compounds, lignin, proteins and tartrates; and a liquid phase rich in ethanol, lactic acid and acetic acid [9].

The legislative situation in Europe governing the by-products produced by the wine industries only indicates that wine lees have to be withdrawn once they have been denatured to make their use in winemaking impossible. Producers with a capacity of less than 25 hL/year of wine may be exempted by the Member State from this obligation. However, most of the Member States have specified the rules for the withdrawal and legal destinations of winery industries by-products [14].

1.1. Grape seed oil

As it was previously commented, grape seed is one of the main by-products of winery industries. Even some investigations have focused on the use of grape seeds as a fuel via pyrolysis [15, 16] or gasification [17] processes, the most interesting activity from an economic point of view consists on the extraction of the interesting compounds prior to the thermal exploitation.

Grape seed oil composition has been studied by different authors, focusing mainly on the fatty acid profile, the phytochemical composition, and antioxidant properties. **Table 1** shows the fatty acid profile for grape seed oil found on a previous study [18]. These values agree with the ones found in literature. For example, Beveridge et al. [19] compared 8 different grape seed oils obtained by two different extraction methods, finding a linoleic acid content between 66.76 and 73.23%. They also reported oleic acid, palmitic acid, and stearic acids as main fatty acids in the range of 12.63–18.95%, 6.28–8.62%, 3.60–5.26%, respectively. However, slight differences on the fatty acid profile can be found due to the use of different grapes varieties and/or extraction methods [20–23].

Fatty acid		Percentage (wt.%)
Palmitic	C16:0	6.9
Palmitoleic	C16:1	0.1
Stearic	C18:0	4.0
Oleic	C18:1	19.0
Linoleic	C18:2	69.1
Arachidic	C20:0	0.3
Gadoleic	C20:1	0.3
Others		0.3

Table 1. Fatty acid composition (wt.%) of grape seed oil.

With respect to vitamin E active compounds, most of the vegetable oils only contain a significant content of tocopherols, meanwhile tocotrienols are seldom. Virgin grape seed oil contains up to 10 mg α-tocopherol/100 g and different tocotrienols with a total amount of around 35 mg/100 g [24]. Crews et al. [25] investigated the total content of tocopherols and tocotrienols of 30 samples of grape seed oil, finding a total content between 63 and 1208 mg/kg. This range is much wider than the values given at the Codex Alimentarius (240–410 mg/kg). Hassanein and Abedel-Razek [26] reported a value of 380 mg/kg of total tocopherol content. These values are relatively low compared with other vegetable oils, indicating that most of vitamin E active compounds remain on the solid phase of the seed (grape seed flour).

Phytosterols are natural sterols which occur in plants and vegetable oils. A content of phytosterols ranged from 2580 to 11,250 mg/kg phytosterols have been reported in literature. However, independently of the author, β-sitosterol is the main phytosterol found in grape seed oil (67–70%) [25, 26].

Different techniques have been proposed for grape seed oil extraction. The traditional method consists of pressing the whole seeds in a hydraulic press or the milled and heated seeds in screw press. Cold-pressing is a method of oil extraction that involves no heat or chemical treatment, and hence might retain more health beneficial components, such as natural antioxidants. The cold-pressed oils may be a better source of beneficial components, such as antioxidative phenolic compounds, as well as other health-beneficial phytochemicals. Although the yield is usually lower than that with conventional solvent extraction, there is no concern about solvent residues in the oil, making for a safer and more consumer-desired product [27]. Using this technique, Lutterodt et al. [20] determined no significant differences on the fatty acid profile of cold-pressed grape seed oil, but an increment on the oxidative stability index, indicating a higher content of antioxidant compounds.

Solvent extraction of grape seed oil is a diffusion process achieved by immersing the seed in solvent or by percolating solvent through a bed of seeds. Solvent is recovered from the oil-solvent mixture (known as micella) by an evaporator and recycled to the process. In order to determine the influence of the solvent, Fernández et al. [28] compared hexane, acetone, pentane, acetonitrile, diethyl ether, methanol, and ethanol. Among them, diethyl ether showed the highest extraction yield (20.8%). However, its use was rejected due to its high flammability and hexane was chosen as an extraction agent. It was also found that the use of polar solvents enhanced the oxidation stability of the extracted grape seed oil, indicating a higher antioxidant content in those cases. In order to combine both effects, solvents mixtures were tested and it was observed a maximum in oil yield (18.5%) and oxidation stability (16.3 h) by using a mixture hexane/acetone 1:1 (v/v). Moreover, Soxhlet and Soxtherm extration technologies were compared, without observing a significant diference on the extraction yield nor on the oxidation stability of the oil. Luque-Rodríguez et al. [29] compared superheated hexane extraction with conventional Soxhlet system, finding that similar yields can be obtained in shorter extraction times.

Supercritical fluid extraction of vegetable oils has been intensively studied in the last decades and the results of the research are protected by several patents. However, this procedure is not applied in large-scale plants because of the difficulties with continuous transport of seeds into, through and out of the high-pressure extractor. Smaller amounts of seeds can be extracted in semi-batch mode, with the supercritical fluid flowing through a fixed bed of material [30]. Molero Gómez et al. [31] compared the supercritical extraction and the non-supercritical one using CO_2 as extraction agent, showing a great increment on oil yield by working at supercritical conditions. Jokić et al. [32] determined that the optimal conditions for obtaining the highest oil yield (14.49%) and antioxidant activity using supercritical CO_2 as extracting agent were 400 bar and 41°C. Fiori [33] determined a break-even value of grape seed oil obtained at 550 bar and 60°C of 5.9 €/kg using the same technology. Freitas et al. [23] pointed that supercritical propane is a more suitable solvent for grape seed oil extraction since a higher extraction yield and faster kinetic was observed. However, most of authors conclude that the grape variety, the harvesting conditions, and the cultivation structure have a major influence on the grape seed oil yield, composition and antioxidant components concentration [19].

As a result of the fatty acid profile of grape seed oil with a high content of linoleic acid and the low amount of nutraceutical compounds compared with other vegetable oils, grape seed oil is not as important as olive or rapeseed oil. Moreover, phenolic compounds, whose concentration is very high in the grape seed, are only extracted into the oil only to a very small extent [24]. From this, it can be drawn that grape seed oil is not as an interesting vegetable oil as others from a nutritional point of view. Hence, in this chapter, the major non-food applications of grape seed oil are discussed, mainly focusing on the transesterification to produce biodiesel and the epoxidation and further ring-opening reactions to produce biobased polyols.

2. Transesterification of grape seed oil for biodiesel production

Biodiesel is an alternative fuel for diesel engines which is made from renewable resources, such as vegetable oils and animal fats. It is biodegradable, nontoxic, sulfur- and benzene-free and, therefore, is environmentally advantageous. The recent concern about the increase in the crude oil price, the limited resources of fossil oil, and the environmental concerns have increased the interest on vegetable oils to make biodiesel [34].

Different alternatives have been proposed on literature to increase the content of vegetable oil-based compounds on diesel fuels. The first approach consisted on directly using vegetable oils without any chemical transformation as direct fuel. Thus, soybean oil [35], sunflower oil [36], canola oil [37] or rapeseed oil [38] have been studied. However, the direct use of vegetable oils or the use of blends with traditional petroleum-based diesel has not been considered a suitable solution due to the high viscosity, the gum formation due to oxidation processes and the deposition of carbon on piston and head of the engines [39]. As an alternative, some authors proposed the pyrolysis of rapeseed oil [40], soybean oil [41] and safflower oil [42] to yield small molecules. Nevertheless, the required equipment is expensive and the obtained results do not justify the use of this technology. In addition, the pyrolysis of vegetable oils implies the remove of oxygen from biofuel, which nowadays is a legal requirement in biobased fuels [43]. Therefore, the most interesting route to produce biodiesel from vegetable oils is the transesterification.

Transesterification reactions consists of the alcoholysis of a triglyceride to form esters and glycerol. To improve the reaction rate and yield, a catalyst is usually used. As the reaction is limited by an equilibrium, excess of alcohol is used to enhance the production of esters. Among the alcohols that can be used during the transesterification process, methanol and ethanol are the most efficient and commonly used ones. With respect to catalysts, alkalis (NaOH, KOH, carbonates, and alkoxides), acids (sulfuric acid, sulfonic acid, and hydrochloric acid), and enzymes (lipases) have been proposed in literature [44]. However, alkali-catalyzed transesterification is much faster and economically viable and, therefore, is the most often used industrial system for producing biodiesel from vegetable oils [34]. However, for alkali-catalyzed transesterification, the glycerides and alcohol must be substantially anhydrous, because water causes a partial reaction change to saponification which produces soaps. Soaps consume the catalyst and reduce the catalytic efficiency, as well as causing an increase in viscosity, the formation of gels and difficulty in achieving separation of glycerol at the end of the reaction [44].

In respect to the transesterification mechanism, it consists of three consecutive reversible reactions where the triglyceride is converted into diglyceride, monoglyceride, and finally, glycerol, liberating a mole of esters at each step (**Figure 1**). After the transesterification reaction, a mixture of esters, glycerol, alcohol, catalyst, and unreacted tri-, di-, and monoglycerides is obtained.

The obtained transesterification product should be purified and characterized to quantify the global biodiesel quality. **Table 2** shows the requirements of the most important regional standards and the analytic methods.

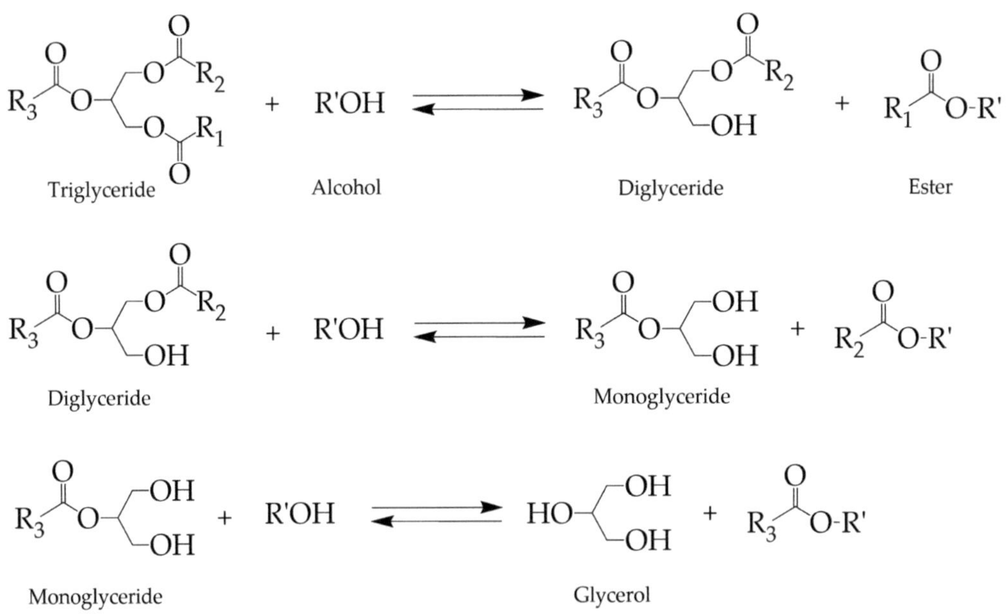

Figure 1. The transesterification reaction of triglycerides with alcohol.

http://dx.doi.org/10.5772/intechopen.71710

		EU (EN 14214:2012)		USA (ASTM D6751-15ce1)	
Parameter	**Units**	**Limit**	**Method**	**Limit**	**Method**
Ester content	% (m/m)	>96.5	EN 14103	–	–
Density at 15°C	kg/m³	860–900	EN ISO 3675 EN ISO 12185	–	–
Viscosity at 40°C	mm²/s	3.50–5.00	EN ISO 3104	1.9–6.0	ASTM D445
Flash point	°C	>101	EN ISO 2719 EN ISO 3679	>93	ASTM D93
Distillation temperature (90% recovered)	°C	–	–	<360	ASTM D1160
Cetane number	–	>51.0	EN ISO 5165	>47	ASTM D613
Copper band corrosion	Classification	1	EN ISO 2160	3	ASTM D130
Oxidation stability	h	>8.0	EN 14112 EN 15751	>3.0	EN 15751
Acid value	mg KOH/g	<0.50	EN 14104	<0.50	ASTM D664
Iodine value	g I/100 g	<120	EN 14111 EN 16300	–	–
Cloud point	°C	*	EN 23015	Report	ASTM D2500
CFFP	°C	*	EN 116	–	–
Methyl linoleate	% (m/m)	<12.0	EN 14103	–	–
Polyunsaturated methyl esters	% (m/m)	<1.00	EN 15779	–	–
Methanol content	% (m/m)	<0.2	EN 14110	<0.2	EN 14110
Monoglyceride content	% (m/m)	<0.70	EN 14105	<0.40	ASTM D6584
Diglyceride content	% (m/m)	<0.20	EN 14105	–	–
Triglyceride content	% (m/m)	<0.20	EN 14105	–	–
Free glycerol	% (m/m)	<0.02	EN 14105 EN 14106	<0.02	ASTM D6584
Total glycerol	% (m/m)	<0.25	EN 14105	≤0.24	ASTM D6584
Water content	mg/kg	<500	EN ISO 12937	–	–
Carbon residue	% (m/m)	–	–	<0.05	ASTM D4530
Water and sediment	% vol.	–	–	<0.05	ASTM D2709
Total contamination	mg/kg	<24	EN 12662	–	–
Sulphated ash content	% (m/m)	<0.02	ISO 3987	<0.02	ASTM D874
Sulfur content	mg/kg	<10.0	EN ISO 20846 EN ISO 20884 EN ISO 13032	<15	ASTM D5453

		EU (EN 14214:2012)		USA (ASTM D6751-15ce1)	
Parameter	Units	Limit	Method	Limit	Method
Cold soak filterability test	s	—	—	<200	ASTM 7501
Group I metals (Na + K)	mg/kg	<5.0	EN 14108	<5.0	EN 14538
			EN 14109		
			EN 14538		
Group II metals (Ca + Mg)	mg/kg	<5.0	EN 14538	<5.0	EN 14538
Phosphorous content	mg/kg	<4.0	EN 14107	<10	ASTM D4951
			EN 16294		

*Differences exist between the national versions of the EN 14214 standard. These differences relate to cold weather requirements and are detailed in the national annex of each standard.

Table 2. Comparison of the European and American biodiesel quality standards.

Among the general parameters for biodiesel, the viscosity controls the characteristics of the injection into the engine. The viscosity of fatty acid methyl esters can reach very high values when the vegetable oil used as raw material is highly unsaturated and, therefore, it should be controlled. The flash point of a fuel is the temperature at which it will ignite when exposed to a flame or a spark. The flash point of biodiesel is higher than the petroleum-based diesel, which is safe for transport purposes. Cetane number measures how easily ignition occurs and the smoothness of combustion. Cetane number is a critical variable because it affects important performance parameters such us combustion, stability, production of white smokes, noise, and emissions of CO and hydrocarbons. The presence of mono-, di-, and triglycerides, and other non-interesting products (alcohol, glycerol, phosphorylated and sulfated compounds, etc.) cause engine problems, such as fuel filter plugging, carbon deposition, and hydrolytic and oxidative reactions of fatty acid methyl esters [34].

In their previous work, Ramos et al. [18] studied the influence of vegetable oils properties and composition on the quality of biodiesel synthesized from different vegetable oils. Among them, grape seed oil was also used as raw material. As a resume, **Table 3** shows a comparison of properties of biodiesel obtained using different vegetable oils.

As can be observed, grape seed oil-based biodiesel shows similar values of ester content, viscosity, flash point, and acid value to those obtained from traditional vegetable oils (soybean oil, sunflower oil, etc.). However, iodine number, cetane number, and oxidative stability exclude the direct use of this biodiesel because their values are out of the range indicated by the major biodiesel standards.

Iodine number is an important parameter on biodiesel quality, since heating highly unsaturated fatty acid methyl esters results in polymerization. This can lead to the formation of deposits or to deterioration of the lubricant properties of biodiesel. It is well known that biodiesel cetane number depends on the feedstock used for its production. The longer and the more saturated the fatty acid carbon chains, the higher the cetane number [45]. Also,

Parameter	Grape	Palm	Olive	Rape	Soybean	Sunflower
Ester content (% mass)	97.8	97.7	99.0	99.5	96.9	97.2
Viscosity at 40°C (mm^2/s)	4.1	4.5	4.5	4.4	4.2	4.2
Flash point (°C)	175	176	178	170	171	177
Cetane number	48	61	57	55	49	50
Oxidative stability at 110°C (h)	0.5	4.0	3.3	2.0	1.3	0.8
Acid value (mg KOH/g)	0.27	0.12	0.13	0.16	0.14	0.15
Iodine value (g I/100 g)	138	57	84	109	128	132
Linoleic acid content (% mass)	0.4	0.2	0.6	7.9	6.3	0.2
CFPP (°C)	−6	10	−6	−10	−5	−3
Methanol content (% mass)	0.00	0.00	0.00	0.00	0.00	0.00
Monoglyceride content (% mass)	0.28	0.17	0.67	0.41	0.21	0.37
Diglyceride content (% mass)	0.08	0.06	0.09	0.08	0.10	0.07
Triglyceride content (% mass)	0.03	0.04	0.03	0.03	0.07	0.04
Free glycerol (% mass)	0.00	0.01	0.00	0.01	0.07	0.00
Total glycerol (% mass)	0.09	0.06	0.19	0.09	0.00	0.11

Table 3. Properties of biodiesel obtained from different vegetable oils via transesterification.

oxidation stability is one of the major issues affecting the use of biodiesel because it is very difficult to meet the 6 h of stability required by most of the main standards, even for many common raw materials. Therefore, since linoleic acid (C18:2) is the majoritarian fatty acid on grape seed oil composition, cetane number, iodine value, and oxidation stability requirements cannot be easily accomplished.

On the other hand, biodiesel obtained from grape seed oil showed some excellent properties, such us a very low cold filter plugging point (CFPP). Certain types of biodiesel present an operability problem based on the wax settling and plugging of filters and fuel lines when overnight temperatures approach. Low-temperature properties depend mostly on the saturated ester content. In this sense, grape seed oil barely contains a 11% of saturated fatty acid methyl esters, so its excellent CFPP is justified [46, 47].

To improve the performance of grape seed oil as biodiesel feedstock, different approaches have been discussed in literature. One of the most common solutions consists of the biodiesel-biodiesel blending from different feedstocks. Fernández et al. [28] studied the feasibility of blending jatropha, grape, rape and palm biodiesel, finding good properties at specific mixing portions. Atabani et al. [48] improved the kinematic viscosity of *Sterculia foetida* methyl esters by blending it with soybean-based biodiesel. Also, biodiesel has been blended with petroleum-based diesel and other fuels to meet the standards requirements [49, 50].

The issue of oxidative stability affects biodiesel mainly during extended storage. Generally, factors such as presence of air, elevated temperatures or presence of metals facilitate oxidation. Several approaches have been discussed in literature to either prevent or decelerate oxidation rate [51, 52]. The most obvious solution consists on preventing contact of the biodiesel with atmospheric air. Another solution is to prevent contact with prooxidative substances and avoid elevated temperatures and light. However, these solutions are not always viable and the use of antioxidants is of significant interest. Different natural and synthetic antioxidants had been used on biodiesel obtained from different vegetable oils. The most interesting result, apart from the obvious increment on oxidative stability, was that properties such us viscosity, CFPP, density, carbon residue and sulphated ash remained constant [53]. Different antioxidants, such as vitamin E [54], tocopherols [55], and synthetic ones (butylated hydroxytoluene, tert-butyl hydroquinone, and pyrogallol) [56], had been studied over biodiesel obtained from different feedstocks.

Nevertheless, grape seed oil-based biodiesel has been used for different applications. For example, Karthikeyan [57] proposed its use as a biofuel in marine engines by blending it with traditional petroleum-based fuel. It was found that flash point, cloud point and pour point where improved and NOx emissions were significantly reduced. Bazooyar et al. [58] compared the behavior of biodiesels based on various vegetable oils in a semi-industrial boiler. Grape seed oil biodiesel showed the lowest CO and CO_2 emissions. Hence, it can be concluded that grape seed oil can be used.

3. Epoxidation of grape seed oil

Epoxidation reaction consists of the addition of an oxygen atom to a carbon-carbon double bond. This reaction has been established as one of the most important methods for the formation of carbon-oxygen bonds [59]. Epoxides are highly reactive and versatile functional groups which are commonly used as precursors for alcohols, glycols, carbonyls, alkanolamines, substituted olefins, polyester polyurethanes, and epoxy resins [60, 61]. Increasing environmental problems related to fossil sources overuse have made plant oils an attractive alternative for the production of epoxy-based materials.

In the industry, epoxidized vegetable oil, and more specifically soybean and linseed oil, are currently used as plasticizers for PVC and related resins, reaching an annual production of about 200,000 tons [62]. Indeed, the epoxides change the solubility and flexibility of the PVC resins and react with hydrochloric acid liberated from the PVC resins under the prolonged action of light and heat. Epoxidized vegetable oils are also used as lubricants [63, 64] and as prepolymer in coating formulations [65–67]. The applicability of an epoxidized oil depends on its purity, oxirane number, and iodine number.

There are three different potential sources of vegetable oils containing epoxy groups. Firstly, there is a variety of natural occurring epoxy vegetable oil, mainly *Vernonia galamensis* [68] and *Euphorbia lagascae* oil [69], which contain up to 70% of vernolic acid (12S,13R-epoxy-9-cis-octadecenoic acid). The second option consists of the production of oil from genetically modified seeds which already contains vernolic acid in order to increase its content [70, 71].

http://dx.doi.org/10.5772/intechopen.71710

However, none of these vegetable oils are commercially available at a competitive cost. Therefore, the only remaining solution is the chemical transformation of unsaturated vegetable oils, such us soybean [72] or grape seed oil [73], on epoxidized vegetable oils.

A wide variety of methods have been proposed for epoxidation of vegetable oils and related products (e.g., unsaturated free fatty acids). The most important ones are listed below [74]:

- Chlorohydrin process: this indirect epoxidation process is based on the reaction of the unsaturation with HOCl, forming the chlorohydrin which subsequently yields the epoxide on alkaline treatment. However, this method is highly unfriendly environmentally.
- Halcon reaction: epoxides can also be prepared by treating the unsaturations with tert-butyl hydroperoxide using vanadium, titanium of molybdenum complexes as catalysts.
- Epoxidation with dioxirane: the enantioselective epoxidation in a neutral medium is made possible by using dioxirane and an optically active manganese (III) salt as catalyst.
- Epoxidation with molecular oxygen: this process is catalyzed by silver and is the cheapest and greenest route to epoxidize low molecular weight molecules (e.g., ethylene and butadiene). However, this process is not efficient to be applied to vegetable oils, but its use is mostly restricted to low molecular weight substrates.

However, the selectivity and workup of these processes are not satisfactory for industrial applications, and currently, the industrial processes are based on the use of peracids. The Prilezhaev reaction consists on the epoxidation of alkenes with peracids, typically performic or peracetic acid, formed *in situ* from hydrogen peroxide and formic or acetic acid, respectively. Among them, peracetic acid is the most used one due to its low price, higher epoxidation efficiency, and safety issues at ordinary temperatures. Acidic catalysts, either strong acids or acidic ion exchange resins, are required on this process. Dinda et al. [75] found that the order of catalyst effectiveness for the vegetable oil epoxidation reaction based on mineral acids was headed by H_2SO_4, followed by H_3PO_4, HNO_3, and HCl.

During the *in situ* epoxidation process, different reactions take place. Firstly, the peracid must be formed from hydrogen peroxide and the carboxylic acid. Secondly, the previously formed peracid attacks the alkene group, generating an oxirane group and the carboxylic acid. Hence, the carboxylic acid is not consumed during the epoxidation reaction. However, two side reaction might take place during the epoxidation process. Since oxirane rings are formed in aqueous acidic media, they can be hydrolysed into hydroxyl groups. Consequently, reactions between hydroxyl groups form different triglycerides led to the formation of oligomers linked by ether groups.

3.1. A case study: influence of temperature on grape seed oil epoxidation

Grape seed oil was epoxidized using peracetic acid formed *in situ* from acetic acid and hydrogen peroxide using H_2SO_4 as catalyst. The epoxidation reactions were carried out at 50, 60, 90, and 100°C [73]. To evaluate the conversion and selectivity of the process, the concentration of different functional groups was determined.

Figure 2 shows the evolution of double bonds, epoxy groups, and hydroxyl groups along the reaction time at the previously commented reaction temperatures.

As can be observed, a gap exists in the double bond conversion (**Figure 2A**), depending on the reaction temperature, with increased conversion at greater temperature. If the temperature is higher than 90°C, conversions greater than 90% can be obtained in less than 1.5 h. Complete conversion of double bonds was only achieved when the reaction temperature and time were 100°C and 4 h, respectively. However, the double bonds conversion decreased much slower when the reaction temperature was 60°C or lower, requiring reaction times longer than 6 h for complete double bond conversion. Nevertheless, high double bond conversion rate cannot be the desired reaction conditions because the presence of secondary reactions can be promoted.

In respect to the oxirane ring formation (**Figure 2B**), the obtained data also indicated an important influence of reaction temperature. The reactions performed at high temperature (90 and 100°C) showed a maximum concentration of epoxide groups at short reaction times, between 1 and 1.5 h. After this, oxirane rings were completely consumed when the reaction time was 6 h. However, when the reaction was performed at lower temperatures (50 and 60°C), the epoxide group concentration increased constantly during the reaction, without observing any depletion.

In the same way, the hydroxyl group concentration (**Figure 2C**) increased abruptly at 100°C, reaching its maximum value in 4 h. At this point, the hydroxyl group concentration decreased in the reaction media, indicating the presence of another secondary reaction. On the other hand, when the reaction was performed at lower temperatures, the hydroxyl group formation was almost negligible.

The diminishing in the concentration of hydroxyl groups at 100°C indicates the presence of another secondary reaction, which mainly takes place when oxirane rings exist. The most common reaction in these cases reported in literature consists of the oligomerization of triglycerides. The generation of oligomers in the reaction bulk was confirmed by Gel Permeation Chromatography (GPC). All chromatograms presented three peaks, indicating the presence of dimers and trimers in addition to the main product.

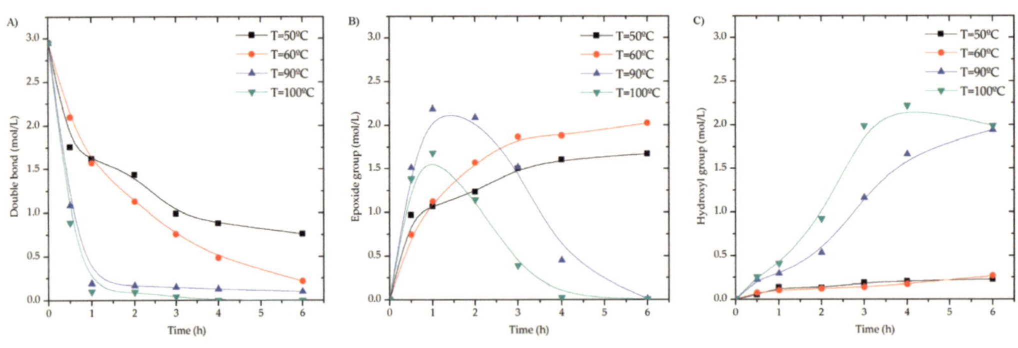

Figure 2. Concentration of different functional groups along the reaction time as function of different temperature used.

The concentration of oligomers (dimers and trimers) could be calculated through the mass balance presented in Eq. (1). This mass balance indicates that the initial content of double bonds ($DB_{t=0}$) is equal to the unreacted double bonds (DB_t), the oxirane rings (OR_t), half of the hydroxyl groups (OH_t), and the oligomers ($OLIG_t$). The portion of dimers (DIM) and trimers (TRIM) was quantified using the area ratio of these compounds from the chromatograms. In this estimation, the same response factor was assumed for both dimers and trimers. The obtained values are presented in **Figure 3**.

$$DB_{t=0} = DB_t + OR_t + \frac{OH_t}{2} + OLIG_t \qquad (1)$$

When the epoxidation reaction was performed at low temperature (50 or 60°C), oligomer formation was almost neglected. However, the oligomer concentration profiles obtained at high temperatures (90 and 100°C) indicate that the formation of dimers (**Figure 3A**) is required to form a trimer (**Figure 3B**) and that a unique reaction step is not taken place. Moreover, a change in the trend can be observed at a reaction time of 1.5 and 2.5 h for the reactions performed at 90 and 100°C, respectively. This result could be explained by the lack of oxirane ring groups in the reactor at this time, suggesting that the formation of oligomers requires the presence of both oxirane rings and hydroxyl groups.

3.2. Kinetic model for the *in situ* epoxidation of grape seed oil

Janković and Sinadinović-Fišer [76] have already postulated a kinetic model for the *in situ* formation of peracetic acid from acetic acid and hydrogen peroxide in the presence of a homogeneous acid catalyst. The mechanism can be summarized in the following steps: (i) formation of peracetic acid in the presence of the catalyst and (ii) reaction between peracetic acid and the double bonds to produce oxirane rings, releasing acetic acid. Moreover, different secondary reactions were observed during the epoxidation of grape seed oil: the hydrolysis of oxirane

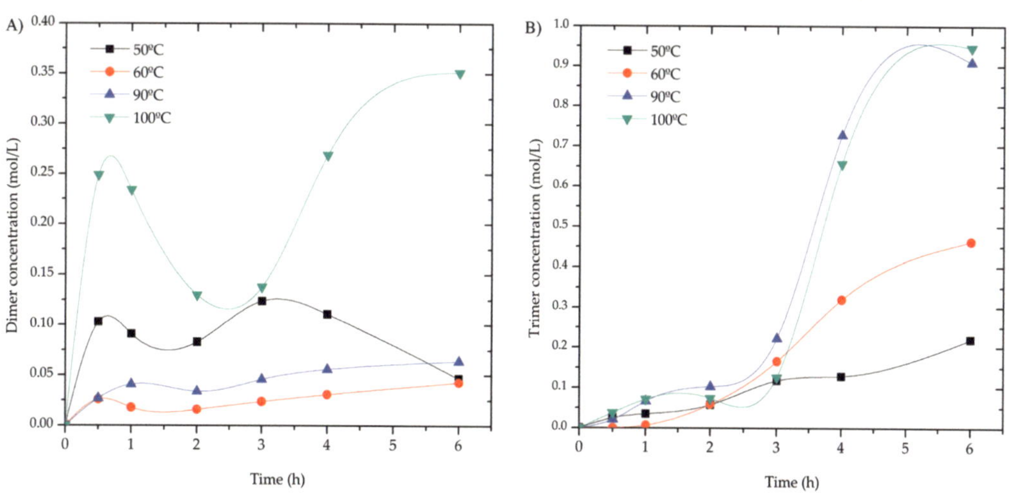

Figure 3. Concentration of dimmers and trimmers along the reaction time as a function of different temperature used.

rings (iii) and the formation and dimers (iv) and trimers (v). Finally, it is important to take into account the degradation of hydrogen peroxide in water (vi) [77].

From the above mechanism, the following reaction system (Eqs. (2)–(7)) can be proposed:

$$AA + H_2O_2 \underset{k_0}{\overset{k_1}{\longleftrightarrow}} PAA + H_2O \tag{2}$$

$$PAA + DB \overset{k_2}{\rightarrow} OR + AA \tag{3}$$

$$OR + H_2O \overset{k_3}{\rightarrow} 2OH \tag{4}$$

$$OR + OH \overset{k_4}{\rightarrow} DIM \tag{5}$$

$$OR + DIM \overset{k_5}{\rightarrow} TRIM \tag{6}$$

$$H_2O_2 \overset{k_6}{\rightarrow} H_2O + 0.5\, O_2 \tag{7}$$

where AA, PAA, DB, OR, OH, DIM, and TRIM represent the acetic acid, peracetic acid, double bond, oxirane ring, hydroxyl groups, dimers, and trimers, respectively.

The kinetic model was simplified assuming that the epoxidation reaction takes place in a unique phase (i.e., pseudo-homogeneous system). This assumption avoids the use of distribution constants, which would be necessary to quantify the concentration of the species in the organic and aqueous phases. To minimize the error caused by this assumption on the kinetic model, a high agitation rate was used. This high agitation rate favors the formation of small oil drops in the aqueous phase, increasing the external surface for mass transfer and minimizing limitations to the mass transfer [73].

Therefore, the corresponding differential equation system (Eqs. (8)–(16)) can be drawn from the proposed reaction set:

$$\frac{d[AA]}{dt} = -k_0[AA][H_2O_2] + k_1[PAA][H_2O] + k_2[PAA][DB] \tag{8}$$

$$\frac{d[H_2O_2]}{dt} = -k_0[AA][H_2O_2] + k_1[PAA][H_2O] - k_6[H_2O_2] \tag{9}$$

$$\frac{d[PAA]}{dt} = k_0[AA][H_2O_2] - k_1[PAA][H_2O] - k_2[PAA][DB] \tag{10}$$

$$\frac{d[H_2O]}{dt} = k_0[AA][H_2O_2] - k_1[PAA][H_2O] - k_3[OR][H_2O] + k_6[H_2O_2] \tag{11}$$

$$\frac{d[DB]}{dt} = -k_2[PAA][DB] \tag{12}$$

$$\frac{d[OR]}{dt} = k_2[PAA][DB] - k_3[OR][H_2O] - k_4[OR][OH] - k_5[OR][DIM] \tag{13}$$

$$\frac{d[OH]}{dt} = 2k_3[OR][H_2O] - k_4[OR][OH] \tag{14}$$

$$\frac{d[DIM]}{dt} = k_4[OR][OH] - k_5[OR][DIM] \tag{15}$$

$$\frac{d[TRIM]}{dt} = k_5[OR][DIM] \tag{16}$$

The differential equation system was solved numerically by using the fourth-order Runge-Kutta method. Hence, all the kinetic constants (k_i) were calculated by fitting the experimental concentrations of the species to those predicted from the equations Eqs. (8)–(16).

A comparison between the experimental values and the theoretical trends for the concentration of DB, OR, OH, DIMS, and TRIMS is shown at **Figure 4**. As can be observed, a good agreement exists among the experimental and predicted values, with an error lower than

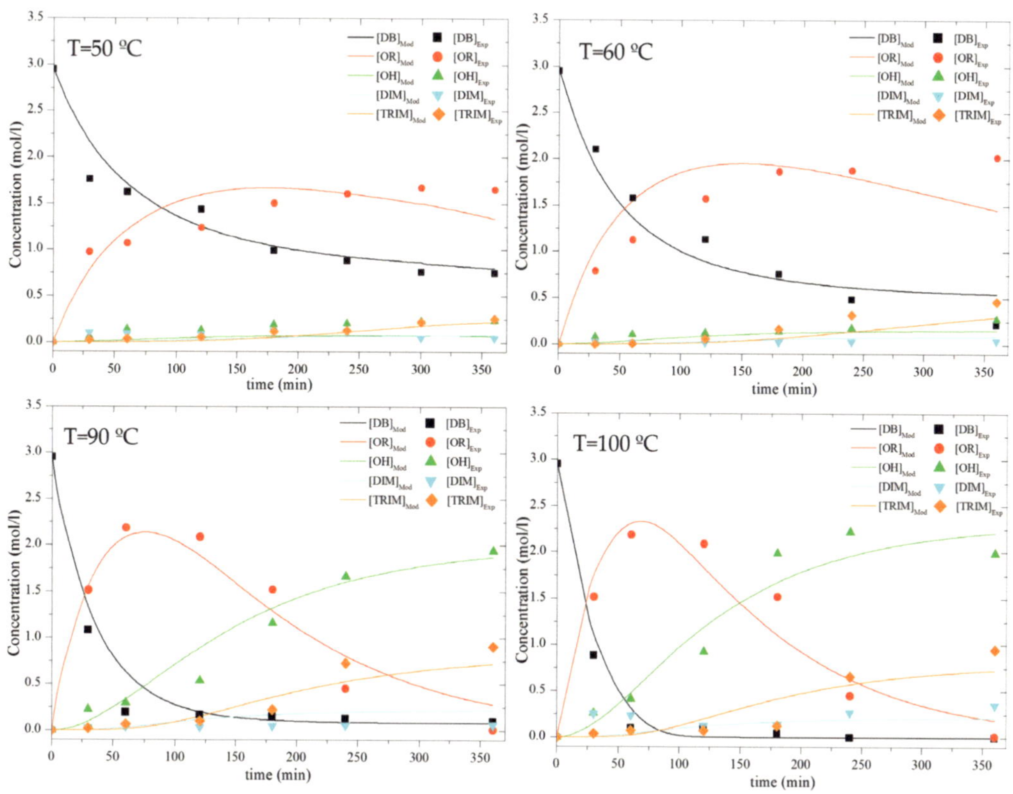

Figure 4. Theoretical and experiential profiles of functional groups along the reaction time for different temperatures.

T (°C)	k_0 (L/mol min)	k_1 (L/mol min)	k_2 (L/mol min)	k_3 (L/mol min)	k_4 (L/mol min)	k_5 (L/mol min)	k_6 (L/min)
50	1.35×10^{-2}	6.56×10^{-2}	0.129	0.247×10^{-4}	0.106×10^{-3}	1.04×10^{-2}	5.85×10^{-3}
60	1.20×10^{-2}	4.02×10^{-2}	0.178	0.288×10^{-4}	5.94×10^{-3}	1.05×10^{-2}	7.10×10^{-3}
90	1.01×10^{-2}	0.102×10^{-2}	0.485	1.38×10^{-4}	3.81×10^{-3}	1.50×10^{-2}	0.14×10^{-3}
100	0.990×10^{-2}	0.573×10^{-2}	0.560	1.75×10^{-4}	3.07×10^{-3}	1.70×10^{-2}	6.30×10^{-3}

Table 4. Kinetic constants for the proposed model.

0.5% in all cases. Student's *t*-test was performed, observing in all cases *t* values lower than the critical one (2.56) [73].

Table 4 shows the obtained values for the kinetic constants. All the kinetic constants presented the same order of magnitude, excepting k_3, which was the lowest. This lower value could be related to the high concentration of water that was considered in this model by neglecting the presence of a secondary aqueous phase. Therefore, it is expected that this value would increase in a two-phase model. Respecting to the values of k_0 and k_1, it indicates that the peracetic formation is not favorable at high temperatures. Moreover, the kinetic constants k_4 and k_5 indicate that the formation of trimers is favored over dimers one.

One of the most traditional representations of kinetic constants is the Arrhenius equation (Eq. (17)). To obtain the pre-exponential factor (A) and the activation energy (E_A), all obtained constants for the chemical reactions were adjusted by non-linear fitting.

$$k = A \cdot e^{\frac{-E_A}{R \cdot T}} \tag{17}$$

The E_A, A, and coefficient of determination values (R^2) are summarized in **Table 5**. A good fit between the experimental kinetic constants and the Arrhenius equation was observed for all reactions, except for the reverse reaction of peracetic acid synthesis (k_1).

Different authors have determined the E_A value of different *in situ* epoxidation system using different vegetable oils as raw materials. For example, Cai et al. [78] and Mungroo et al. [79] found a value of 10.3 and 10.7 kcal/mol, using soybean and canola oil as raw material, respectively. There is a small difference among these values and our determined one (7.3 kcal/mol). Janković and Sinadinović-Fišer [76] concluded that the number and the position of the double bonds, their position with respect to the carboxylic group, and the presence of *cis*- or *trans*-isomers have

Kinetic constant	k_0	k_1	k_2	k_3	k_4	k_5	k_6
E_A (kcal/mol)	−1.46	−16.51	7.30	10.19	−5.33	2.47	5.33
A	1.35×10^{-3}	4.35×10^{-13}	1.12×10^{4}	171	2.27×10^{-6}	0.468	23.18
R^2	0.972	0.738	0.994	0.979	0.937	0.966	0.996

Table 5. Arrhenius equation parameters.

a strong influence on the activation energies. Moreover, E_A values are affected by the simplifications and assumptions considered along the proposed model. Therefore, slight differences in the activation energies values are expected.

4. Synthesis of biopolyols

The uncertainty of petroleum price and its availability, combined with the global and institutional tendencies toward the principles of green chemistry, have forced the chemical industry to explore the use of renewable resources. Polyurethanes (PUs), with a global production of 12.28 Mt in 2010, are the sixth most widely used polymer [80]. This is due to the high versatility of use in mostly all the fields of polymer applications: foams, elastomers, adhesives, coatings, sealants, fibers, etc. PUs are obtained by the reaction of a polyol and a polyisocyanate, being both of these raw materials typically obtained from petroleum. However, the chemical industry is paying a great interest to the production of biobased polyols (biopolyols).

The most interesting alternative to produce biopolyols consists of the chemical transformation of unsaturated vegetable oils through different alternatives to produce hydroxyl groups, which enable the reaction with polyisocyanates to get PUs. Some of the proposed routes in literature are the ozonolysis-hydrogenation process [81], hydroformylation [82], dimerization of fatty acids [83], thiol-ene coupling [84], and formaldehyde addition [85]. However, the most used one consists of the epoxidation of double bonds and its further epoxide ring opening.

4.1. Ring-opening reaction with glycerol

To obtain desired biopolyols, the oxirane rings can be opened with compounds containing active hydrogen atoms, such as monoalcohols [86], amines [87, 88] or carboxylic acids [89], among others.

In addition, if diols or triols are used as nucleophile for epoxide ring opening, then both primary and secondary hydroxyl functions could be inserted on each epoxide group. Thus, although the production of polyols from different vegetable oils has been reported, there is a lack of information about the yield of this reaction toward hydroxyl groups and the viability of using grape seed oil as unsaturated vegetable oil.

Therefore, in this work, the ring-opening reaction was carried out using glycerol as ring-opening agent and double metal cyanide (DMC) complex as catalyst (**Figure 5**) [90].

The epoxidized grape seed oil previously produced, with an oxirane oxygen concentration of 5.87%, a hydroxyl value of 31.98 mg KOH/g, and an average molecular weight of 1151.75, was stirred at 900 rpm during 2 h at 80°C. **Figure 6** shows the infrared spectrum of the grape seed oil, the epoxidized oil, and the green-polyol obtained after ring-opening reaction with glycerol. It can be observed a drop in the intensity of the bands at 3009 and 1660 cm^{-1} related to unsaturations during the epoxidation reaction. Moreover, a big increase in the intensity area at the stretching band of 3500 cm^{-1} is observed, which confirms the rupture of the oxirane ring and the incorporation of a large number of terminal hydroxyl groups from the glycerol molecule.

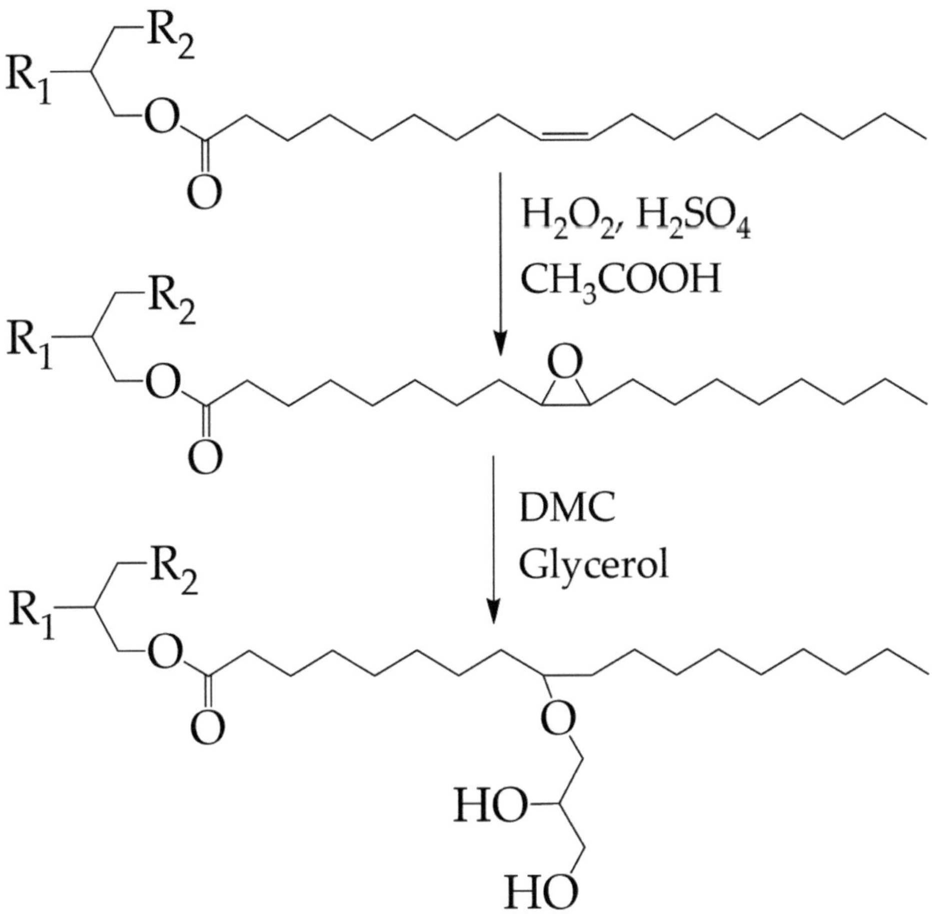

Figure 5. Reaction scheme of the ring-opening process in presence of glycerol.

The developed renewable green-polyol presents the following characteristics: an oxirane oxygen concentration of 1.06%, a hydroxyl value of 80.84 mg KOH/g, and an average molecular weight of 1800.94 g/mol. These values indicate that 82% of the oxirane rings were opened by the glycerol.

4.2. Ring-opening reaction with NaN_3

As it was previously commented, biopolyols synthetized from epoxidized vegetable oils can be functionalized by different groups, when the epoxide group is hydrolysed. Among the possible pathways, the ring-opening reaction with sodium azide [91] is one of the most interesting alternatives for obtaining very interesting compounds (**Figure 7**) [92].

The presence of nitrogen atoms in the structure of the biopolyol resulting from the incorporation of azide groups enables the polyol to act as thermal stabilizer compound in polyurethane foams [93]. Additionally, the azide group is highly polar, and when it is linked to a hydrocarbon chain, the hydrophobic character is modified, favoring its use as a surfactant [94].

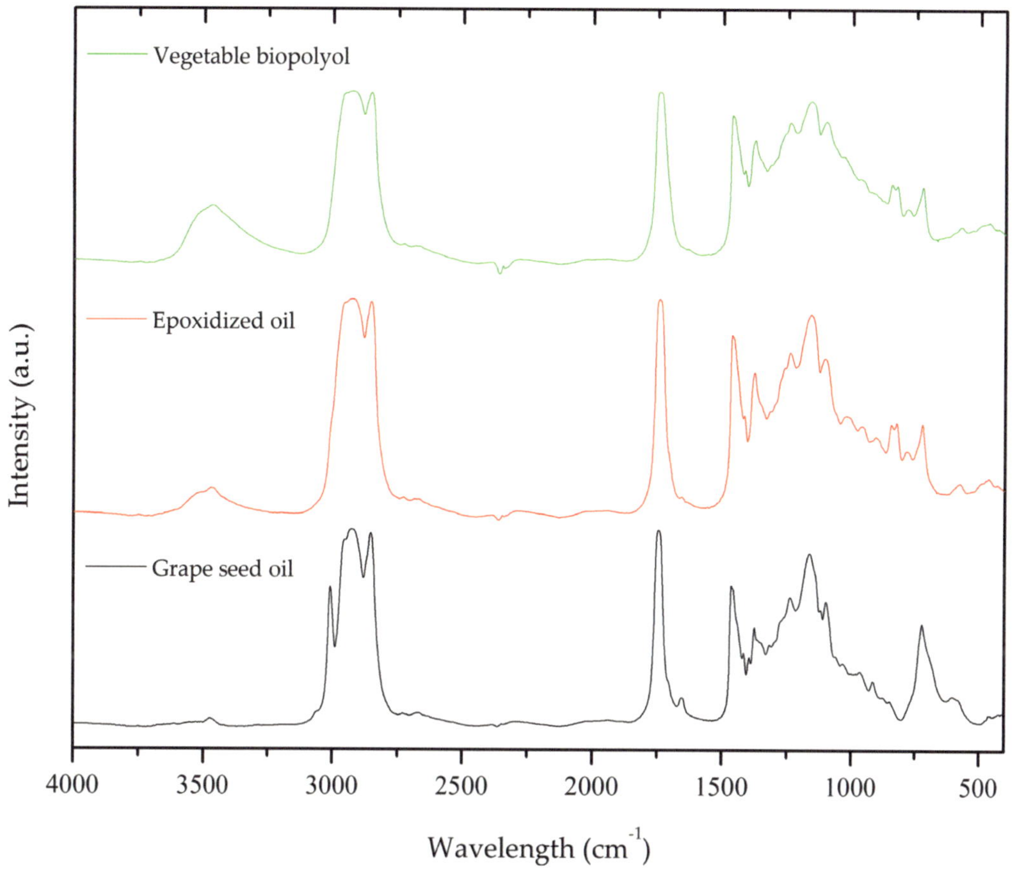

Figure 6. Infrared spectrum of the grape seed oil, the epoxidized oil, and the vegetable biopolyol.

Moreover, the azide group is one of the two main parts for performing the 1,3-dipolar cycloaddition reaction among azides and alkynes in the field of click chemistry [95]. The latter one is of the most actively investigated approaches for preparing tailor-made bioactive substances at this time.

The azidification reaction was performed at three different temperatures (50, 60, and 70°C) for 24 h. Samples were extracted at four points: 1, 2, 4, and 24 h [92]. The hydroxyl value and oxirane content of the synthesized azidified biopolyols are presented in **Figure 8**.

As can be observed, high reaction times and temperatures favor the transformation of oxirane rings into hydroxyl groups. The influence of temperature was determined considering that the reactions follow a pseudo-first-order kinetics. Hence, the observed kinetic constant (k_{obs}) can be determined from the slope of the representation ln(C_{epox}) vs. reaction time. The high correlation coefficient ($R^2 = 0.990$) indicated a good fit between the experimental and theoretical values. Moreover, it is also remarkable that longer reaction times than 24 h are required to achieve full conversion of epoxide groups. However, the polyol obtained after 24 h at 70°C was used for the subsequent foaming process.

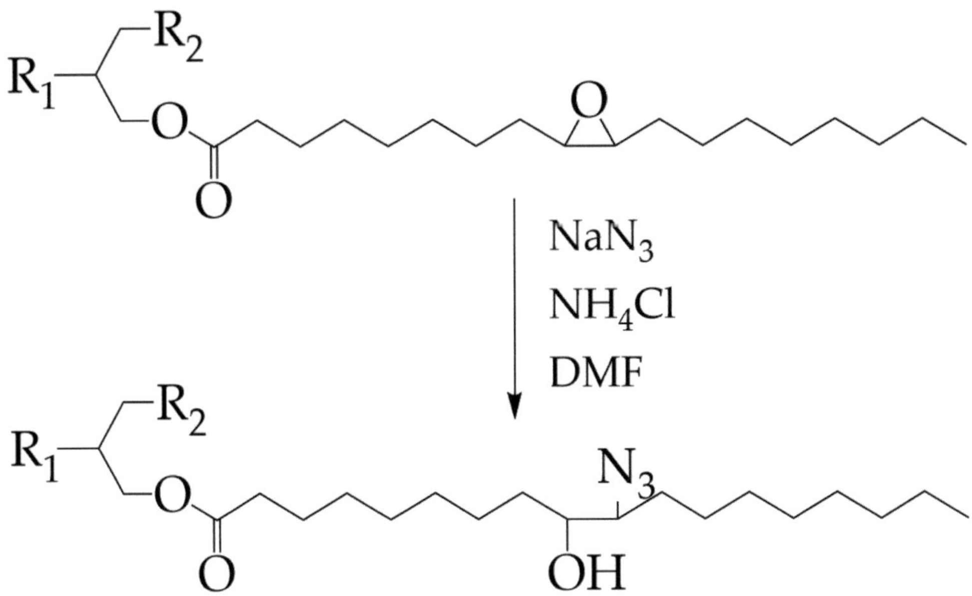

Figure 7. Reaction scheme of the ring-opening process in presence of sodium azide.

Kinetic constant values of 2.70×10^{-2}, 5.05×10^{-2}, and 7.18×10^{-2} s^{-1} were obtained at 50, 60, and 70°C, respectively. As expected, higher reaction temperatures favored the ring-opening reaction. Arrhenius equation (Eq. (17)) was used to determine the E_A and A for the proposed reaction system. Also in this case, a good correlation ($R^2 = 0.987$) was observed. E_A and A was calculated from the slope and y-intercept of its linearized form, obtaining values of $E_A = 0.341$ J/mol and A = 0.121.

4.3. Rigid polyurethane foam synthesis

As was stated previously, polyurethanes constitute a large family of polymers which is characterized for their versatility. Depending on the characteristics and structures of polyols and polyisocyanates, polyurethanes can be obtained as thermoplastics, adhesives, coatings, foams, etc. Rigid polyurethane foams are widely used in different applications, such as thermal [96] and acoustic [97] insulation.

The versatility of these products and the great consumption of polyurethanes (PUs) in the new emerging economies, which are believed to increase in incoming years, are forcing researchers to find alternative raw materials to avoid the petroleum dependence. Related to this, use and disposal of petroleum-based PUs is to be taken into account. Therefore, since polyol components usually are around half the weight of the final product, the use of greener polyols in PU synthesis has attracted much interest during the last decade [98].

In this section, different rigid PU foams were synthesized from the previously synthesized grape-seed-oil-based polyols. The commercial polyol Alcupol R4520 was used as a reference

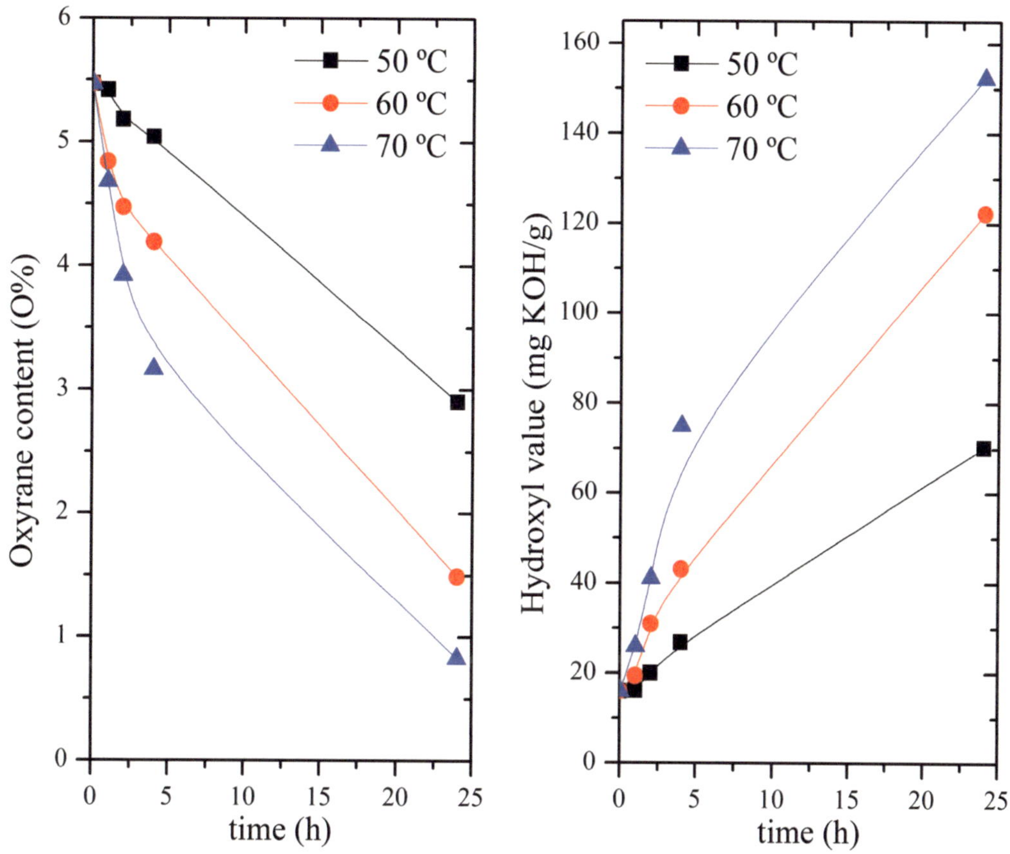

Figure 8. Oxirane content and hydroxyl values for the azidified biopolyol obtained at different reaction temperatures during 24 h.

	Foam A	Foam B	Foam C
Type of polyol	R4520	Vegetable biopolyol	Azidified biopolyol
Polyol (wt.%)	47.75	48.43	47.75
Water (wt.%)	0.75	1.21	0.75
Tegostab B8404 (wt.%)	3	0.73	3
Tegoamin 33 (wt.%)	0.37	0.37	0.37
Tegoamin BDE (wt.%)	0.37	1.20	0.37
MDI (wt.%)	47.76	48.06	47.76

Table 6. Weight percentages of raw materials for the synthesis of rigid PU foams.

material to compare the internal structure and properties of the final product. **Table 6** shows the formulations used to synthesize the PU foams. The required quantities of these reactants were calculated based on the hydroxyl number of the polyol used to synthesize the rigid PU foam, as indicated elsewhere [99].

No differences during the foaming process (growth start time, growth rate, and final aspect of the foam) were observed between the two polyols. The only appreciable difference that was observed between the three PU foams was that the azidified foam was slightly darker than the commercial one due to the higher coloration of the azidified biopolyol [91].

The internal morphology of all the rigid PU foams was observed using a scanning electron microscope (SEM) (**Figure 9**). In all three foams, a polyhedral cell structure can be observed. A slight increase in the cell size of Foams B and C with respect to that of Foam A can be observed.

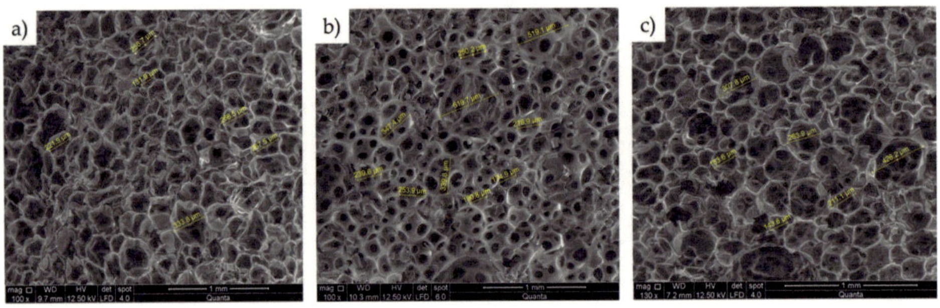

Figure 9. SEM images of (a) Foam A (synthesized from commercial polyol) (b) Foam B (synthesized from the biopolyol obtained using glycol as ring-opening agent) and (c) Foam C (synthesized from the azidified biopolyol).

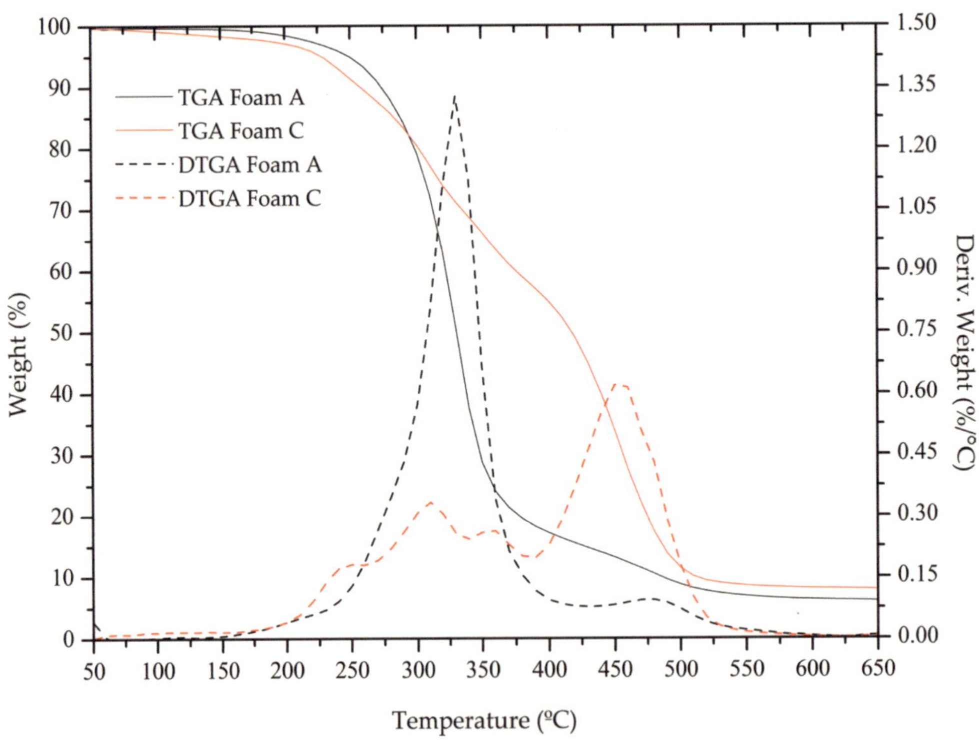

Figure 10. Comparison between thermogravimetric curves of Foam A (commercial foam) and Foam C (azidified foam).

This increase can primarily be attributed to two effects: the different volumes of pendant groups and the unadjusted foam formulation, which modify the nucleation and foaming rate.

Finally, thermogravimetric analyses (TGA) were used to investigate the effect of the incorporation of azide groups on the thermal stability of the biopolyols. **Figure 10** compares the TGA and DTG curves for Foam A and Foam C under an air atmosphere.

Two degradation steps can be observed for both polyurethane foams. The first one is assigned to the thermal degradation of the crosslinked polymer, which can be divided into hard and soft segments. The second step is attributed to the decomposition of the hydrocarbon chains formed by the main components of the foam (polyol and isocyanate) [100, 101]. As can be observed in the thermogravimetric curve, the incorporation of 0.09 g of nitrogen per gram of Foam C increased the maximum degradation temperature from 330 to 450°C. Moreover, the residue yield observed at 650°C improved from 6.28 to 8.11, indicating that Foam C is more thermally stable than Foam A. These results confirm the enhancement in the thermal stability of the PU foams because of incorporating nitrogen groups into their chemical structures by using azidified biopolyols from grape seed oil as raw materials.

5. Other non-food applications

Cosmetics are substances or products used to enhance or modify the aspect or smell of the body. During the last years, a wide sector of cosmetic consumers is expressing their interest in natural products due to the appearance of allergies and skin irritations due to the synthetic preservatives (mainly parabens), colorants, stabilizers, etc. [102]. Grape seed oil has been presented by Fiume et al. [103], as a safe alternative to these synthetic compounds, which are to be incorporated in cosmetics. Mbah [104] compared different vegetable oils with grape seed to act as lipophilic substance, indicating that it can be a potential dermal permeation enhancer. Attending to micro- and nano-structured cosmetic systems, Contri et al. [105] determined the viability of preparing grape seed oil nanocapsules for cutaneous applications, which enhanced the antioxidant activity. Moreover, Glampedaki and Dutschk [102] studied the preparation of oil-in-water emulsions using wine and grape seed oil, observing a good stability, when using glycerol monostearate as emulsifying agent.

Grape seed oil was compared with soybean and rapeseed oil for producing renewable crosslinked polyester resins by Clark et al. [106]. They concluded that grape seed oil presented a lower crosslink density due to its higher linoleic acid content but a higher thermal stability, being this product a viable substitute of traditional polyester resins in low stress or high-demanding thermal applications.

6. Conclusions

Grape seed oil is one of the main by-products of the winery industry. The fatty acid profile of this oil showed that it is mainly composed by linoleic and oleic acid, but the total amount and proportion of this depend on the grape variety, the weather, and the extraction method. However, grape seed oil has a limited application from a nutritional point of view, since most nutraceutical compounds (vitamin E active compounds, phytosterols, etc.) remain on other by-products, mainly on the grape skin and grape seed flour. Hence, different non-food applications have been proposed and discussed along with this book chapter.

The transesterification of grape seed oil leads to the production of a biodiesel with excellent low-temperature properties. On the other hand, other quality requirements, such as cetane number and oxidation stability, cannot be easily accomplished using grape seed oil as raw material due to its high content of polyunsaturated fatty acids. However, this can be easily solved by blending the grape seed oil based biodiesel with other fuels.

The epoxidation of grape seed oil using peracetic acid formed *in situ* from acetic acid and hydrogen peroxide is a feasible process for obtaining a highly epoxidated vegetable oil. The negative effect of the secondary reactions of this process, namely the hydrolysis and further oligomerization, can be diminished by working at high temperature (90°C) and short reaction times (1 h).

Epoxidized vegetable oils, such as grape seed oil, are of high interest for producing biobased materials that nowadays are produced from petroleum. So, biopolyols can be produced from epoxidized grape seed oil using different ring-opening agents such us glycerol and sodium azide. Different rigid polyurethane foams were synthesized from these biopolyols, using a commercial petroleum-based polyol as reference material to compare the internal structure and their properties. No internal differences were observed between the foams. However, it was observed an increase on the thermal stability of the foam which incorporated N_3 groups.

Acknowledgements

Authors gratefully acknowledge the financial support from Spanish Ministry of Education, Culture and Sport for the fellowship for PhD studies FPU014/00009.

Author details

Juan Carlos de Haro*, Juan Francisco Rodríguez, Manuel Carmona and Ángel Pérez

*Address all correspondence to: jcarlos.haro@uclm.es

Department of Chemical Engineering, Institute of Chemical and Environmental Technology, University of Castilla-La Mancha, Ciudad Real, Spain

References

[1] FAOSTAT [Internet]. 2014. Available from: http://www.fao.org/faostat [Accessed: 2017]

[2] Jin B, Kelly JM. Wine industry residues. In: Biotechnology for Agro-Industrial Residues Utilisation: Utilisation of Agro-Residues. Dordrecht, Netherlands: Springer; 2009. pp. 293-311. DOI: 10.1007/978-1-4020-9942-7_15

[3] Manzone M, Paravidino E, Bonifacino G, Balsari P. Biomass availability and quality produced by vineyard management during a period of 15 years. Renewable Energy. 2016;**99**:465-471. DOI: 10.1016/j.renene.2016.07.031

[4] Max B, Salgado JM, Cortés S, Domínguez JM. Extraction of phenolic acids by alkaline hydrolysis from the solid residue obtained after prehydrolysis of trimming vine shoots. Journal of Agricultural and Food Chemistry. 2010;**58**(3):1909-1917. DOI: 10.1021/jf903441d

[5] Bustos G, Moldes AB, Cruz JM, Domínguez JM. Influence of the metabolism pathway on lactic acid production from hemicellulosic trimming vine shoots hydrolyzates using Lactobacillus pentosus. Biotechnology Progress. 2005;**21**(3):793-798. DOI: 10.1021/bp049603v

[6] Cortés-Camargo S, Pérez-Rodríguez N, Oliveira RPDS, Huerta BEB, Domínguez JM. Production of biosurfactants from vine-trimming shoots using the halotolerant strain Bacillus tequilensis ZSB10. Industrial Crops and Products. 2016;**79**:258-266. DOI: 10.1016/j.indcrop.2015.11.003

[7] Mediavilla I, Fernández MJ, Esteban LS. Optimization of pelletisation and combustion in a boiler of 17.5 kWth for vine shoots and industrial cork residue. Fuel Processing Technology. 2009;**90**(4):621-628. DOI: 10.1016/j.fuproc.2008.12.009

[8] Gañán J, Al-Kassir Abdulla A, Cuerda Correa EM, Macías-García A. Energetic exploitation of vine shoot by gasification processes. A preliminary study. Fuel Processing Technology. 2006;**87**(10):891-897. DOI: 10.1016/j.fuproc.2006.06.004

[9] Dávila I, Robles E, Egüés I, Labidi J, Gullón P. The biorefinery concept for the industrial valorization of grape processing by-products. In: Handbook of Grape Processing By-Products. Academic Press; 2017. pp. 29-53. DOI: 10.1016/B978-0-12-809870-7.00002-8

[10] Yu J, Ahmedna M. Functional components of grape pomace: Their composition, biological properties and potential applications. International Journal of Food Science and Technology. 2013;**48**(2):221-237. DOI: 10.1111/j.1365-2621.2012.03197.x

[11] Pinelo M, Arnous A, Meyer AS. Upgrading of grape skins: Significance of plant cell-wall structural components and extraction techniques for phenol release. Trends in Food Science and Technology. 2006;**17**(11):579-590. DOI: 10.1016/j.tifs.2006.05.003

[12] Prozil SO, Evtuguin DV, Lopes LPC. Chemical composition of grape stalks of Vitis vinifera L. from red grape pomaces. Industrial Crops and Products. 2012;**35**(1):178-184. DOI: 10.1016/j.indcrop.2011.06.035

[13] Jordão AM, Ricardo-da-Silva JM, Laureano O. Evolution of proanthocyanidins in bunch stems during berry development (*Vitis vinifera* L.). Vitis. 2001;**40**(1):17-22

[14] Spigno G, Marinoni L, Garrido GD. State of the art in grape processing by-products. In: Handbook of Grape Processing By-Products. Academic Press; 2017. pp. 1-27. DOI: 10.1016/B978-0-12-809870-7.00001-6

[15] Părpăriță E, Brebu M, Azhar Uddin M, Yanik J, Vasile C. Pyrolysis behaviors of various biomasses. Polymer Degradation and Stability. 2014;**100**(1):1-9. DOI: 10.1016/j.polymdegradstab.2014.01.005

[16] Ferreira CIA, Calisto V, Cuerda-Correa EM, Otero M, Nadais H, Esteves VI. Comparative valorisation of agricultural and industrial biowastes by combustion and pyrolysis. Bioresource Technology. 2016;**218**:918-925. DOI: 10.1016/j.biortech.2016.07.047

[17] Fidalgo B, Berrueco C, Millan M. Chars from agricultural wastes as greener fuels for electric arc furnaces. Journal of Analytical and Applied Pyrolysis. 2015;**113**:274-280. DOI: 10.1016/j.jaap.2015.01.027

[18] Ramos MJ, Fernández CM, Casas A, Rodríguez L, Pérez A. Influence of fatty acid composition of raw materials on biodiesel properties. Bioresource Technology. 2009;**100**(1):261-268. DOI: 10.1016/j.biortech.2008.06.039

[19] Beveridge THJ, Girard B, Kopp T, Drover JCG. Yield and composition of grape seed oils extracted by supercritical carbon dioxide and petroleum ether: Varietal effects. Journal of Agricultural and Food Chemistry. 2005;**53**(5):1799-1804. DOI: 10.1021/jf040295q

[20] Lutterodt H, Slavin M, Whent M, Turner E, Yu L. Fatty acid composition, oxidative stability, antioxidant and antiproliferative properties of selected cold-pressed grape seed oils and flours. Food Chemistry. 2011;**128**(2):391-399. DOI: 10.1016/j.foodchem.2011.03.040

[21] Passos CP, Silva RM, Da Silva FA, Coimbra MA, Silva CM. Supercritical fluid extraction of grape seed (*Vitis vinifera* L.) oil. Effect of the operating conditions upon oil composition and antioxidant capacity. Chemical Engineering Journal. 2010;**160**(2):634-640. DOI: 10.1016/j.cej.2010.03.087

[22] Rubio M, Alvarez-Ortí M, Andrés A, Fernández E, Pardo JE. Characterization of oil obtained from grape seeds collected during berry development. Journal of Agricultural and Food Chemistry. 2009;**57**(7):2812-2815. DOI: 10.1021/jf803627t

[23] Freitas LDS, De Oliveira JV, Dariva C, Jacques RA, Caramão EB. Extraction of grape seed oil using compressed carbon dioxide and propane: Extraction yields and characterization of free glycerol compounds. Journal of Agricultural and Food Chemistry. 2008;**56**(8):2558-2564. DOI: 10.1021/jf0732096

[24] Matthäus B. Virgin grape seed oil: Is it really a nutritional highlight? European Journal of Lipid Science and Technology. 2008;**110**(7):645-650. DOI: 10.1002/ejlt.200700276

[25] Crews C, Hough P, Godward J, Brereton P, Lees M, Guiet S, et al. Quantitation of the main constituents of some authentic grape-seed oils of different origin. Journal of Agricultural and Food Chemistry. 2006;**54**(17):6261-6265. DOI: 10.1021/jf060338y

[26] Hassanein MMM, Abedel-Razek AG. Chromatographic quantitation of some bioactive minor components in oils of wheat germ and grape seeds produced as by-products. Journal of Oleo Science. 2009;**58**(5):227-233. DOI: 10.5650/jos.58.227

[27] Shinagawa FB, de Santana FC, Torres LRO, Mancini-Filho J. Grape seed oil: A potential functional food? Food Science and Technology. 2015;**35**(3):399-406. DOI: 10.1590/1678-457X.6826

[28] Fernández CM, Solana M, Fiori L, Rodríguez JF, Ramos MJ, Pérez Á. From seeds to biodiesel: Extraction, esterification, transesterification and blending of *Jatropha curcas* oil. Environmental Engineering and Management Journal. 2015;**14**(12):2855-2864

[29] Luque-Rodríguez JM, De Castro MDL, Pérez-Juan P. Extraction of fatty acids from grape seed by superheated hexane. Talanta. 2005;**68**(1):126-130. DOI: 10.1016/j.talanta.2005.04.054

[30] Sovová H, Kučera J, Jež J. Rate of the vegetable oil extraction with supercritical CO2-II. Extraction of grape oil. Chemical Engineering Science. 1994;**49**(3):415-420. DOI: 10.1016/0009-2509(94)87013-6

[31] Molero Gómez A, Pereyra López C, De La Martínez Ossa E. Recovery of grape seed oil by liquid and supercritical carbon dioxide extraction: A comparison with conventional solvent extraction. Chemical Engineering Journal and the Biochemical Engineering Journal. 1996;**61**(3):227-231. DOI: 10.1016/0923-0467(95)03040-9

[32] Jokić S, Bijuk M, Aladić K, Bilić M, Molnar M. Optimisation of supercritical CO_2 extraction of grape seed oil using response surface methodology. International Journal of Food Science and Technology. 2016;**51**(2):403-410. DOI: 10.1111/ijfs.12986

[33] Fiori L. Supercritical extraction of grape seed oil at industrial-scale: Plant and process design, modeling, economic feasibility. Chemical Engineering and Processing: Process Intensification. 2010;**49**(8):866-872. DOI: 10.1016/j.cep.2010.06.001

[34] Meher LC, Vidya Sagar D, Naik SN. Technical aspects of biodiesel production by transesterification—A review. Renewable and Sustainable Energy Reviews. 2006;**10**(3):248-268. DOI: 10.1016/j.rser.2004.09.002

[35] Adams C, Peters JF, Rand MC, Schroer BJ, Ziemke MC. Investigation of soybean oil as a diesel fuel extender: Endurance tests. Journal of the American Oil Chemists' Society. 1983;**60**(8):1574-1579. DOI: 10.1007/BF02666588

[36] Engler CR, Johnson LA, Lepori WA, Yarbrough CM. Effects of processing and chemical characteristics of plant oils on performance of an indirect-injection diesel engine. Journal of the American Oil Chemists' Society. 1983;**60**(8):1592-1596. DOI: 10.1007/BF02666591

[37] Strayer RC, Blake JA, Craig WK. Canola and high erucic rapeseed oil as substitutes for diesel fuel: Preliminary tests. Journal of the American Oil Chemists' Society. 1983;**60**(8):1587-1592. DOI: 10.1007/BF02666590

[38] Peterson CL, Auld DL, Korus RA. Winter rape oil fuel for diesel engines: Recovery and utilization. Journal of the American Oil Chemists' Society. 1983;**60**(8):1579-1587. DOI: 10.1007/BF02666589

[39] Schlick ML, Hanna MA, Schinstock JL. Soybean and sunflower oil performance in a diesel engine. Transactions of the American Society of Agricultural Engineers. 1988;**31**(5):1345-1349. DOI: 10.13031/2013.30868

[40] Billaud F, Dominguez V, Broutin P, Busson C. Production of hydrocarbons by pyrolysis of methyl esters from rapeseed oil. Journal of the American Oil Chemists' Society. 1995;**72**(10):1149-1154. DOI: 10.1007/BF02540981

[41] Niehaus RA, Goering CE, Savage LD Jr, Sorenson SC. Cracked soybean oil as a fuel for a diesel engine. Transactions of the American Society of Agricultural Engineers. 1986;**29**(3):683-689. DOI: 10.13031/2013.30213

[42] Schwab AW, Bagby MO, Freedman B. Preparation and properties of diesel fuels from vegetable oils. Fuel. 1987;**66**(10):1372-1378. DOI: 10.1016/0016-2361(87)90184-0

[43] Ma F, Hanna MA. Biodiesel production: A review. Bioresource Technology. 1999;**70**(1): 1-15. DOI: 10.1016/S0960-8524(99)00025-5

[44] Fukuda H, Kondo A, Noda H. Biodiesel fuel production by transesterification of oils. Journal of Bioscience and Bioengineering. 2001;**92**(5):405-416. DOI: 10.1016/S1389-1723(01)80288-7

[45] Demirbas A. Biodiesel production from vegetable oils via catalytic and non-catalytic supercritical methanol transesterification methods. Progress in Energy and Combustion Science. 2005;**31**(5-6):466-487. DOI: 10.1016/j.pecs.2005.09.001

[46] González Gómez ME, Howard-Hildige R, Leahy JJ, Rice B. Winterisation of waste cooking oil methyl ester to improve cold temperature fuel properties. Fuel. 2002;**81**(1):33-39. DOI: 10.1016/S0016-2361(01)00117-X

[47] Wu M, Wu G, Han L, Wang J. Low-temperature fluidity of bio-diesel fuel prepared from edible vegetable oil. Petroleum Processing and Petrochemicals. 2005;**36**(4):57-60

[48] Atabani AE, Mahlia TMI, Masjuki HH, Badruddin IA, Yussof HW, Chong WT, et al. A comparative evaluation of physical and chemical properties of biodiesel synthesized from edible and non-edible oils and study on the effect of biodiesel blending. Energy. 2013;**58**:296-304. DOI: 10.1016/j.energy.2013.05.040

[49] Pramanik K. Properties and use of jatropha curcas oil and diesel fuel blends in compression ignition engine. Renewable Energy. 2003;**28**(2):239-248. DOI: 10.1016/S0960-1481(02)00027-7

[50] Aydin H, Ilkiliç C. Effect of ethanol blending with biodiesel on engine performance and exhaust emissions in a CI engine. Applied Thermal Engineering. 2010;**30**(10):1199-1204. DOI: 10.1016/j.applthermaleng.2010.01.037

http://dx.doi.org/10.5772/intechopen.70861

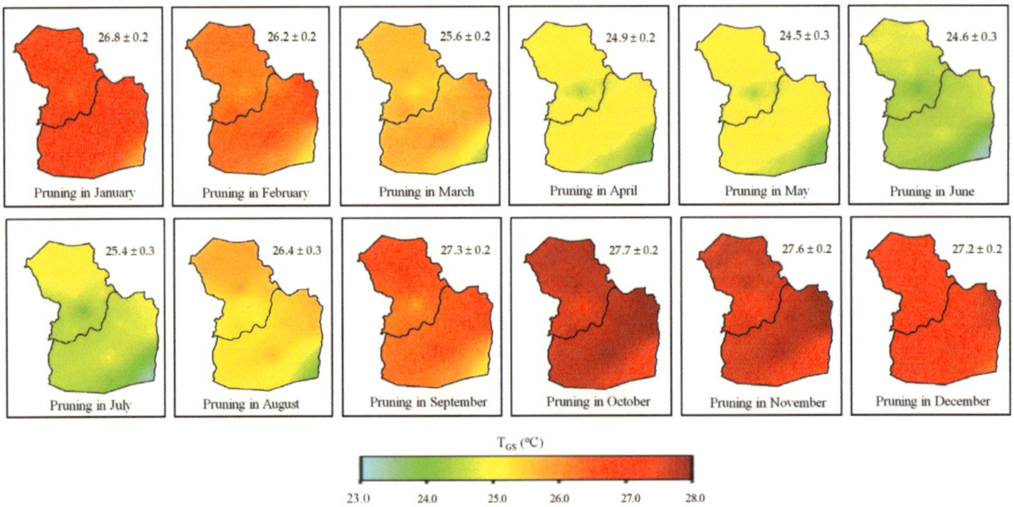

http://dx.doi.org/10.5772/intechopen.72751

http://dx.doi.org/10.5772/intechopen.72799

Treatment*	Lipid peroxidation nmol MDA g^{-1} FW	Superoxide dismutase (SOD) U mg^{-1} prot	Peroxidase (POX) U mg^{-1} prot	Conyferilalcohol peroxidase (CA-POX) U mg^{-1} prot	Polyphenoloxidase (PPO) U mg^{-1} prot	Total Phenols µg caffeic acid g^{-1} FW	Total Phenylpropanoid glycosides µg verbascoside g^{-1} FW	Total flavonoids µg rutín g^{-1} FW
C	94.6 ± 20.5^{b}	89.3 ± 21.7^{b}	4.9 ± 1.2^{b}	53.6 ± 16.5^{b}	1415.5 ± 615.4^{b}	988.6 ± 142.2^{a}	1689.9 ± 290.5^{b}	1122.3 ± 211.0^{b}
CT	130.1 ± 23.0^{a}	71.6 ± 18.0^{b}	4.2 ± 1.5^{b}	66.2 ± 19.3^{ab}	900.3 ± 189.0^{a}	1108.9 ± 158.0^{a}	2013.5 ± 188.3^{b}	1589.7 ± 132.4^{a}
R100C	88.3 ± 18.4^{b}	158.7 ± 22.8^{a}	5.2 ± 1.2^{b}	78.2 ± 20.8^{a}	$1510–3 \pm 312.5^{b}$	1045.6 ± 100.1^{a}	1865.0 ± 366.0^{ab}	1325.6 ± 126.1^{ab}
R100T	91.4 ± 24.3^{b}	81.5 ± 20.1^{b}	2.6 ± 1.0^{a}	79.0 ± 22.1^{a}	1389.0 ± 340.0^{b}	966.0 ± 144.5^{a}	1924.4 ± 171.3^{b}	1159.0 ± 206.6^{b}

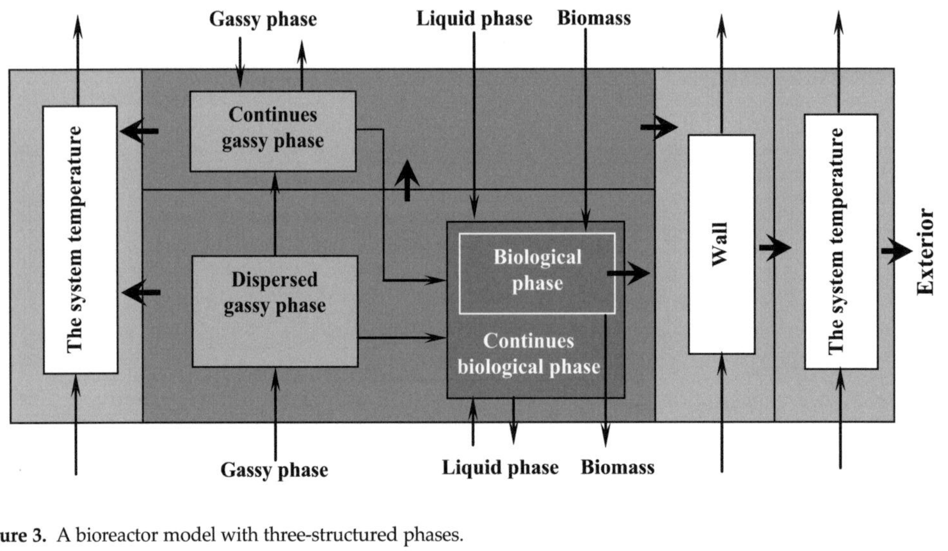

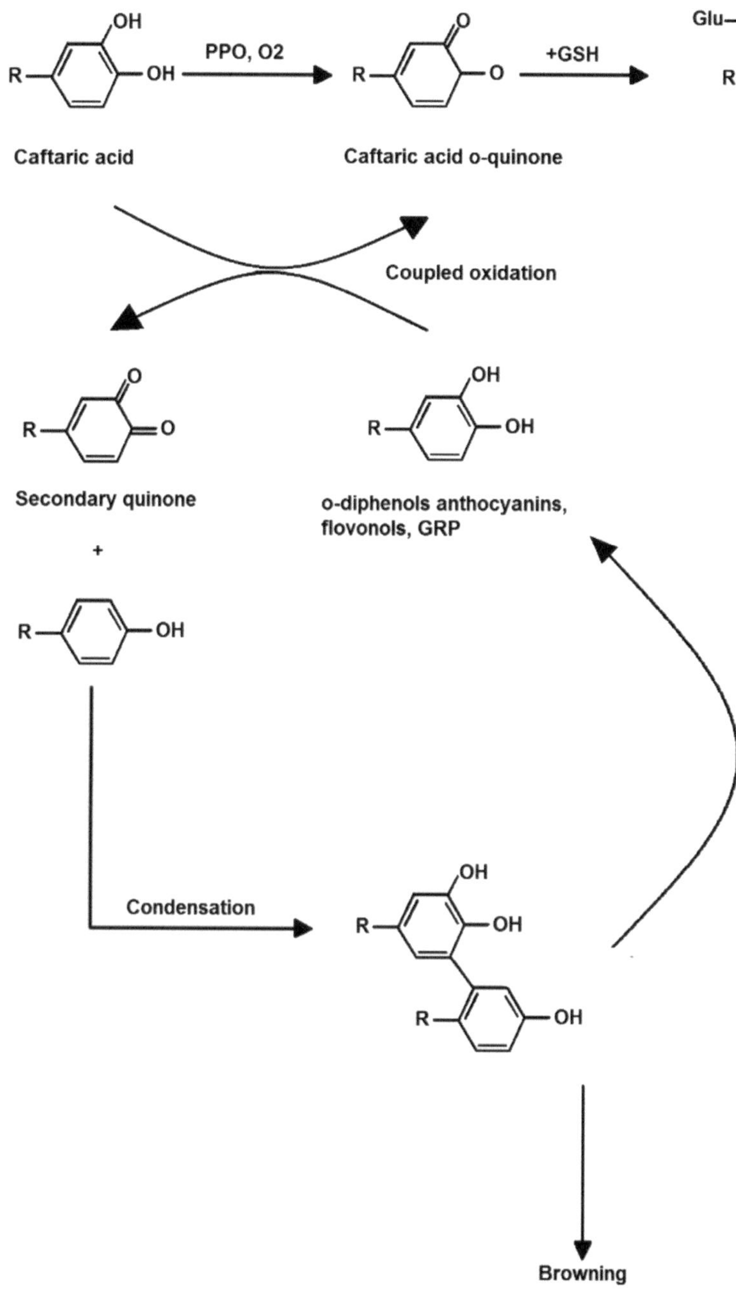

INTECH
open science | open minds

http://dx.doi.org/10.5772/intechopen.73039

http://dx.doi.org/10.5772/intechopen.73039

[51] El Boulifi N, Bouaid A, Martinez M, Aracil J. Optimization and oxidative stability of biodiesel production from rice bran oil. Renewable Energy. 2013;**53**:141-147. DOI: 10.1016/j.renene.2012.11.005

[52] Lepak GS, Moser BR, Bakota EL, Sharp J, Thornton CD, Walker T. Improved oxidative stability of biodiesel via alternative processing methods using cottonseed oil. International Journal of Sustainable Engineering. 2017;**10**(2):105-114. DOI: 10.1080/19397038.2016.1250839

[53] Knothe G. Some aspects of biodiesel oxidative stability. Fuel Processing Technology. 2007;**88**(7):669-677. DOI: 10.1016/j.fuproc.2007.01.005

[54] Liang YC, May CY, Foon CS, Ngan MA, Hock CC, Basiron Y. The effect of natural and synthetic antioxidants on the oxidative stability of palm diesel. Fuel. 2006;**85**(5-6):867-870. DOI: 10.1016/j.fuel.2005.09.003

[55] Mittelbach M, Schober S. The influence of antioxidants on the oxidation stability of biodiesel. JAOCS, Journal of the American Oil Chemists' Society. 2003;**80**(8):817-823. DOI: 10.1007/s11746-003-0778-x

[56] Polavka J, Paligová J, Cvengroš J, Šimon P. Oxidation stability of methyl esters studied by differential thermal analysis and rancimat. JAOCS, Journal of the American Oil Chemists' Society. 2005;**82**(7):519-524. DOI: 10.1007/s11746-005-1103-4

[57] Karthikeyan S. An environmental effect of *Vitis vinifera* biofuel blends in a marine engine. Energy Sources, Part A: Recovery, Utilization and Environmental Effects. 2016; **38**(21):3262-3267. DOI: 10.1080/15567036.2016.1179362

[58] Bazooyar B, Ghorbani A, Shariati A. Combustion performance and emissions of petrodiesel and biodiesels based on various vegetable oils in a semi industrial boiler. Fuel. 2011;**90**(10):3078-3092. DOI: 10.1016/j.fuel.2011.05.025

[59] Linker T. The Jacobsen-Katsuki epoxidation and its controversial mechanism. Angewandte Chemie (International Edition in English). 1997;**36**(19):2060-2062. DOI: 10.1002/anie.199720601

[60] Petrović ZS. Polyurethanes from vegetable oils. Polymer Reviews. 2008;**48**(1):109-155. DOI: 10.1080/15583720701834224

[61] Yadav GD, Manjula Devi K. A kinetic model for the enzyme-catalyzed self-epoxidation of oleic acid. JAOCS, Journal of the American Oil Chemists' Society. 2001;**78**(4):347-351. DOI: 10.1007/s11746-001-0267-2

[62] Corma Canos A, Iborra S, Velty A. Chemical routes for the transformation of biomass into chemicals. Chemical Reviews. 2007;**107**(6):2411-2502. DOI: 10.1021/cr050989d

[63] Adhvaryu A, Erhan SZ. Epoxidized soybean oil as a potential source of high-temperature lubricants. Industrial Crops and Products. 2002;**15**(3):247-254. DOI: 10.1016/S0926-6690(01)00120-0

[64] Campanella A, Rustoy E, Baldessari A, Baltanás MA. Lubricants from chemically modified vegetable oils. Bioresource Technology. 2010;**101**(1):245-254. DOI: 10.1016/j.biortech.2009.08.035

[65] Wan Rosli WD, Kumar RN, Mek Zah S, Hilmi MM. UV radiation curing of epoxidized palm oil-cycloaliphatic diepoxide system induced by cationic photoinitiators for surface coatings. European Polymer Journal. 2003;**39**(3):593-600. DOI: 10.1016/S0014-3057(02)00241-0

[66] Shaker NO, Kandeel EM, Badr EE, El-Sawy MM. Synthesis and properties of renewable environment-friendly epoxy resins for surface coatings. Journal of Dispersion Science and Technology. 2008;**29**(3):421-425. DOI: 10.1080/01932690701718800

[67] Thames SF, Yu H. Cationic UV-cured coatings of epoxide-containing vegetable oils. Surface and Coatings Technology. 1999;**115**(2-3):208-214. DOI: 10.1016/S0257-8972(99)00244-3

[68] Perdue RE, Carlson KD, Gilbert MG. *Vernonia galamensis*, potential new crop source of epoxy acid. Economic Botany. 1986;**40**(1):54-68. DOI: 10.1007/BF02858947

[69] Krewson CF, Scott WE. *Euphorbia lagascae* Spreng., an abundant source of epoxyoleic acid; seed extraction and oil composition. Journal of the American Oil Chemists Society. 1966;**43**(3):171-174. DOI: 10.1007/BF02646296

[70] Cahoon EB, Ripp KG, Hall SE, McGonigle B. Transgenic production of epoxy fatty acids by expression of a cytochrome P450 enzyme from *Euphorbia lagascae* seed. Plant Physiology. 2002;**128**(2):615-624. DOI: 10.1104/pp.128.2.615

[71] Yu K, Li R, Hatanaka T, Hildebrand D. Cloning and functional analysis of two type 1 diacylglycerol acyltransferases from *Vernonia galamensis*. Phytochemistry. 2008;**69**(5):1119-1127. DOI: 10.1016/j.phytochem.2007.11.015

[72] Petrović ZS, Zlatanić A, Lava CC, Sinadinović-Fišer S. Epoxidation of soybean oil in toluene with peroxoacetic and peroxoformic acids—Kinetics and side reactions. European Journal of Lipid Science and Technology. 2002;**104**(5):293-299. DOI: 10.1002/1438-9312(200205)104:5<293::AID-EJLT293>3.0.CO;2-W

[73] de Haro JC, Izarra I, Rodríguez JF, Pérez Á, Carmona M. Modelling the epoxidation reaction of grape seed oil by peracetic acid. Journal of Cleaner Production. 2016;**138**:70-76. DOI: 10.1016/j.jclepro.2016.05.015

[74] Armylisasa AHN, Hazirahb MFS, Yeonga SK, Hazimaha AH. Modification of olefinic double bonds of unsaturated fatty acids and other vegetable oil derivatives via epoxidation: A review. Grasas y Aceites. 2017;**68**(1):1-11. DOI: 10.3989/gya.0684161

[75] Dinda S, Patwardhan AV, Goud VV, Pradhan NC. Epoxidation of cottonseed oil by aqueous hydrogen peroxide catalysed by liquid inorganic acids. Bioresource Technology. 2008;**99**(9):3737-3744. DOI: 10.1016/j.biortech.2007.07.015

[76] Janković MR, Sinadinović-Fišer S. Kinetic models of reaction system for the in situ epoxidation of unsaturated fatty acid esters and triglycerydes. Chemistry & Industry. 2004;**58**(12):569-576. DOI: 10.2298/HEMIND0412569J

[77] Miller CM, Valentine RL. Hydrogen peroxide decomposition and quinoline degradation in the presence of aquifer material. Water Research. 1995;**29**(10):2353-2359. DOI: 10.1016/0043-1354(95)00059-T

[78] Cai C, Dai H, Chen R, Su C, Xu X, Zhang S, et al. Studies on the kinetics of in situ epoxidation of vegetable oils. European Journal of Lipid Science and Technology. 2008;**110**(4):341-346. DOI: 10.1002/ejlt.200700104

[79] Mungroo R, Goud VV, Pradhan NC, Dalai AK. Modification of epoxidised canola oil. Asia-Pacific Journal of Chemical Engineering. 2011;**6**(1):14-22. DOI: 10.1002/apj.448

[80] Shen L, Worrell E, Patel M. Present and future development in plastics from biomass. Biofuels, Bioproducts and Biorefining. 2010;**4**(1):25-40. DOI: 10.1002/bbb.189

[81] Petrović ZS, Zhang W, Javni I. Structure and properties of polyurethanes prepared from triglyceride polyols by ozonolysis. Biomacromolecules. 2005;**6**(2):713-719. DOI: 10.1021/bm049451s

[82] Guo A, Demydov D, Zhang W, Petrovic ZS. Polyols and polyurethanes from hydroformylation of soybean oil. Journal of Polymers and the Environment. 2002;**10**(1-2):49-52. DOI: 10.1023/A:1021022123733

[83] Tolvanen P, Mäki-Arvela P, Kumar N, Eränen K, Sjöholm R, Hemming J, et al. Thermal and catalytic oligomerisation of fatty acids. Applied Catalysis A: General. 2007;**330**(1-2):1-11. DOI: 10.1016/j.apcata.2007.06.012

[84] Samuelsson J, Jonsson M, Brinck T, Johtansson M. Thiol-ene coupling reaction of fatty acid monomers. Journal of Polymer Science, Part A: Polymer Chemistry. 2004;**42**(24):6346-6352. DOI: 10.1002/pola.20468

[85] Biermann U, Friedt W, Lang S, Lühs W, Machmüller G, Metzger JO, et al. New syntheses with oils and fats as renewable raw materials for the chemical industry. Angewandte Chemie—International Edition. 2000;**39**(13):2207-2224. DOI: 10.1002/1521-3773(20000703)39:13<2206::AID-ANIE2206>3.0.CO;2-P

[86] Dahlke B, Hellbardt S, Paetow M, Zech WH. Polyhydroxy fatty acids and their derivatives from plant oils. Journal of the American Oil Chemists' Society. 1995;**72**(3):349-353. DOI: 10.1007/BF02541095

[87] Biswas A, Adhvaryu A, Gordon SH, Erhan SZ, Willett JL. Synthesis of diethylamine-functionalized soybean oil. Journal of Agricultural and Food Chemistry. 2005;**53**(24):9485-9490. DOI: 10.1021/jf050731o

[88] Lee KW, Hailan C, Yinhua J, Kim YW, Chung KW. Modification of soybean oil for intermediates by epoxidation, alcoholysis and amidation. Korean Journal of Chemical Engineering. 2008;**25**(3):474-482. DOI: 10.1007/s11814-008-0081-7

[89] Miao S, Zhang S, Zhiguo S, Wang P. A novel vegetable oil-lactate hybrid monomer for synthesis of high-T g polyurethanes. Journal of Polymer Science, Part A: Polymer Chemistry. 2010;**48**(1):243-250. DOI: 10.1002/pola.23759

[90] Díaz-Medino A, Pérez A, Rodríguez JF, De Lucas A, Carmona M. Rigid polyurethane foam from grape seed oil. Research Journal of Chemistry and Environment. 2014; **18**(8):22-28

[91] de Haro JC, Rodríguez JF, Pérez Á, Carmona M. Incorporation of azide groups into bio-polyols. Journal of Cleaner Production. 2016;**138**:77-82. DOI: 10.1016/j.jclepro.2016.05.012

[92] Akintayo ET, Thomas Z, Akintayo CO. Characterization of the reaction products of epoxidised Adenopus breviflorus Benth. Seed oil with sodium azide and trimethylsilyl azide. Industrial Crops and Products. 2006;**24**(2):166-170. DOI: 10.1016/j.indcrop.2006.03.008

[93] Lu SY, Hamerton I. Recent developments in the chemistry of halogen-free flame retardant polymers. Progress in Polymer Science (Oxford). 2002;**27**(8):1661-1712. DOI: 10.1016/S0079-6700(02)00018-7

[94] Zhong Q, Steinhurst DA, Carpenter EE, Owrutsky JC. Fourier transform infrared spectroscopy of azide ion in reverse micelles. Langmuir. 2002;**18**(20):7401-7408. DOI: 10.1021/la0260234

[95] Kolb HC, Finn MG, Sharpless KB. Click chemistry: Diverse chemical function from a few good reactions. Angewandte Chemie - International Edition. 2001;**40**(11):2004-2021. DOI: 10.1002/1521-3773(20010601)40:11<2004::AID-ANIE2004>3.0.CO;2-5

[96] Serrano A, Borreguero AM, Garrido I, Rodríguez JF, Carmona M. Reducing heat loss through the building envelope by using polyurethane foams containing thermoregulating microcapsules. Applied Thermal Engineering. 2016;**103**:226-232. DOI: 10.1016/j.applthermaleng.2016.04.098

[97] Oliviero M, Verdolotti L, Stanzione M, Lavorgna M, Iannace S, Tarello M, et al. Bio-based flexible polyurethane foams derived from succinic polyol: Mechanical and acoustic performances. Journal of Applied Polymer Science. 2017;**134**:1-12. DOI: 10.1002/app.45113

[98] Ugarte L, Gómez-Fernández S, Peña-Rodríuez C, Prociak A, Corcuera MA, Eceiza A. Tailoring mechanical properties of rigid polyurethane foams by sorbitol and corn derived biopolyol mixtures. ACS Sustainable Chemistry and Engineering. 2015;**3**(12):3382-3387. DOI: 10.1021/acssuschemeng.5b01094

[99] Simón D, Borreguero AM, De Lucas A, Rodríguez JF. Glycolysis of viscoelastic flexible polyurethane foam wastes. Polymer Degradation and Stability. 2015;**116**:23-35. DOI: 10.1016/j.polymdegradstab.2015.03.008

[100] Chattopadhyay DK, Webster DC. Thermal stability and flame retardancy of polyurethanes. Progress in Polymer Science (Oxford). 2009;**34**(10):1068-1133. DOI: 10.1016/j.progpolymsci.2009.06.002

[101] Sacristán M, Ronda JC, Galià M, Cádiz V. Phosphorus-containing soybean-oil copolymers: Cross-metathesis of fatty acid derivatives as an alternative to phosphorus-containing reactive flame retardants. Journal of Applied Polymer Science. 2011;**122**(3): 1649-1658. DOI: 10.1002/app.34011

[102] Glampedaki P, Dutschk V. Stability studies of cosmetic emulsions prepared from natural products such as wine, grape seed oil and mastic resin. Colloids and Surfaces A: Physicochemical and Engineering Aspects. 2014;**460**:306-311. DOI: 10.1016/j.colsurfa.2014.02.048

[103] Fiume MM, Bergfeld WF, Belsito DV, Hill RA, Klaassen CD, Liebler DC, et al. Safety assessment of vitis vinifera (grape)-derived ingredients as used in cosmetics. International Journal of Toxicology. 2014;**33**(6):48S-83S. DOI: 10.1177/1091581814545247

[104] Mbah CJ. Studies on the lipophilicity of vehicles (or co-vehicles) and botanical oils used in cosmetic products. Die Pharmazie. 2007;**62**(5):351-353. DOI: 10.1691/ph.2007.5.6184

[105] Contri RV, Ribeiro KLF, Fiel LA, Pohlmann AR, Guterres SS. Vegetable oils as core of cationic polymeric nanocapsules: Influence on the physicochemical properties. Journal of Experimental Nanoscience. 2013;**8**(7-8):913-924. DOI: 10.1080/17458080.2011.620019

[106] Clark AJ, Ross AH, Bon SAF. Synthesis and properties of polyesters from waste grapeseed oil: Comparison with soybean and rapeseed oils. Journal of Polymers and the Environment. 2017;**25**(1):1-10. DOI: 10.1007/s10924-016-0883-3